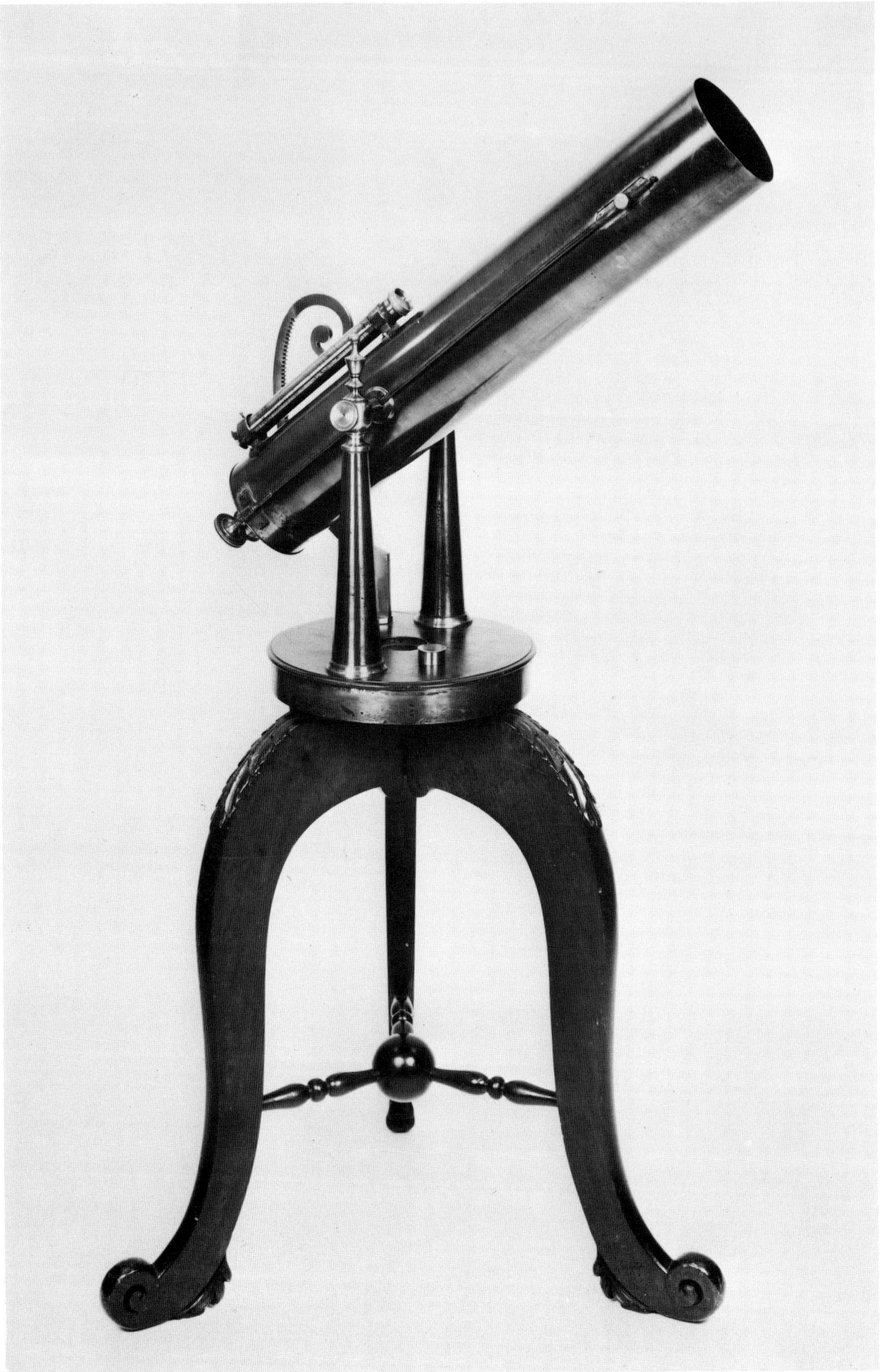

BRASS & GLASS

*Scientific Instrument Making Workshops
in Scotland as illustrated by instruments
from the Arthur Frank Collection
at the Royal Museum
of Scotland*

T.N. CLARKE, A.D. MORRISON-LOW,
A.D.C. SIMPSON

NATIONAL MUSEUMS OF SCOTLAND
1989

© Trustees of the National Museums of Scotland, 1989

British Library Cataloguing in Publication Data
Clarke, T.N.
 Brass & glass: scientific instrument making workshops in Scotland as
 illustrated by instruments from the Arthur Frank Collection at the Royal
 Museum of Scotland.
 1. Scotland. Scientific instrument industries & trades
 I. Title II. Morrison-Low, A.D. III. Simpson, A.D.C. (Allen David
 Cumming), *1947-* IV. National Museums of Scotland
 338..4'768175'09411

ISBN 0-948636-06-8

Cover illustration:
Horizontal sundial by Alexander Adie of Edinburgh, with the coat of arms of Agnew of Lochnaw, *c.*1825 (catalogue 18)

Frontispiece:
Gregorian reflecting telescope by Thomas Morton of Kilmarnock, *c.*1860 (catalogue 96)

CONTENTS

Preface		vii
Introduction		ix
1.	James Short 'the Optic'	1
2.	Peter Hill	11
3.	James Veitch of Inchbonny	17
4.	The Adie Business	25
	Patrick Adie	75
	George Hutchison	85
5.	John and Thomas Dunn	89
6.	Retailing Instrument Makers	96
	John Davis	99
	Zenone & Butti	102
7.	Leith Nautical Instrument Makers	104
	David Laird	106
	David Stalker	108
	W. Craig	111
8.	The Bryson Family	112
9.	The Lennie Business	123
	Turnbull & Co.	130
10.	William Hume	133
11.	Dundee Nautical Instrument Makers	138
	Alexander Cameron	140
	Peter Airth Feathers	143
12.	George Lowdon of Dundee	146
13.	Aberdeen Instrument Makers	152
	Peter McMillan	154
	Berry & Mackay	156
	A. & J. Smith	162
14.	The Gardners of Glasgow	164
	David Carlaw	180
	James Brown	183
15.	Thomas Morton and Andrew Barclay, Engineers	188
	Thomas Morton	190
	Andrew Barclay	197
16.	'Carvers and Gilders'	203
	Antoni and John Galletti	205
	Charles Gerletti	208
	J. & M. Riva	209
17.	Richard Melville, Maker of Sundials	210
18.	Nautical Instrument Makers	217
	Robert Park	219
	Macalister & Fyfe	220
	Samuel Holborn Fyfe	222
	Henry Bell	224
	Cameron & Blakeney	225
19.	Alexander Dobbie	228
20.	Duncan McGregor	235
	Christie & Wilson	250
21.	James White and Lord Kelvin	252
	Mathew Edwards	276
	James More & Co.	278
	Reid & Young	280
22.	Whyte and Thomson	281
23.	Baird & Tatlock	289
24.	Retailing and Manufacturing Instrument Makers	292
	John Spencer	293
	W.A.C. Smith	294
	Baird & Son	296
	Abraham & Co.	298
	John Lizars	301
	Norton & Gregory	306
Concordance		309
Index of names		311

DEDICATED TO
MICHAEL ARNOLD CRAWFORTH
1932-1988

I should like to record my gratitude and appreciation to all those who contributed to the production of this catalogue. Nine years of painstaking research have resulted in a work which undoubtedly will be recognised by future generations of scholars to be a definitive study covering the history of instrument making in Scotland.

Special praise is also due to the Department of Conservation and Analytical Research for its role in researching, conserving and restoring material from the collection.

A skilled and dedicated restoration programme was certainly needed for in my search for Scottish material, I was content to accept items in distressed or incomplete state. I was in no way deterred by lack of perfection and my criteria for accepting instruments were a Scottish provenance and a hope that the item would contribute to our knowledge of the Scottish instrument making trade.

Arthur Frank
Jersey, Channel Islands
September, 1989

PREFACE

Perhaps not surprisingly, the National Museums of Scotland possess the largest and most comprehensive collection of Scottish-made scientific instruments. An already strong collection was supplemented by that amassed by Mr Arthur Frank which was part purchased, part generously donated to the Museums. This book documents that collection and provides hitherto ungathered material about the workshops which constructed the instruments.

Most scientific instruments are neither obscure nor esoteric. They are often of vital everyday importance. Seamen need sextants for navigation, surveyors need theodolites to build railways, roads and bridges. Medical students need microscopes. All were important during the second half of the eighteenth and the nineteenth centuries when developing education, trade, communication and industrialization produced significant wealth for Scotland.

The authors have illuminated much economic and social history, beyond the conventional confines of the history of science. Their research has brought together the unique kinds of evidence which are to be uncovered in museums with archival and published materials, the usual historical sources. Here is a major contribution to our knowledge of the detailed structure of the instrument-making trade in Scotland in the nineteenth century.

R.G.W. Anderson
Director
National Museums of Scotland

INTRODUCTION

'The stock-in-trade of this old gentleman comprised chronometers, barometers, telescopes, compasses, charts, maps, sextants, quadrants, and specimens of every kind of instrument used in the working of a ship's course, or the keeping of a ship's reckoning, or the prosecuting of a ship's discoveries. Objects in brass and glass were in his drawers and on his shelves, which none but the initiated could have found the tops of, or guessed the use of, or having once examined, could have ever got back again into their mahogany nests without assistance.'

Charles Dickens, *Dombey and Son* (London, 1848)

Dickens' fictional Solomon Gills, whose shop was managed by Captain Cuttle, may be taken as representative of the scores of small mid-nineteenth century instrument makers and retailers in London and in ports around the country, who provided for the needs of navigators and other professional users of scientific instruments. The description of Gills' 'stock-in-trade' conveys at once something of the romance of nautical adventure and the mystery of things technical. In some ways this mirrors the clientele of the instrument maker. Ostensibly he catered principally for the initiated elite of professionals, such as navigators, surveyors, draughtsmen and engineers. However, his instruments were also marketed to appeal to enquiring and more affluent clients, for whom a fine telescope or a microscope may have been a key to the wonders of science but was also a prestige object of pride and beauty. The aesthetics of precision engineering - the fascination of 'brass and glass' - had an appeal in Victorian times that is shared by today's collectors of early scientific instruments.

The instrument maker's shop in Dickens' novel was in London, which in this field as in so many others was the dominant centre of the British retail trade, and at least until the mid-nineteenth century also dominated manufacture. It is therefore natural that analysis of the composition and structure of the instrument trade has concentrated on London. The first detailed study to address a geographically separate area within Britain was D.J. Bryden's *Scottish Scientific Instrument-Makers 1600-1900* published in the Royal Scottish Museum's Information Series in 1972. This study was conceived simply as a directory of Scottish instrument makers, providing more reliable and comprehensive details on practitioners than were available from existing London-centred sources, and was intended primarily as a guide to the dating of instruments from makers' inscriptions. However, the work developed to include the first considered discussion of the structure and operation of a market that can loosely be categorised as 'provincial'. In practice, there are characteristics of a separate Scottish patronage and Scottish market which are likely to make the relationship with London somewhat different from that of an English region.

Subsequent work by Bryden and others has further illuminated aspects of Scottish scientific instrument making, but his 1972 study has yet to be superseded. It is certainly not our intention that the present volume should be seen as replacing it: we have examined only a proportion of the makers previously surveyed, although we have tried to examine their businesses in much greater depth. In doing so we have attempted to highlight features of the operational and financial stability of firms that we feel may be significant for the trade as a whole.

The selection of the firms examined has been determined entirely by the content of the Arthur Frank Collection, for which this volume also provides a descriptive catalogue. Despite the strong bias of this collection towards optical instruments, the scope is such that it is at least geographically representative of the Scottish instrument trade. No major firm is excluded, and the composition of minor makers has allowed us to explore a range of types of trading operation as well as the training and dynastic succession in several more substantial workshops. The Frank collection has provided an appropriate framework for discussing the fortunes of a very diverse group of instrument producers and retailers in a way that provides a more rounded view of their activities than has been available before.

Such an assemblage of business biographies makes no pretence at being balanced in coverage or exhaustive in detail. Some of the makers discussed are well known and prominent figures, others are artisans who have left only sufficient evidence for comparatively superficial accounts to be prepared. It might be argued that others have made such slight contributions that they deserve equally superficial treatment, but we have felt it important to attempt a critical view of their trade, even where this led only to the bankruptcy courts. Where suitable archival information is available we have tried to use this to make points about the economics of their markets.

To some extent the lengths of the individual entries reflect the importance we ascribe to the craftsmen and businesses concerned, but it also takes into account the size of the existing secondary literature. Thus, for example, we have not said much about the celebrated eighteenth century Scots telescope maker James Short, who has attracted critical interest, especially over the last twenty years, although we have introduced some additional information relating to patronage and manufacturing practice. Similarly, there is now an extensive literature on the scientific work on Lord Kelvin, and to a more limited extent on his commercial enterprises. This is reflected in our treatment of Kelvin & James White, which is concerned largely with the development of James White's Glasgow workshops and in the nature of Kelvin's financial involvement - aspects which have previously received scant attention.

By way of contrast, very little has been published about the Adie business in Edinburgh, which played such a distinguished role in the scientific and technical life of the capital. We make no excuse, therefore, for having devoted by far the largest section in this volume to the work of this influential dynasty of instrument makers, probably the most significant to have worked outside London.

Nevertheless this group of business studies is littered with unresolved problems. Part of our purpose in publishing them in this form is the hope that more source material will be drawn to our attention as a result. We would welcome any information that would contribute to our understanding of the business of instrument making in Scotland, and we would be particularly pleased to learn of surviving instruments by Scottish makers.

Sources
In the sixteen years since the publication of Bryden's survey, a great deal of Scottish material has been acquired by the Museum and further material has been studied in collections elsewhere. Although this material, and the files on instrument makers that we continue to develop, has been mined for this project, an extended research effort has been necessary which has been very much a collaborative undertaking.

As with Bryden's survey, we have made use of published street directories where these are available. We are acutely aware of the limitations of this type of information in the absence of any adequate understanding of the methods used to compile and revise the directories. A more cautious approach has led to minor revisions of Bryden's dates, and wherever possible we have sought to confirm key dates in more reliable sources.

We have for example normally secured death dates and established family relationships from original registers, census returns and wills. Public records such as these and the Court records that will be discussed below are cited with permission of the Registrar General of Scotland and the Keeper of the Records of Scotland. Partnership details have been extracted from the legal notices in the *Edinburgh Gazette*, which unfortunately note only the terminations and changes in partnerships and the cessation of trading.

Perhaps the single most valuable category of source material used has been the proceedings of bankruptcy cases. Under Scots law, the bankruptcy process involved voluntary petition or petition by creditors to the Court of Session or the Sheriff Court, followed immediately by sequestration of assets. A Trustee was appointed by the Court to administer the estate, to summon meetings of creditors, and to determine with the bankrupt the level of composition which would be offered to the creditors to clear the outstanding claims. The bankruptcy would then be formally discharged. Again, the legal notices covering the stages in this process were published in the *Edinburgh Gazette*, where they were no doubt scanned by anxious creditors.

In the majority of cases the Trustee's papers themselves survive. This virtually untapped archive, held at the Scottish Record Office, is complex to use, and we must acknowledge our debt to Michael Moss of Glasgow University Archive, who willingly gave us access to the indices to the sequestration papers he is developing. In some instances a *verbatim* manuscript record is preserved in the Sederunt Book for the case, in which the bankrupts outlined the history of their business activities, including the relationship between partners, and were required to account for the errors that had led to sequestration. The creditors are all listed, with the amounts of the outstanding debt, providing information about the suppliers and wholesalers used.

The proportion of instrument firms that went bankrupt at some stage appears to be fairly high. However, we have no comparative figures to suggest that it was different from that in other trading areas or that the instrument business was inherently more or less risky. What is, on the face of it, surprising is the frequency with which some firms were sequestrated, and the fact that it was clearly possible to remain actively in trade before being discharged. It is conceivable that sequestration was used to some extent as a method of wiping the slate clean, and that the process may have been used to provide an element of liability protection. Although sequestration cases can provide valuable information on business failure, it might be argued that the record we present is therefore distorted because we have comparatively little information of this sort about successful firms. Against this it must be pointed out that even James White went bankrupt (and indeed this had a significant effect on his subsequent practice) and the Gardners were manifestly successful even though each generation of the firm was subject to sequestration.

The other substantial manuscript source that has been extensively used is the archive of the Royal Scottish Society of Arts (founded in 1821 as the Society of Arts for Scotland). This collection is held at the National Library of Scotland, and we are grateful to the Society's Council and to the Librarian of the National Library of Scotland for permission to use it.

Colleagues at museums, libraries and archives in many parts of Britain and Ireland have helped us with advice and information, often responding patiently to strings of questions and requests, and we have been particularly glad of their active help. We have not given separate acknowledgements at the end of each chapter, but have acknowledged the sources of individual references in the relevant footnotes. We would, however, wish to thank Dr Anita McConnell for her continued and much appreciated support; and to D.J. Bryden, now Keeper of the Department of Science, Technology and Working Life in the National Museums of Scotland, for his detailed criticism of the manuscript. Our thanks too, go to our colleagues in the Photographic Studio, for their perserverance.

One aspect that we have found particularly gratifying has been meeting the descendents of several of the craftsmen whose work is discussed here. They have generously provided information which we could not have obtained otherwise, and have shared family traditions and recollections with us.

The Frank Collection
As befits a study conducted in a museum, the instruments themselves have formed a valuable part of our evidence. In addition to the Royal Museum of Scotland's own collection, comparative material in other collections has been used, and we have also cited material in the published catalogues of auction houses and dealers. The instruments in the Frank collection have been studied minutely in the course of an extended conservation programme, undertaken by our colleagues in the Department of Conservation and Analytical Research, and this has been particularly useful in distinguishing manufactured and retailed pieces and in drawing comparisons between instruments that appear to have common origins in the workshops of wholesale manufacturers.

The Frank Collection of Scottish Scientific Instruments in the Royal Museum of Scotland formed part of a larger collection formed by Arthur Frank, F.B.O.A., F.B.C.O., formerly of Glasgow. Mr Frank trained as an optician and optical instrument maker, and until his retirement in 1977 he directed the optical firm established in Glasgow at the turn of the century by his father, Charles Frank.

Mr Frank's interest in early instruments began when he took over the business on his return from service in the Army in 1945, and was faced with sorting the material that had accumulated in the cellar of the shop. He determined then that some of these early pieces would form the basis of a collection of scientific instruments, and this was later to become an absorbing interest. Over the years it grew in size, scope and importance, and it was perhaps natural that early optical instruments of all sorts should have attracted his particular attention.

In 1980 the Royal Scottish Museum, which from 1985 has been a part of the new National Museums of Scotland, acquired a substantial number of scientific instruments with Scottish associations from Mr Frank's collection. This material, which is described in the present catalogue, was a valued addition to the Museum's growing research holdings of scientific apparatus. The criterion for selection had not necessarily been the quality of individual pieces (although there are items of considerable importance) but rather a Scottish instrument-trade provenance, and therefore the potential for illuminating issues of manufacturing, retailing and marketing practice, mainly in the nineteenth century.

The material in the Arthur Frank Collection of Scottish Scientific Instruments was drawn from two sources: first, a loan collection at Glasgow Museums and Art Galleries, Kelvingrove, previously exhibited as part of that museum's 'Tools of Science' exhibition in 1973, which was transferred in its entirety; and secondly, additional Scottish material acquired since that date and forming part of Mr Frank's main collection on Jersey.

This was the second occasion that the Royal Scottish Museum had received sections of Mr Frank's collection. In 1979 an important group of microscopes illustrating the introduction of achromatic optics in the period 1800 to 1860 had been acquired. In both instances the material was obtained by a combination of gift and purchase, and in the acquisition of the Scottish material we must acknowledge substantial assistance from the Heritage Fund for Scotland (a precursor of the National Heritage Memorial Fund). Although for administrative reasons the material was accessioned in two stages in 1980 and 1981, for grant-aiding purposes it was considered as a single transaction. We must acknowledge Mr Frank's enthusiasm for preserving this important aspect of our culture and thank him for the assistance he has provided in ensuring that it is preserved for public benefit. Material from the collection was first displayed in the Royal Museum of Scotland's new 'Instruments of Science' gallery, which opened in March 1986. The Royal Museum of Scotland is part of the National Museums of Scotland, but is located at two sites in Edinburgh: all references to the Royal Museum of Scotland in this volume are to the Chambers Street site.

Cataloguing Conventions

Certain conventions have been used in compiling the catalogue entries, and it remains to describe these briefly. Five-year dating intervals have been used to estimate the dates of instruments (e.g. in the form c.1855 or c.1860), except for those that can be dated exactly: this approach has been adopted because in general the error margins for individual pieces are small and of comparable size. Normally the dimensions given include key measurements, such as the clear aperture of objective lenses for telescopes. Where the instrument has draw tubes or extension tubes, the length dimension is the unextended length. In the descriptive entries the term 'bronze' is used for all bell-metal, gun metal and similar alloys, and the term 'oxidised' is used in preference to 'bronzed'. Similarly, 'ivory' is used for convenience to describe whalebone as well as ivory. The use of deep coloured lacquers is noted but not the use of light lacquers.

Alison Morrison-Low
Allen Simpson
National Museums of Scotland

Tristram Clarke
Scottish Record Office

February 1989

1 James Short, M.A., F.R.S. (1710-68). Portrait attributed to
Benjamin Wilson, in the Royal Observatory, Edinburgh. *The
Astronomer Royal for Scotland*

1

JAMES SHORT 'THE OPTIC'

James Short (1710-1768) is the earliest of the Scottish instrument makers whose work is represented in this collection, and he is one of the most celebrated, for by the time of his death he was acknowledged throughout Europe and beyond as the past master of the reflecting telescope. His career presents the almost unblemished success story of the archetypal lad o' pairts. His early difficulties were surmounted by talent and dedication, and with the help of influential patrons, these qualities brought him great prosperity. Short's stature as a telescope maker and consequent significance for the history of astronomy have led to a number of brief but detailed studies of his life and work, and the present account therefore principally draws from these.[1]

Short, who was born in 1710 the son of a burgess wright of Edinburgh, was orphaned at the age of ten and was consequently placed at George Heriot's Hospital where he received his basic education. Early academic promise led to four years at the High School, and then to his enrolment in 1726 as an arts student at Edinburgh University and to the completion of the normal courses.[2] His aptitude for the classics and his grandmother's wishes pointed him towards the church as his career, but he abandoned divinity studies in 1731 or 1732, apparently as a result of attending the classes of the dynamic professor of mathematics, Colin Maclaurin (1698-1746).[3] Maclaurin was the first major exponent of Newtonian mathematics and natural philosophy in Scotland, and through his Fellowship of the Royal Society of London, and his friendships with several other Fellows, was aware of the resurgent European interest in the reflecting telescope which had occurred by 1720. He included discussion of the prevailing theories of astronomical optics in his lectures and classes.[4]

When Short 'made so great a figure in the class',[5] Maclaurin was then a good position to draw out his precocious mechanical and mathematical talent, and allowed him the use of his rooms in the College in 1732 for his early attempts at telescope construction.[6] Perhaps stimulated by the attempts of two Edinburgh makers to make and sell reflecting instruments, Short soon acquired great proficiency in grinding specula, and within two years was producing Gregorian reflectors far superior, in Maclaurin's opinion, to others available at the time. One measure of their performance was that a fifteen inch focus reflector by Short belonging to the Professor of Scots Law enabled the *Philosophical Transactions* to be read at the then considerable distance of 500 feet. By December 1734 Short had produced some thirty pieces, taking 'vast pains to make the instruments as perfect as possible'.[7]

Short's precocity as a telescope maker was drawn to the attention of the Royal Society in June 1735 by Maclaurin, and a series of tests of his telescopes was conducted during 1736 by Fellows of the Society.[8] That summer Short was in London at the summons of Queen Caroline, to instruct her son William, Duke of Cumberland in mathematics, but by November he was back in Edinburgh. The following month as a result of the London tests Short's skill as a maker was recognised by his election as a Fellow of the Royal Society.

In February 1737 Scottish astronomers observed an annular eclipse, and later that year a comet was visible in Scottish skies. These astronomical events were discussed amongst an interested group, amongst them Short, all of whom were friends of Maclaurin. Short was later associated with Maclaurin, Alexander Monro *primus*, Sir John Clerk of Penicuik and the Earl of Morton in the formation in Edinburgh of the Society for Improving Philosophy and Natural Knowledge, a precursor of the Royal Society of Edinburgh.[9] The broadly-based Society, with members drawn from across the professions and the aristocracy, was to provide Short with a framework of patronage and influence which extended across Europe. He was the only one of the forty-six founder-members to earn his living by a trade, but it is clear that his education and technical ability raised him above the level of an artisan, led to a readier acceptance in scientific circles, and at the same time provided him with access to potential patrons and customers.[10] The Earls of Macclesfield and Morton were reportedly among his best patrons.[11]

In spite of his increasing status in Edinburgh, the attraction of London as the market-place for scientific instruments in Europe was too strong for Short, and he quit Edinburgh early in 1738. His earnings in Edinburgh had not, however, been negligible, for he is said to have deposited £500 in the Bank of Scotland by summer 1736.[12] In addition, in August that year he loaned 2000 merks Scots to another member of the Philosophical Society, Hugh Murray Kynynmound of Melgund and Kynynmound, an advocate in Edinburgh

A TABLE

Shewing the Focal Lengths, Magnifying Powers, and Prices of Reflecting Telescopes, constructed after the *Gregorian* Form, by Mr. SHORT, in *Surry-street,* in the *Strand,* LONDON.

Number.	Focal Lengths in Inches.	Magnifying Powers.	Prices.
1	3	1 Power of 18 Times.	3 Guineas.
2	4½	1 — — 25	4
3	7	1 — — 40	6
4	9½	2 — 40 and 60	8
5	12	2 — 55 and 85	10
6	12	4 — 35, 55, 85 and 110	14
7	18	4 — 55, 95, 130 and 200	20
8	24	4 — 90, 150, 230 and 300	35
9	36	4 — 100, 200, 300 and 400	75
10	48	4 — 120, 260, 380 and 500	100
11	72	4 — 200, 400, 600, and 800	300
12	144	4 — 300, 600, 900, and 1200	800

N.B. The first five Telescopes are moved by plain Joints and the rest by Rack-work or Screws.

2 Price list for James Short's telescopes, undated. Private collection

who was borrowing heavily at this time. Short was still receiving interest on the outstanding bond in December 1740, through his factor in Edinburgh.[13] Once in London Short's earnings became very high and placed him above most fellow Scots who moved there to seek a living in literature or commerce; men such as Tobias Smollett the novelist, who went to London in 1739, could not command such a steady patronage or high income, although some merchants exceeded Short's fortune by massive commercial undertakings.[14] Indeed, part of Short's unique achievement is that he accrued his small fortune by working alone on a single type of scientific instrument. How was this possible?

The reasons were three-fold. First, Short's technique of grinding the metal specula was in advance of other makers' skills. He had begun in 1732 by working his mirrors in glass, as Newton had recommended in his *Opticks*, but soon abandoned this method in favour of grinding them in 'speculum metal' following the success of London optical workers in the early 1730s.[15] His special skill lay in the very difficult conversion of the spherical concave shape of the metal that could be achieved in the conventional grinding process, to the theoretically required parabolic shape, a process which at the time only Short appeared able to master. His exact method, though guessed at by contemporaries, remained a secret which died with him.

Secondly, Short's success lay in his system of production. Whereas to begin with he constructed all the parts of the telescopes himself, by the time he set up in London he was using out-workers, who provided the constituent parts, while he concentrated on the mirrors. By keeping the volume of work to within his own capacity he avoided the need for assistants or apprentices, and therefore kept his technique to himself. He was sufficiently successful with this continued specialisation in reflecting telescopes (a most unusual feature in itself), for his instruments to command a price double that asked by the other leading London makers. His rate of production has been analysed to show that he could buy in completed tubes and stands by other makers, fit only his own mirrors and still clear a substantial profit.[16] In fact, it can be shown that this was the method used by Short, and that even before his move to London he had negotiated with London suppliers for the brasswork of his telescopes.[17]

Such a degree of successful specialisation would not have been possible without a strong demand for his telescopes, not only for serious work but also as prestige items for recreational use. To some extent this demand was led by the growing importance of astronomical problems in scientific thought at this time, particularly in pressing areas such as precision cartography and long-distance navigation. Short's career spanned the beginnings of wide interest in the scientific possibilities of the reflecting telescope[18] and its perfection at his hands by mid-century coincided happily with two opportunities to make simultaneous observations across the globe, of the rarely occurring transits of the planet Venus over the Sun's disc. Short was associated with the scientific and computing work of the Board of Longitude, and as a member of the Council of the Royal Society was involved in the preparations for both the 1761 and 1769 transits of Venus. These were seen at the time as of the most fundamental importance for astronomical calculation: namely, the measurement of the absolute distance of the Earth from the Sun and hence the size of the known solar system.[19] Indeed, such was the reputation of Short's prestigious reflectors that most of the observatories in Europe possessed one, and large numbers of both serious astronomers and dilettanti purchased them. Short's reputation in London was thus high, not only as a master optician but also as an astronomer.[20]

After he had established himself in London his contacts with Scotland were maintained not only through his Edinburgh lawyer, but also through his patrons and his brother Thomas Short, who practised as an instrument maker in Edinburgh and Leith from about 1737. Of his patrons the most important was James, 14th Earl of Morton, a cultured politician interested like many of his contemporaries in agricultural improvement, but also (less commonly) in science. Short's relationship with him was deferential and intimate, at least in the early years, and it is likely that he was rewarded with several commissions by and through his patron. In addition to providing Morton with his services as a telescope maker, Short was called on to provide the baseline for a survey of the Earl's Orkney estates in 1739 during his progress there.[21] As Morton's protégé Short was entrusted with the care of his eldest son Sholto Charles Douglas, Lord Aberdour, to whom he appears to have been a tutor in the summer and autumn of 1742, and he also performed several tasks in London on the Earl's behalf.[22]

Short's letters to Morton referring to these activities are interesting for the light they shed on the mechanism of the Earl's patronage and Short's related business activities. In September 1742 he was responsible for sending pictures, dyed silk and 'your cases of Instruments made by Mr Bird' to Morton. This consignment of instruments included 'several sizes of compasses', and it is clear that Short was acting as a sort of agent between the two men, for John Bird asked him to tell Morton that he had had to meet extra packing charges for the compasses above the price agreed with the Earl.[23] Short mentioned that he himself owned a brass beam compass by Bird, and amongst the consignment was probably the duplicate which Morton had ordered for himself. George Graham was another maker who worked for Morton through Short.[24] It is interesting to note that although Short's working methods and apparent ease in making his living might be thought to have set him apart from his fellow instrument makers, he had close links with several of them, and was generous to at least one craftsman who was in a sense his rival. This was John Dollond, whose work on achromatic refracting telescopes he communicated to the Royal Society.[25]

Short's links with the Earl of Morton obviously went deeper than acting as agent on his behalf, as the same correspondence shows. In autumn 1742 Short was setting up house in London, and along with a housewarming present of linens, Morton sent him an equinoctial dial to be altered by adding a twelve inch focus telescope to it, presumably so that it would act as a small equatorial instrument. Short quoted the price as ten or twelve guineas, stating that he had adjusted its three feet to work as well as with four, but that it would be 'impossible to alter the needle without being at a good deal of expense'. He had been busy with 'the Great Telescope', which he promised to send to Morton by sea, and added revealingly,

> 'but as to the altering of the machinery it would occasion a great expence to me, and besides could not be done before Candlemass, but this I promise to your Lordship, that if I have another of that size bespoke, if you please I shall take that back again, and make you another altered in the way you propose.'[26]

Short was evidently busy at this period, but was at the same time reluctant to dampen his patron's wishes. This also emerges in an episode which provides the only known instance of Short's work on anything other than optical instruments. Morton's agricultural interests appear to have led him to commission Short to produce a working model of a straw-cutting machine in summer 1742. James Short's letter is ambiguous about whether he or his brother Thomas was asked, but it was James who executed the one-fifth scale model in brass, which he sent to Morton along with the fabric and compasses noted above. As he explained:

'my Brother was so busy, I was obliged to make it myself, you'll please therefore excuse the roughness of it, and I wish you may be able to understand what is meant by it; at least by it you'll be able to remember the machine you saw.'[27]

The likely explanation of this unusual piece of work is that James' attempts to pass on the commission to his brother in Scotland fell through because of Thomas' workload, or indeed, his incompetence, something commented upon by Maclaurin in 1743.[28] Apart from exercising Short's brassworking skill in a way which seems to have become increasing less likely in later life, the model evinces Short's links with his brother. At the same date James promised Morton to obtain information about the working of manchineel, a West Indian wood, from Thomas who knew of 'those that have wrought it'.[29] Of actual trade links between the two brothers there is no hint, which is not surprising given James' solitary method of production. However, Thomas Short altered the stand of a four-inch Gregorian telescope belonging to the natural philosophy class at the University of St. Andrews, originally supplied by James in 1736. Later, after his brother's death he ran his business in London for eight years before returning to Edinburgh in 1776 to found the Calton Hill Observatory.[30]

James Short's links with Scotland lasted until his later years, although it is not known the extent to which he found buyers for his pieces there. He represented for

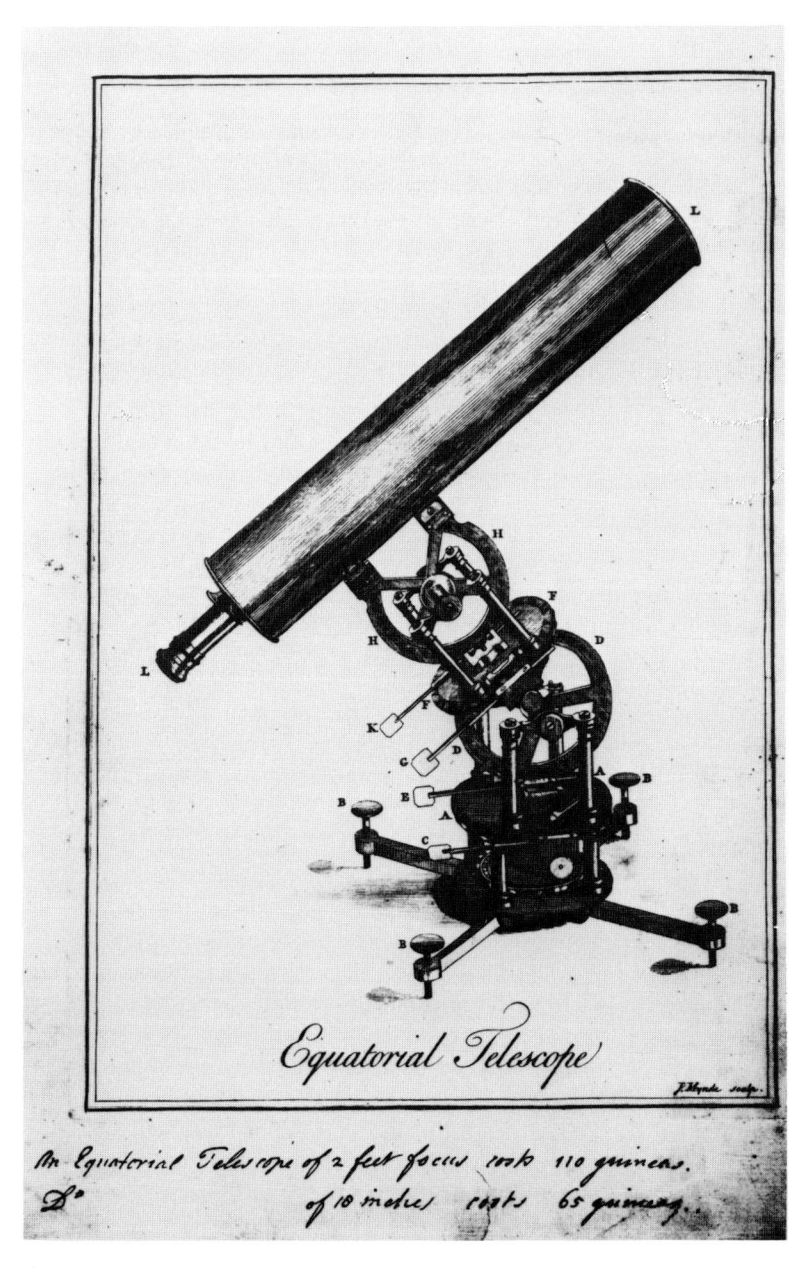

3 Engraving of Short's Equatorial Telescope. A pull from the engraved plate used to illustrate Short's paper 'Description and Uses of the Equatorial Telescope, or Portable Observatory', *Philosophical Transactions* 46 (1749-50). Short has added the prices for his 24-inch and 18-inch instruments in his own handwriting. *Museum of the History of Science, Oxford*

like-minded Scots a contact who was well placed in London's scientific community, and it is in this capacity that James Watt was sent to him in 1755 with a letter of introduction. As has been suggested, Short's method of working was perhaps one reason why he found another maker to take on Watt as a journeyman.[31] The same may have been true of other putative assistants, whether Scots or English. He also kept up with Professor Maclaurin, who was himself a correspondent of the Earl of Morton.[32] Short's last known link with Scotland was his creation as an honorary Burgess and Guildbrother by Edinburgh Town Council in August 1766.[33] He had already been honoured by his election as a foreign member of the Royal Swedish Academy of Science in 1758, but the fact that his freedom of Edinburgh came belatedly did not prevent him from mortifying £200 to the Town's College for a mathematics bursary.[34]

The 1760s, which marked one of his busiest periods both as astronomer and telescope maker, also saw the estrangement of Short from his erstwhile patron the Earl of Morton. The rift allegedly came when Short, with characteristic generosity, supported the claim of the chronometer maker John Harrison for the Government's £20,000 longitude prize in 1764. He is said to have pressed it too hard for the liking of the Earl of Morton, who was then President of the Royal Society, and influential in the filling of the vacant post of Astronomer Royal that year. Although Short was widely considered as a likely candidate, he failed to obtain the coveted office because of Morton's opposition.[35] His reputation as a telescope maker nonetheless continued up to and after his death in 1768, despite the increasing challenge provided by the newly improved refracting telescopes, such as those made by John Dollond. At his death he is reckoned to have left a fortune of about £20,00, and a corpus of some 1,370 telescopes. As one writer has observed, he made his money 'chiefly out of polishing speculum metal',[36] and this is perhaps a just assessment of his achievement as a telescope maker.

REFERENCES

1. D.J. Bryden, *James Short and his Telescopes* (Edinburgh, 1968): this handbook, prepared for the bicentennial exhibition of Short's work at the Royal Scottish Museum in 1968, is a comprehensive study and the most easily-available one. G.L'E. Turner, 'James Short, F.R.S., and his Contribution to the Construction of Reflectiong Telescopes', *Notes and Records of the Royal Society of London* 24 (1969), 91-108: contains the best account of his life and work in London. D.J. Bryden, 'James Short, M.A., F.R.S., Optician Solely for Reflecting Telescopes', *University of Edinburgh Journal* 24 (1970), 251-261: is largely a recasting of his 1968 work but draws on Turner. A.D.C. Simpson, 'The Early Development of the Reflecting Telescope in Britain', unpublished Ph.D. thesis (University of Edinburgh, 1981), 314-326: discusses Short's telescope production to about 1740.

2. Bryden (1968), *op. cit.* (1), 3. He did not, however, take a degree, which was not untypical of the period. Short was awarded the M.A. degree in 1753 by St. Andrews University, for services rendered, but only when that institution was happy that he had completed the course at Edinburgh (Bryden (1970), *op. cit.* (1), 251-252).

3. David Stuart Erskine, 11th Earl of Buchan, 'Life of Mr James Short, Optician', *Transactions of the Society of Antiquaries of Scotland* 1 (1792), 251-256. This is the earliest account of Short, but although drawn from personal knowledge at first or second hand, it contains errors corrected in the recent articles noted above.

4. Bryden (1970), *op. cit.* (1), 252-254.

5. Buchan, *op. cit.* (3), 252.

6. *Ibid.*, 253.

7. Bryden, (1970), *op. cit.* (1), 254.

8. Simpson, *op. cit.* (1), 317.

9. Bryden (1970), *op. cit.* (1), 255. For an account of the Society, see Roger L. Emerson, 'The Philosophical Society of Edinburgh 1737-1747', *British Journal for the History of Science* 12 (1979), 154-191. For a discussion of the Society after its revival in the 1740s see *idem*, 'The Philosophical Society of Edinburgh, 1748-1768', *ibid.* 14 (1981), 133-176. The later phase of the Society's work provided a context for the Adie business: see Chapter 4.

10. Simpson, *op. cit.* (1), 319; Emerson (1979), *op. cit.* (9).

11. Buchan, *op. cit.* (3), 254. In 1739 Short was commencing a survey of Orkney for Morton during the Earl's progress round his Orkney estates (*ibid.*, 254-255; Bryden (1970), *op. cit.* (1), 256).

12. Buchan, *op. cit.* (3), 254. Bryden (1970) mistakes this date for 1738 (*op. cit.* (1), 255).

13. National Library of Scotland, MS 13,728, ff. 133, 207. Discharges by Short's factor, 13 December 1738, 11 December 1740. He had been empowered 8 April 1738, presumably just before Short's departure for London. The sterling equivalent then of 2000 merks Scots was about £110.

14. A coppersmith called Forbes who operated in both Aberdeen and London bought the estate of Callendar and Almond near Falkirk for £85,850 in the 1780s (Scottish Record Office, GD 150/3504/29).

15. On the context to this, see Simpson, *op. cit.* (1), 303-311.

16. Turner *op.cit* (1), 100-101.

17. Royal Society, London, MS. Fo.4.3. At end [in a postscript] of letter, Colin Maclaurin to Sir Martin Folkes, 28 July 1736: 'Mr Short goes for London in a few days to settle a correspondence for his brasswork which is complained of'; published in Stella Mills (ed.), *The Collected Letters of Colin MacLaurin* (Nantwich, Cheshire, 1982), 263.

18. Short's arrival in London in 1738 coincided with the publication of Robert Smith's highly influential *Compleat System of Opticks* (London, 1738), which highlighted Short as the best commercial maker (Simpson, *op. cit.* (1), 319-320).

19. Bryden, *op. cit.* (1), 260-261. Cf. Harry Woolf, *Transits of Venus* (Princeton, New Jersey, 1959) for the world-wide interest

in the transits and Short's activities concerning them. One of Short's fellow members of the Royal Society committee for the 1769 transit was James Ferguson (1710-1776), his exact contemporary, and like him a talented and learned mechanic who came to London from Scotland to earn his living (E. Henderson, *Life of James Ferguson, F.R.S., in a brief Autobiographical Account, and Further extended Memoir* (London, 1867); John R. Millburn, *Wheelwright of the Heavens: the Life & Work of James Ferguson, F.R.S.* (London, 1988)).

20. However, he was supposed to have altered data when calculating his published figures for the value of the solar parallax in 1763, which involved him in controversy (Woolf, *op. cit.* (19), 145-147).

21. Buchan, *op. cit.* (3), 254-255; Bryden (1970), *op. cit.* (1), 256-257. The survey of Orkney was undertaken by Murdoch Mackenzie, senior, during the 1740s, and published in 1750; he also had Colin Maclaurin as a patron (D.G. Moir, 'A History of Scottish Maps' in D.G. Moir, *et al.* (eds.), *The Early Maps of Scotland to 1850* 3rd edition, I (Edinburgh, 1973), 89-90; Diana C.F. Smith, 'The Progress of the *Orcades* Survey, with Biographical Notes on Murdoch Mackenzie Senior (1712-1797)', *Annals of Science* 44 (1987), 277-288).

22. S.R.O., Morton papers, GD 150/3485/10, James Short to Earl of Morton, 9 September 1742.

23. S.R.O., GD 150/3485/15, James Short to Earl of Morton, 25 September 1742. John Bird (1709-1776), was one of the most prominent London instrument makers with an international reputation for major divided instruments (cf. C.D. Hellman, 'John Bird (1709-1776) Mathematical Instrument Maker in the Strand', *Isis*, 17 (1932), 127-153).

24. S.R.O., GD 150/3485/35, James Short to Earl of Morton, 25 November 1742. George Graham (1673-1751) was one of the group of contemporary instrument makers who became Fellows of the Royal Society; his encouragement of Short on his first visit to London, and his later friendship with him, meant that through Graham Short was introduced to a whole network of makers and scientists in the metropolis (cf. Edwin A. Battison, 'George Graham', in C.C. Gillispie (ed.), *Dictionary of Scientific Biography* V (New York, 1972), 490-492; Olivia Brown, 'The Instrument-Making Trade', in R. Porter, *et al.*, *Science and Profit in 18thCentury London* (Cambridge, 1985), 19-20).

25. Turner, *op. cit.* (1), 92.

26. S.R.O., GD 150/3485/35. In 1742 Short made a twelve foot focus reflector for the Duke of Marlborough (Bryden, (1970), *op. cit.* (1) 259), but the reference here is to a different instrument, presumably. Colin Maclaurin thanked Morton for the use of it in January 1744 (S.R.O., GD 150/3486/2, Maclaurin to Morton, 27 March 1744).

27. S.R.O., GD 150/3485/15.

28. British Library, Add. MS. 6861 f.35, Colin Maclaurin to Sir Andrew Mitchell, 31 March 1743: '... I ask liberty to inclose a letter from Mr Short because it is too bulky for ye common post. A'propos I must caution you against *Shorts*. His brother has proved the completest Villain has ever fall'n in my way. It would amaze you to tell how he has used me but *non est tanti*.' [i.e. 'it matters little']; published in Mills, *op. cit.* (17), 101.

29. *Ibid.*

30. D.J. Bryden, *Scottish Scientific Instrument-Makers 1600-1900* (Edinburgh, 1972), 8.

31. Turner, *op. cit.* (1), 101-102.

32. See Maclaurin's letters to Morton 1742-1744 (S.R.O., GD 150/3485/38, 3486/1-10).

33. Bryden (1970), *op. cit.* (1), 256.

34. *Ibid.*, 26; Turner, *op. cit.* (1), 92.

35. Bryden (1970), *op. cit.* (1), 257-258. Morton's reported reason for opposing Short's candidacy, '... that Mr Short is a Scotch man, though he acknowledges that he is the fittest for it of any Man...' (*ibid.*), perhaps reflects Morton's pragmatic recognition of the unpopularity of the Scots during and after the ministry of the Earl of Bute, who was forced to resign from office in April 1763.

36. Turner, *op. cit.* (1), 102.

---

1. Reflecting telescope: James Short, Edinburgh, 1735 (T1981.27)

9½" focus mounted Gregorian reflecting telescope in brass, with speculum metal optics (the secondary remounted). Engraved on the mirror retention plate 'JAMES SHORT EDINBURGH 1735.' Modified external focusing rod, the threaded end working in a boss which projects through a slot in the barrel and locates in a hole in the sliding mount for th secondary speculum, being retained by an internal clip. A small scratch on the boss and barrel (not the normal punch mark) indicates the setting for focus at infinity. Eyepiece assembly with single eye-lens, but lacking the external eye stop. The cast primary speculum with turned finish on the outer and inner edges, and with a painted vertical orientation mark on the reverse; retained, by a separate triangular spring plate pierced at the centre, against a flange in the mirror cell.

Screw-fit end cover. Pillar-mounted for attachment to a fitted wooden case, now lacking its lid. Replacement knuckle-joint for altitude adjustment and free azimuth motion. The pillar pierced for a tommy bar and with the thread of the lower attachment bolt tapered to form a wood screw.

Length of barrel: 330mm.
Barrel diameter: 61mm.
Speculum diameter: 60mm.
Case size: 405 x 130 x 95mm.

The instrument shows signs of 19th century modification to the mounting and to the focus mechanism. The knuckle-joint on the stand has been renewed, and soldered to the top of the pillar: the upper retaining ring of the azimuth motion has been re-used but reduced in diameter. A long slotted attachment plate to the barrel is fitted, but the original plugged attachment holes in the barrel are visible. The sliding plate to which the secondary is attached has a plugged hole at its centre, and a fingernail recess at the inner edge suggests that it was originally mounted the other way round and was adjusted manually. A split or cut at the barrel seam has been repaired by inserting a narrow plate. There is an extended metal fracture in the barrel.

Provenance: Purchased (with item 6) from David Letham Ltd., Edinburgh, 1967, who acquired it at Lyon & Turnbull, Edinburgh, 20 September 1967, Lot 73. This instrument is unusual in having no serial number added after the date, and no other examples of this are known. Serial numbers are first recorded in 1734: the earliest noted is 1734.7.14, (i.e. the 14th instrument produced and the 7th of this size) which is in a private collection (see D.J. Bryden, *James Short and his Telescopes* (Edinburgh, 1968), 4, item 1), and 1734.3.27 is in the National Maritime Museum (inventory OO/Rs27). Several other 1735 instruments have been recorded, but all have serial numbers: 1735.11.44 (Royal Museum of Scotland, NMS T1893.452); 12.45.1735 (The Antiquarian Scientist, Catalog 20 (May 1988), item 405); 1735.22.60 (Jena Optical Museum, V.E.B. Carl Zeiss, Jena, D.D.R.); 1735.27.65 (Sotheby's, 23 October 1985, Lot 233); 1735.30.68 (Royal Museum of Scotland, NMS T1967.93).

2. Reflecting telescope: James Short, Edinburgh, 1737 (T1981.28)

9½″ focus mounted Gregorian reflecting telescope in brass, with speculum metal optics. Engraved on the mirror retention plate 'JAMES SHORT EDINBURGH 1737 55/117'. External focusing rod, the threaded end working on a projection from the internal sliding mount for the secondary speculum, and with defining marks to indicate the setting for focus at infinity. Eyepiece assembly with single eye-lens and external sleeve carrying an eye stop. The cast primary speculum with turned finish on the outer and inner edges, and a painted vertical orientation mark on the reverse; retained by leaf springs against 3 small lugs in the mirror cell. Plain sights, with both vanes lacking. Pillar-mounted, originally for attachment to the lid of a fitted wooden case (now lacking), with a knuckle-joint for altitude adjustment, and free azimuth motion. The pillar pierced for a tommy bar.

Length of barrel: 347mm.
Barrel diameter: 64mm.
Speculum diameter: 57mm.

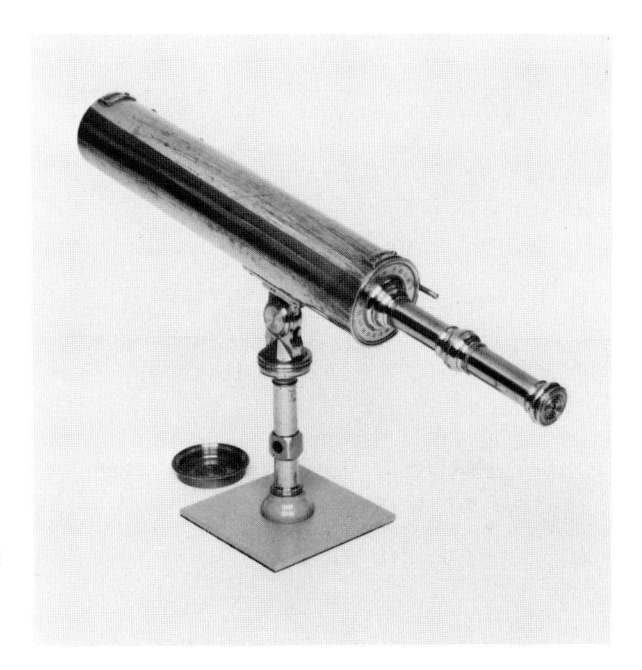

The instrument shows signs of 19th century modification to the focus mechanism, and the chamfered runners for the secondary speculum support have been replaced. The iron wood screw which originally projected from the base of the piller has been turned off. The wing nut for adjusting the knuckle-joint on the stand is damaged. Lacking sights and focus adjusting knob. The pushfit end cover is not original.

Provenance: Purchased at Sotheby's, 24 May 1971, Lot 81 (previously offered at Sotheby's, 22 February 1971, Lot 8).

3. Reflecting telescope: James Short, London, 1743 (T1981.29)

12″ focus mounted Gregorian reflecting telescope in brass, with speculum metal optics. Engraved on the mirror retention plate 'JAMES SHORT LONDON J743 25/353 = 12.' External focusing rod, the threaded end working on a projection from the internal sliding mount for the secondary speculum, and with defining points to indicate the setting for focus at infinity. The original eyepiece lacking, but with a threaded cover plate in the eyepiece aperture. The cast primary speculum with turned finish on the outer and inner edges. Residual painted vertical orientation mark on the reverse; retained by leaf springs against 3 lugs in the mirror cell. Plain sights (the front sight remounted). Push-fit end cover. Pillar-mounted on a fitted wooden case. Knuckle-joint for altitude adjustment and free azimuth motion. The pillar pierced for a tommy bar and the thread of the lower attachment bolt tapered to form a wood screw.

Length of barrel: 415mm.
Barrel diameter: 78mm.
Speculum diameter: 71mm.
Case size: 495 x 195 x 125mm.

The secondary may not be original to this instrument: the mounting slide has a plugged threaded hole apparently for an attachment point of the type used by Short at a later date, and a second focus defining mark has been added.

Provenance: Acquired privately in Coventry, 1972, and understood to have been in the collection of a small private observatory. It was recorded by Thomas Court in 1929, but with no further details given: T.H. Court and M. von Rohr, 'A History of the Development of the Telescope from about 1675 to 1830 based on documents in the Court Collection', *Transactions of the Optical Society* 30 (1929), 224. The serial number was printed inaccurately as 25/354 = 12, and the error has been corrected in Court's copy of the off-print: Science Museum Library, London, MS Court U (we are grateful to Jane Insley for providing this information).

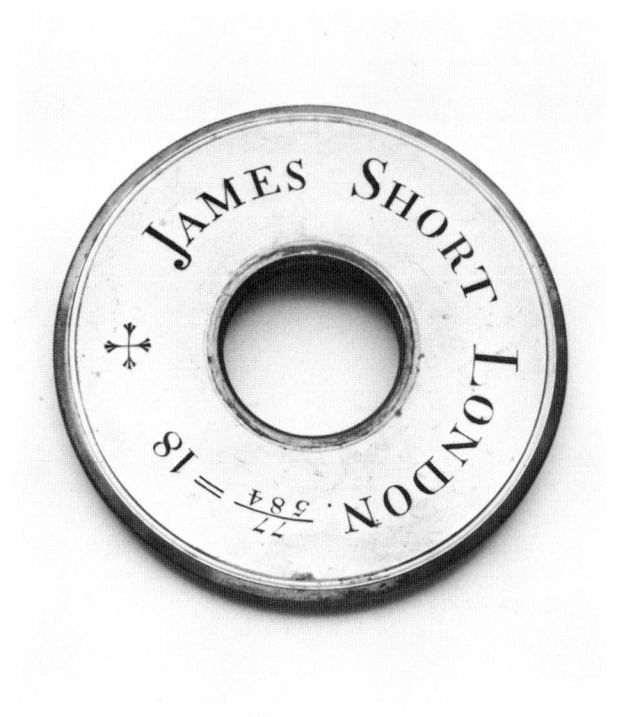

The tube has been badly distressed and the finish destroyed by inexpert restoration. The mirror cell is of the correct diameter for the tube and the presence of a focus defining point indicates that it is a Short tube. However, all the fitments are attached by modern screws, but neither the mounting nor the finder telescope are aligned with the tube axis. Components surface mounted on the tube are mainly made to conform with a larger diameter tube. The position of the focusing attachment on the mounting plate for the secondary has been moved. A reinforcing plate behind a patch in the barrel has holes corresponding to the four attachment points for the stand and may be from another tube. The two eyepieces (one lacking optics) differ in style and are not an adequate fit in the mirror plate. The tripod legs, although stamped with identification numbers for screw attachment, are riveted in place. An acrylic lacquer finish had been applied.

Provenance: Acquired in 1968.

4. Reflecting telescope: James Short, London, c.1750 (T1981.30)

18" focus mounted Gregorian reflecting telescope in brass (restored), lacking primary speculum, and apparently a composite of two or more instruments. Engraved on the mirror retention plate 'JAMES SHORT LONDON 77/584 = 18'. External focusing rod, the threaded end working on an internal sliding plate adapted to take either of two removeable secondary mirrors, and with defining marks (one obscured) to indicate the setting for focus at infinity. Telescopic sight (incomplete). Pillar-mounted on a folding table tripod stand; adjustable in altitude by a tangent screw engaging in the perimeter of a semi-circular arc with a clamp release, and with a clamped screw-operated azimuth motion. With fitted case.

Length of barrel: 598mm.
Barrel diameter: 105mm.
Provision for speculum of 98mm diameter.
Case size: 680 x 295 x 165mm.

5. Reflecting telescope: James Short, London, c.1750 (T1981.31)

12" focus mounted Gregorian reflecting telescope in brass, with speculum metal optics. Engraved on the mirror retention plate 'JAMES SHORT LONDON 93/706 = 12.' External focusing rod, the threaded end working on a projection from an internal slide carrying the secondary speculum, and with defining marks to indicate the setting for focus at infinity. The eyepiece assembly with field lens, eye-lens and external eye stop. The primary speculum cast with chamfered outer and inner edges, filed-down edge sprue and a turned finish on the inner edge. Residual painted vertical orientation mark on the reverse; retained by leaf springs against 3 lugs in the mirror cell. Plain sights. Pillar-mounted on a fitted wooden case; the pillar end and insert on the case top with matching punch marks. Knuckle-joint for altitude adjustment and free azimuth motion. The pillar pierced for a tommy bar and the thread of the lower attachment bolt tapered to form a wood screw.

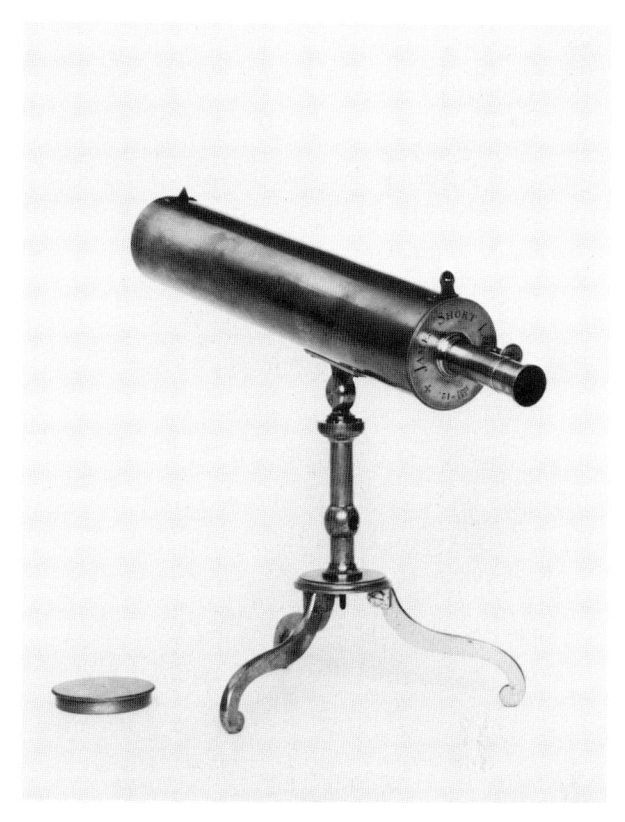

Length of barrel: 445mm.
Barrel diameter: 82mm.
Speculum diameter: 75mm.
Case size: 525 x 195 x 130mm.

The instrument shows signs of being a composite. The mirror plate internal thread and the eyepiece assembly thread have been re-cut and the flange on the eyepiece refaced; the diameter of the field lens mount has had to be turned down to avoid fouling on the mirror plate thread. The relief of the eye stop is insufficient to define the image area. The slide mount for the secondary is a poor match to its chamfered guide and may be a replacement.

Provenance: Acquired in 1972.

6. Reflecting telescope: James Short, London, c.1755 (T1981.32)

12″ focus mounted Gregorian reflecting telescope in brass, with speculum metal optics. Engraved on the mirror retention plate 'JAMES SHORT LONDON 182/1001 = 12.' Short external focusing rod, the threaded end working on a projection from a long internal sliding plate slotted for an attachment at its upper end and carrying the secondary speculum on a 3-point adjustment, and with defining marks to indicate the setting for focus at infinity. The eyepiece assembly with field and eye lenses and external eye stop. The primary speculum cast with chamfered outer and inner edges, filed-down edge sprue and turned finishing on the inner edge, and with a painted vertical orientation mark on the reverse; retained by leaf springs against 3 lugs in the mirror cell. Plain sights. Push-fit end cover. The edges of the mirror plate and end cover with assembly numbering marks of 5 strokes. Pillar-mounted on a folding table tripod stand, with a knuckle-joint for altitude adjustment and free azimuth motion. The pillar pierced for a tommy bar and the screw attachment to the tripod finished as a tapering wood screw.

Length of barrel: 488mm.
Barrel diameter: 82mm.
Speculum diameter: 75mm.

The knuckle-joint was found to be damaged and the thread stripped: the outer leaves of the joint were replaced in the Museum in 1983. The mirror cell has been re-soldered within the barrel at some stage, requiring the re-positioning of 5 screw holes at the barrel end.

Provenance: Purchased (with item 1) from David Letham Ltd., Edinburgh, 1967, who acquired it at Lyon & Turnbull, Edinburgh, 20 September 1967, Lot 73.

7. Reflecting telescope: James Short, London c.1760 (T1981.33)

18″ focus mounted Gregorian reflecting telescope in brass, but lacking mirrors and lenses. Engraved on the mirror retention plate 'JAMES SHORT LONDON 180/1098 = 18.' External focusing rod, the threaded end working on a projection from an internal sliding mount adapted to carry removeable secondary specula, and with defining marks to indicate the setting for focus at infinity. Telescopic sight. The edge of the mirror plate with an assembly numbering mark of 5 strokes. Pillar-mounted on a folding table tripod stand. Adjustable in altitude by a tangent screw engaging in the perimeter of a semi-circular arc with a clamp release; the azimuth slow-motion work lacking.

Length of barrel: 615mm.
Barrel diameter: 105mm.
Provision for speculum of 97mm diameter.

The field lens and eye lens are lacking from the eyepiece assembly. The eye stops of the telescope and finder are reproductions. Metal surfaces have been abraded and are finished with acetate lacquer.

Provenance: Acquired in 1967.

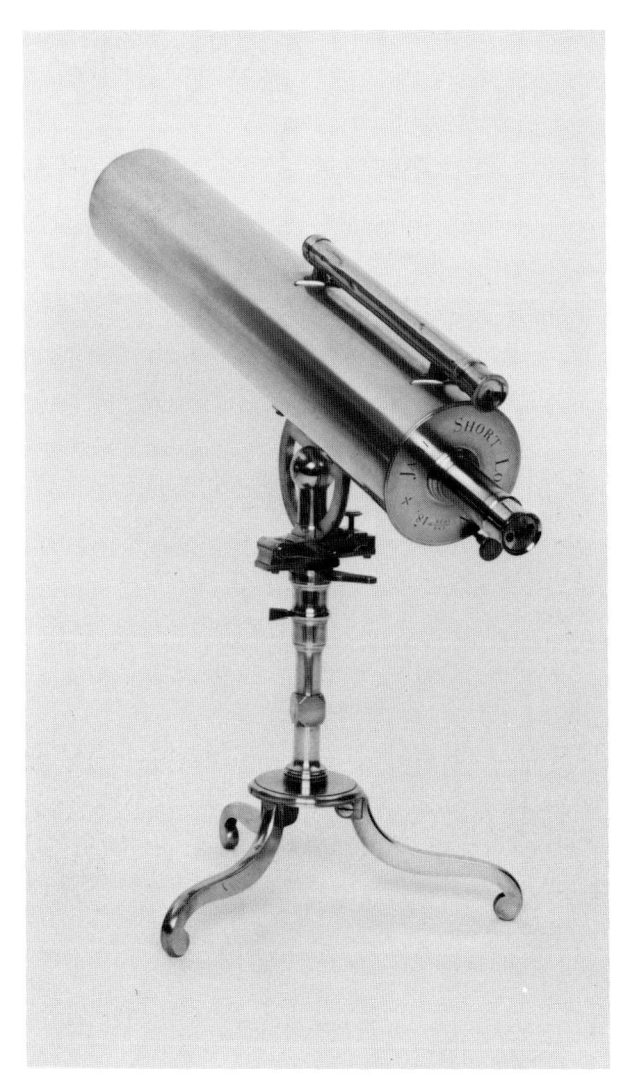

8. Reflecting telescope: James Short, London, c.1765 (T1981.34)

12" focus Gregorian reflecting telescope in brass, with speculum metal optics. Engraved on the mirror retention plate 'JAMES SHORT LONDON 244/1305 = 12.' Short external focusing rod, the threaded end working on a long internal sliding plate slotted for an attachment at its upper end and carrying the secondary speculum on a 3-point adjustment, and with one of the two defining marks to indicate the setting for focus at infinity. The eyepiece assembly contains the eye lens and has an external eye stop, but lacks the field lens. The primary speculum cast with chamfered outer and inner edges, filed-down edge sprue and turned finish on the inner edge, and with a painted vertical orientation mark on the reverse; retained by leaf springs against 3 lugs in the mirror cell. With push-fit end cover, but lacking the original stand. The edges of the mirror plate and end cover, and both ends of the tube, with assembly numbering marks of 2 strokes.

Length of barrel: 455mm.
Barrel diameter: 83mm.
Speculum diameter: 75mm.

The four original mounting holes for the stand are present. The attachment point for the focusing plate has been modified by widening the slot in the barrel (removing the focus defining mark) and fitting a new threaded collar on a radiused washer attached to the plate by a slotted internal keeper.

Provenance: Acquired in 1967.

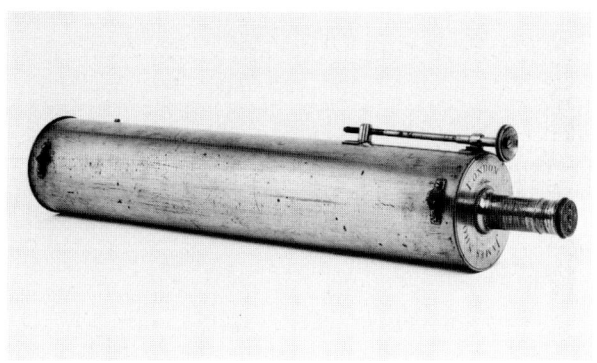

2
PETER HILL

Little is known of the working life or biography of Peter Hill, a mathematical instrument maker established on the South Side of the City in the early years of the nineteenth century. No details of an apprenticeship are provided in the Burgess Roll, but the quality of his work is sufficiently good for one to believe that he might have trained under a master of the competence of John Miller, uncle of Alexander Adie (q.v.).

The difficulty in tracing Peter Hill arises from two areas of uncertainty. Firstly, he formed part of what appears to have been an extended family in which there were at least three other members working as smiths in the same neighbourhood. Secondly, the Nicolson Street and Richmond Street area in which they lived was still being developed during this period, leading to confusion about the renumbering of houses and even occasional changes in street names and divisions.

Peter Hill was born in 1778, the son of another Peter Hill, smith, who was first recorded in Thomas Aitchison's *Edinburgh Directory* for 1796 in Richmond Street, which runs east to west between the Pleasance and Nicolson Street.[1] This Peter Hill, described as smith of Chapel Street or Richmond Street, together with his wife Barbara White, is recorded in the Register of Sasines, the official register of heritable property, as living in the ground floor of a tenement house in 'Nicolson's Park' in April 1794.[2] Lady Nicolson's former parkland, through which Nicolson Street was built in the 1790s as the new southern approach to the city, included Richmond Street, and this is presumably the address recorded by Aitchison in 1796.

Peter Hill 'junior', optician, was included in a section of Aitchison's 1801 *Directory* devoted to 'those who have

4 Peter Hill's trade card for his address between 1804 and 1810

moved since Whitsunday 1801'.[3] The distinction 'junior' was promptly dropped and the two Peter Hills have adjacent entries in the directories for a number of years at Richmond Street, East Richmond Street (from 1804), 7 East Richmond Street (from 1806), 9 East Richmond Street (from 1811) and finally 9 Richmond Street (from 1812),[4] during which time they presumably lived or worked in the same tenement; and if any physical move was involved, they moved together. Peter Hill junior appears to have acquired a flat in this tenement in his own name in 1810.[5]

In the 1813 directory, the instrument maker is entered at Union Place, at the top of Leith Walk on the west side, whereas the smith remained at Richmond Street and was last recorded there in the 1821 directory. His New Town address was given more specifically as 6 Union Place from 1818, and 7 Union Place from 1823. The expanded *Post Office Directory* of 1824 also lists his house at 2 Greenside Place, off the other side of Leith Walk, but in the directories from 1825 he is listed only at the Greenside Place address until the entry lapses after 1828.[6]

He clearly retained property interests in Richmond Street, because in 1815 a 'back house and piece of ground' in Nicholson's Park was acquired by Peter Hill senior and his wife 'and Peter and James Hill their Sons'.[7] In 1826 he sold a flat in a tenement in Richmond Street (presumably the one purchase in 1810) to George Hill 'Smith and Beam Maker' of Richmond Street.[8] George Hill was presumably a relative (and most probably his brother George) because Peter Hill senior is recorded as lending him £200 in 1808, gaining title to property in Richmond Street as security.[9]

Aitchison's *Directory*, and subsequently the *Post Office Directory*, described Hill only as 'optician' until 1813 when he was termed optician and mathematical instrument maker. As early as 1804, however, Denovan's *Directory*[10] described him as a mathematical instrument maker, and on his own trade card for the period to 1811 he advertised himself as 'Optician and Mathematical Instrument Maker'.[11]

Some of the surviving instruments which carry his name are clearly retailed pieces, typical of the bought-in stock of an instrument maker. The signature on the small telescope in this collection is an addition;[12] and another Hill telescope in the Museum collection, although of better quality, also appears to have been retailed.[13]

An altogether different picture is presented by the pantograph in the Frank collection, which reveals Hill as an accomplished and original maker.[14] This somewhat enigmatic piece, probably of about 1805-10, has several interesting features. Firstly, the arms are not arranged in the conventional fashion, where the two long arms form adjacent sides of a hinged parallelogram. Here two opposite sides are extended to become the long arms, a layout not normally found until the 1820s. Secondly, much greater flexibility of use has been incorporated by allowing all three of the tracing and pivot points to be moveable. Finally, although the instrument has been signed, the engraving of Hill's name matching closely in style that on the trade card in the box and showing characteristic features of Edinburgh engraving practice, the arms of the pantograph are not divided or calibrated with the conventional reduction ratios. This last point alone appears to rule out the pantograph having been supplied by a manufacturing wholesaler as well as the possibility that the instrument was never completed.

The attractive alternative is that it was made as a special commission for a client who had particular copying and reduction requirements not readily met by stock commercial instruments. This is perhaps reinforced by the presence of a single short extension piece for one of the long arms, for which special provision is made in the fitted case. Use of this extension allows accurate copying at ratios of around 1:1 with the instrument set up in the so-called reverse manner, and far more nearly in balance than could be achieved with a conventional pantograph which is lop-sided at this ratio range. The only restriction to be made in adjusting the instrument manually and without the assistance of scales would be the mathematical requirement that all three of the pivot points should be in line and this can readily be arranged.

If it was indeed Hill's intention to adapt the pantograph and provide an instrument with better mechanical characteristics for approximately full-sized copying, then to some extent he (or his commissioning client) was anticipating the thinking behind Professor William Wallace's development of the eidograph some ten or fifteen years later.[15] It may be that Hill's client was one of the draughtsmen or engravers involved in cartographic work or the assembly of composite encyclopaedia plates who later employed the eidograph to such good effect.[16]

There was always repair work. For example, in 1815 William Kyle, a Glasgow land-surveyor, sent his travelling telescope to Hill stating:

> 'You will observe that the Skrew which tightens the instrument on the cone is broken. This is the second time that it has given way;
> therefore, instead of soldering it, as was done when it first failed, I request you will make an entirely-new skrew, at as short an interval as your convenience will permit, for it is an instrument which I have very frequent occasion to use.'[17]

He continued his letter with strict specifications for a brass parallel ruler. Makers like Hill would have found ready customers from professionals like Kyle. An ivory rule, possibly used by an architect, or in an office, has been noted.[18] A brass compass with a silvered dial in a mahogany case,[19] presumably made for the surveying market is recorded, as is a surveyor's level with a 15-inch telescope.[20]

That Hill could be regarded as amongst the first rank of Edinburgh instrument-makers is confirmed by the writings of Sir David Brewster (1787-1868) who commissioned four Scottish opticians - Peter Hill, Alexander Adie (q.v.), James Veitch (q.v.) and William Blackie[21] - to construct microscope lenses from precious and semi-precious gems (which have higher refractive indices and lower dispersion than glass) in order to reduce the image problems encountered in microscopes used before the advent of corrected microscope lenses.[22]

This suggestion had first been mooted by Brewster in 1813,[23] and in an anonymous review in 1829 he wrote:

> '... About ten years ago, Mr Peter Hill, an ingenious optician in Edinburgh, executed for him [i.e. Brewster] two single lenses of ruby and garnet, which were used both as single microscopes, and as the object glasses of a compound microscope. Mr Sivright of Meggetland had also executed for him, we believe by the same artist, a single plane convex lens, of the colourless topaz of New Holland.'[24]

Later, Brewster commented that Hill's ruby and garnet lenses

> 'performed admirably, in consequence of their producing, with surfaces of inferior curvature, the same magnifying power as a glass lens; and the distinctness of the image was increased by their absorbing the extreme blue rays of the spectrum.'[25]

It has been suggested[26] that the garnet lens Hill made for Brewster survives in an unsigned microscope now in the British Museum.[27] No other microscopes by Hill are known.

Hill was apparently a competent glass-worker too, and a mercury Fahrenheit thermometer signed by him is in the Royal Museum of Scotland[28] and barometers with his signature have been recorded.[29] On 27 December 1822 he communicated to the Society of Arts for Scotland (later the Royal Scottish Society of Arts) 'An Improved Saccharometer', but nothing further about the device is known.[30] It is possible that Hill had connections with Alexander Allan, an instrument maker who was active in Edinburgh between 1806 and 1835[31] and whose workshop produced Allan's saccharometer, designed by Thomas Thomson (1773-1852), a device reportedly in exclusive use (but without legal authority) for testing the strength of spirits by the Scottish Excise from 1805 to 1816.[32] So far, attempts to uncover any relationship between the two makers, or between Hill and Thomson, have proved unfruitful.

Although Peter Hill disappeared from the *Post Office Directory* in 1829, he resurfaced in the late 1830s in the last two issues of a short run of Edinburgh directories published by John Gray, where he is described in 1836 and 1837 as Peter Hill, optician, at 13 Abbey.[33] The usual reason for a move to the Abbey Strand was to take advantage of the ancient right to escape creditors in the city by claiming sanctuary within the precincts of Holyrood Abbey. Debtors thereby avoided sequestration and were even able to continue trading, although they were liable to arrest if they left the sanctuary, except on Sundays.[34] Hill is indeed recorded in the Register of Protection maintained by the Bailie of Holyroodhouse as having been granted sanctuary, and therefore having been declared bankrupt in the eyes of the law, on 16 December 1834.[35] He was originally in the lodging run by A. Petrie who was Keeper of the Sanctuary Records and whose wife was Keeper of the Chapel Royal in the Abbey.[36] It is possible that when he was later at 13 Abbey Strand he was associated in business with David Hume, smith, who appears at this address and whose wife took lodgers.[37]

It appears that the official registration of the 1826 sale of his property in Richmond Street to George Hill was not completed at the time. It was entered in the Register of Sasines only in 1834, presumably to prevent the possibility of it being sequestrated as part of Peter Hill's assets. Hill was at this stage described as 'sometime Optician and Mathematical Instrument Maker'.[38]

It is not known when Peter Hill cleared his debts and was able to return to Edinburgh, but it may have been fairly quickly because he was not found in the 1841 Census returns at Holyrood. The only other reference to him that we have found is an entry in the street directory for the single year 1845 of Peter Hill, smith and philosophical instrument maker, at the address of the same George Hill, smith and beam maker, at 12 Richmond Place.[39] Although there are entries for a Peter Hill, smith, at a newly built house in neighbouring Roxburgh Street from 1857, we cannot say whether this is the same man.[40]

REFERENCES

1. T. Aitchison, *The Edinburgh and Leith Directory* (Edinburgh, 1796). The International Genealogical Index, British Isles, Scotland, fiche F0402, records the marriage of Peter Hill [senior] and Barbara White on 11 July 1773 (p.8,124) and the baptisms of their sons Peter on 20 September 1778 (p.8,124), George on 30 July 1782 (p.8,106) and James on 15 March 1788 (p.8,111).

2. Scottish Record Office, Particular Register of Sasines, Edinburgh, RS 27/389, ff. 197-205.

3. Aitchison's *Directory* 1801, 46.

4. *Ibid.* 1803; *Denovan and Co's Edinburgh and Leith Directory, from July 1804 to July 1805* (Edinburgh, 1804); *Edinburgh Post Office Directory* 1806-1812.

5. S.R.O., Particular Register of Sasines, Edinburgh, RS 27/648, ff. 243-248.

6. *Edinburgh Post Office Directories* 1813-1828.

7. S.R.O., Abridgements of Particular Register of Sasines, Edinburgh, 1781-1820, no. 19384.

8. S.R.O., Abridgements of Particular Register of Sasines, Edinburgh, 1831-36, no. 3580.

9. S.R.O., Particular Register of Sasines, Edinburgh, RS 27/618, ff. 108-116. The baptism of George Hill, son of Peter Hill and Barbara White, is noted at reference 1 above.

10. Denovan's *Directory* 1804.

11. With the pantograph NMS T1980.117.

12. NMS T1980.245.

13. NMS T1988.22.

14. NMS T1980.177.

15. The eidograph is discussed in Chapter 4.

16. Discussed in an unpublished paper by A.D.C. Simpson, 'Brewster's Society of Arts and the Pantograph Dispute', read at the Second Greenwich Scientific Instrument Symposium, September 1982.

17. Strathclyde Regional Archives, T-KF 6/3, pp. 86-87, Register of William Kyle, copy letter, William Kyle to Peter Hill, 20 October 1815.

18. Offered of sale at Sotheby's Belgravia, 9 November 1977, Lot 59.

19. Offered for sale at Sotheby's Belgravia, 7 September 1979, Lot 116; another offered at Christie's South Kensington, 7 January 1982, Lot 283.

20. At Birmingham Museum of Science and Industry (inventory 53.325).

21. Details of William Blackie (1808-1838) and his short career are to be found in John Coldstream, 'Memoir of the late Mr William Blackie, Optician', *Transactions of the Royal Scottish Society of Arts* 2 (1844), 315-320; and in the *Scotsman*, 31 January 1838, an excerpt taken from the *Caledonian Mercury*.

22. For a fuller account of this episode in the history of microscope optics, see G.L'E. Turner, 'The Rise and Fall of the Jewel Microscope 1824-1837', *Microscopy* 31 (1968), 85-94 and R.H. Nuttall and A. Frank, 'Makers of Jewel Lenses in Scotland in the Early Nineteenth Century', *Annals of Science* 30 (1973), 407-416.

23. David Brewster, *A Treatise on New Philosophical Instruments* (Edinburgh, 1813), 402.

24. Anon. [David Brewster, editor], 'Account of Mr Pritchard's Single Lens Microscope', *Edinburgh Journal of Science* 10 (1829), 328.

25. David Brewster, *Treatise on Optics* (London, 1831), 337.

26. Nuttall and Frank, *op. cit.* (22), 409.

27. Discussed as item 17 (chest microscope with garnet lens) by A.D. Morrison-Low, 'Scientific Apparatus associated with Sir David Brewster', in A.D. Morrison-Low and J.R.R. Christie (eds.), *'Martyr of Science': Sir David Brewster 1781-1868* (Edinburgh, 1984), 90-91; British Museum, inventory 97 12-23.1.

28. In the Royal Museum of Scotland (NMS T1981.L11.3).

29. Nicholas Goodison, *English Barometers 1680-1860* 2nd edition (Woodbridge, 1977), 330, records a stick barometer by Hill. Another was offered for sale by Phillips, 26 January 1983, Lot 1.

30. Anon., 'Proceedings of the Society of Arts for Scotland', *Edinburgh Philosophical Journal* 9 (1824), 190. No examples of this instrument have been recorded.

31. D.J. Bryden, *Scottish Scientific Instrument-Makers 1600-1900* (Edinburgh, 1972), 43.

32. *Ibid.*, 15. There is an example in the Royal Museum of Scotland (NMS T1975.152).

33. *Gray's Annual Directory ... of Edinburgh, Leith, and Suburbs* (Edinburgh, 1832).

34. For discussions of the operation of the Holyrood Sanctuary see E.F. Catford, *Edinburgh The Story of a City* (London, 1975), 96-108, and H. Hannay, 'The Sanctuary of Holyrood', *Book of the Old Edinburgh Club* 15 (1927), 55-98.

35. S.R.O., RH 2/8/20, [copy] Register of Protections of Sanctuary of Holyroodhouse, vol. 6, 1822-1880.

36. Gray's *Directory* 1832, 1833.

37. *Ibid.* 1837; *Post Office Directory* 1834-1840.

38. S.R.O., Abridgements of Particular Register of Sasines, Edinburgh, 1831-36, no. 3580.

39. *Edinburgh Post Office Directory* 1845. The entry is misleadingly given as 'Hill, Peter Smith, philosophical instrument maker'; no entry is given in the classified section.

40. *Ibid.* 1857.

9. Pantograph: P. Hill, Edinburgh, c.1810 (T1980.177)

Copying pantograph in brass, comprising four arms hinged in a horizontal plane, 3 of them with sliding vertical sockets, and supported on ivory wheels; and with a leather-covered lead anchor weight. Engraved on one arm 'P Hill Richmond Street Edin.ʳ'. Although some lines have been scribed across the arms carrying the sliding sockets, they are otherwise undivided; and there is no obvious relationship between these lines. With an extension piece for one limb, but lacking other accessories. With fitted case containing trade card for P. Hill, 7 Richmond Street, Edinburgh.

Length closed: 685mm.
Case size: 700 x 95 x 80mm.

Lacking 5 of the 7 wheel units, the tracing point and pencil holder.

10. Refracting telescope: P. Hill, Edinburgh, c.1820 (T1980.254)

1½″ aperture 3-draw hand-held refracting telescope with doublet objective and 2-component eyepiece and erecting assemblies; in brass, with wooden barrel. Engraved on the first draw tube 'P Hill / Edinburgh'. The objective ferrule and the draw slides scratched with assembly marks 'III', and the barrel similarly marked at one end.

Length closed: 240mm.
Tube diameter: 47mm.
Aperture: 40mm.

Lacking the external eye stop and objective cover.

5   James Veitch (1771-1838) aged 56, by J. Wilson, c.1825.
    W. H. Veitch, Esq.

3

JAMES VEITCH OF INCHBONNY

James Veitch was born at Inchbonny, just south of Jedburgh in the Scottish borders, in 1771.[1] His grandfather had settled first at Mossburnford in Jedwater in 1734, where he was a country joiner and wood merchant, and then in 1738 he had purchased the smallholding of Inchbonny Braes from Sir John Rutherfurd of Hunthill. James Veitch's father William carried on the family business of wright work at Inchbonny, and in 1775 he succeeded with his elder brother James, who farmed there, to an equal share of the property. The heirs of James subsequently sold out to William Veitch, whose family thus became sole owners.[2]

James Veitch was the third child and eldest son of William Veitch and his wife Isabel Miller. His son wrote:

> 'My father was born 1771 and was at first educated at a womans school at Fernihirst afterwards at the school of Mr Clark Jedburgh where he learned to write well and also to be a good arithmetician the study of which with the higher branches of mathematics he took much pleasure in[.] he did not remain long at school but his determination to acquire knowledge increased with his strength.'[3]

At the age of twelve he started to work with his father, using the 'long saw[,]' a kind of hard labour now little known[;] meanwhile he was careful in using his spare time in reading scientific books of every description especially studying mathematics and astronomy.'[4] He was fortunate to have a father who was himself particularly fond of reading self-improving books and conversing with like-minded people in the neighbourhood. For the young James

> 'mechanical knowledge was not neglected he soon began to apply what he knew of mathematics to his da[i]ly employmint [sic] by finding the best curv[e]s for the mouldboard of the plough for diminishing the friction ... a great improvement was made on the plough about that time by James Small of Pat[h]head[.] my father went for sometime and wrought with him[,] where no doubt he acquired a good deal of the practical method of plough making which was beneficial afterwards.'[5]

It appears that Veitch was apprenticed to James Small (c.1740-1793), whose modifications to the plough were to be considered a significant contribution by the agricultural improvers of the Scottish Enlightenment.[6] On his return to Inchbonny, where he set up business as a ploughwright:

> 'my father's shop was frequented by many scientific men both old and young who continued to keep up acquaintance and correspondence with him throughout their lives some of them in very distant places[.] one great patron of his was Admiral Elliot of Mounteviot famous for defeating Thurot of[f] the coast of Ireland[.] he not only gave him encouragement by causing him to make ploughs in order to test his improvements but introduced him to many of the nobility and gentry who were the most scientific agriculturists of the day and most active members of the Highland society many of whom were visitors at Mounteviot and before a deputation of which a public trial was made at Timpendean of my father's plough with many of the best that could be got[.] the superiority over them all was plainly seen both in regard to the work done the much smaller force required to draw it and the care with which it was held by the man[.] among those who interested themselves in the improvements Sir John Riddel of Riddel Lord Cathcart Graham of Balgowan Sir John Sinclair and many others who were visitors of the Minto family[.] In a short time his ploughs were extensively used through all Britain[.] he put them all together with his own hands to insure good workmanship.'[7]

The meeting of the Highland Society was described in their *Transactions*: 'the Committee having met at Admiral Elliot's at Mount Tiviot, on the 8th July 1808, proceeded to examine ... a Dynamometer, made by Mr Veitch'. This was derived from a description of Regnier's dynamometer, and the Committee found it 'to be correctly made, and of sufficient strength to render it particularly well adapted to the comparative trial of ploughs ... [and] appears, to them a mark of genius and industry rarely to be met with, and deserving the notice of the Highland Society.' In the plough trials, Veitch's pattern 'was evidently neater'. The report concluded:

> 'Mr Veitch makes his ploughs himself, with the assistance of a boy and occasionally of a sawyer. He delivers, upon average throughout the year, two ploughs per week. He has hitherto made

them complete for 2l. 16s.; but, from the increased price of wood, and of expense of carriage of his cast-iron mouldboards, he cannot afford to continue to make them under the ordinary price of ploughs of Small's pattern, to the best of which they are fully equal in value of materials, and neatness of workmanship, besides the improvement in construction.'[8]

Veitch described his innovations in a paper published by the Highland Society,[9] and a model of his new plough was deposited in the Highland Society's collection of agricultural scale models.[10] As the portable technology of a largely illiterate society, its impact can be seen in the following letter dated 8 July 1808, addressed to the schoolmaster in Jedburgh:

> 'I had the pleasure of travelling with you on the stage-coach between Hawick and Langholm rather more than twelve months ago, when you showed me the model of a plough made by one Veich in the neighbourhood of Jedburgh. I am in want of a plough at present and I will take it as an obligation if you would have the goodness to desire him to send me one immediately ...'[11]

Ploughmaking remained the main means of Veitch's livelihood until 1826, and was presumably sufficiently renumerative to keep him and his growing family. His son William commented:

> '... my memory of them [his family] began about the year 1810[.] My father was then very busy with the plough trade through the day and at night was either reading or calculating ... about this time my father's business of ploughmaking was at its best and his ploughs were in much request throughout both England and Scotland[.] those made of iron were not in use for some years after[.] many of the nobility and gentry throughout the country about that time took a great interest in agriculture and the improvement of the implements used.'[12]

However, despite this busy working life making agricultural implements, Veitch made time to continue studying. William recorded that his father was first influenced by a local man some years his senior, Alexander Scott of Fala, who appears to have been interested in mechanics, and who gave James Veitch and his own younger brother Thomas each 'a speculum of about 2½ inches diameter for a Gregorian telescope setting them both to work to make a telescope about 15 inches long which my father accomplished ... [it was the] first he had made'.[13] A number of men with similar interests and enquiring minds were mentioned by William Veitch as being in the habit of gathering at his father's shop to discuss astronomy, mathematics and other branches of knowledge: Robert Easton, a surveyor; James Fair of Langton; James Scott, later minister at Dalkeith; Robert Hall, who also became a minister; Amos Reid; James Anderson, schoolmaster; George Noble, poet; and David Brewster - 'all having a desire of acquiring a knowledge of natural philosophy and communicating what they had learned at College or otherwise and some of them being very clever in many branches of science made their meetings very entertaining'.[14]

Of those named, perhaps the one who became most famous was the natural philosopher David Brewster (1781-1868): 'he was the very youngest of the quaint and varied group', wrote his daughter. 'When he began his visits I do not know, but we find that at the age of ten he finished the construction of a telescope at Inchbonny, which had engaged his attention at a very early period, and at which he worked indefatigably, visiting the workshop daily, and often remaining till the dark hours of midnight, to see the starry wonders and test the powers of the telescopes they had been making.'[15] Brewster was a son of a Jedburgh schooolmaster and, after a spell at the University of Edinburgh, was destined for a career in the Church. However, he found that this did not suit him temperamentally, and he opted for a less financially secure profession, as an editor of scientific journals and encyclopaedias.[16] While mostly living away from the Borders after his childhood, he maintained a lifelong correspondence with his old friend Veitch on optical and astronomical matters.[17]

His letters are filled with allusions to science and instrument making, besides mention of the Edinburgh luminaries whom he met and his thoughts on divinity when still a student. Whilst attending the University of Edinburgh Brewster wrote regularly for the *Edinburgh Magazine*, a periodical combining science and literature. By 1802 he had abandoned ideas of the ministry and become its editor.[18] In 1805 he wrote to Veitch to tell him that

> 'You will probably have seen that I noticed [i.e. discussed] your new Plough in one of the late numbers of the *Edinburgh Magazine*. On account of my distance from Leith, however, it has hitherto been out of my power to see your models at the Foundry.'[19]

By 1808 Brewster was immersed in another great project, one which was to absorb his energies until 1830, the *Edinburgh Encyclopaedia*. Again, he enlisted the assistance of his friend James Veitch:

> 'We are now at the article AGRICULTURE, and are about to send the drawings of the ploughs to the engraver. It occured to me, however, that it might be of use to you to have a drawing and description of your new plough inserted ... If, therefore, you could send me by post, as soon as possible, a drawing and description of your plough, I shall with pleasure publish them in our Encyclopaedia.'[20]

Brewster was not by any means Veitch's sole patron, although he was possibly the most influential in the long term as far as scientific activities were concerned. The network of patronage in early nineteenth century agricultural and scientific affairs appears to have had extended ramifications which Veitch's son William attempted to explain:

> 'Through the influence of the neighbouring gentry chiefly Admiral Elliot Lord Minto and the Minto family my father was introduced to most of the leading agriculturists of the time Though then busily engaged in hard work through the day[,] at night he was equally busy with the study of Astronomy mathematics and every branch of natural philosophy especially Optics[.] He made many reflecting Telescopes for gentlemen who often visited him when with their friend[s] in the neighbourhood many of whom were fond of scientific pursuits such as Lord Minto Rutherfurd of Edgerton Sir Henry MakDougal Sir Thomas Brisbane Sir Walter Scott Dr Sommerville [*sic*] &c who were always very kind to my father and introduced him to other visitors who were some of the most scientific and learned men in the kingdom[.] from them he was kept informed of any improvements or new discoveries made by the scientific men of the time[.] he was also presented with many books of which they were the authors and kept up a considerable correspondence with many of them (which may be seen by his library and manuscripts preserved)'[21]

Veitch's astronomical studies brought him some fame. Through his correspondence with Brewster his observations of the eclipse of the sun in February 1804 were published.[22] He claimed to be the first person in Britain to have observed the great comet of 1811,[23] and he observed the occultation of Jupiter.[24] In 1836 he assisted the astronomer Francis Baily in the observation of the annular solar eclipse of 15 May, the ground track of which passed over Inchbonny, a site conveniently close to Sir Thomas Makdougall Brisbane's observatory at Makerstoun, where extra instruments could be obtained and the chronometers rated. The term 'Baily's Beads', describing the apparently discontinuous ring exhibited by the Sun at the centre of the eclipse, was coined from the published description of this eclipse.[25]

However, Veitch's instrument making brought him further renown. His wooden ploughs began to lose their former popularity as iron ones displaced them, and he appears to have stopped making them altogether about 1826. This was the year that Parliament brought in the Imperial system of standard weights and measures, which in theory replaced the old Scottish local standards, and which had been established by an 1824 Act of Parliament. Through Lord Minto

> 'my father was engaged in making the comparisons between the old standards and the new in the counties of Roxburgh Berwick and Selkirk and in making out Tables for calculating their different proportions in these Counties which were published in each County by order of the Sheriffs[.] He was also appointed inspector of weights and measures for the County of Roxburgh and the Burgh of Jedburgh[.] He was to get a salary of Ten pounds a year from the County and two pounds yearly from the Burgh and if sent to the different towns such as Kelso Berwick Melrose &c one guinea for each day so engaged[.] he also got new Bushel measures for the whole County[.] This was a very good job for a while the people getting soon supplied[.] he also got the fees for stamping which did not amount to much[.] This with occasionally making Telescopes microscopes Barometers &c was all the work he now did.'[26]

As Weights and Measures Inspector for the area, Veitch drew up and published the tables for converting the old system to the new.[27] This brought him into contact with the novelist Sir Walter Scott, as Sheriff of Selkirkshire.[28] Scott, too, recommended Veitch to friends,[29] and he himself owned a clock and a telescope made by Veitch.[30]

Besides Scott, the list of Veitch's known customers for telescopes is impressive: Mrs Mary Somerville, scientific authoress;[31] Sir Thomas Makdougall Brisbane, soldier and astronomer;[32] local landowners Sir Henry McDougall and Mr Rutherfurd of Edgerston, and the minister at Jedburgh, Dr Thomas Somerville;[33] Charles Hope, Lord President of the Court of Session;[34] Sir Alexander Muir Mackenzie of Delvine;[35] Professor John Playfair;[36] Lord Renton;[37] the Earl of Hopetoun;[38] Professor Heinrich Schumacher of Altona Observatory;[39] the Earl of Minto;[40] Robert Shortrede of Jedburgh;[41] John Shortrede, the latter's son.[42] Besides these telescope sales, he sold an electrical machine to a Mr Charlton for fifteen shillings in 1814,[43] a magic lantern to James Douglas of Cavers for £11 in 1830,[44] and a microscope to Major Elliot for a guinea in 1817, and another for two guineas to a Mr Selby of Twizel Castle in 1834.[45]

James Veitch kept meticulous accounts throughout his working life, and these reveal his three main sources of income: as a ploughwright, as an instrument maker and as a weights and measures inspector. The accounts also demonstrate that the emphasis on each of these activities changed through different periods in his life.[46] The first mention of a sale of an instrument (as opposed to making one for himself) was on 12 February 1811, to Lord Hermand, of a reflecting telescope costing eight guineas, and 3s 6d for its box.[47] These sales gradually increased, and by 1821 there was a marked diminution in sales of agricultural implements; however, by 1826, the account book makes it quite clear that most of Veitch's income was due to his weights and measures work. Even so, the proportion of his wealth made by instrument sales was not negligible.

Telescopes were, without doubt, Veitch's main line in instrument making, although he did construct microscopes, barometers and other philosophical apparatus too. The telescope in this collection sadly lacks a provenance,[48] although another, dated 1820, which has recently come to light may well have once belonged to one of Veitch's ploughing patrons as it is inscribed 'Admiral Elliot' on the barrel. This is unlikely to have been his early patron Admiral John Elliot of Mount Teviot, who died in 1808; but more probably Admiral Sir George Elliot (1784-1863), second son of the first Earl of Minto.[49] David Brewster mentioned at least four different telescopes made by Veitch in his letters,[50] and disparaged the optical quality of those he found for sale in Edinburgh compared with those made at Inchbonny. However, the exterior appearance of those made in the capital led to the rueful comparison that a particularly poor instrument was 'fitted up in a fine brass tube & mounted on an excellent Stand, whereas ours bear a greater resemblance to *coffins* or *waterspouts* than anything else'.[51] In due course, Brewster was to mention Veitch's telescope making prowess in the same breath as that of James Hadley and James Short (q.v.).[52]

For an important series of investigations into the refractive index of various materials, Brewster asked Veitch to make him a replacement microscope lens in 1812,[53] and although Veitch's workmanship was not wholly successful,[54] Brewster turned to his old mentor again for some delicate grinding work when he began experimenting with the optical properties of precious minerals.[55] In 1828 he sent Veitch a piece of garnet from his mineral collection to be made into microscope lenses.[56] Although no microscope with jewel lenses by Veitch appears to be extant, it is possible that an example, now in the Royal Museum of Scotland, is of the type for which garnet lenses were intended.[57]

Another protégé of Brewster, James David Forbes, who was to become professor of Natural Philosophy at the University of Edinburgh in direct competition with his patron,[58] wrote to Brewster:

> '... When at Jedburgh I saw Veitch. He has a most exquisite single garnet lens: I saw the lenticular structure of the eye you showed me quite as well (you know that it was not a remarkably favourable opportunity when I saw yours) as through the Sapphire lens; looked as well *as possible*. The two garnets Gen. Leslie spoke of were lenses of different power. Could you give me any idea of what Veitch would expect for repolishing a 4 or 5 inch metal of a four feet reflector?...'[59]

Indeed, by 1830 Veitch was receiving some attention in the local press for his ability in the construction of microscopes with garnet lenses.[60]

Several instruments have remained in the possession of Veitch's descendants, and illustrate the diversity of the material which he was able to produce.[61] An undated example of a stick barometer signed by Veitch is held by the Royal Museum of Scotland.[62]

Our knowledge of the extent of his instrument sales must raise Veitch's status from one of amateur instrument maker to that of a professional. What makes his case so unusual is that it is documented. Few makers were as isolated as he was from colleagues, training, and ancillary suppliers; he was not based in a city, where he could fit into a ready-made market and he had no workshop, in the sense that he is not known to have had outworkers or apprentices of his own. It is apparent from his son's account that he did not encourage any of his children to imitate his scientific interests, although the youngest of them, John, was apprenticed to a Jedburgh watchmaker named Shairp and 'excelled making Telescopes microscopes Dials &c'.[63] Unfortunately, he died before the age of eighteen.

Perhaps James Veitch was not a particularly unusual figure for his times, although rare enough to be remarked upon by contemporaries. He appeared to lack the money-making incentive that our own age finds so peculiarly attractive, and he made sufficient from his ploughs and his smallholding to keep his family from want and himself scientifically stimulated. The figure he most resembles is John Gibson of Kelso (d.1795),[64] a clockmaker and instrument maker, who was a generation older than Veitch, but may have had some influence over him.[65] Of the businesses discussed in this work, his is perhaps closest to that of Thomas Morton of Kilmarnock (q.v.) in that telescope making was not his main livelihood; he was far removed in circumstance and outlook from the thriving businesses of the Adies (q.v.) or the Gardners (q.v.), or the later financial concerns of James White (q.v.). He was, perhaps, particularly fortunate in his patrons, especially the young David Brewster. He died 'between 9 and 10 in the morning of the 10th June 1838' aged 67.[66]

REFERENCES

1. The main published account of Veitch's life has appeared in a number of places: 'Astron', 'James Veitch, Philosopher, Inchbonny', *Jedburgh Post*, 29 September and 6 October 1899; reprinted as George Watson, *James Veitch, Philosopher, Inchbonny, Jedburgh* (Jedburgh, 1899) and idem., *The Border Magazine* 5 (1900), 15-18, 34-36 and 45-48.

2. George Tancred, *Rulewater and its People* (Edinburgh, 1907), 220-221.

3. Veitch Papers: William Veitch, manuscript autobiography, 11-12. We are grateful to W.H. Veitch Esq., for permission to consult and quote from items held in the Veitch papers, which are listed by the National Register of Archives (Scotland), survey 0337, 11-14.

4. *Ibid.*, 12.

5. *Ibid.*, 12-13.

6. For a comprehensive discussion of the agricultural background to James Veitch's ploughmaking activities, see Hugh Cheape, 'James Small and Plough Innovation', *Acta Museorum Agriculturae Pragae* 18 (forthcoming). We are grateful to Hugh Cheape for letting us read his unpublished paper. An abstract of this paper appeared as Hugh Cheape, 'James Small och den agrare revolutionen i England', *Folkets Historia* 15 (1987), 30-41, with a summary in English, 57.

7. Veitch, *op. cit.* (3), 15-16.

8. 'Report of a Committee of the Society to whom it was referred to examine particularly and to report upon a new Plough stated to be of an Improved Construction made by James Veitch of Inchbonny near Jedburgh', *Prize Essays and Transactions of the Highland Society* 4 (1816), 243-247.

9. James Veitch, 'Description of the Improvements made on the Original and Ordinary Construction of Small's Plough', *ibid.*, 248-252.

10. The Veitch model plough from the Highland and Agricultural Society's collection is now in the Royal Museum of Scotland, NMS 1865.79.14.

11. National Museums of Scotland, NMAS MS Archive 1981/59. Our thanks to Hugh Cheape for bringing this manuscript to our attention.

12. Veitch, *op. cit.* (3), 21-22.

13. *Ibid.*, 18-19.

14. *Ibid.*, 19.

15. M.M. Gordon, *The Home Life of Sir David Brewster* (Edinburgh, 1869), 30.

16. W.H. Brock, 'Brewster as a Scientific Journalist', in A.D. Morrison-Low and J.R.R. Christie (eds.), *'Martyr of Science': Sir David Brewster 1781-1868* (Edinburgh, 1984), 37-42.

17. N.R.A.(S.) survey 337; the letters from Brewster to Veitch have been microfilmed and can be consulted at the Scottish Record Office, RH/4/21, and we are grateful to W. H. Veitch for permission to quote from them; some of these are reproduced by Mrs Gordon, *op. cit.* (15), 37-66. Letters from Veitch to Brewster must be presumed destroyed in the fire at the Brewster family home at the turn of the century.

18. A.D. Morrison-Low, 'Published Writings of Sir David Brewster: a Bibliography', in Morrison-Low and Christie, *op. cit.* (16), 107-108.

19. S.R.O., RH/4/21.45, Brewster to Veitch, 9 December 1805. D[avid] B[rewster], 'Memoirs of the Progress of Manufactures, Chemistry, Science and the Fine Arts', *Scots Magazine* 67 (1805), 735; Gordon, *op. cit.* (15), 58. The rather complicated history of Brewster's editorship of various Edinburgh periodocals is outlined by Brock, *op. cit.* (16), 37 and especially his note 3, where he discusses the *Edinburgh Magazine* and the *Scots Magazine*.

20. Gordon, *op. cit.* (15), 65-66. Veitch's contribution appears in the article 'Agriculture', *Edinburgh Encyclopedia* (Edinburgh, [1808]-1830), I, 252-253 and is illustrated Plate VI, fig 11.

21. Veitch, *op. cit.* (3), 22-23.

22. D[avid] B[rewster], 'Memoirs of the Progress of Manufactures, Chemistry, Science and the Fine Arts', *Scots Magazine* 66 (1804), 203.

23. Mentioned in T. Dick, *The Diffusion of Knowledge* (London, 1833), 73; T. Dick, *Sidereal Heavens* (London, 1840), 461; and by Martha Somerville (ed.), *Personal Recollections from Early Life to Old Age of Mary Somerville* (London, 1874), 100; the letter from Veitch to Mrs Somerville dated 12 October 1836 describing his discovery, 101-102. All of these references are cited by Watson, *op. cit.* (1).

24. *Edinburgh Journal of Science* 1 (1824), 179.

25. Francis Baily, 'On a Remarkable Phenomenon that occurs in Total and Annular Eclipses of the Sun', *Memoirs of the Royal Astronomical Society* 10 (1836), 1-40; *Monthly Notices of the Royal Astronomical Society* 3 (1836), 199. Discussed by Watson, *op. cit.* (1), and by J. N. McKie, 'James Veitch, 1771-1838', *Journal of the British Astronomical Association* 87 (1976), 48.

26. Veitch, *op. cit.* (3), 40-41.

27. James Veitch, *Tables for converting the Weights and Measures, hitherto used in Roxburghshire into the Imperial Standards, as established by Act 5, Geo. IV. Cap.74...* (Jedburgh, 1826); the only copy of this pamphlet in a public collection traced so far is in Hawick Area Library. Our thanks to Hugh K. Mackay, Deputy Librarian, for his assistance. James Veitch, *Tables for converting the Weights and Measures hitherto used in Selkirkshire into the Imperial Standards ...* (Edinburgh, 1828); the only copy of this pamphlet traced so far is in the Borders Regional Library Headquarters, Selkirk. Our thanks to Alan Carter, Regional Librarian, for his assistance. The third pamphlet, in which Veitch converted the weights and measures of Berwickshire into the Imperial standards, has not been located, although it is described by Watson, *op. cit.* (1), and Gordon, *op. cit.* (15), 24. Veitch's appointment was announced in the *Kelso Mail*, 13 and 27 April 1826; the Roxburghshire *Tables* were announced on 4 May 1826, and advertised on 6 July 1826. The MS. of this advertisement is in the Veitch Papers, annotated with a list of names of people who purchased the pamphlet from James Veitch.

28. Mrs Gordon recounts how Sir Walter, initially ignorant of the technicalities of the experiments, soon mastered the problem (Gordon, *op. cit.* (15), 24).

29. Scott wrote to James Ellis of Otterburn: 'I heard these particulars from James Veitch, a very remarkable man, a self-taught philosopher, astronomer, and mathematician,

residing at Inchbonny, and certainly one of the most extraordinary persons I ever knew ... James Veitch is one of the very best makers of telescopes, and all optical and philosophical instruments, now living, but prefers working at his own business as a ploughwright, excepting at vacant hours. If you cross the Border, you must see him as one of our curiosities; and the quiet, simple, unpretending manners of a man who has, by dint of private and unaided study, made himself intimate with the abstruse sciences of astronomy and mathematics, are as edifying as the observation of his genius is interesting.' (quoted by Gordon, *op. cit.* (15), 25-26; from Willis' *Current Notes*, 25 January, 1856). This letter is quoted in full in H.J.C. Grierson (ed.), *The Letters of Sir Walter Scott* (London, 1932-37), IV, 221.

30. Veitch wrote to Scott on 12 October 1820 to say that the telescope was ready: 'I think the best way of carrying it will be a basket with a belt round a mans shoulder. I would have sent it long ago with the Coach but was afraid of getting it spoiled. If you have a man to spare you can send him on a work horse which will carry it safely. I will make your Clock as soon as possible. Jupiter and Saturn is both seen at night. I see Jupiters moons and his belts with your telescope and likewise Saturns ring.' (*ibid.*, VI, 227). Scott replied on 20 October: 'I send my piper for the telescope with a basket. He has charges to be particular in his care of it and I think will do better than a lad on horseback. I am dear James with regard Yours &c. Walter Scott' (*ibid.*, VI, 277-278). Another anecdote linking Veitch and Scott is worth quoting: 'There is a watchmaker at Jedburgh who has a very extraordinary genius for mechanics. His telescopes, watches, and clocks are of the very first description, and he is nearly self-taught. One day he ran into the street calling aloud to a friend: "I'm the happiest man in a' the world. I've just brought Lord Minto and Sir Walter to agree." "I did not know they had quarrelled," said his friend. It proved to be two timepieces which he was making for them.' (Horace G. Hutchinson, (ed.), *Letters and Recollections of Sir Walter Scott by Mrs Hughes of Uffington* (London, 1904), 274-275). The clock had to wait at Inchbonny until it had a proper home: 'As I am about to build at Abbotsford, I will not trouble you to fetch over the clock till that job is finished; I will then have a better and more distinguished situation for the work of your hands.' (Gordon, *op. cit.* (15), 26). The grandfather clock by Veitch is still at Abbotsford: we are grateful to Mrs Patricia Maxwell-Scott for this information. Another clock with Veitch's signature is also in a private collection.

31. Watson, *op. cit.* (1). '...James Veitch ... made excellent telescopes, of which I bought a very small one; it was the only one I ever possessed' (Somerville, *op. cit.* (23), 99).

32. Watson, *op. cit.* (1). 'Reflecting telescope, by James Veitch, Inchbonny, mounted equatorially, by Adie & Son, diameter of reflection 3 inches', Lot 48 in *Catalogue of the Valuable Astronomical and Philosophical Instruments of the late General Sir Thos. Makdougall Brisbane Bart... which will be sold by auction by Mr T. Nisbet in his Great Room No 11 Hanover Street Edinburgh, ... April 4, 1860* (Edinburgh, 1860), 3: Royal Greenwich Observatory, RGO 6/170 Section 38, 268. Our thanks to Dr John Chaldecott for drawing our attention to this pamphlet.

33. Watson, *op. cit.* (1). An expanded revised manuscript version of Watson's biography of James Veitch remains with the Veitch Papers and contains further references to customers and shall henceforth be referred to as the 'Veitch Biography'.

34. Veitch sold a telescope and a plough to Hope for £20 in 1812 (Veitch Biography, *op. cit.* (33), 37).

35. Veitch sold Delvine a telescope for £21 in 1815 (*ibid.*).

36. Veitch supplied Playfair with 'two specula and two glasses for a reflector, at the price of £6 15/-' (*ibid.*). These were destined for the Natural Philosophy Class of the University of Edinburgh; but by the time the first systematic inventory was made in 1831, these had become 'CU 7 Single lens by Veitch'; this survival is no longer extant.

37. 'In 1819 Veitch made two telescopes for Lord Renton (*ibid.*).

38. *Ibid.*

39. 'The price of this splendid instrument was twenty-six guineas' (Gordon, *op. cit.* (15), 102; also mentioned in Veitch Biography, *op. cit.* (33)). This instrument was ordered through Brewster, mentioned by him in a letter dated 9 October 1821 and was a '... best reflecting telescope 2 feet 8 inches focal length and 5 inches aperture, completely mounted on a stand of Brass &c', costing about £25; it was to be delivered to Altona, Hamburg (S.R.O., RH/4/21.58). Brewster mentions that he is pleased to hear the instrument is almost complete on 23 April 1823, as Schumacher is becoming impatient (S.R.O., RH/4/21.59). For Schumacher, see Sister Maureen Farrell, F.C.J., 'Heinrich Christian Schumacher' *Dictionary of Scientific Biography* XII (New York, 1975), 234-235.

40. 'A 3-foot telescope was constructed in 1821 on the same plan [as for Professor Schumacher] for the Earl of Minto, at the price of £26 5/-' (*ibid.*; Gordon, *op. cit.* (15), 102); for the account of the clock which Veitch made for the Earl of Minto, see reference 30 above.

41. 'In 1822 he supplied a telescope to Robert Shortreed of Jedburgh, the friend of Scott, for £5' (*ibid.*, 38).

42. 'Veitch executed a Newtonian reflector with a magnifying power of 500, in the year 1835, at the price of 35 guineas' (*ibid.*).

43. *Ibid.*, 38-39.

44. *Ibid.*, 41.

45. *Ibid.*, 43.

46. There are four account books in the Veitch Papers; one dated 1799-1819, contains plough repairs and farm rents; a second, although dated 1796-1815, is concerned mainly with later weights and measures, their prices and the price of adjustment. A third account book dated 1805-1811 deals with ploughs. A fourth account book starts with agricultural implement sales from 12 June 1812, and, interspersed with instrument sales, finishes with the inspection of a weighing machine on 17 October 1837. These document quite clearly Veitch's transition from a ploughwright to an instrument maker, through to the financial security of the weights and measures appointment of his old age.

47. Veitch Papers, Account Book 1809-1837. Lord Hermand to Veitch 8 October 1810, mentions a telescope; Lord Hermand to Veitch 21 February 1811: 'The telescope came last week, & with the letter being addressed to Hermand did not reach me till some days later ...'; the account written on the same letter shows that Lord Hermand paid Veitch £3 12s for a plough and spare sock (ploughshare) in October 1810, and a total of £12 3s 6d in February 1811 for the reflecting telescope and its box. Lord Hermand was a law lord, and was mentioned in

John Kay, *Series of Original Portraits ...* (Edinburgh, 1842), II, 380.

48. NMS T1981.37; discussed by A.D. Morrison-Low, 'Scientific Apparatus associated with Sir David Brewster: An Illustrated Catalogue of the Bicentenary Display at the Royal Scottish Museum 21 November 1981-9 April 1982', in Morrison-Low and Christie, *op. cit.* (16), 82-83.

49. NMS T1987.316. This 5 inch Gregorian telescope is signed on the brass band around the lower end of the octagonal wooden tube: JAMES VEITCH INCHBONNY 1820. The tube is 48 inches long and it has no stand. It was offered for sale by Sotheby's, 28 October 1986, Lot 190.

50. S.R.O., RH4/21, letters of Brewster to Veitch, 29 March 1799 (RH4/21/9) '... your Gregorian Telescope of 3 feet'; 21 March 1800 (RH4/21/25) '... your 25 Inch Newtonian reflector'; 26 December 1800 (RH4/21/31) 'I am so happy to hear that you have got such a good Speculum for your Seven feet Reflector, & that it shews so distinctly'; and the two letters referring to Professor Schumacher of Copenhagen's order for a best reflecting telescope discussed in reference 39.

51. S.R.O., RH4/21/30, Brewster to Veitch, 2 October 1800; Gordon, *op. cit.* (15), 39.

52. 'Looking back from the present advanced state of practical science, how great is the contrast between the loose specula of Gregory and the fine Gregorian telescopes of Hadley, Short, and Veitch, - between the humble six inch tube of Newton and the gigantic instruments of Herschel and Ramage' (David Brewster, *The Life of Sir Isaac Newton* (London, 1831), 28).

53. S.R.O., RH4/21/52, Brewster to Veitch, 7 August 1812, 'In consequence of having completely broken the object glass of the microscope with which I have made all my experiments in Refractive powers, I write to you at present to beg that you would have the goodness to grind me a lens equally convex on both sides, about ⅓ of an inch in diameter, & having the radius of each surface 1.16 inches - I have sought thro' all the opticians here and cannot get a glass of this kind, & they are such bunglers in the grinding of lenses that I could not trust to them for one on which so much depends. As the lens must have the same focal length & be made of glass of the same refractive power as the one which I broke, & as I do not know whether it was made of crown glass or plate glass, you would oblige me very much if you could grind me two, one of each kind of glass ...'

54. S.R.O., RH4/21/53, Brewster to Veitch, 19 September 1812, 'I am much obliged to you for the two lenses which you sent me, & for the telescope which you propose to make for me, for which I shall take care to have a good stand provided. The lenses, tho' very good & nearly of the proper focal lengths, were unequally convex from which circumstance they would not suit my experiments ...'

55. The subject of jewel lenses has received extensive treatment: G. L'E. Turner, 'The Rise and Fall of the Jewel Microscope', *Microscopy*, 31 (1968), 85-94, and R.H. Nuttall and A. Frank, 'Makers of Jewel Lenses in Scotland in the Early Nineteenth Century', *Annals of Science* 30 (1973), 407-416. Brewster first suggested the idea of a lens with a high refractive index in his *Treatise on New Philosophical Instruments...* (Edinburgh, 1813), 402.

56. S.R.O., RH4/21/63, Brewster to Veitch, 12 October 1828, 'I send you from my collection of Minerals a fragment of Garnet sufficiently large for your purpose, but I do not see how you can get it cut at Jedburgh with the pieces you require. This can only be done under your own eye in Edinburgh ...' Nuttall and Frank, *op. cit.* (55), 410, explain that the garnet came from Sir Charles Giesecké (1761-1833) who had collected it on a six year expedition to Greenland (obituary of Charles Giesecké, *London and Edinburgh Philosophical Magazine* 3rd series, 4 (1834), 445-446). Brewster commented on this in his *Treatise on Optics* (London, 1891, 337-338, and the subsequent edition, *Optics* (London, 1838), 337-338, and in his article on the 'Microscope' produced for the seventh edition of *Encyclopaedia Britannica* XV (Edinburgh, 1837), 31. The remaining pieces of garnet have been described by Morrison-Low, *op. cit.* (48), 89.

57. NMS T1979.96; this item has no provenance. It was offered for sale by Christie's South Kensington, 31 May 1979, Lot 143. Described by Morrison-Low, *op. cit.* (48), 90, and discussed more fully by R.H. Nuttall, 'A Simple Microscope by James Veitch of Inchbonny', *Microscopy* 34 (1983), 569-573.

58. J.B. Morrell, 'Brewster and the early British Association for the Advancement of Science', in Morrison-Low and Christie, *op.cit.* (16), 25-29.

59. St. Andrews University Library, Forbes Correspondence, Letterbook I, 198-199, J.D. Forbes to Sir David Brewster, 11 September 1830.

60. 'From a correspondent - I think that it is due to the merits of a most deserving artist to call the attention of the scientific and the curious to a microscope of very astonishing power lately made by Mr James Veitch of Inchbonny. The lens is formed of garnet, and so perfect is the polish and configuration, that, though it magnifies to the astonishing amount of three hundred times, it gives neither an obscure nor imperfect, nor double vision. This has long been a desideratum in the garnet lens; but the difficulties which have hitherto frustrated the labours of so many artists of eminence have at last been completely surmounted by the skill and indefatigable perseverance of Mr Veitch. On the same frame he has fixed another lens, of the same material, and magnifying one hundred and fifty times, and both together are so fitted up as to be, with the greatest facility, adjusted by anyone who is familiar with the use of instruments of great power and delicacy - Jedburgh September 4, 1830' (*Kelso Mail*, 13 September 1830; reprinted *ibid.*, 10 August 1945).

61. Now on display at the Visitor Centre, Royal Observatory, Edinburgh; these were exhibited at Inchbonny in 1924 and described in 'Relics at Inchbonny', *History of the Berwickshire Naturalists' Club* 25 (1924), 216-217. Items not restricted by entail were sold at Christie's, 9 April 1975, Lots 38 to 46.

62. NMS T1984.2; it was offered for sale by Christie's, 16 December 1982, Lot 21. It has no provenance.

63. Veitch, *op. cit.* (3), 39.

64. Sir John Sinclair Bt. (ed.), *The Statistical Account of Scotland* X (Edinburgh, 1794), 591.

65. That Veitch knew Gibson is revealed in a letter in the Veitch Papers sent by his brother William Veitch to James Veitch 23 April 1813, describing William's visit to the shop of the instrument maker Charles Tulley in Islington on James' behalf: '... he is not backward either to speak or shew me anything I ask him he is very lick [i.e. 'like'] John Gibson ...'

66. Veitch, *op. cit.* (3), 66, gives a harrowing deathbed scene; he suffered from severe chest pains, and vomited blood.

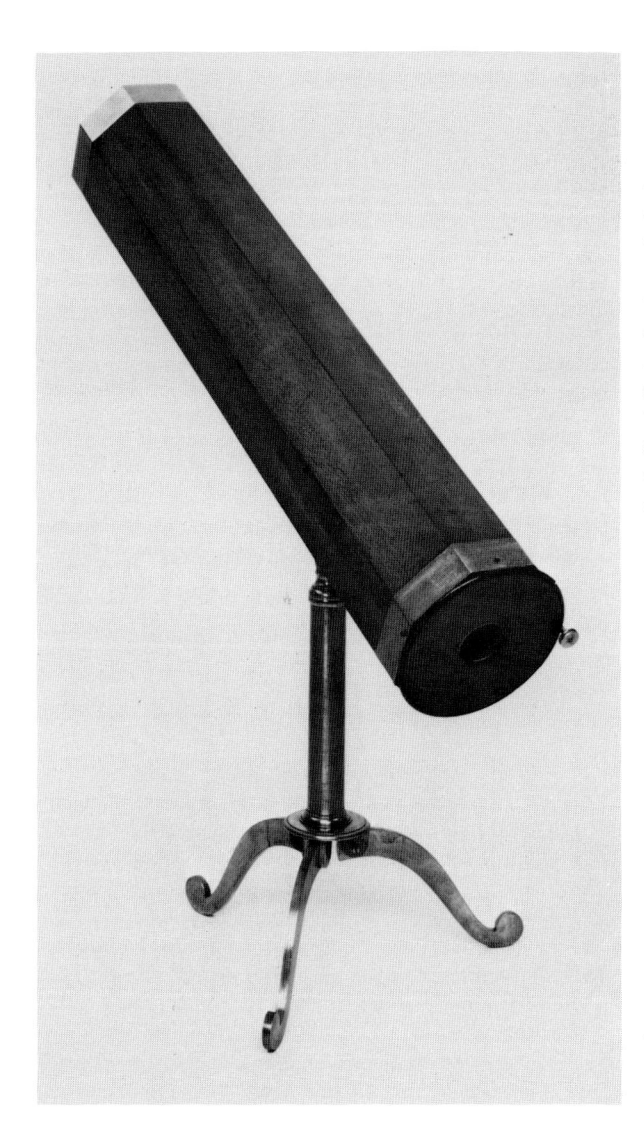

11. Reflecting telescope: James Veitch, Jedburgh, c.1820 (T1981.37)

18" focus mounted Gregorian reflecting telescope, with octagonal brass-bound wooden barrel and speculum metal optics, but lacking eyepiece. Engraved on the brass band at the lower end of the tube 'JAMES VEITCH / INCHBONNY'. External focusing rod, the threaded end working on a projection from an internal sliding iron mount carrying the secondary speculum on a 3-point adjustment. The primary speculum cast with a chamfered outer edge, a filed-down edge sprue, and some spotting on the mirror surface; retained between two sets of 3 brass brackets within the tube, and with a turned wooden end plate with a threaded aperture for an eyepiece. On a brass pillar mount with geared altitude motion (defective) and folding table tripod stand (the legs replaced).

Length of barrel: 630mm.
Barrel external width: 124mm.
Speculum diameter: 104mm.

The upper brass band on the barrel is a replacement, constructed in the Museum in 1981; the lower band has split along its original join. The instrument lacks its eyepiece, the centre support for the focusing rod and the tangent screw for altitude adjustment. The legs are not compatible with the tripod base.

Provenance: Acquired privately in Glasgow, c.1975. It was exhibited in a special exhibition commemmorating the bicentenary of the birth of Sir David Brewster in 1981: A.D. Morrison-Low, 'Scientific Apparatus associated with Sir David Brewster: An Illustrated Catalogue of the Bicentenary Display at the Royal Scottish Museum 21 November 1981 - 9 April 1982', in A.D. Morrison-Low and J.R.R. Christie (eds.), *'Martyr of Science': Sir David Brewster 1781-1868* (Edinburgh, 1984), 82-83. It is not known for whom Veitch constructed this instrument.

4

THE ADIE BUSINESS

John Miller, his nephew Alexander Adie and Adie's four sons, together form the most talented and creative of the family businesses discussed in this volume. Both their commercial success and their technical and scientific achievements place them in a wider context than merely that of highly successful provincial instrument makers. Although it may ultimately be concluded that Edinburgh was no more than a provincial centre of instrument production dominated from England, the history of the Adie business makes it clear that such a judgement needs to be heavily qualified.[1] The Adies flourished in the midst of a vital scientific community which, while it kept abreast of developments outside Scotland, valued its own independence and prided itself on its evident superiority in many fields.

John Miller (1746-1815), who came to rank among the foremost eighteenth century Scottish instrument makers, was the son of John Miller, an Edinburgh turner.[2] His early career remains a matter of conjecture, although two reliable sources agree that he worked for a period in London in the large and prestigious work-

6 Level, by John Miller. From a plate in John Francis Erskine of Mar, *General View of the Agriculture of the County of Clackmannan ...* (Edinburgh, 1795)

shops of George Adams, instrument maker to George III. Miller first re-appears in Edinburgh in 1769 when he was described as '... from this place, and bred by Adams in Fleet Street',[3] and a second much later account states that he 'was educated as an optician by Adams of London, where he resided for some years'.[4] Neither of these references suggests that Miller entered a formal apprenticeship with Adams, for whom he probably worked as a qualified journeyman, and certainly no apprenticeship was recorded in the books of Adams' London guild company.[5] It remains possible of course that the arrangement was a more informal one in which a talented but unqualified craftsman was taken on for an enhanced financial premium. However, the likelihood that he had already served an apprenticeship in Edinburgh gives more credence to the conjecture (probably based on a tradition in the firm) that he served under John Yeaman.[6]

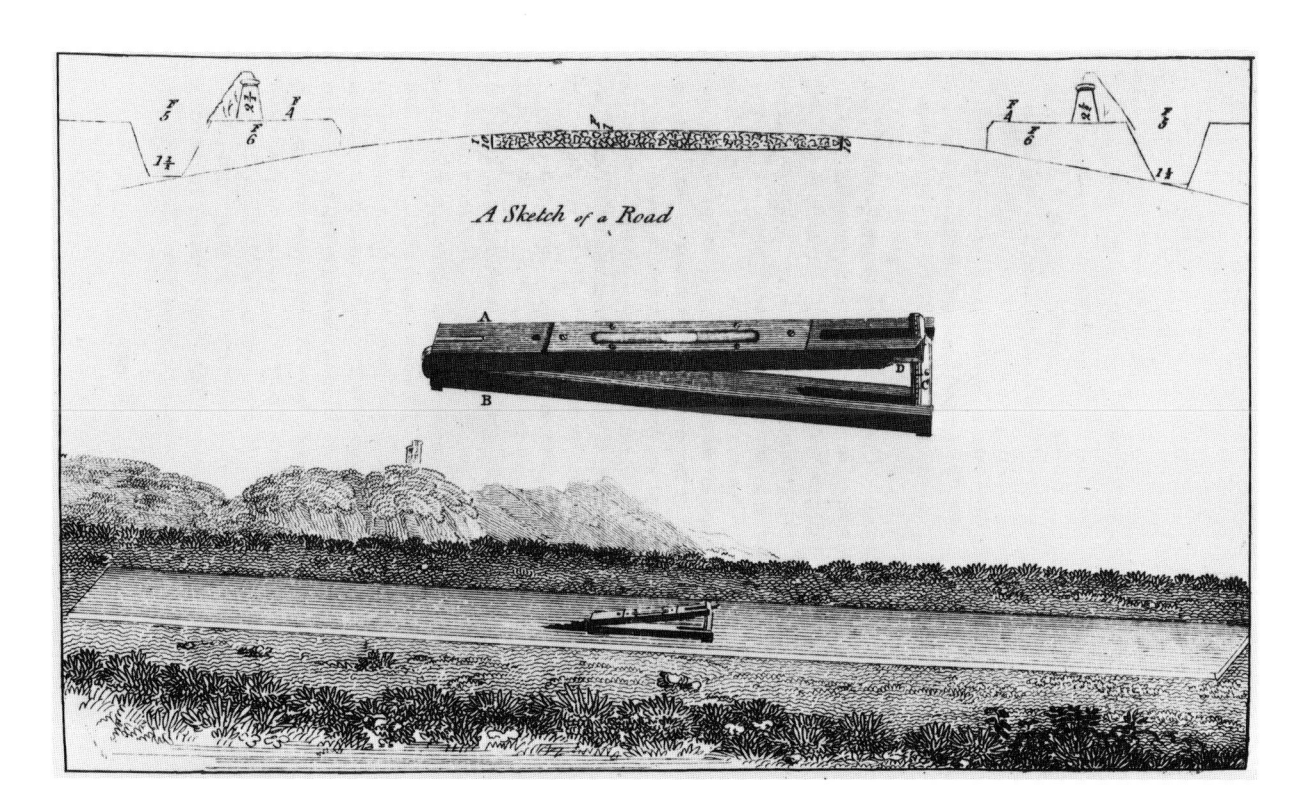

Yeaman was one of the few instrument makers working during the likely period of Miller's apprenticeship in the 1760s. He is recorded as a mathematical instrument maker in Edinburgh between about 1752 and 1780.[7] However, he can now be identified with the John Yeaman previously listed as a watch and clock maker between 1734 and 1749 in the separate burgh of the Canongate, immediately to the east of the old Edinburgh city wall, but describing himself from at least 1745 as a 'mathematical instrument maker'.[8] Yeaman developed a reputation for the quality of his levels, supplied to the rapidly growing number of land surveyors employed in the division of Scotland's common grazing land in the 1750s and 1760s.[9] Indeed, the only surviving Yeaman instrument known to us is a fine level, now in the Royal Museum of Scotland.[10]

It is also possible that Miller may have gained his initial experience under his father, who was himself involved in instrument engineering in the 1750s. John Miller senior was one of three Edinburgh craftsmen who in 1754 manufactured and adjusted two large high-precision capacity measures for the County of Stirling, raised from the ancient Scots pint.[11] The work was conducted under the supervision of Dr John Stewart (1715-1759), Professor of Natural Philosophy at Edinburgh from 1742 to 1759.[12] Although these are the only pieces so far identified with John Miller senior, the association with John Stewart may indicate a closer working relationship, perhaps even akin to that with one of Stewart's successors, John Robison, who held the chair from 1774 to 1805, and for whom Miller senior acted as experimental assistant.[13]

The 1754 work on the firlot measures provides an important context for John Miller's senior's work. In the supervision of this project John Stewart was joined by James Gray, Master of the Dalkeith Ironmill, who like Stewart was a member of the Philosophical Society of Edinburgh, which with the award of its Charter in 1783 became the Royal Society of Edinburgh. The Philosophical Society had its origins in an earlier grouping in which James Short (q.v.) had been active.[14] Co-ordinated solar eclipse observations in 1748 provided a stimulus to revive the Society, and it was again in vigorous operation by 1752.[15] The editors of the Society's new proceedings, the anatomist Alexander Monro *primus* and the philosopher David Hume, published in their first volume in 1754 a paper by Gray describing the Stirling standard pint measure.[16] This had recently been re-discovered by a third member of the Society, the Rev. Alexander Bryce (1713-1786), Minister at Kirknewton, Midlothian, who had been another of the 1748 observers: Bryce's measurements of the pint's capacity were used by Gray, and in 1754 Bryce also did metrological work for the City of Edinburgh.[17] In his reconstruction of the Philosophical Society's activities, Roger Emerson has noted that metrology figured in its business at this time, and that this work involved further members.[18] On this occasion, at least, Miller serviced a project of concern to the Society, and it is reasonable to assume that he came to the notice of those active in the Society's experimental programme. Bryce's involvement with instrumental matters is also seen in 1752, when he acted for the Blair Drummond Estate in the repair by John Yeaman

of a theodolite and the supply of a level.[19] If anything, this tends to reinforce the association between Yeaman and the Millers.

Very little can be discovered of John Miller senior's working life. When he married Elizabeth Gregg in April 1745 he was described as a 'journeyman wright', whereas by 1772 he was a 'turner'.[20] He obtained his Burgess ticket, giving him formal entitlement to trade, only in the City's purge of 1782.[21] He is listed as 'turner' in Libberton's Wynd, off the Lawnmarket, in Peter Williamson's first street directory for Edinburgh in 1773, and then as 'silver-turner' at the same location from 1774 to 1788.[22] A facility at working in metal as well as in wood is clearly of interest in terms of his likely role as an instrument maker, and it suggests he may perhaps be identified with the John Millar apprenticed to William Ayton, goldsmith, in 1732.[23] In 1780 he was again described merely as a turner, but now at the foot of Libberton's Wynd, apparently to distinguish him from another turner, Charles Miller, who had first been listed at the head of Libbeton's Wynd in the previous year. After a gap of several years in which there is no entry, he is listed at Old Excise Court in 1786 and 1790, but at Merchant Court, off Candlemaker Row in 1788 and from 1793 to 1801.[24] The date of Miller's death has not been found, but it was probably at about this time.

At the top of the Merchant Court house a small astronomical observatory was established, which will be mentioned later in connection with meteorological observations made there by Alexander Adie. In later life Adie recalled the interest the observatory had created in the city's scientific community. A prominent observer was Thomas Brisbane (1773-1860), later General Sir Thomas Makdougall Brisbane, Bt., who was to become an important patron of astronomy:[25]

> 'General Brisbane was a frequent visitor at the Merchant Court observatory and when a very young man would leave gay society to make his observations in full ball dress.'[26]

Another observer was undoubtedly John Robison (1739-1805), who had been appointed Professor of Natural Philosophy in 1774. It has been concluded that John Miller senior was the 'elderly person, a carpenter by occupation, who was employed by the celebrated Dr John Robison ... as an assistant in his class experiments', as recounted much later by William Wallace, who held the mathematics chair at Edinburgh from 1819 and whose initial introduction to Robison in about 1792 came through this assistant.[27] Wallace's obituarist patronisingly described how:

> 'This man, though a great reader of books, was no mathematician; but he had sat too near the feet of Gamaliel not to have imbibed a respect for the science, and for the pursuits of his young friend [Wallace]. With an excusable vanity he was in the habit of boasting of his intimacy with

the professor, to whom he proposed to introduce Mr Wallace ...'[28]

Although a period of apprenticeship in Edinburgh to a maker such as John Yeaman, or to his own father, would have left John Miller junior well grounded in his craft, the extensive London workshops of George Adams would have provided opportunities of a different order for developing expertise. Indeed, one can readily appreciate that Adams' reputation as a mechanic would lead to the conclusion that he had in effect 'bred' and 'educated' Miller, by introducing him to the advanced techniques of the centre of the instrument trade, and by finishing him as a journeyman. Certainly Miller became a trusted assistant of Adams, as can be inferred from the episode when he demonstrated to George III the classic 'guinea and feather' experiment, where a coin and a feather are seen to fall at the same rate in a glass vessel evacuated by an air pump.[29]

It is not known how John Miller junior came to work under George Adams. However, Emerson has made a persuasive case for the effectiveness of the Philosophical Society as a patronage mechanism, particularly in contacts with London, so the move is perhaps not surprising.[30] On the assumption that he was booked as an apprentice in Edinburgh on reaching the age of fourteen and served six years, he would have qualified as a journeyman in mid-1766.[31] A possible route of influence at this time would have been through John Stewart's colleague Joseph Black, the respected chemist and a member of the Society. Black taught at Edinburgh in the mid-1750s and after a period at Glasgow returned to Edinburgh as Professor of Chemistry in 1766.[32] His protégé John Robison, for whom he secured his old post in Glasgow in 1766 and who later obtained the natural philosophy chair at Edinburgh, was involved in instrument-making circles in London in about 1760 and in 1762 was appointed by the Board of Longitude to assist with trials of John Harrison's famous marine chronometer H4.[33] Robison and Black were on close terms with James Watt, who had been introduced to James Short in London in 1755 and had subsequently been placed with the instrument maker John Morgan. John Miller senior may also have been in touch with Short, directly or through his brother Thomas Short in Leith. It is even possible that some support was provided at a higher level: James Douglas, 14th Earl of Morton, was patron both of James Short and Alexander Bryce, and in addition to being President of the Philosophical Society was President of the London Royal Society from 1764 to 1768.[34] His recommendation would undoubtedly carry weight with the King's instrument maker.

Having completed his period as Adams's assistant 'for some years' (probably three years), Miller returned to Edinburgh some time before summer 1769.[35] He first appeared in the Edinburgh street directory in 1774 at the back of the Fountain Well in the Netherbow; subsequently, from 1775 until 1794 he was in Parliament Square, latterly given as 7 Parliament Close, and from 1795 until 1801 or 1802 at 38 South Bridge and 86 South Bridge in 1803.[36] The street directory then shows him moving to 94 Nicolson Street in 1804 shortly after taking his nephew Alexander Adie into partnership, as will be discussed below; they both appeared there between 1804 and 1809.[37] Miller received his burgess ticket entitling him to trade as a freeman in the Burgh, only in 1782, but this is of little significance in the context of his development as an instrument maker,[38] and it is from other scattered sources that a picture of his activities can be built up.

As far as can be judged from documentary evidence and from surviving instruments from his hand, Miller's business was healthy and the quality of his pieces well-regarded. The polish he had acquired under Adams brought him attention and work shortly after his return to Scotland, and significantly it was in the field of astronomy that he gained his first known commissions. Telescopes were the vogue scientific instrument of the 1760s because of the interest generated in the rare transits of the planet Venus across the Sun's disc in 1761 and 1769,[39] and it is therefore not surprising to find that on his return from work under George Adams, John Miller was constructing instruments for observing the second transit. His patron was the Edinburgh physician Dr James Lind (1736-1812), who had a strong interest in astronomy, and whose letters to Lord Loudoun provide the main descriptive source for the early phase of his protégé's Edinburgh business. Lind was the key figure in Miller's three earliest known commissions, and entrusted him with later work also; in June 1769 he was referring to him warmly as 'my Mathematical Instrument Maker here, one Miller a very cliver young man', and the inference is that Lind had employed Miller on previous occasions.[40]

Perhaps the most significant of these commissions was Miller's construction of some 'very good instruments for me for observing the late Transit of Venus', as Lind explained to Lord Loudoun on 23 June 1769.[41] The observations had taken place earlier that month at Hawkhill near Edinburgh, the seat of Andrew Pringle (1715-1776), a judge who sat in the Court of Session as Lord Alemoor, and like Lind was a member of the Philosophical Society.[42] Lind, who lived at Restalrig, was his neighbour, and was on intimate terms with him; and as Henry Mackenzie recalled, Alemoor was not only a *bon viveur* but also 'a lover of science and had an observatory at Hawkhill'.[43] The principal instruments used were Alemoor's 3½-foot triple-objective achromatic refracting telescope used by James Hoy 'our young observer', Lind's own 2-foot triple achromat, which was probably also by Dollond of London and had probably been supplied to Lind by Miller, and thirdly the 18 inch focus reflecting telescope used by Alemoor himself which is likely to

7 John Miller's shop beside St. Giles Cathedral. A detail from 'The Old Parliament Close, and Public Characters of Edinburgh nearly a century since', by T. Dobbie and John Le Conte, c.1844

have been by James Short (q.v.), possibly indicating a longer standing interest in astronomy on Alemoor's part: in his observations, Lind was assisted by 'a mathematical instrument maker [undoubtedly Miller] who counted seconds from the clock'.[45] Henry Mackenzie's more comical eyewitness account of the observation of the Transit of Venus in 1769 is interesting for the anomalies it contains, which can perhaps be attributed to imperfect recollection, for he was only fourteen at the time and was writing many years later. Referring to the observations made by Alexander Bryce in his Kirknewton parish (at which he incorrectly claimed Lind was present), he recalled that 'Dr Lind, who was a great mechanic, had made an instrument for assisting them in the accuracy of the observation'.[46] Almost certainly this refers to the sophisticated equatorial stand for the telescope, which appears to have been Miller's principal contribution, and which was perhaps originally made to a design by Lind. Other examples of these are recorded: a surviving stand by Miller at the Science Museum, London, supports an achromatic telescope signed by Ramsden, a 1774 example was owned by James Stuart Mackenzie, brother of the 3rd Earl of Bute, who was Prime Minister 1762-3, and another was acquired for Edinburgh University in the 1770s.[47] Mackenzie referred to Bryce as 'another zealous astronomer' and acquaintance of Lind's, and it was Lind who communicated Bryce's observations of the 1769 transit to the Astronomer Royal, Nevil Maskelyne.[48] Bryce's astronomically derived latitudes and longitudes were used by Andrew and Mostyn Armstrong in their 1773 large-scale 'Map of the Three Lothians'.[49] Reminiscent of the large equatorial stands on which Miller's early reputation rests, is a small theodolite recently acquired by the Royal Museum of Scotland; the universal equatorial mount of this converts it into a miniature portable observatory.[50] Although its small size makes it difficult to imagine whether it would have fulfilled a useful purpose, it is undoubtedly an instrument whose quality reflects the standards of the Adams workshop and the design of his instruments.

It is clear that Lind thought highly of Miller's instrument work, and was passing commissions his way. In June 1769, on Lord Loudoun's behalf, he ordered from him 'the Glass and Head of a Camera Obscuro', and within six weeks Miller had supplied the required type to Loudoun for 15s, probably having constructed it himself.[51] At about this time Lind's attention was turned to surveying instruments, and when he designed a clinometer, it was to Miller that he turned for its manufacture.[52] Lind's interest in practical surveying was extensive, for he supplied the great English tourist Thomas Pennant with a map of Islay,[53] and he is known to have assisted William Roy with experiments in the early 1770s to measure height barometrically in Scotland.[54] His scientific interest also extended to improving meteorological instruments, and he described in 1774 a new form of anemometer for measuring wind pressure, which incorporated a bent tube containing liquid and which became known as Lind's Anemometer.[55] Miller almost certainly produced this for him initially, although we have not recorded examples:[56] he is known to have had an interest in meteorological instruments and devised for instance the ingenious and 'simple contrivance' of rods and gearing which allowed the rotation of a wind vane to be shown on a vertical plate.[57]

By the time that Lind settled in Windsor in about 1777, and his connection with Miller probably ceased, Miller was well established as a leading maker in Edinburgh, with a diverse and flourishing business. Extant instruments show that he could turn his hand to all types of work and that he catered for professional as well as amateur customers. An elegant and unexceptional stick barometer indicates the fact that at least part of his production was geared to the domestic barometer market;[58] and like other makers he produced garden sundials.[59] His stock also included material bought-in from the south: telescopes, for example, were imported from Spencer, Browning & Rust of London, a firm with whom the Edinburgh business maintained links for many years, with Spencer, Browning & Rust also acting as agents for Adie Instruments.[60] There is also evidence of Miller undertaking specialist work for instruments signed by other Scottish makers.[61] Telescopes, barometers and sundials were standard items in an instrument maker's repertoire, but what proportion of Miller's output they formed is uncertain. On top of these were the prestigious individual commissions such as those which Lind had provided, but it now appears likely that a significant part of his business was the production of surveying instruments.

Miller's commencement in business coincided with the end of a period of rapid agricultural change in Scotland which had gathered pace since about 1750, and during which the number of land surveyors working in Scotland increased from about ten to about seventy.[62] But although the work of enclosure and the division of common land slackened between 1770 and 1800, the number of active surveyors remained high, and it is likely that some of them were Miller's customers. If Yeaman was responsible for his intitial training, Miller must have been well versed in the skill of making levels. A fine example by him is included in the Frank collection, and it is known from another source that Miller frequently constructed spirit levels: in about 1790, when Alexander Keith tested the accuracy of a mercurial level he had devised (and which Miller probably constructed for him), he and Miller used 'a line drawn upon the opposite side of Parliament Square, fronting his shop, by which he has been in use to adjust his spirit levels'.[63] An example of the way in which Miller modified his products to satisfy practical requirements is special purpose level developed to help road engineers and inspectors determine gradients accurately, of which an account was published in 1795.[64] It is clear that a wide range of surveying instruments such as theodolites were also being produced, and it is a significant indication of the quality and volume of Miller's work that by the early date of 1793 he was using a circular dividing engine, probably the only one of its kind in Scotland.[65]

In 1776, at about the time of Lind's departure from Edinburgh, Miller became involved in a project which demonstrates both his concern to cater for the amateur or dilettante market, and his links with the world of the land surveyor. He proposed publishing terrestrial and celestial globes on an advance subscription basis in association with the surveyor and engraver John Ainslie. This was the first attempt to manufacture globes in Scotland in commercial competition with readily available London-made products, and although this particular project fell through by 1777, Miller was producing a pocket globe on his own account in 1793.[66]

Perhaps the most significant early commissions to have been placed with Miller were from John Robison, who in 1774 took up the chair of natural philosophy at Edinburgh University. Because the demonstration apparatus used by a professor was normally his personal property, bought out of his class fees, it was not necessarily available to his successor. Robison had to re-equip, but was fortunate in being permitted by the Town Council, as patron of the University, to use £300 which had been bequeathed by Dr Charles Stewart of London in 1770 and which Robison was authorised to spend in four stages in the period 1777 to 1781.[67] Some of the instruments were purchased from London makers, but sizeable orders went to Miller. The arrangement, however, was not without some financial difficulty for Miller, and Robison had to write to the Council in August 1779 to say that

> 'Mr Miller having pressing occasion for money came to me and begged that I would procure him payment for an Air pump which he had made for my Class, and execute intirely to my Satisfaction.'[68]

The two most impressive survivals from this collection, both of which are at the Royal Museum of Scotland, are astronomical demonstration devices. One is a 'cometarium' showing the variation of orbital speed of a body in an elliptic orbit, and based on a design by J.T. Desaguliers.[69] The other is a large orrery which is unusual in having the planetary orbits in a vertical plane and is clearly intended for classroom use.[70] Both these items are mounted on elegant mahogany stands, which were perhaps contributed by John Miller senior, but it is also possible that he may have produced the numerous demonstration experiments for mechanics, dynamics, etc, which must have been largely made of wood and for which no maker's name is known. The wheelwork and internal construction of orreries has much in common with clocks, and it is possible that such work was done for John Miller by his younger brother Alexander who had completed a clockmaking apprenticeship in Edinburgh in 1772.[71] Alexander appeared in the street directory at what had previously been listed as John Miller's address in Parliament Close only in the year 1783, but he is not recorded after this date.[72] Possibly he died or left Edinburgh, because in 1789 John Miller was advertising for the assistance of a locksmith or clockmaker in constructing some unspecified apparatus.[73] Perhaps it was no coincidence that on 1 January 1790 Miller advertised an orrery 'of a very elegant construction' for sale at his shop, the works of which were driven by a timekeeper which ran for one month.[74] In type, at least, this example can be linked to the orrery commissioned earlier by Robison, and Miller's production of them typifies the dual capacity of the skilled instrument-maker in designing a piece of philosophical apparatus which was at once a tool for higher learning and an object for instructive amusement in Edinburgh drawing rooms.

Miller appears to have remained single, but in April 1772 his sister Betty married John Adie, who owned property in the parish of Torrieburn near Dunfermline, Fife.[75] Adie became a printer, was a proprietor of the *Edinburgh Evening Courant*, and was said to have contributed the foreign articles in the *Scots Magazine*.[76] He died about three months before his second boy Alexander James Adie was born in January 1775.[77] His widow remarried, but died soon after,[78] and subsequently Alexander Adie was adopted by his uncle John Miller, living with him until Miller's death in 1815.

Although Adie was brought up by one of the most important Scottish scientific instrument makers of the day, he was not apparently originally intended to follow his uncle's trade, since he was set to work at the age of twelve to be a stocking maker with William Coulter, later Lord Provost of Edinburgh.[79] It was only when his elder brother John died in about 1787 that he was apprenticed to his uncle as an optician. As his biographer tells us, because he had left school at twelve his education

> 'was of necessity very limited; but his connection with his uncle was good for stimulating self-improvement. John Miller was a well informed caustic but kind hearted old scotchman fond of books and philosophy whose chiefest crony was David Herd an antiquary of some note in Edinr.[80] These two met almost nightly for discussions of various subjects the length of the sittings being regulated by the time it took to finish a moderate quantity of toddy ... So far therefore as an instructor in his profession went Mr Adie had the advice of a good master, but it was some years after he became an optician when fully able to regret the loss of a more extended course of instruction in his early youth that at his own cost he attended lectures and got

teachers for himself in the evening for mathematics algebra, chemistry and other branches of the rudiments of philosophy.'[81]

We know little of the details of this period of Adie's life, for his biographer moves on quickly to his later activities. However, quite apart from some non-professional pursuits in the Volunteer Reserve and so forth, the period of his assistance of, and subsequent partnership with John Miller is of interest in itself and in shaping his later independent work. His apprenticeship probably lasted until about 1796, and from its expiry until 1803 Adie worked as his uncle's assistant. Their partnership began that year as 'Miller & Adie', at 86 South Bridge. They moved soon after in 1804 to 94/96 Nicolson Street, working there until 1809; they moved in 1810 to 8 Nicolson Street, renumbered to 15 Nicolson Street from 1811, and the firm remained at that address until after it ceased to trade under the name 'Miller & Adie' in 1822.[82] In November 1804 Adie married Marion Ritchie, the seventeen year old daughter of John Ritchie, a Burgess slater, and Janet Sibbald, daughter of the prominent family of Edinburgh smiths.[83] From 1807 the street directory also lists the Adies' house at 11 Lothian Street,[84] but, by this time they also had a property in the country. In 1802 John Miller and Alexander Adie each purchased ground in the newly feued Canaan estate at Morningside, to the south of the city; Adie's plot was subsequently sold, but on the larger plot they built Canaan Cottage, a fine house in a large suburban garden, which still stands today.[85] It is likely that this was made possible by the earlier sale of Adie family land in Fife.[86]

An impressively broad range of instruments marketed under the trade name 'Miller & Adie', has survived or been documented; the Royal Museum of Scotland's collection includes three in the Frank collection, namely a level, a telescope and a microscope. Of the known instruments by the partnership the earliest is probably a portable hydrometer signed by them and dateable to late 1803, or the first half of 1804.[87] Related instruments also designed for rugged outdoor use, are a mountaineering barometer and a box sextant.[88] The level in this collection is representative of the continued production of the surveying instruments that were probably a mainstay of the firm's work.[89] Theodolites from the workshop were sturdily constructed: one such instrument sold in about 1799 to William Kyle, a land surveyor from Glasgow, lasted for seventeen years of heavy use before it was worn out and useful only for work where accuracy was not needed.[90] A small 'Miller & Adie' theodolite of this period in the Museum's collection is understood still to have been in use in about 1860 in the Ordnance tertiary survey of west Perthshire.[91] For experimental and demonstration work at least two air pumps are known to have been made; one was supplied to Dr Andrew Ure for Anderson's Institution, Glasgow,[92] and a second for the natural philosophy classroom at the University of St. Andrews in 1812.[93] The other two pieces in the Frank collection, the telescope and the microscope, are of interest as examples of the two most popular optical instruments, illustrating the increasing diversity of the business's market at this period: the Martin-type universal microscope is of a conventional pattern and representative of the type of stock obtained from wholesale suppliers, whereas the signed telescope mount is presumably for a more substantial commission.[94] Some indication of the range of their stock can be gained from an advertisement of 1808, for a type of camera obscura termed a camera guido, which noted that they supplied

> '... all kinds of instruments for surveying land, viz. Theodolites of various constructions, Telescopic Levels, Measuring Chains, Pentographs, &c. Barometers and Thermometers of all kinds, Achromatics, Telescopes; Spectacles, in Silver, Tortois-shell, and Steel Frames, the best kind; and any article in the Mathematical, Optical, and Philosophical line, executed agreeable to order.'[95]

A clear indication of their willingness to buy in stock is seen in a later advertisement of 1818 in which it was noted that a large supply of Brewster patent kaleidoscopes had been obtained.[96] Miller & Adie were not one of the initial suppliers licenced by Brewster to sell the instrument, and they appear to have been operating in competition with Brewster's official Edinburgh agent, John Ruthven: however, whether these were instruments produced by one of the approved manufacturers or cheaper pirated versions from France or elsewhere cannot be said.[97]

While these examples and others suggest a picture of fruitful activity in the partnership of uncle and nephew, some doubts remain about the relative part each played in the manufacture of pieces bearing their joint trade name. These arise mainly from the fact that although the business continued until 1822 as Miller & Adie, John Miller himself died on 4 February 1815,[98] so that for about eight years of the eighteen during which the name was used, Alexander Adie can be assumed to have been the maker. Setting aside (because there is no evidence) the question of whether there were journeymen working under Miller and Adie during this period, who might have provided a continuity in the workshop, the problem of dating the signed pieces remains. Without further documentary evidence this is difficult to resolve. Arguably items such as the telescope and the air pumps would have been standard productions for an optician trained under George Adams, and might therefore be attributable to the period when Miller's expertise was available to his clients, although it is also clear that Adie was capable of constructing both these instruments. However, on the evidence of Adie's known activities after his uncle's death, an item such as the mountain barometer probably belongs to that later period, when as we shall see, Adie was actively working on capillary instruments. From the period up to 1822 one instrument, at least, can be attributed with some certainty to Adie: this was the statical hydrometer, described favourably by David Brewster as 'one of the neatest and most correct instruments that we have seen', and first constructed by Adie, we are told, in 1799.[99] If on the one hand Adie was already an inventive craftsman at the age of twenty-four, his uncle on the other hand appears to have been active in the business until the end of his life. This is implied by his obituarist in the *Scots Magazine*, who was in no doubt as to his merits both as a shrewd, benevolent and generally esteemed man in public and private life, and as an optician 'who for upwards of 48 years, held the first rank in his profession: his excellence, as a workman, was admitted by his contemporaries, while the fertility of his genius added many improvements to our instruments of science.'[100]

One person at least who seems not to have been of the same unqualified opinion was the natural philosopher David Brewster, who was later to dominate the field of optics in Britain. Although Brewster complained of the quality of Miller's work in his private correspondence in 1799,[101] it must be uncertain whether his criticism was really directed at Adie. From his student days or from after his graduation at Edinburgh University in 1800, Brewster must have known Miller & Adie's shop, but he is not known to have commissioned an instrument from them until 1806. He had sent a specification for an eye-piece micrometer to William Cary, the London instrument maker, and the following year had an example of the instrument made for him by Miller & Adie.[102] Three years later, in April 1809, Brewster had Alexander Adie modify a newly-made goniometer by William Harris of London, to enable it to perform an additional type of measurement.[103] A telescope circular micrometer of a type first proposed by Brewster was made for him by Adie in 1810 or 1811 and incorporated a circular degree scale in the focal plane minutely divided on mother-of-pearl.[104] These commissions are interesting not only in the context of Brewster's development as an experimental scientist, but also as early examples of his patronage of Adie, for he was perhaps partly responsible for the good reputation which Adie acquired. In actual fact his opinion of Adie's work was qualified and in print at least even modest praise was sparely given.[105] Brewster's role as Adie's patron can be distorted, because the several instances where his influence, and probable friendship, secured for Adie commissions and recognition of his status, came at a later date. Moreover, Brewster's apparent ubiquity in the published transactions of Edinburgh scientists from 1800 onwards,[106] has perhaps obscured the role which other experimenters played in stimulating Adie's mechanical talent. But it remains true that Brewster, for ever publicising his own doings, was partly responsible for bringing Adie's qualities to the attention of fellow scientists.

Arguably, of course, in the 1800s Alexander Adie did not stand in need of the good opinion of the precocious Brewster, who, although he wielded editorial influence, was only one among many customers. Adie's own disposition as an instrument maker also counted for much towards his growing reputation, as his biographer explained:

> 'At his business as an optician Mr Adie worked with great assiduity from early in the morning till far into the night, and from his skill as a mechanic, quickness of inventive powers and sound judgement he was much applied to by all kinds of inventors to work out schemes and to reduce them if possible to a practical or at least a working shape.'[107]

Other commissions for experimental work in which Adie was involved in the early years of the century were also important in this respect because of the attention they attracted at the time. They came from the geologist Sir James Hall of Dunglass (1761-1832) and from Professor John Leslie (1766-1832), who was the controversial choice for the chair of mathematics at Edinburgh University in 1805. In terms of the advancement of scientific knowledge Hall's work was the more important, for he was the first to provide successful experimental demonstration of James Hutton's argument for the igneous origin of basalt and other types of rock. Between 1798 and 1805 he conducted more than 500 experiments on weighed amounts of rock at very high temperature and under extreme pressure, using an *ad hoc* arrangement of gun barrels as crucibles, weights to pressurise the contents, and Wedgwood pyrometers to indicate the temperatures.[108] Although there is no direct reference to Adie in any of the published papers which detailed the series of experiments, Adie's biographer states 'Mr Adie took great interest and assisted in the preparation of the necessary instruments for the experiments by which Sir James Hall supported the igneous theory in Geology.'[109] From the trial and error nature of the experiments it would appear that Adie's technical expertise was important to Hall.

In the case of work executed for John Leslie, the scientific achievements were less spectacular, but the close association which grew up between Leslie and Adie was more significant in commercial terms. The two men had more in common, too. While Adie and Sir James Hall both attended lectures at the University as outside students, their social standing and age separated them. Leslie, on the other hand, like Adie, was of humble origin and had his roots in Fife. A rapport existed between them which led Leslie to entrust to Miller & Adie the Scottish manufacture of the thermometric instruments he devised, and to enlist

8 Sir John Leslie (1766-1832), engraving by John Horsburgh after a painting by Sir David Wilkie, in which Leslie's differential thermometer, made locally by Miller & Adie, is placed prominently beside the professor's left hand

Adie's technical assistance in experimental work. The dating of the development and production of the instruments is not known exactly, but an advertisement by Miller & Adie of 1808 lists:

> 'an assortment of Professor Leslie's Differential Thermometers, under the several forms of Photometer, Hygrometer, and *Pyroscope*. The accuracy of them may be relied upon, being regulated under the eye of the Professor.'[110]

Whether or not a formal agreement was reached for their manufacture, Leslie was clearly overseeing the production in some way.

By 1813 Leslie's instruments also included the atmometer and hygroscope, and he described his entire family of capillary instruments in a short book that year. This included the revealing note that all the instruments were to be had 'of the most accurate and perfect construction' from Miller & Adie in Edinburgh, and 'Mr Cary' of London.[111] This joint manufacture of the six instruments described in the 1813 work was still in force when Leslie published a sequel to it in 1820, which described in addition a modified pyroscope known as the aeturioscope. A price list, apparently common to both Cary and Adie, lists the range of instruments according to the style of mounting, from a differential thermometer at £1 5s to an aeturioscope at £5 5s.[112]

Also included are the prices of air pumps 'for congelation' or the freezing of liquids, ranging from £42 to £84, depending on the number of plates included.[113] Leslie's experimental work on freezing was done using air pumps and with Adie's help he succeeded in freezing water and mercury. He froze water in 1810, and succeeded in freezing mercury some time before 1813, after one failed attempt using a defective pump. As the press reported: 'This remarkable experiment was performed in the shop of Mr Adie, Optician, here, with an air pump of a new and improved construction made by that skilful artist.'[114] Adie's important role in this episode suggests that he had assumed the lead in the partnership, and that the commissions for pumps which date from this period can be attributed to him. The pump commissioned for the natural philosophy classroom at St. Andrews in 1812 was possibly as a result of his work with Leslie.[115]

Another association with John Leslie at about this time came with a commission from the civil engineer Robert Stevenson, a friend of Adie's who was Engineer to the Commissioners for the Northern Lighthouses. In about 1810 Stevenson designed a greatly improved reflector lamp for use in lighthouses and this was to see very widespread use around British and foreign coasts. The silvered copper parabolic reflectors were manufactured in Edinburgh by the brassfounder James Milne: to ensure accuracy these were formed on special moulds constructed by Adie from curves drawn by Leslie.[116]

By 1815, therefore, Adie was well known in his own right, but he retained the name of the partnership either out of caution or modesty. He was kept busy, as we learn from a letter of William Kyle the land surveyor who owned the theodolite by Miller mentioned above. In 1817 Kyle requested Adie to repair his main theodolite (originally made by Jones of London) as soon as possible, although he realised that 'the extensive nature of your business' might prevent this.[117] Allowing for a degree of flattering persuasiveness, the remark indicates that Adie's skill as a repairer of instruments was in demand; the letter details the required repairs and also asks for his advice about lubricating the moving parts. Adie's grasp of the physical and chemical properties of the materials of his trade was also evident in areas not obviously connected with his instrument work. In 1815, for instance, he was called in to prepare the platinum inscription plates, coins and papers for preservation in the foundation stones of the Regent's Bridge, then about to be constructed to the designs of Archibald Elliot to carry the eastern approach to Princes Street, Edinburgh.[118]

The surveying instruments produced by Adie in the first half of the century seem to represent a mainstay of the firm's business, as it had under Miller. In general, these were conventional instruments designed for the use of professional land surveyors involved in land division and field enclosure work. One device however deserves particular mention. In 1821 James Hunter of Thurston, East Lothian, described a simple and ingenious 'odometer' or distance measuring wheel to the Highland and Agricultural Society; this was designed to enable land owners to undertake speedy field measuring and especially to check survey measurements which had been performed by chaining.[119] The odometer was made by Adie and marketed in conjunction with a watchmaker friend James Howden, and it was apparently popular, because it remained available for an extended period.[120]

The years after John Miller's death saw Adie increasingly turning his attention to meteorological instruments. His professional interest in the weather was no new thing, as his biographer explains:

> 'Meteorological observations attracted his early attention & with a view both to have regular registers of weather and also to astronomical observations he erected a small private observatory on the top of his residence in Merchant Court Edinr. This was long before the establishment of a public observatory in Edinr & Mr. Adie has often been heard to tell in his very quiet way how much interest it created among scientific men at the time.'[121]

It is not at all clear whether the astronomical observatory had its origins with Miller or Adie, but the meteorological station can probably be dated to 1795, or shortly before, from a register of his observations made at Merchant Court, covering the period 1795-1805.[122] Professor James David Forbes (who later became closely connected with the Adies), speculated in 1861 that Adie was 'Stimulated probably by Professor [John] PLAYFAIR'S example and advice' into beginning 'what appears to have been a very careful register of the thermometer, barometer, wind, and rain, on the 1st January 1795'.[123] The quartet of instruments he used were all made by himself and were 'certainly equal in accuracy to any then constructed'.[124] Adie was barely twenty when he embarked on his observations, which proved to be the first in a long series.

His biographer continues:

> 'Arising out of his barometrical observations Mr Adie devoted much consideration to the possibility of constructing an instrument sufficiently sensitive to measure the tides in the atmosphere. In this he did not succeed but the invention of the sympiesometer was the important practical result of these investigations.'[125]

The reference to atmospheric 'tides' is interesting, because it echoes strongly the claims made by John Leslie in 1813 about the indispensibility of the hygrometer in meteorological observations. He argued that it might make an essential contribution towards laying 'the foundation of a juster and more comprehensive knowledge of the various modifications which take place in the lower regions of our atmosphere'.[126] The common aim of precise meteorology underlay both the work jointly undertaken by Leslie and Adie, and Adie's own independent work which in the 1810s came to centre on barometry. The sympiesometer, or 'new air barometer', for which Adie obtained a patent on 23 December 1818,[127] was a variant on the then standard type of mercurial barometer, in which oil replaced mercury as the hydrostatic fluid and a column of gas was used. It was especially well suited for its intended use as a marine barometer, because its sensitivity provided earlier warning of weather changes at sea, and its consistency under rugged conditions gave it a distinct advantage over conventional marine barometers.

The sympiesometer's genesis remains obscure, but it is clear from the passage quoted above, and from his own account, that Adie had been carrying out 'investigations' in increasing the sensitivity of the barometer, which involved the construction of experimental instruments. An improved barometer, of which there are examples in the collection, belongs to this 'Miller & Adie' period, and indicates that one or both of the partners were concerned with the problem of barometer adjustment.[128] This was a developed version of the English maker Edward Troughton's barometer, and was referred to as 'Miller & Adie's Barometer' in a source of 1810 although the date of its construction is uncertain.[129] It incorporated two concentric adjusting screws to facilitate and improve the barometer's performance, but it was criticised for 'falling short of the desideratum of self-adjustment'.[130]

Adie was only too well aware of the shortcomings of conventional barometers, and after unwittingly repeating Robert Hooke's use of the compression of an enclosed column of air to measure atmospheric pressure, concluded experimentally that a column of hydrogen with tinted almond oil as the fluid were the best hydrostatic materials for the instrument.[131] For normal use the sympiesometer was adjusted with a thermometer, but for survey work it was necessary also to use a hygrometer, to measure atmospheric humidity at the required locations. Adie states he was experimentally ascertaining the hygrometrical properties of various materials in the winter of 1816-7, and he patented his own hygrometer simultaneously with the sympiesometer.[132]

One of the first sympiesometers had been sent on board a Greenock ship bound for the East Indies in 1816, and the readings produced on the voyage compared very favourably with those of a common marine barometer.[133] More favourable still were the results obtained from comparative readings of a sympiesometer and a marine barometer between 24 April and 11 November 1818 on H.M.S. *Isabella* during the Arctic expedition to discover the North West Passage. The expedition commander Captain John Ross predicted that the sympiesometer would supersede the marine barometer 'when it is better known',[134] and further plaudits came from Adie's friend Robert Stevenson, who tried it against a barometer on coastal voyages in the Lighthouse Board's main supply vessel for two years. By December 1820 the sympiesometer was in regular service for the Board on this vessel and the smaller *Pharos*.[135] Reports in the Scottish press were naturally complimentary: one typically referred to Adie as 'our very ingenious townsman', and noting that the sympiesometer could be made pocket-sized, argued that 'it is likely to become a valuable acquisition to the geologist.'[136] The *Scots Magazine* reported the successful trials on the Indies and Arctic voyages, referred to a portable version specially constructed by Adie for measuring heights, and thought that his hygrometer promised to be 'a valuable addition to our stock of philosophical instruments'.[137]

Exactly how widely used the sympiesometer became is not known, but it was produced in considerable numbers, and placed Adie's work before a wider public. Responding to, or anticipating demand in England, he had by October 1819 authorised the London maker Thomas Jones to make and sell both the patent sympiesometer and the patent hygrometer.[138] Other agencies were established, apparently solely as retail

9 Alexander Adie's trade card for his address between 1823 and 1829

10 Alexander Adie's trade card for his address between 1830 and 1834

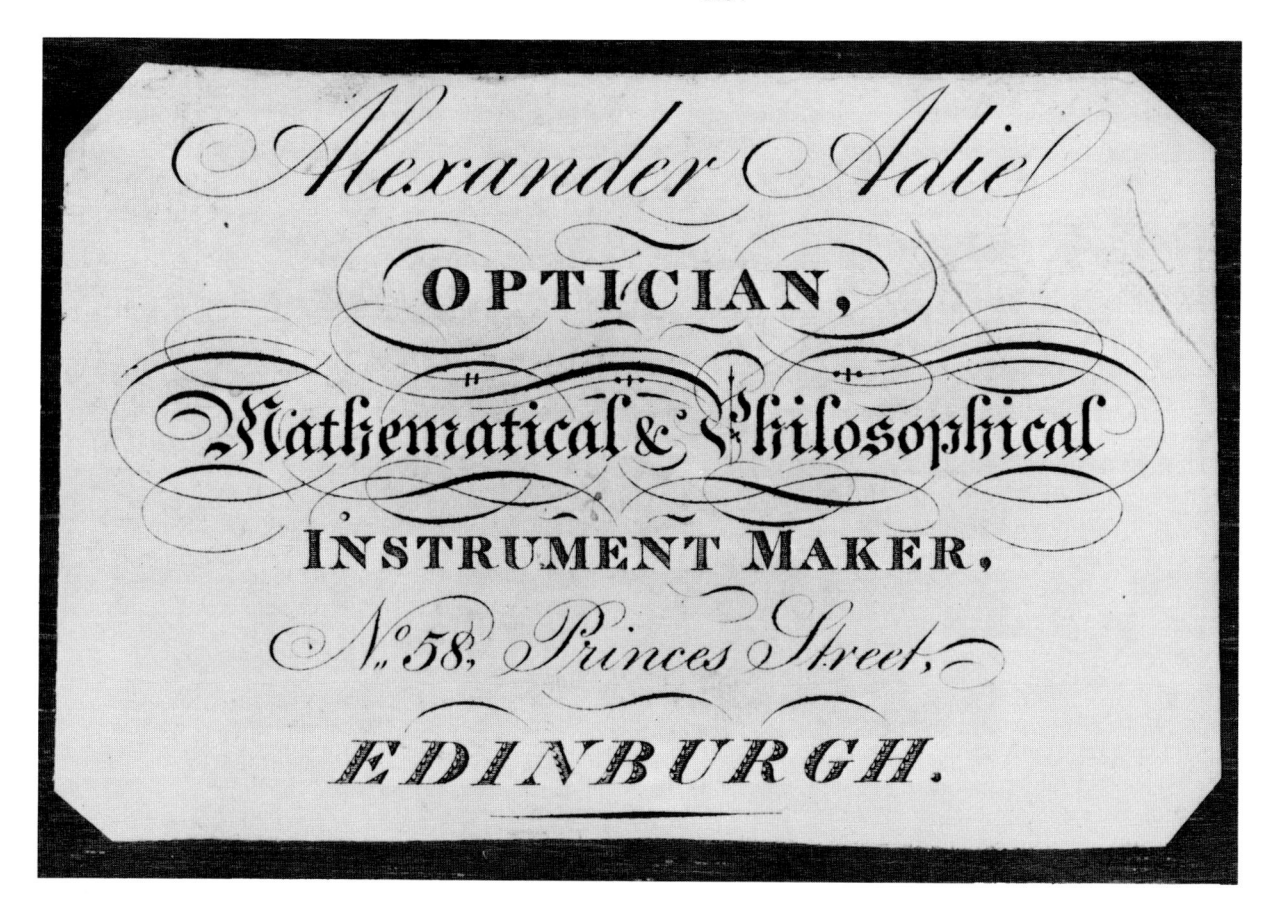

outlets. The earliest sympiesometer presently recorded, No 383, bears in addition to Adie's plate the name of 'Wm. Heron Agent Greenock'.[139] The location of this agency indicated the importance for Adie's successful marketing of the sympiesometer of an outlet in one of Scotland's main shipbuilding and trading centres. The instrument's number suggests it dates from the early to mid-1820s. Another instrument with a low number in the Museums' collection, No. 579, is also engraved 'Spencer Browning & Rust Agents London' and appears to date from the late 1820s.[140] Dealings between the Edinburgh and London firms were mutual, with each acting as specialist supplier of wholesaled items to the other: mariner's octants by Spencer, Browning & Rust were retailed by Adie stamped with his own name.[141]

The sympiesometer in this collection falls approximately one third of the way along the known production run of about 3,200 instruments, and was made in about 1830-35.[142] It has been estimated that about ninety instruments were made each year during the fourteen years of monopoly afforded by the patent, and about thirty thereafter, although these figures rest on the approximate dating of only a handful of pieces,[143] and assume for simplicity that production of the sympiesometer continued throughout the period up to 1880 when the name of Adie & Son was dropped. However, it is misleading to assume that the instrument with the highest known serial number, No 3262, which is signed 'Adie & Son, Edinburgh Adie Strand London',[144] necessarily falls towards the latter end of the period: the most that can be said at present is that it dates from after 1844, when Alexander Adie's son Patrick (q.v.) set up in London.

Adie's biographer, commenting on the sympiesometer's popularity, noted that in about 1860 'The public after a trial of above forty years still demand increasing supplies of this instrument.'[145] It is perhaps a matter of comment then that although 'the quiet investigations of an unobtrusive citizen in Edinr produced an instrument which has greatly contributed to the safety and comfort of the ships of the greatest mercantile nation in the world',[146] the instrument should not have been manufactured by Adie in even greater numbers. In practice, the local demand may have been lower, a factor perhaps being the innate conservatism of sailing captains who were reluctant to be guided by any instrument more scientifically exact than a marine barometer. Lieutenant Robertson of H.M.S. *Isabella* thought the sympiesometer totally superior to it, but added 'If it has any fault, it is that of being too sensible of small changes, which might frighten a reef in when there was no occasion for it.'[147] The most telling factor was that rival products, including sympiesometers and 'improved sympiesometers' produced by other manufacturers, came to compete for the same market, and that in particular the sympiesometer was overshadowed from the 1850s by the increasingly sophisicated aneroid barometer which became the standard all-purpose type.[148]

Shortly after Alexander Adie patented the sympiesometer and hygrometer he was elected a Fellow of the Royal Society of Edinburgh on 25 January 1819.[149] It is hard to avoid the conclusion that this unprecedented election of an instrument maker to the august body of scientists, doctors, literati and men of affairs, was both a recognition of Adie's scientific inventiveness and investigative ability, and of his proven mechanical skill as a maker. This is reinforced by the fact that one of Adie's three sponsors for his election was David Brewster,[150] the Society's secretary, who had already called on Adie to help with his optical investigations and whose classical knowledge had been applied in choosing the name of sympiesometer.[151] Adie's election, coinciding as it did with the emergence of the sympiesometer before an impressed public, almost certainly tightened his connections with the professional and amateur scientists who were his customers, and it ushered in the most prolific period of his working life, remarkable both for the variety of commissions he undertook and for his own energetic scientific activity. In explaining the secret of his success in terms of his intelligence, dexterity and capacity for hard work, his biographer took for granted the unique status which Adie's membership of the Royal Society of Edinburgh conferred on him. It was undoubtedly a boost both for his business and for his own experimental work.

Also of great importance, particularly in the latter respect, was his membership of the Society for the Encouragement of the Useful Arts in Scotland (later known as the Royal Scottish Society of Arts). This body was instituted in 1821 by David Brewster as an improving society and forum for new inventions and was established under George IV's patronage in the following year. The inference to be drawn from Adie's presence at the first meeting of the Society's council to consider practical questions of assessing technical communications is that Brewster looked to him as precisely the kind of expert whose judgement would be needed both to evaluate submitted designs and to encourage further work.[152]. Brewster's faith in Adie was well-founded and for his part Adie profited from the contacts and stimuli which the prestigious and lively activities of the Society created.

The field in which Adie seems to have become more involved in the 1820s was optics. The commissions for Brewster in 1806 and 1809 form two related elements in the obscure pattern of Adie's work, but two commissions dating from 1815 provide further indications of his optical instrument work. C.R. Goring, then a medical student with private means at Edinburgh University, was inspired by the article on

'The Telescope' by the Edinburgh natural philosopher John Robison in the *Encyclopaedia Britannica* to become a self-confessed 'reformer of microscopes' in the search for a true achromatic instrument.[153] He commissioned 'that distinguished artist, Mr Adie, of Edinburgh to execute for me a microscope similar to that recommended by the Professor [Robison], on which no expense was spared'. Goring was unhappy with the result because the 'too complicated' construction and inconvenient length caused by the lens arrangement had not elimated the chromatism. Nor was a second instrument commissioned from Adie 'on the plan of the erecting eye-piece of the ordinary construction' and fitted with Huyghenian eye-pieces truly achromatic, although it was agreed that the image was very distinct. Having attributed the defects of the first microscope to Adie's 'imperfect execution', Goring then 'set Mr Adie down, at the time, for a bungler, who could not adjust the foci and intervals of the glasses in a proper manner': his description of Adie echoes Brewster's stricture on all the Edinburgh opticians in 1812. Goring subsequently turned to Charles Tulley of Islington, but it was only after his inadequate optical theory had been improved that Goring felt he had achieved some success.[154]

Perhaps as the result of this episode, and because of his preoccupation with the development of the sympiesometer in the years following, Adie is not known to have executed microscope work until the early to mid 1820s. This was at a period when his links with Brewster were strong, and when Brewster was attempting to put into practice his published proposals of 1813, for the use of exotic optical media such as garnet, ruby, sapphire and diamond in lenses. He had argued that single lens microscopes of much greater power and resolution could be produced from highly refractive materials with lower dispersive powers than optical glasses.[155] The performance of these simple microscopes was felt to justify the undoubted cost and difficulty of constructing such lenses and this appeared to be a valid alternative route to the empirical correction of the multiple optics of compound microscopes.[156] Brewster employed the optician Peter Hill (q.v.), his early associate the Jedburgh instrument-maker James Veitch (q.v.) and later the Edinburgh lapidary William Blackie, to make lenses of these materials; and in addition, he commissioned Adie in about 1824 to grind two garnet lenses, which he judged to be better than any solid lens he had seen.[157] Although these last items are not known to survive it is possible to evaluate the quality of Adie's lenses from other sources.

His biographer gives an account of the apparently crude but ingeniously effective method which Adie devised to make moulds for grinding the minute lenses.[158] He probably only made a few sets of these painstakingly constructed lenses: their expense prevented them from being widely used and the

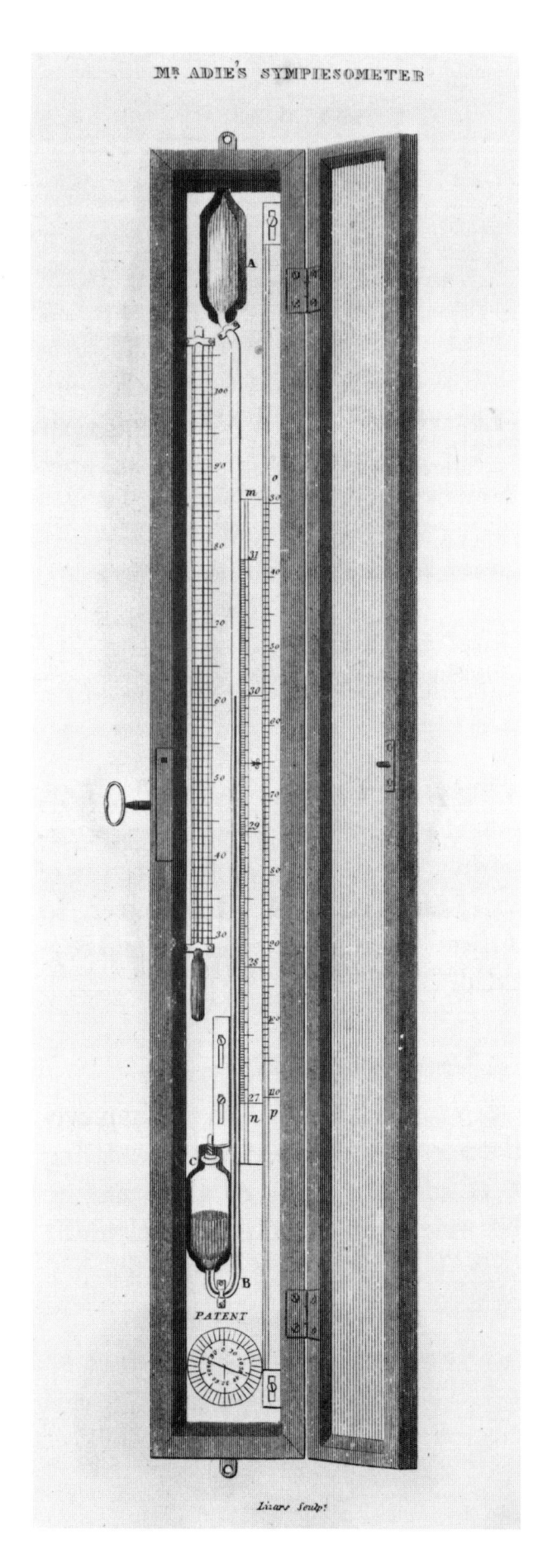

11 Alexander Adie's Sympiesometer, 1819; engraving by W. H. Lizars

predominance of the achromatic type by 1840 in effect removed the market for jewel lenses. However, it has been shown that the optical performance of jewel lenses was equal to contemporary achromatic lenses and that C.R. Goring in conjunction with Andrew Pritchard in London successfully marketed jewel lens microscopes.[159] The question therefore arises as to why Adie was not comparably successful, given both his association with Brewster, the main proponent of jewel lenses, and his evident skill and established position as a main supplier to the Edinburgh scientific community.

In the first place, the circumstances of the construction of the two garnets for Brewster are unknown: their date of about 1824 is inexact and it is not known how they relate to the chronological development of Brewster's ideas in microscopy. Equally problematic is the group of six lenses attributed to Adie, five of them jewel lenses, which forms one of the most interesting items in this collection. The attribution made by Nuttall and Frank in 1973 rests on two cryptic notes, one of which is apparently in Adie's hand, found in the box, and on the supposition that the lenses were unsold stock which passed into the hands of J. R. Hutchison, an Edinburgh optician and instrument collector, whence they came to Arthur Frank.[160] The first manuscript note, if it indicates the prices of the items (and this has not been established), shows them to have cost from £4 to £9 5s, and it has been suggested that the correspondence of this list with the six lenses demonstrates that they remained unsold.[161] Such prices would have been higher than those charged by Pritchard for his sapphire lenses, and they would presumably have appealed to a strictly limited market. A group of three garnet lenses of $1/50$ inch focal length in the Wellcome Collection has also been confidently but inconclusively attributed to Adie on the basis of its very close similarity to the set in this collection.[162] Although the excellence of the lenses demonstrated in modern tests[163] tallies with Brewster's high opinion of the garnets which Adie made for him, the attribution must still be treated with some caution.

Comparatively few microscopes from Adie's hand are known, but extant examples impressivly support his reputation. Two pieces in the Royal Museum of Scotland's collection are dateable to the period 1823-29. The first is an extremely large microscope, which is only superficially of standard eighteenth century conception: its most novel feature is a substage polariser, making it easily the earliest recorded polarising microscope.[164] It was constructed for the Royal Society of Edinburgh at some time after June 1823 when Brewster proposed the establishment of a 'Physical Cabinet of Instruments', and before 1829 when its existence is first recorded.[165] The second instrument is a reflecting microscope, similar to the instruments produced by G.B. Amici of Modena in Italy from about 1818. A reflecting microscope was also amongst items to be commissioned by Brewster for the Society's Cabinet, although in this case it is not known to have been delivered; this may be the very instrument intended for the Royal Society or another made by Adie at the same time.[166] The mirrors are of speculum metal, but not of the same composition as Amici's. It is possible that they may be by John Cuthbert of London who produced an improved reflecting microscope in 1826 and who had corresponded on microscope mirrors with Brewster, whose description of them given to the Society of Arts was published in his *Edinburgh Journal of Science*.[167] The instrument is by no means a slavish copy of Amici's microscope. The detailed modifications reveal a grasp of engineering principles later apparent in a precision chemical balance made to the pattern developed by Thomas Charles Robinson.[168]

Another microscope, of later date, is an example of the only model of a simple microscope by Adie & Son (and therefore post-1834) in the collection, and incorporates several features which distinguish it from the Pritchard model from which it is derived.[169] In particular the standard set of optics has an additional garnet lens of the very short focal length of $1/100$ inch (which limited its use to uncovered objects), and for this reason the microscope is an important example of Adie's workmanship.[170]

The optical work which Adie secured through Brewster appears to have been confined to a few commissions such as for the garnet lenses, and other recommendations such as the microscopes for the Royal Society of Edinburgh. Although Brewster was turning increasingly to London specialists such as Andrew Ross for optical instruments, Adie's link with him was sustained through their common interest in meteorology. In 1821 Adie recommenced his weather observations (which had ceased abruptly in about the middle of 1805) at his new house, Canaan Cottage, at Morningside, to the south of Edinburgh. From 1824, abstracts of the thermometer readings were printed in Brewster's periodical the *Edinburgh Journal of Science*,[171] and in 1825 Adie published an article on atmospheric phenomena in the *Journal*, following it in 1829 with a second on the temperature of Bombay.[172] In 1826 we find him filling in the questionnaire circulated on behalf of the Royal Society of Edinburgh by Brewster, which called for hourly readings from specified instruments throughout the day of 17 July 1827.[173] A typical project of Brewster's, it was a modest variation on a scheme which had involved continuous hourly readings throughout the course of the years 1824-7 inclusive. Adie had constructed a 'large and accurate thermometer' specially for the observations, which were begun at Leith Fort by the garrison on 1 January 1824, and which continued for the next four years. The few errors in the mass of data which was gathered were, to Adie's credit, human rather then technical.[174]

The massive subsidised undertaking at Leith Fort was

balanced by a multiplicity of private observations, some of which appeared in print in the scientific journals and acknowledged the quality of Adie's meteorological instruments. For example, in 1825 the Rev. Wastell of Newbrough, near Hexham, published his rain-gauge readings from 'one of Mr Adie's gauges' which he owned,[175] an illustration of the type of free publicity which Adie gained at a time when meteorological observation was apparently becoming a vogue, and evidently holding a particular attraction for clergymen and schoolmasters. The other well-known application of the barometer to surveying was also publicised when the Rector of Perth Academy, Mr Adam Anderson, claimed to have calculated a corrected formula for calculating heights with a barometer by Adie, and in doing so to have read off the thousandth part of an inch.[176] The manifest excellence of Adie's products was underlined by readings taken with a sympiesometer in Corfu in 1821[177] and Adie's business profited from the interest in that and other instruments. A price list of meteorological instruments sold by Adie dating from 1820[178] enables some comparisons to be made with another of about 1826.[179]

Commissions for research purposes, of the type Professor Leslie had made in the previous two decades, continued in the 1820s. Adie adjusted the thermometers used by W.T. Hay Craft about 1825 in his experiments on the specific heat of gases,[180] for example, and barometers were also being worked on in the late 1820s.

Brewster was responsible for effecting an important introduction in 1829 when he recommended Adie to the young James David Forbes (1809-1868) as an experimenter on capillary instruments in touch with new developments.[181] Forbes, who became Professor of Natural Philosophy at Edinburgh University in 1833, seems to have been one of Adie's most important patrons in the 1830s. He instigated a project funded by the British Association for the Advancement of Science to investigate heat flow in soil and bedrock at three sites in and around Edinburgh, including the extinct volcanic plug which formed Calton Hill, on the summit of which was the Royal Observatory. Observations began in February 1837, with three series of thermometers constructed by Alexander Adie 'under Mr Forbes's directions'. Each series was nearly of the same extent and range, and consisted of thermometers of lengths 3, 6, 12 and 24 French feet.[182] Their construction represented a considerable technical achievement which makes another commission secured through Forbes in 1838 seem almost trival by comparison: on this later occasion the British Association was again the sponsor, this time of straight-forward meteorological observations by barometer and thermometer at Inverness and Kingussie: 'Mr Adie, of Edinburgh ... [having made] the necessary instruments ... under the superintendence of Professor Forbes ...'.[183] The first of this pair of commissions clearly excited interest, for in about 1840 Adie supplied a John Caldecott at Trevandrum, India, with three thermometers sunk to depths of 3, 6 and 12 French feet for soil temperature observations.[184]

Meanwhile, Alexander Adie had been continuing his own twice daily meteorological observations. From 1821 until May 1831 they were made at Canaan Cottage, from mid-May 1831 until mid-May 1838 at the back of Adie's house at 10 Regent Terrace, finally resuming at Canaan Cottage in 1838 and ending in 1850.[185] He had apparently ceased his first observations in 1805, not from lack of interest, but because he believed he had accumulated sufficient material to assess the Edinburgh climate, to do which he reduced the observations and projected them in graph form.[186] In 1834 he performed a similar exercise for the benefit of a subsection of Section A of the British Association meeting in Edinburgh, with a view of Edinburgh weather during the previous decade.[187] When in 1850 he finally ended his observations, made with the help of his family, they filled seven paper-bound quarto volumes. After Adie's death J.D. Forbes, who had long contemplated the reduction of Adie's thermometer readings, obtained funds from the Royal Society of Edinburgh to employ computors, and with the help of a clerk in the business, James Grassick, worked through the manuscripts which the family had placed in his hands. The result was his classic paper delivered to the Society in 1861, based mainly on the labour of the 'zealous and careful observer of meteorological instruments'.[188]

The 1820s were among Alexander Adie's most active and fruitful years, to judge by the abundant evidence of his activities both in and out of his workshop. His active membership of the Royal Society of Edinburgh, the Society of Arts for Scotland and the Wernerian Natural History Society indicate the extent to which he was a highly respected instrument-maker, a shrewd judge of both the principles and practice of scientific experiment, and an assiduous experimenter in his own right.

Not surprisingly, Adie's interests extended into areas of physical and chemical science which were only just beginning to be explored. In December 1823 and January 1824 Adie demonstrated the experiments of the German chemist Johann Wolfgang Döbereiner of producing light by the action on platinum of hydrogen, and of forming water by the actions on pulverised platinum of oxygen and hydrogen, before the Royal Society of Edinburgh.[189] It is clear that these were not merely for show, because by late 1824 Adie had implemented the first of Döbereiner's theories and constructed an instant light-giving machine: it was less complicated than a version by the London maker Alexander Garden, but was prohibitively expensive,[190] and nothing further is known of the piece.

Adie was also interested in steam boilers, and accompanied his friend Robert Stevenson, the civil engineer, to inspect the exploded boiler at the Lochrin Distillery, Edinburgh in April 1821.[191] In 1832 and 1839 he sat on committees of the Society of Arts which examined proposals submitted to the Society for the prevention of explosions and methods of feeding boilers.[192] Although Adie was probably involved from time to time in technical consulting work, little information about this has come down to us. One such significant occasion, however, was the scientific gauging with James Jardine for the county reports to the Exchequer of the local standard weights and measures which were derived from the pre-Union Scots standards, against the new Imperial standards enacted by Parliament in 1824.[193]

Further indications of Adie's wide mechanical knowledge can be found in the minute books of the Society of Arts. In the year 1829 alone he assessed the merits of several clock pendulums, a lathe, a rain gauge, a magnetic needle, two hydrometers, an orrery and a mangle,[194] and many other similar examples in subsequent years could be quoted. Adie was a councillor of the Society 1828-31, its foreign secretary 1835-38, and sat on the committee of accounts, in addition to his principal work as a mainstay of the Mechanical Department.[195]

In 1823 when Alexander Adie dropped his uncle's name and began trading only under his own name, he remained at 15 Nicolson Street. In June 1828 he moved to much more fashionable premises in the New Town at 58 Princes Street, where the business remained until 1843, advertising their Royal Warrant as Instrument Makers to the King after 1835.[196] In due course Alexander's son John was brought into partnership, and the business traded as Adie & Son from 1835 until 1880. Between 1844 and 1876 the shop was at 50 Princes Street, and from 1877 until 1880 at 37 Hanover Street.[197] By his marriage to Marion Ritchie, Adie had four sons and seven daughters.[198] Their eldest son John (1805-1857) became his father's partner and is discussed in more detail below. Alexander James Adie (1808-1879), their second son, served an apprenticeship with the civil engineer James Jardine and became a respected railway and bridge engineer.[199] Richard (1810-1881), who was baptised Ritchie after his mother, and Patrick (1821-1886), the younger sons, were also instrument makers and were the agents for expanding the business into Liverpool and London, with Richard returning to run the Edinburgh firm after the deaths of his father and brother John.

Although John Adie's early years remain obscure it can reasonably be assumed that he served his apprenticeship in Edinburgh, either under his father or under another master optician or brassfounder. The earliest reference to him is in a pass made out for him to leave the Isle of Man in May 1826,[200] and this would coincide with the likely date of the end of his apprenticeship. By the summer of 1828 John had commenced significant work alongside his father. William Galbraith, a teacher of mathematics in Edinburgh, who had drawn up tables designed to facilitate barometric measurements, enlisted John Adie's cordial support for his project of verifying the tables by observation. Adie provided 'two mountain barometers of the best construction' (of the syphon type), and an easily-used and accurate hygrometer. The two set out from Edinburgh 'on a short tour, by way of Stirling, Callander, the Trossachs and Loch Cathrine so as to combine a little amusement with our more important pursuits ...', and on 27 August took simultaneous readings at the summit and foot of Ben Lomond: Adie, already perhaps rather overweight, stayed at the level of the loch.[201]

Galbraith's interest in surveying resulted in several small commissions from the Adie workshop in subsequent years. In September 1829 he used an Adie sympiesometer to calculate the height of Allermuir in the Pentland Hills, and in August 1830 he used an Adie barometer 'of the best construction' to measure Ben Nevis, referring in the process to the working of the latest sympiesometers which were fitted with an engraved table, enabling rapid calculations to be made.[202] The usefulness of the instrument for mountain work had been questioned in 1829 by J. D. Forbes, who distinguished sharply between its performance as a marine barometer, which he praised, and on survey work, which he critised strongly.[203] Galbraith's publicised use of the sympiesometer and mountain barometers therefore made him a useful ally as well as a valued customer, and a close professional relationship developed between him and John Adie. In 1831 Galbraith calculated the magnetic deviation on Arthur's Seat, accompanied by his friend Mr John Trotter, who used 'a new surveying compass by Mr Adie': Adie also helped with the calculations.[204]

The following year Galbraith measured the height of Cheviot barometrically, which involved simultaneous readings by Sir Thomas Makdougall Brisbane (Alexander Adie's old acquaintance, then living at Holy Island) and by Richard Adie at 58 Princes Street.[205] An interesting and apparently unique venture of Alexander Adie's occurred in 1833, when he was a co-publisher of a volume of barometric tables compiled by Galbraith.[206] The last of Galbraith's commissions from this period was a 'New Pocket Box Circle' constructed by John Adie to his designs, and exhibited to the Society of Arts on 13 April 1836 in an incomplete state, 'Mr Adie not having yet divided the Circle'. It gained the Society's Silver Medal.[207]

Enough has been said of John Adie's known work in the period up to 1836 to indicate that his mechanical

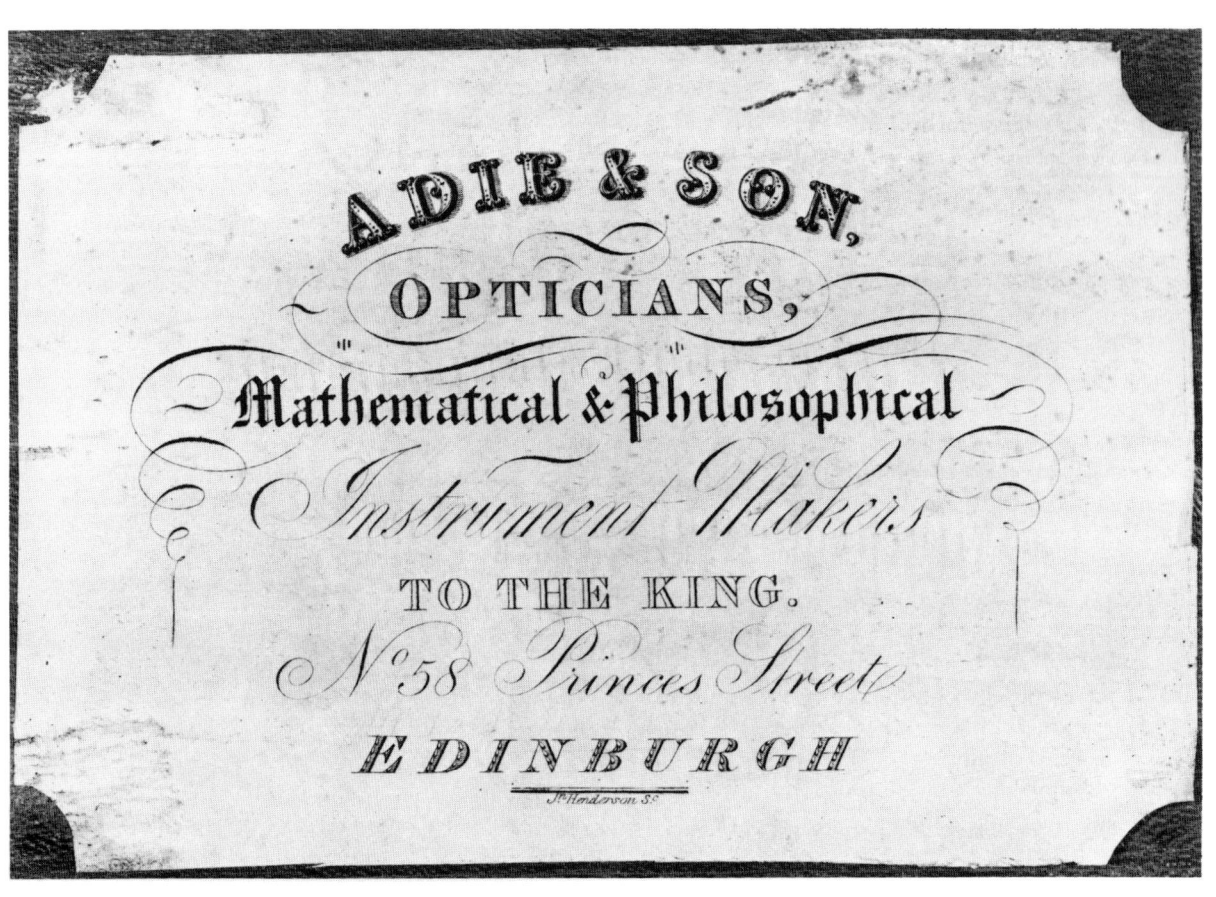

12 Adie & Son's trade card for their address between 1835 and 1837, when they held the Royal Warrant as Instrument Makers to William IV in Edinburgh

13 Adie & Son's trade card for their address between 1838 and 1843, when they held the Royal Warrant as Instrument Makers to Queen Victoria in Edinburgh

and scientific ability was attracting attention by his early twenties. This poses a problem parallel to that discussed above concerning Alexander Adie's contribution to his uncle's business: how far can separate specialisations and activities within the workshop be discerned? Difficulties of interpreting the relative contribution of father and son obviously begin from the moment when John can reasonably be supposed to have executed commissioned work in his own right. As has been shown this could be earlier than 1828, and certainly by the time the business became Adie & Son in about 1835 John's contribution was already considerable. Later, the point at which Alexander retired remains uncertain. Evidence of commissions provides valuable clues as to their activies, but acknowledgements and indeed references of all sorts all too often refer unhelpfully to 'Mr Adie' as the maker of a particular instrument. Again, as with the Miller & Adie period, the dating of pieces to a specific time within the forty-five years during which the name Adie & Son was used, is made more critical by the fact that both Alexander and John were dead by 1860. To some extent continuity during the remaining phase of Adie & Son up to 1880 was maintained by Richard who continued the business, and by the existing staff (who are discussed below). In general, however, the problems of attribution remain interesting and vexed.

An indication of the importance John had assumed in the workshop by 1829 is given by his execution of a 'large and delicate' pressure pump on the plan of H.C. Oerstedt for Professor Leslie. The interest here lies not only in Leslie's glowing reference to Adie as 'our ingenious young optician', but also in the fact that Leslie's commissions, previously executed by Alexander, were being undertaken by John.[208] As noted above, the earlier commissions were principally for meteorological instruments, and it has been shown that one of Alexander Adie's great strengths lay in his excellent capillary work, in all applications of the thermometer, barometer and sympiesometer. It is therefore instructive to see the way in which John was responsible for some improvements made to these types of instruments from the late 1820s. At the same time that he was helping Galbraith with his barometrical work, John Adie was experimenting with dewpoint instruments which were used alongside barometers for measurement, and he read a paper on a new version he had constructed to the Society of Arts in February 1829.[209] In April 1829 he described his improved adjustment mechanism for the cistern barometer.[210] The sympiesometer also received the Adies' attention, probably because they took J.D. Forbes's strictures on the instrument's use for mountain work to heart. Forbes was sufficiently persuaded of its improved state to take at least two with him for his fieldwork on glaciers near Monte Rosa, in the Alps, in August 1842. One was a small version and was 'universally admired', as Forbes wrote to a friend; he continued, 'Tell him

[John Adie] however that the improved Sympiesometer does not yet work well, & that he must exercise his ingenuity upon it again.'[211] John Adie was clearly something of an expert in glass working for different types of instruments of which the sympiesometer was only one. In 1835 he described to the Society of Arts a new method originating in France, which he had learnt recently in London, of cutting and working glass with turpentine as a lubricant.[212] To Adie goes the credit of providing a solution to the fouling of large-bore barometer tubes, and in about 1853, when Adie was in London, John Welsh of the Kew Observatory successfully followed Adie's advice on the problem, which had been holding up the construction of a standard barometer for the observatory. Welsh's account refers to a remark by Adie that he had experienced the same problem, although it is not known when or for whom he worked on large-bore instruments.[213]

Another category of instruments sold by Adie & Son was telescopes, and although Alexander Adie was not noted for telescope work, John appears to have executed some commissions and to have made observations and experiments.[214] One well-documented commission which unexpectedly fell through was for a refracting telescope and camera obscura for the Dumfries and Maxwelltown Astronomical Society in 1835. Contact was made through James Jardine, the Adies' engineer friend, and Alexander Adie agreed in January 1835 to supply a telescope for £73 10s, and a camera obscura 'of the same description as that furnished by him to the Edinbr. observatory for £60'.[215] Meanwhile, Thomas Morton of Kilmarnock (q.v.) had offered to supply the instruments, the telescope at £73 (almost the same as Adie's price) and the camera obscura for £27 10s. The deciding factor was the recommendation of Morton's work by the astronomer and explorer Sir John Ross in April 1835, with the result that Adie was dropped by the Society, although he had gone to the length of preparing drawings of the observation tower.[216]

The episode confirms that the business did not in any sense have a dominant hold over the market, and suggests the idea that Alexander and John Adie's specialisations led to weak spots in their repertoire of instruments. However, a later reference shows that John was highly capable when it came to the mechanics of astronomical instruments. He constructed a Y-mounting for the transit telescope at the Royal Observatory, for Professor Charles Piazzi Smyth. Smyth had attempted to get the telescope's German maker to construct his new and unusual design, but found him 'far too fearful of leaving the old beaten path of instrument-making to attempt any improvement'. When applied to, John Adie constructed the massive cast-iron stablising blocks with their finely-fitted pivot grooves 'in a perfectly satisfactory manner'.[217] When the Adies exhibited at the Universal

14 Four medals awarded by the Royal Scottish Society of Arts to sons of Alexander Adie: a. John Adie's dewpoint instrument and barometer cistern, 1830; b. A. J. Adie junior's experiments on apparatus for heating hothouses, 1832; c. R. Adie's wind-meter, 1844; d. R. Adie's new hermetic barometer, 1860

Exhibition in Paris in 1855 the pieces were astronomical instruments entered under the name Adie, which suggests that John was responsible for them.[218] A telescope used by William Swan, an experimenter in optics closely associated with the Adies, for observing an eclipse in 1851 is described as furnished by Mr Adie of Edinburgh, as were two thermometers accurate to one-tenth of a degree. These items can be attributed safely to John who published an account of the observations.[219] That he was preoccupied with optical instruments in the early to mid 1850s is also suggested by a paper of 1850 on the marine telescope.[220] Their services to astronomers included the provision of observatory regulator clocks. An Adie & Son example of about 1840, with a mercury compensation pendulum, in the Royal Museum of Scotland exhibits detailed constructional features which are more in keeping with instrument making practice than with horological practice, indicating that the movement was made by the firm rather than bought in from one of the London specialists.[221] There was a sidereal regulator by Adie & Son at the private observatory of Sir William Keith Murray (1801-1861) at Ochtertyre, near Crieff, Perthshire in the 1850s.[222]

William Swan's association with John Adie shows the close ties between customer and maker which could arise from an initial commission. The telescope and thermometer were followed by a discussion between Swan and Adie which resulted in John Adie's new variation compass, for which he acknowledged that he was 'indebted to my friend Mr Swan for a suggestion ...';[223] and two years later Swan published his own simplified version, which was constructed for him by John Adie.[224] This was followed by a commission for 'a small collimating magnet by Mr Adie', described by Swan in April 1855.[225] For his experiments on the spectra of carbon and hydrogen compounds, described in April 1856 (again to the Royal Society of Edinburgh), Swan borrowed 'an excellent theodolite by Adie' from John Adie.[226] This may be the same instrument or a similar one described later by Swan in the year of his appointment to the chair of natural philosophy at St. Andrews University as 'an excellent theodolite, constructed expressly for observations of prismatic spectra, by the late Mr John Adie'.[227] At any rate it suggests that John Adie made a theodolite for his own use, which he was prepared to lend. Other evidence suggests that he shared Swan's experimental interest in the optics of the human eye and that this strengthened their scientific association. On 19 March 1849 Adie presented to the Royal Society of Edinburgh a paper on the relationship between the density of the aqueous humour and the achromatism of the eye, and at the same meeting Swan talked on luminous impressions on the eye, measured with a special instrument, the selaometer, which Adie may have constructed for him.[228] That the subject was arousing the interest of the circle of which Adie formed a part is shown by J.D. Forbes's paper on the eye in December 1849.[229]

John Adie's interest in optics dates from an earlier period, although examples of his work are few. In about 1834 he helped the civil engineer Alan Stevenson, son of Robert Stevenson the Engineer to the Northern Lighthouse Board, with a lengthy series of observations to evaluate three experimental lighthouse lenses belonging to the Commissioners of the Northern Lighthouses,[230] an instance not only of the close professional relationship between the Adie and Stevenson families, but also John's capabilities as an optician. For it is clear that John Adie was thoroughly familiar with the principles of optics and recent literature on the subject, and it made him, like his father, an indispensable member of the committee of the Society of Arts, to which he was elected in May 1838.[231] In 1841 he was evaluating an improved camera obscura by the pioneer of photography, Thomas Davidson, in 1842 Davidson's oxyhydrogen microscope and polariscope, and in 1843 his improved camera lucida.[232] Among the many other instances of his work as a technical critic one stands out as a clue to the origins of his association with William Swan, with whom, as shown above, he may have been working in 1849. In 1843 Adie was appointed convener of a committee on Swan's method of determining the angle of prisms, and also sat on the prize committee that year, which awarded the Society's Gold Medal for the session 1842-3 to Swan for his paper.[233] John Adie's approval was clearly worth having. The success of the camera lucida had created a demand for small and finely worked prisms and Adie developed some

expertise in grinding these: another of the Society's prize-winners described a prismatic auriscope for medical use in 1844 in which the 'interior prism especially must possess that highly-finished surface which Messrs Adie & Son ... are so competent to supply'.[234]

Another area of optical instrument work which preoccupied Adie was the mirror arrangement on the sextant. He appears to have been interested in the instrument purely in its surveying capacity rather than for its marine navigational use, and this arguably reflects the overall bias of the Adie business, with the notable exception of the sympiesometer, which of course was intended for both land and sea use. In the mid 1840s John Adie was testing the advantage of metallic reflectors for sextants, and concluded they were more accurate and measured greater angles than conventional mirrors. He exhibited an example to the Royal Society of Edinburgh in 1845 which had been used by a Mr Mossman for a season of surveying in the north of Scotland, and read Mossman's (presumable complimentary) remarks on its performance.[235]

It is not known if the business subsequently produced the metallic mirror sextant as a speciality, but surveying instruments as a whole continued to be a staple product for the firm, like others in Scotland. If in the latter part of the eighteenth century agricultural improvements stimulated a demand for surveyors and the tools of their trade, from the early nineteenth century increasing urbanisation likewise increased demand, not only in conventional surveying, but also in specialised areas such as water and sewage systems created by public health legislation and in civil engineering undertakings such as canal and railway construction, and mine surveying.[236] Although no evidence exists to decide whether the Adie business catered on a large scale to the demand for surveying instruments, it is reasonable to suppose that they had a good share of the market. Examples of individual commissions suggest that in this area too, John Adie was coming to dominate the business by the mid-1840s. It is no surprise to find the civil engineer Thomas Stevenson commissioning from the firm a ball-and-socket levelling instrument, 'an improved portable Levelling Instrument and Rod', and an improved spirit level. The latter was made for Stevenson by 'Mr Adie' in 1840, and the others by 'Messrs Adie', but the likelihood is that John was chiefly responsible for them.[237] In 1844 the land-surveyor John Sang, brother of the polymath Edward Sang, described a sophisticated two-screw levelling head for theodolites; and although Adie is not credited by Sang with its construction, the only example recorded is signed by Adie & Son and is presumably also the work of John.[238]

That the above discussion of John Adie's work in the

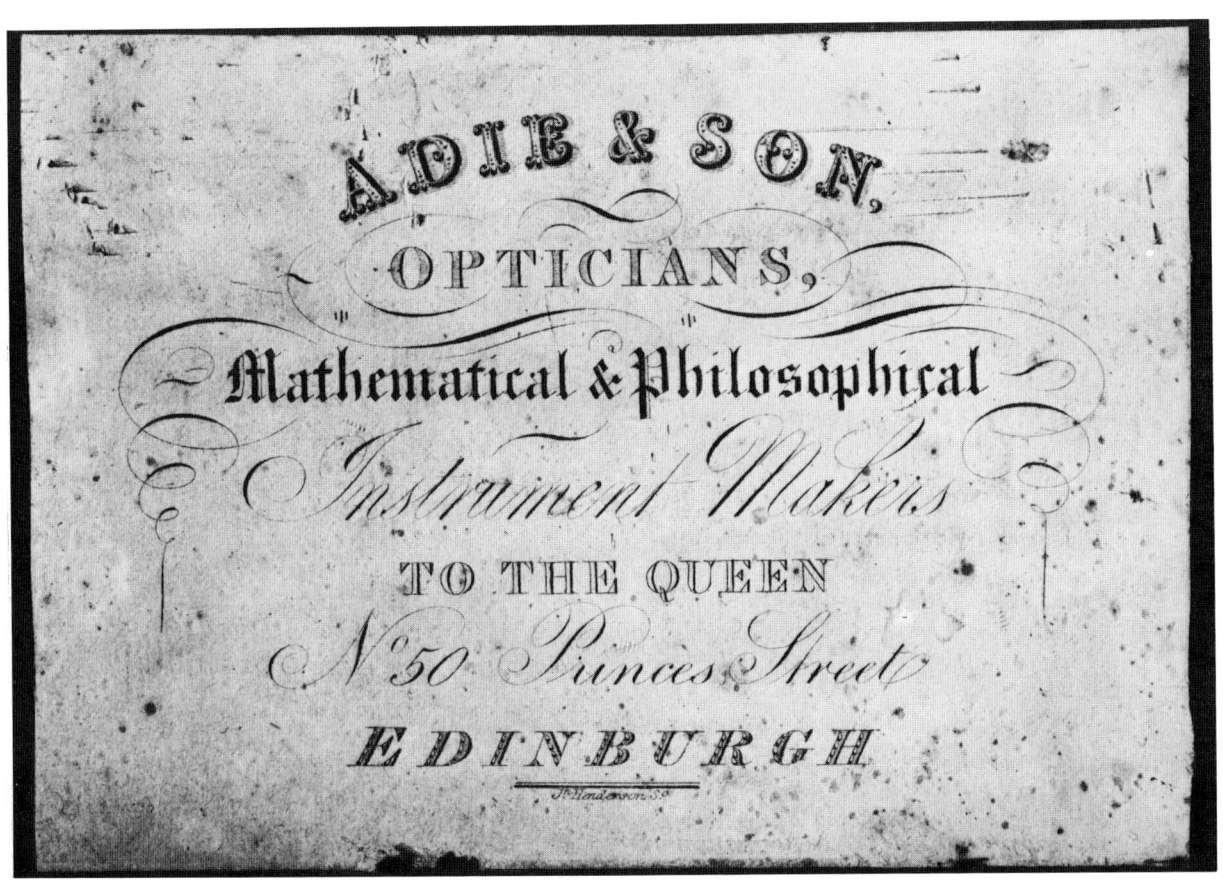

15 Adie & Son's trade card for their address between 1844 and 1876

16 Adie & Son's trade card for their address for 1881-2

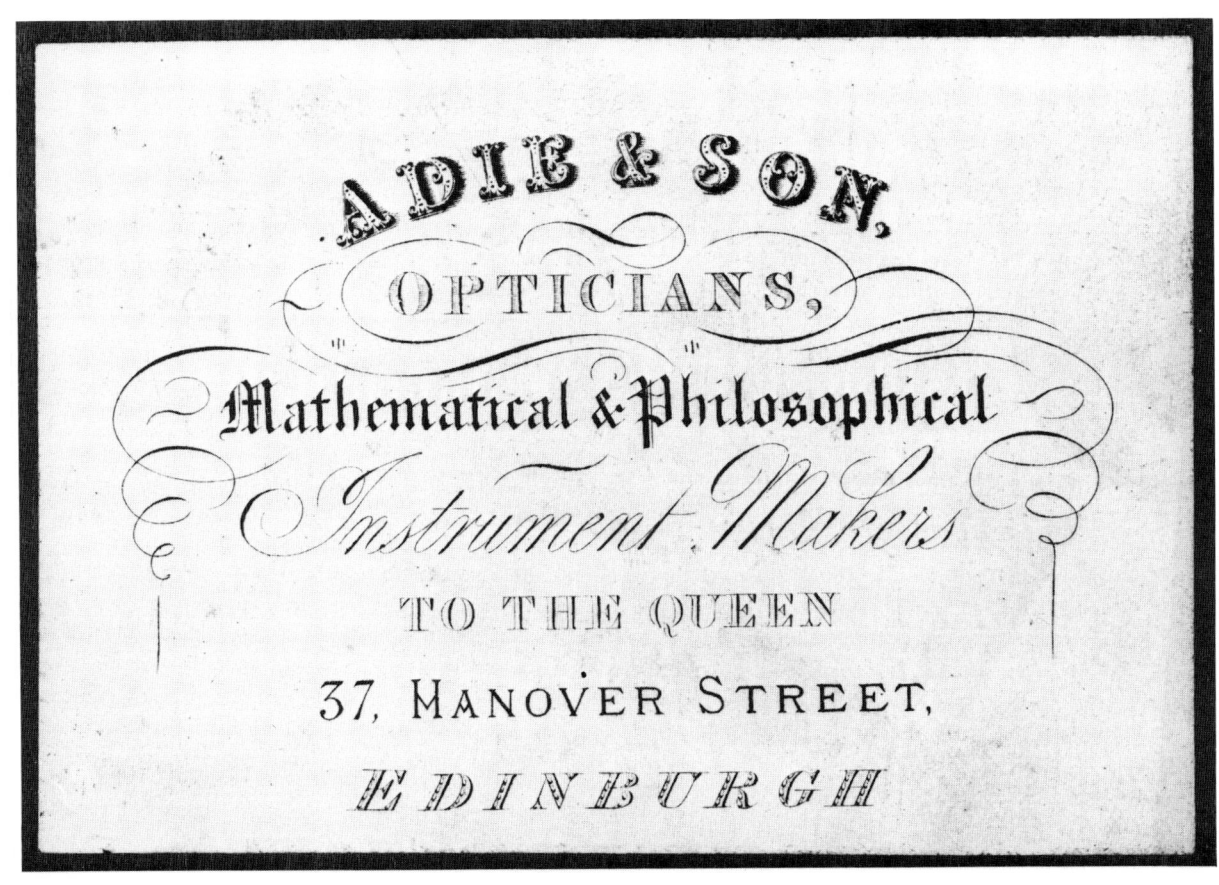

thirty years or so of his known activity may appear to be synonymous with a discussion of the operation of the firm as a whole, is due to the weight of the evidence which points to him as more productive than his father. This is equally true of their activities in the workshop and in the Society of Arts, but while the more taxing and prestigious commissions came to be executed by John it would be misleading to disregard his father's role in the workshop in his later years. According to his biographer, the large thermometers for Professor J.D. Forbes, constructed 1836-37, were 'the last work of much difficulty in which he took much personal trouble'.[239] However, it is not known when he ceased to work altogether. He was in the shop regularly in July 1840, but Richard Adie reported in November 1841 that he had had 'another of those disagreeable attacks in the morning that made him stagger and feel languid all day after'.[240] It is likely that from this time onwards he spent most of his time at Canaan Cottage.

His garden absorbed much of his attention, as his biographer explained:

> 'In after life he was a most assiduous gardener trying many experiments, raising new apples and beyond his eightyth year deriving the greatest pleasure from his garden works and in the culturation of his plants and flowers.'[241]

Typically, his interest was in part scientific, an approach which he shared with Professor Leslie: in 1840 we find him eager to begin observing the decomposition of specially prepared pieces of wood to be buried in his garden.[242] In his advancing years Adie was ministered to by his wife and daughters, and enjoyed the fruits of his early years of hard toil. In addition to his wife and children, the household at Canaan included lizards and birds, hens, pigs and a cow, all of which are recorded in the cheerful letters of Adie's daughters to their brother A.J. Adie. The 'Canaanites' as they dubbed themselves, appear to have lived a tranquil life as 'a highly favoured family'.[243] Three of the five daughters married: Elizabeth married the lawyer George Barron W.S., Janet married Thomas Henderson (1798-1844), Astronomer Royal for Scotland, and Helen married firstly James Marshall and secondly William John Menzies W.S. None of the daughters, however, was physically strong, and one after another they died young. Adie is said to have born up against these losses with a 'manly fortitude':[244] on one occasion, after the death of Jane in 1839, he worked off his grief by busying himself 'contriving a newly invented Bee hive' to divert the family.[245]

Meanwhile, John had set up house in 1838 in the New Town, and enjoyed a state of mildly luxurious bachelorhood, much to his sisters' amusement. Their letters are full of references to his expensive furnishings, his supper parties, and his fondness for food and (especially) drink.[246] Apart from the light their remarks shed on John's character, they indicate that his earnings from the business were enough to support him in some style.[247] Other personal details emerge from the letters. He had a good voice, and entertained his friends with his singing. He was fond of opera, and when in London purchased scores in a shop near the optician George Dollond's premises. He disliked walking, perhaps because of his corpulence, and instead preferred driving fast in a gig; despite several tumbles he seems to have avoided serious injury.[248] A small glimpse of his political views is afforded by his subscription of £1 towards the cause of the Anti-Corn Law League in 1844.[249] Copies of pamphlets generated during the Anti-Corn Law agitation in the 1840s, which have found their way into the family scrapbook, may well reflect Alexander Adie's interest in the inflammatory issue, as much as his son's.[250] It is likely that John shared his father's enthusiasm for reform in 1832, but their respective political interests, like their work in the shop, remains obscure.

John married into the Barron family (into which his sister Elizabeth had also been married) but his wife died childless in the 1850s and he appears to have resumed a semi-independent existence, busying himself with his work in the years following her death. However, by the mid 1850s his health was 'in a delicate condition', and by the latter part of 1856 he was prone to occasional 'fits of despondency'. It therefore came as less of a shock to those who knew him closest, when on the evening of 5 January 1857, John Adie shot himself through the head in his own house.[251] He was fifty-two years old. His death caused a small sensation in Edinburgh because, by a macabre coincidence, the widely-known and respected geologist Hugh Miller had committed suicide at his house in Portobello, in an almost identical manner only twelve days before.[252] Miller killed himself while in a state of paranoia and extreme fatigue from overwork, but no dramatic symptoms appear in Adie's case. Rather, the apparent normality of his last few hours, spent playing whist and having supper with family and friends, belied his depressed mental state, which the doctor who certified his death had noted in December.[253] If there were things going badly wrong in his personal life or with business at the shop, few clues have survived.

However, it is known that Alexander Adie was becoming senile. According to one account, Adie at the age of eighty-two had a severe illness from which he recovered physically

> 'but which wrought a sad change on his memory and perceptive powers. From this time his usual pursuits were abandoned and his bodily frame appeared gradually to change ...'[254]

In November 1858 a cold destroyed his remaining strength, and he died on 4 December 1858 at Canaan Cottage, aged eighty-four.[255] Whether his decay exacerbated John's condition, or whether the shock of John's suicide affected his father, is not known.

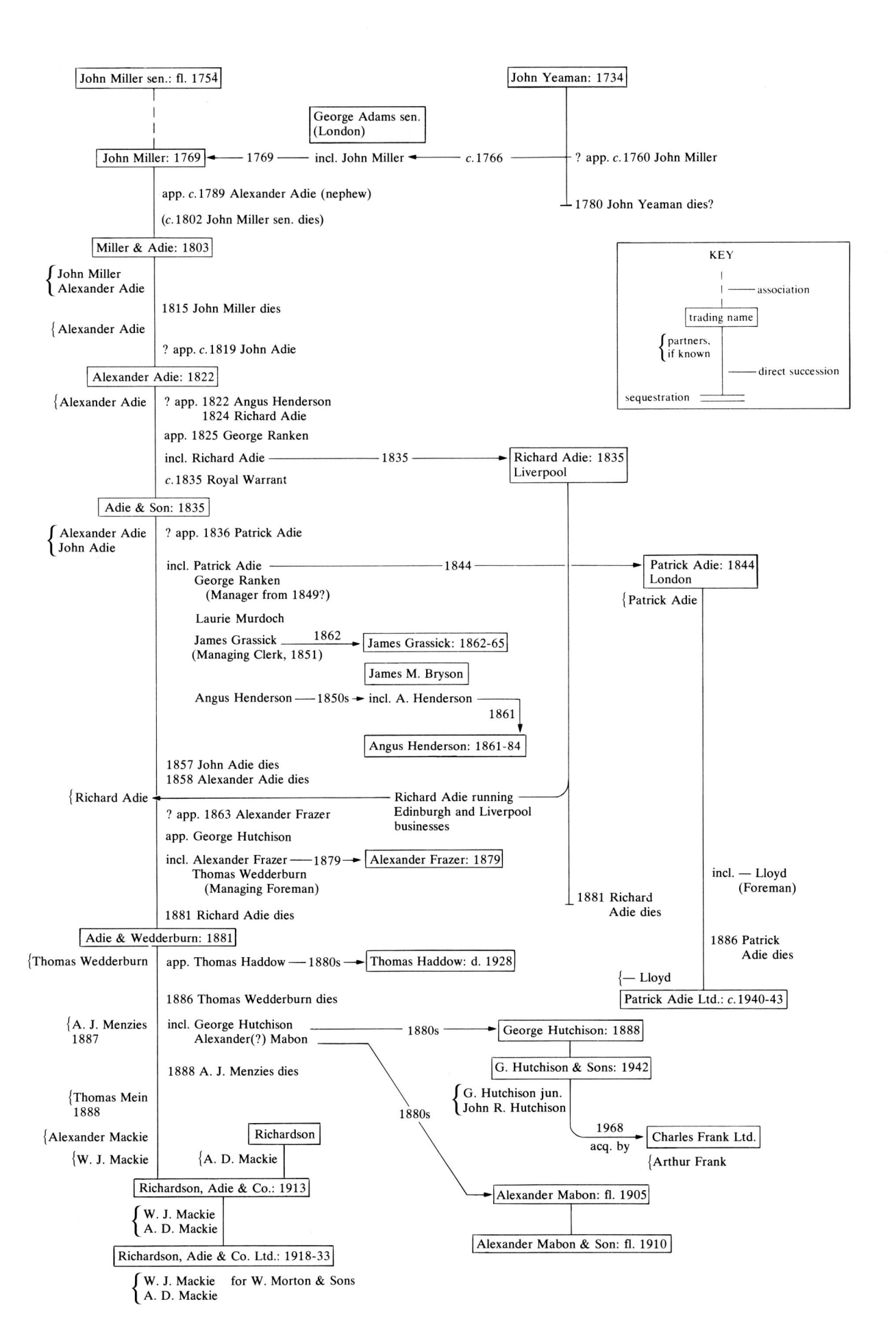

```
John Miller sen.: fl. 1754                                          John Yeaman: 1734
        |                    George Adams sen.
        |                    (London)
 John Miller: 1769 ◄── 1769 ── incl. John Miller ◄── c.1766 ── ? app. c.1760 John Miller
        |                                                           |
  app. c.1789 Alexander Adie (nephew)                              └─ 1780 John Yeaman dies?
  (c.1802 John Miller sen. dies)
 ┌──────────────────┐
 │ Miller & Adie: 1803 │                                 ┌──────────────────────────────┐
 └──────────────────┘                                    │              KEY             │
{John Miller                                             │         |                    │
 Alexander Adie                                          │         |── association      │
                                                         │    ┌─────────┐              │
         1815 John Miller dies                           │    │trading name│            │
                                                         │    └─────────┘              │
{Alexander Adie                                          │ ┌ partners,                  │
         ? app. c.1819 John Adie                         │ ┤ if known                   │
                                                         │ └           — direct succession│
 ┌────────────────────┐                                  │                              │
 │ Alexander Adie: 1822 │                                │ sequestration ▬▬▬            │
 └────────────────────┘                                  └──────────────────────────────┘
{Alexander Adie    ? app. 1822 Angus Henderson
                      1824 Richard Adie
                   app. 1825 George Ranken
                   incl. Richard Adie ─────── 1835 ──────►┌──────────────────┐
                                                          │ Richard Adie: 1835 │
                   c.1835 Royal Warrant                   │    Liverpool      │
                                                          └──────────────────┘
 ┌──────────────┐
 │ Adie & Son: 1835 │
 └──────────────┘
{Alexander Adie   ? app. 1836 Patrick Adie
 John Adie
                  incl. Patrick Adie ──────────── 1844 ──────────────►┌──────────────────┐
                     George Ranken                                     │ Patrick Adie: 1844 │
                       (Manager from 1849?)                            │     London        │
                     Laurie Murdoch                                    └──────────────────┘
                     James Grassick ──── 1862 ──►┌───────────────────┐ {Patrick Adie
                      (Managing Clerk, 1851)     │James Grassick: 1862-65│
                                                 └───────────────────┘
                                                 ┌─────────────┐
                                                 │James M. Bryson│
                                                 └─────────────┘
                     Angus Henderson ── 1850s ──► incl. A. Henderson
                                                          │
                                                         1861
                                                          ▼
                                                 ┌───────────────────┐
                                                 │Angus Henderson: 1861-84│
                                                 └───────────────────┘
                     1857 John Adie dies
                     1858 Alexander Adie dies
{Richard Adie ◄───────────────────────── Richard Adie running ─────────┐
                                          Edinburgh and Liverpool
                     ? app. 1863 Alexander Frazer  businesses
                     app. George Hutchison
                     incl. Alexander Frazer ─ 1879 ─►┌──────────────────┐
                         Thomas Wedderburn          │Alexander Frazer: 1879│
                          (Managing Foreman)        └──────────────────┘       incl. — Lloyd
                                                                                (Foreman)
                     1881 Richard Adie dies                        1881 Richard
                                                                        Adie dies
 ┌──────────────────────┐
 │ Adie & Wedderburn: 1881 │
 └──────────────────────┘
{Thomas Wedderburn  app. Thomas Haddow ── 1880s ──►┌──────────────────┐
                                                    │Thomas Haddow: d. 1928│
                                                    └──────────────────┘
                                                                               ┌─ Lloyd
                    1886 Thomas Wedderburn dies                                ┌──────────────────┐
{A. J. Menzies                                                                 │Patrick Adie Ltd.: c.1940-43│
 1887               incl. George Hutchison ───── 1880s ─────►┌────────────────┐└──────────────────┘
                      Alexander(?) Mabon                      │George Hutchison: 1888│
                    1888 A. J. Menzies dies                   └────────────────┘
{Thomas Mein                                                  ┌───────────────────┐
 1888                                                         │G. Hutchison & Sons: 1942│
                                         ┌──────────┐         └───────────────────┘
{Alexander Mackie                        │Richardson│        {G. Hutchison jun.
                                         └──────────┘         John R. Hutchison
{W. J. Mackie       {A. D. Mackie                                      │   1968   ┌──────────────┐
                                                                       │  acq. by │Charles Frank Ltd.│
 ┌───────────────────────────┐                                         └─────────►└──────────────┘
 │ Richardson, Adie & Co.: 1913 │                                                 {Arthur Frank
 └───────────────────────────┘                           1880s
                     {W. J. Mackie                            ┌──────────────────┐
                      A. D. Mackie                            │Alexander Mabon: fl. 1905│
                                                              └──────────────────┘
 ┌──────────────────────────────┐                                      │
 │ Richardson, Adie & Co. Ltd.: 1918-33 │                              ┌──────────────────────┐
 └──────────────────────────────┘                                      │Alexander Mabon & Son: fl. 1910│
                     {W. J. Mackie    for W. Morton & Sons             └──────────────────────┘
                      A. D. Mackie
```

Diagram 1. Schematic development of the Adie business.

Both men were unstintingly praised in opening addresses to the Royal Society of Edinburgh in 1857 and 1859, and the judgements passed on their technical skills may be taken to represent in some sense general informed opinion as to their work, and provide them with fitting epitaphs. In giving a conspectus of Alexander's life and work, apparently drawn from information supplied by the family or a close friend, it was said:

> 'His attention to business, with his skill as a mechanic, his quick inventive powers, and his sound judgement, led him to his being much employed by all kinds of inventors, to give their schemes a practical form, and in this way he acquired great readiness and experience in the higher parts of his profession.'[256]

Higher praise was reserved for John whose standing as a Fellow of the Royal Society was recalled by the President Dr Robert Christison in 1857:

> 'Mr Adie's enrolment among us is a sufficient proof that he successfully followed his calling as [i.e. 'which is'] one of the scientific arts; and by those by whom he must be better appreciated than by myself, he was greatly esteemed as a man conversant with the highest branches of his profession, and who has left behind him in that respect scarcely an equal, certainly no superior, in Edinburgh, or perhaps even in London itself.'[257]

The evidence of his instruments and his scientific experimentation suggest that this claim was not mere hyperbole, and it is interesting to note that his technical achievements and reputation appear to have eclipsed his father's. It was small wonder that when visiting the steam vessel *Benbow* at Leith in 1839, John should have been received 'as Mr Adie inventor of the sympiesometer'.[258] Confusion as to which of them was which clearly existed even during their lifetime and has persisted. In view of this, perhaps the most appropriate epitaph for their partnership was J.D. Forbes's ambiguous reference, in recalling prominent Fellows who had regularly attended the meetings of the Royal Society of Edinburgh, to 'the accurate Mr Adie'.[259]

After his father's death the shop was overseen by the third son Richard, who was by that time long-established in Liverpool. However, the practical management of the workshop apparently fell on the existing staff, and before examining this phase of its history, it is therefore worth reviewing the surviving evidence concerning the structure of the workshop and the production of instruments while Alexander and John Adie were active. Because no business records survive the known facts are scattered and do not form a complete picture. The workshop equipment is a case in point. John Miller possessed an up-to-date dividing engine in 1793. It appears that Alexander and John Adie were using a dividing engine from the 1830s if not before and this was presumably a more recent model than the Ramsden type in use in Miller's workshop. It has been suggested on the basis of a link between Adie & Son and their Edinburgh contemporary John Dunn (q.v.) that the Adies were perhaps unusual among Scottish instrument makers in mid-century in owning a dividing engine: Dunn undertook the construction of a protracting table to the design of the engineer George Buchanan, and got as far as the main parts of the instrument; the division of the circles was performed by Adie & Son.[260] The immediate reason for this may well have been that Dunn had died in July 1841, although it may nevertheless be true that Dunn did not possess his own dividing engine and would have had to contract the work out to the Adies.[261] Other examples show that the dividing engine was in frequent use for particular commissions. In 1836-7 it was used for Galbraith's pocket box circle, and for the division of the head of a lathe belonging to Edward Sang into minutes of arc, for its use as a goniometer.[262] Later, John Adie executed division work for an improved form of marine reflecting circle constructed for C. Piazzi Smyth.[263] A less sophisticated piece of equipment which was probably used by Adie & Son from an unknown date until about 1880 was a foot-lathe for turning metal, wood or ivory.[264] The existence of instruments divided into grades (400 to the revolution) rather than degrees suggests that the firm had equipment that could handle division in this system also.[265] Linear division was presumably also done mechanically. In 1839 Sir John Robison recommended to a friend who needed both imperial and metric graduations for a barometer, to have a scale laid down by 'Adie' showing both.[266] At some stage, probably in the 1840s, a divided yard scale was produced by the Adies as a standard for Leith Docks Commission.[267]

The link with Dunn in 1841-2 introduces another problematic area in the Adies' history, for little is known of their commercial relationship with other instrument-makers and related craftsmen either in Scotland or further afield. The arrangement for the joint production of Leslie's instruments with Cary of London which lasted between 1813 and 1820 (if not longer) has been noted above. There is no evidence to suggest that either of them retailed the instruments produced by the other, although this was a common arrangement in the trade when one maker had the manufacturing rights or the skill and resources which another lacked. The only known cases in which the Adies were in this position were firstly with the sympiesometer during the early phase of its production, and secondly with the eidograph. This ingenious instrument was an improved version of the simple pantograph (a device for copying, reducing or enlarging illustrations) which was developed from 1821 by William Wallace, Professor of Mathematics at Edinburgh. Wallace, like Alexander Adie, was a protégé of Brewster and a member of both the Society of Arts and the Wernerian Natural History Society. His improvements to the pantograph were encouraged by his friend James Jardine, the engineer who was also closely associated with Adie from the mid-1820s. It is not surprising therefore to find that a redesigned version of the eidograph which emerged in about 1827 from a collusion of Wallace and Adie (and almost certainly the master brassfounder James Milne), was thenceforward manufactured by Adie.[268] A further collaborative project with Wallace was the production of his 1839 chorograph, an improved form of station pointer, which 'may be obtained from ADIE and Son, Opticians, Prince's Street, Edinburgh; or from TROUGHTON and SIMMS, Fleet Street, London'.[269] Milne's name is one of the few that may be associated with the Adies, through their common links with the Stevensons and the Northern Lighthouse Board, and it is likely that as the most prestigious brassfounder in Edinburgh in the first half of the century his firm executed brass work for the Adies.[270] The eidograph was promoted through the Society of Arts and the Royal Society of Edinburgh, and its reliability in a wide variety of graphic work ensured its popularity for many years longer than its rivals. Those produced by the Adies were of high quality, and it appears that they dominated the not inconsiderable Scottish market for the instrument. This is suggested by the fact that they wholesaled it to other makers; an example sold by the Gardners of Glasgow (q.v.) under their name, is clearly of Adie provenance.[271] This is the only known instance of wholesale by the Adies. As an instance of the reverse process, the London instrument maker James Smith is known to have supplied Adie & Son with microscopes.[272] From an examination of student microscopes signed by Adie & Son it is clear that they also bought in separate optics and sub-assemblies from which complete instruments were engineered.[273]

Finally, a third area of uncertainty in the workshop's history, and one directly related to the capacity to produce wholesale goods, is the size of the workforce. The identities of any journeymen and apprentices John Miller may have had (apart from Alexander Adie and George Ranken, who will be discussed below) have not been traced. The same is true of Adie's mature years, for his first known assistant is his son John, probably apprenticed for the period 1819 to 1826, and active thereafter. The second recorded name is that of Adie's fourth son Patrick, who appears to have begun an apprenticeship in the shop in summer 1840.[274] However, probably for some time prior to this date two journeymen had been working for Adie, for in April 1843 George Ranken and Laurie Murdoch acted as witnesses to a codicil of his will.[275] George Ranken, a native of Edinburgh, rose to be foreman of the workshop by the time he was forty in 1851.[276] He had been apprenticed to Adie in January 1825, although this was not recorded until he claimed his Burgess status in July 1849.[277] This may mark the time when Alexander Adie bowed out of the firm in favour of his son John, and Ranken was promoted to foreman. The Census records for 1851 also provide the only known figure for the size of the workshop, where John Adie is described as a master optician employing ten men.[278] Nevertheless, the capacities in which the journeymen were principally employed remains obscure.

By 1854 a young man called James Grassick from Kincardine O'Neil in Aberdeenshire was working as a clerk with Adie & Son.[279] Although his training is not known, he was described in 1860 as a clerk in the firm by J.D. Forbes, whom he had been assisting in reducing Alexander Adie's meteorological data.[280] At about this date he was promoted to be managing clerk, and he is recorded in this post in April 1861.[281] This promotion in the aftermath of the deaths of Alexander and John Adie suggest that he was helping to fill the gap they left, and he appears in the street directories between 1862 and 1865 as an optician.[282] It was thus presumably into the hands of men such as Grassick and George Ranken that the daily running of the firm initially devolved. Richard Adie, who appears to have inherited the firm, retained its name for obvious reasons, and devoted his time to running both it and his Liverpool business simultaneously, with, as he stated in 1870, 'my time nearly equally divided between the two places'.[283] His Liverpool business, which appears to have dealt in similar types of instruments to Adie & Son, was a successful one from its beginning in 1835.[284] Far more than either his father, or any of his brothers (even John), Richard Adie was an experimental scientist. The twenty-seven papers published between 1837 and 1868, and other unpublished researches, attest to the vigour of his mind and the breadth of his investigative interest.[285] The first paper he is known to have delivered concerned his design for an anemometer, which he described to the Society of Arts in Edinburgh on 25 May 1836, and for which he was awarded the Honorary Silver Medal.[286] Although by this date he was established in Liverpool, his design dated from 1834. In 1844 he gained the same award for his adaptation of the gas-meter as an anemometer.[287] His attempts in 1841 to apply Graham and Dalton's laws for the diffusion of gases to the atmosphere of town and countryside, must rank as one of the earliest realisations of the relationship between atmospheric pollution in a city such as Liverpool and its relative humidity.[289]

Richard Adie's reputation as an authority on physical science probably helped sustain the image of Adie & Son, although to a lesser extent than John's scientific standing had done, because he was by no means as mechanically gifted as his brother. Although he kept himself before the attention of the Society of Arts in the 1840s with his occasional contributions, and was elected a Fellow in 1860, his papers in general appear not to have been so numerous by the time he assumed control of the business in the late 1850s. Nonetheless, his interest in meteorology meant that improvements continued to be made to the items which Adie & Son could offer for sale. The statical barometer supplied to the University of Edinburgh in 1859[290] was probably a standard demonstration model, but in 1860 Richard Adie exhibited his new hermetic barometer to the Society of Arts and published a description of it; it was well received.[291]

Although it is not known if this type of barometer was produced on any significant scale, another piece of evidence suggests that the firm continued its line in surveying instruments with some success. A type of levelling pole with fishing-rod joints, was the form 'most generally used in this country' according to a source of 1877, and was known as 'Adie's levelling-rod'.[292] Although its origin is not known, it could equally have been devised by Alexander, John, Richard or A. J. Adie. Specialist commissions also continued, but were notably few in the period after 1856. Some of the apparatus taken by C. Piazzi Smyth to the Great Pyramid in 1864-1865 was constructed by Adie & Son before John's death, but one interesting item appears to have been made in March 1866 by Adie & Son. It was a gun-metal circle divided every 20 minutes of arc, with which Piazzi Smyth measured the angles of the fragments of casing stones he brought back to Edinburgh, a measurement crucial to his theories of the Pyramid's purpose.[293]

Thus the deaths of Alexander and John Adie did not cause a seizure of the workshop, although it is clear that commissions, based as they had been, on a personal-cum-professional relationship between the Adies and their customers, dwindled. Nevertheless, trade continued, and it is possible that the Edinburgh workshop produced instruments for Patrick Adie in London.[294]

At least two of the Adie journeymen set up on their own during this period. In the case of Angus Henderson, it is not known precisely when he had been in the business, but when he began trading independently as a 'Practical Optician, Microscope, Mathematical and Philosophical Instrument maker' in Hanover Street in 1861 he advertised himself as having been 'for many years in the Establishment of Messrs. Adie & Son and Mr James Bryson'.[295] At the 1851 census he was recorded as a journeyman optician, aged 42.[296] It is possible, therefore, that he was apprenticed to the Adies, remaining after completing his apprenticeship in about 1828, and either transferring to J.M. Bryson's (q.v.) shop at the outset in 1850 or when Richard took over the Adie business. He was listed at various addresses until 1884.[297] The only microscopes signed by him which have been recorded have been Continental pieces, and indeed he described himself in 1874 as 'Agent for Oberhauser & Nachet's celebrated Microscopes' and in 1877 as 'importer of E. Hartnack & Nachet's microscopes.'[298] However, his name is inscribed on a plotting protractor divided in grades,[299] and he is credited with devising a differential barometer,[300] which implies that his skills were more than that of retailing.

Alexander Frazer set up independently as a scientific instrument maker in 1879, describing himself as 'optician', and he also is understood to have trained in the Adie business.[301] That same year he was elected to the Society of Arts, and in the following year to the Scottish Meteorological Society, and from the number and range of his contributions to both societies he was clearly an intelligent and active investigator.[302] However, he was most unusual amongst instrument makers in having a formal university qualification, and the award of his M.A. degree by Edinburgh University in 1879 seems to mark the start of his independent career. The fact that his university studies lasted for nine years and were preceded by six sessions at the School of Arts, begun at the age of fourteen, suggests that they were attended on a part-time basis while he was working as a journeyman. His attendance at the School of Arts, which corresponded with the period of his apprenticeship, probably reflects an enlightened attitude on the part of his employers.[303] He gained the School's Diploma in 1869 and had previously gained a prize in the mathematics class.

Richard Adie died of apoplexy at Bonnyrigg near Edinburgh on 25 January 1881, aged seventy.[304] In the terms of his will he left to 'the foreman in charge of my business' in Edinburgh £50, to the foreman's 'shop assistant or second foreman' £25, and to each of 'the other workmen connected with me in business' in Edinburgh and Liverpool, £5 each.[305] Unfortunately neither their names nor their total number has survived, except in the case of the foreman, Thomas Wedderburn. Only a few months after Richard's death, Wedderburn was listed as a master optician,[306] and it is reasonable to assume that at some date, even perhaps before Adie's death, he assumed control of the firm.[307] He changed its name to Adie & Wedderburn, under which it traded until 1913. At about the time of the change the business moved from 50 Princes Street, to 37 Hanover Street where it traded 1881-1882, and then at 17 Hanover Street 1883-1902. In 1903 it was at 33 Hanover Street, and from 1909 until 1913 at 52 George Street.[308]

Thomas Wedderburn followed the Adie tradition in being elected to the Society of Arts in February 1884, and was presumably responsible for having the firm elected as a member of the Scottish Meteorological Society in June 1882.[309] He did not survive Richard Adie long and died of typhoid aged forty-nine on 9 August 1886.[310] If little is known of the firm's activities in the last years of the Adie & Son phase, still less is known of the period when it traded as Adie & Wedderburn. The evidence of the street directories indicates a shift towards retail of a wider range of

> **Telephone 2489.**   **Telegrams: 'LEVELS, EDINBURGH.'**
>
> # ADIE & WEDDERBURN,
> ## *Opticians, Photographic Dealers,*
> ## Mathematical and Surveying Instrument Makers.
>
> THEODOLITES and LEVELS, STAVES, MEASURING RODS, LAND CHAINS, DRAWING INSTRUMENTS, BOARDS, SQUARES, SCALES, Etc.
>
> *Barometers, Barographs, Thermometers.*
>
> ### SPECTACLES AND EYE-GLASSES
> in Gold Frames from 21s.; in Steel Frames from 2s. 6d.
>
> SPECIAL ATTENTION GIVEN TO OCULISTS' PRESCRIPTIONS.
>
> **CAMERAS.**   Kodaks, Sanderson and Magazine Plate Cameras.
> Our Half-Plate Set, *70s.,* is best value in the City.
>
> Developing, Printing, and Enlarging. Prompt Delivery, Best Results, Moderate Prices.
>
> *Please note Address—*
>
> ## 33 HANOVER STREET, EDINBURGH.

17 Advertisement for Adie & Wedderburn, 1905

goods, including photographic materials, in which the Adies are not known to have dealt.[311] This shift reflects similar tendencies in Glasgow firms. Instruments from the period do not point to any remarkable activity in the workshop, and the likelihood is that they represent the bought-in stock of a once great firm now becoming dominated by the wholesale trade from outside Scotland. A continued specialisation in surveying instruments is hinted at in their telegram codeword 'Levels', and apparatus supplied to the Northern Lighthouse Commissioners indicates that the shop still held a respected place as a supplier, if no longer as a maker across the full range of instrumentation.[312] However, another advertisement stated that they were 'practical manufacturers of surveying, mathematical, and meteorological instruments' and offered 'engine dividing to the trade'.[313] In one instance at least, Adie & Wedderburn attempted to market a surveying instrument devised by a local factor, but his patent was subsequently abandoned, and it appears that few examples of Mackenzie's dendrometer were produced.[314]

Nonetheless, there are indications that Wedderburn's death constitutes a logical *terminus ad quem* for any detailed discussion of the Adies' history. He was probably among the last to have worked under the Adies,[315] and it has been shown that talented journeymen who had also done so, such as Angus Henderson and Alexander Fraser, had already left the firm. Moreover, hearsay evidence suggests that after Wedderburn's death several of the 'workers' in the firm dispersed elsewhere. Amongst them was one Mabon, who is said to have gone to Glasgow.[316] The Edinburgh photographer and dealer Thomas Haddow who set up at the end of the century, was apprenticed to Adie & Wedderburn; he was the first to introduce cinematography to Scotland.[317]

The main reason for this change appears to have been the acquisition of the business first by Alexander James Menzies in February 1887, and then following his death a year later, it was acquired by an optician named Thomas Mein.[318] Subsequently the firm was bought by Alexander Mackie, a toolmaker with a shop in Victoria Street, who retained the name Adie & Wedderburn. The firm moved from Hanover Street to 52 George Street in 1909, and in 1913 one of his sons, A. D. Mackie, who owned Richardsons, a cutlery shop, also at 52 George Street, amalgamated with a second son, William James Mackie, an optician who then owned Adie & Wedderburn, to form Richardson, Adie & Co. The new firm dealt in cutlery, jewellery, clocks, firearms and sporting goods, but also advertised as instrument makers,[319] and continued to do so after it was bought in 1918 by the Sheffield cutlers, William Morton & Sons, who formed it into a limited company under the directorship of the two Mackie brothers. It ceased business in 1933 and although the Mackies hoped to revive it, they withdrew the name from the Register of Companies in 1949.[320]

REFERENCES

1. The extent to which the Scottish market for apparatus (as opposed to the means for supplying it) should not be considered as provincial has been discussed by A.D.C. Simpson, 'The Adies of Edinburgh: Satisfying the Scottish Market', a paper delivered at the 'Business of Instruments' conference held at the National Museums of Scotland, 22 March 1986. The Adie business is discussed in D.J. Bryden, *Scottish Scientific Instrument-Makers 1600-1900* (Edinburgh, 1972).

2. In the period 1780-82 the City authorities mounted a purge of un-free tradesmen, requiring those who were eligible, to become burgesses. John Miller claimed his burgess ticket, entitling him to trade, on 8 February 1782 by right of the freedom of his father, John Miller senior, turner, whose ticket (presumably purchased) was awarded two weeks earlier (C.B.B. Watson (ed.), *Roll of Edinburgh Burgesses and Guild-Brethren, 1761-1841* (Edinburgh, 1933), 111). It is known that John Miller, turner, also had a son Alexander, apprenticed to the clockmaker James Duff in March 1766 (John Smith, *Old Scottish Clockmakers* 2nd edition (Edinburgh, 1921), 262, citing the records of the Edinburgh Hammermen's Guild), probably at the customary age of 14, and a daughter Betty, who married John Adie in April 1772 (F.J. Grant (ed.), *Register of Marriages for the City of Edinburgh, 1751-1800* (Edinburgh, 1922), 5). The only relationship that satisfies this (deduced from the International Genealogical Index, British Isles, Scotland, fiche F0412, May 1988) is John Miller or Millar marrying Elizabeth Gregg or Greig on 21 April 1745 (p. 12,573), with children John, baptised 2 April 1746 (p. 12,573), Elizabeth, baptised 17 September 1747 (p. 12,506) and Alexander, baptised 26 April 1752 (p. 12,465), all at Edinburgh.

3. Mount Stuart, Marquis of Bute's Papers, National Register of Archives (Scotland) survey 631, Box 1768-1771: 1769/Bundle 6, letter of Dr James Lind to Lord Loudoun, 23 June 1769.

4. Scottish Record Office, Henderson Papers, GD 76/464/4, Notes on the late A.J. Adie, 1859, p.3. This account throws interesting sidelights on Miller and his relationship with his nephew Alexander Adie. As D.J. Bryden notes (*op. cit.* (1), 9 n.37) the manuscript was apparently prepared by a member of Alexander Adie's family after his death in 1859 (not 1858, as he states) or by someone close to him; parts of it were included in an obituary of Adie in Lord Neave's 'Opening Address' to the Royal Society of Edinburgh in December 1859 (*Proceedings of the Royal Society of Edinburgh* 4 (1857-62), 225-227). An abridged version of the manuscript is located at GD 76/464/6-9. A third manuscript account of Adie's life, concentrating briefly on his forebears, was written by his third son Richard Adie, but is too incomplete and discursive to be of importance as a source for his father's work (GD 76/476 'Historical sketch of the Adie family so far as it is known to Richard Adie'). A fourth account resembles the first two in content and approximate date, and includes important references to Adie's optical work (GD 76/464/10-11).

5. Apprentices booked by George Adams *senior* are listed in Joyce Brown, *Mathematical Instrument Makers in the Grocers Company 1688-1800* (London, 1979), 36. Before the publication of these records, D.J. Bryden had assumed that a formal apprenticeship was involved (*op. cit.* (1), 31).

6. J.R. Hutchison, 'Notable Opticians of Auld Reekie', *Ninth Annual Conference [of] The Scottish Association of Optical Practitioners* (Edinburgh, 1939), 9, 17 and 17. On the involvement of Hutchison's father in the Adie firm see the final section of this chapter.

7. Bryden, *op. cit.* (1), 59. He is not however recorded in J. Gilhooley, *A Directory of Edinburgh in 1752* (Edinburgh, 1988), which was compiled from window and annuity tax rolls.

8. Smith, *op. cit.* (2), 399. On 21 April 1745 'John Yeoman mathematical instrument maker', married Jean Robertson in Canongate Parish (F.J. Grant (ed.), *Parish of Holyroodhouse or Canongate, Register of Marriages, 1564-1800* (Edinburgh, 1915), 584). He may be related to John Yeaman, gunsmith, who became a Burgess of the Canongate by purchase in July 1718 (H. Armet (ed.), *Register of the Burgesses of the Burgh of the Canongate, 1622-1733* (Edinburgh, 1951), 71). This may be the John Yeoman whose son John was baptised in July 1716 in the Parish of St. Cuthberts (I.G.I., British Isles, Scotland, fiche F0428, May 1988, p.20,105).

9. Ian Adams, 'The Land Surveyor and his influence on the Scottish Rural Landscape', *Scottish Geographical Magazine* 84 (1968), 248-255; *idem.*, 'Economic Process and the Scottish Surveyor', *Imago Mundi* 27 (1975), 13-18.

10. NMS T1977.133.

11. The two standard firlot measures for the County of Stirling are inscribed 'John Miller Turner Edin[r]', 'David Robertson Smith Edin[r]' and 'Ja[s] Stark Joiner at Bristow nr Edin[r]' (Smith Art Gallery and Museum, Stirling, inventory B1239, B9136): they will be discussed by A.D.C. Simpson and R.D. Connor in a history of Scottish metrology in progress.

12. On John Stewart, 5th Bt., see 'Steuart or Stewart of Coltress' in G.E. C[okayne], *Complete Baronetage* IV 1665-1707 (Exeter, 1904), 376.

13. See below, reference 27.

14. R.L. Emerson, 'The Philosophical Society of Edinburgh 1737-1747', *British Journal for the History of Science* 12 (1979), 154-191.

15. *Idem*, 'The Philosophical Society of Edinburgh 1748-1768', *ibid.* 14 (1981), 133-176.

16. James Gray, 'Of the Measures of Scotland, compared with those of England', *Essays and Observations, Physical and Literary, read before a Society in Edinburgh ...* 1 (1754), 200-204.

17. H. Scott, *Fasti Ecclesiae Scoticanae* new edition, vol. 1, Synod of Lothian and Tweeddale (Edinburgh, 1915), 151.

18. Emerson, *op. cit.* (15), 160.

19. S.R.O., GD 24/5/4(130), receipted account of John Yeaman, Edinburgh, 9 May 1752.

20. H. Paton (ed.), *The Registers of Marriages for the Parish of Edinburgh, 1701-1750* (Edinburgh, 1908), 377; Smith, *op. cit.* (2), 262.

21. Watson, *op. cit.* (2), 111.

22. *Williamson's Directory for the City of Edinburgh, Canongate, Leith, and Suburbs, from the 25th May 1773, to 25th May 1774* (Edinburgh, 1773); Williamson's *Directory* 1774-1788.

23. John Millar, son of George Millar, merchant Burgess, put to Wm Aiton, goldsmith and Jeweller 29 November 1732 (C.B.B. Watson (ed.), *Register of Edinburgh Apprentices, 1701-1755* (Edinburgh, 1929), 61). It has not been possible to confirm this against the records of the Incorporation of Goldsmiths of Edinburgh (S.R.O. GD 1/482): no indentures are noted in the minute book to 1738, a second minute book to 1738 includes some bookings, the separate record of apprentice indentures is incomplete, and a gap in the minutes 1738-1743 covers the period when the apprenticeship would have been completed.

24. Williamson's *Directory* 1778-1794; T. Aitchison, *The Edinburgh Directory from July 1793 to July 1794* (Edinburgh, 1793); Aitchison's *Directory* 1794-1801.

25. Obituary by Alexander Bryson, *Transactions of the Royal Society of Edinburgh* 22 (1861), 655-680.

26. S.R.O., GD 76/464/1, f.5.

27. Anon., *Memoir of Professor Wallace* (London, 1844), 4. The identification of Miller senior as Robison's assistant is based on the location of the observatory (otherwise only noted in connection with Adie) at Miller senior's address, and the close and early association between Robison and Miller junior over the supply of Robison's teaching apparatus.

28. *Idem.*

29. 'Mr Miller ... used to tell that he was desired to explain the airpump experiment of the guinea and feather to Geo: III. In performing the experiment the young optician provided the feather the King supplied the guinea and at the conclusion the King complimented the young man on his skill as an experimenter but frugally returned the guinea to his waistcoat pocket.' (S.R.O., GD 76/464/3-4).

30. Emerson, *op. cit.* (15), 161-163.

31. An analysis of the apprenticeship period for clockmakers booked in the Incorporation of Hammermen, presumed to be an equivalent training, shows 6 years to be the minimum period (see Smith, *op. cit.* (2), 12, 64, 82, 169, 172, etc.). This date corresponds well with the claim made by Miller's obituarist in 1815 that Miller had been a leading optician for over 48 years (*Scots Magazine* 77 (1815), 239-240).

32. R.G.W. Anderson, 'Joseph Black: an Outline Biography', in A.D.C. Simpson (ed.), *Joseph Black 1728-1799* (Edinburgh, 1982), 7-11.

33. John Playfair, 'Biographical Account of the late John Robison', *Transactions of the Royal Society of Edinburgh* 7 (1815), 501-508.

34. Emerson, *op. cit.* (15), 161, 163.

35. See above, reference 3.

36. Williamson's *Directory* 1774-1780, 1784-1794; Aitchison's *Directory* 1793-1803. There is no entry for the years 1782 or 1783, but Alexander Millar, watchmaker, presumably Miller's brother who was free of the Hammermen in 1782, appeared at Parliament Square for the single year 1783. It is possible that with the move to South Bridge in 1794, Miller also moved his living accomodation: from 1794 to 1803 a John Miller with no occupation noted is listed at Ritchie's Land, Nicolson Street. In 1804 the shop moved to Nicolson Street, but also Alexander Adie, who stayed with his uncle, married the daughter of John Ritchie, Burgess slater.

37. Denovan & Co.'s *Directory* 1804; Campbell's *Directory* 1804, with entries both for John Miller at 86 South Bridge and Miller & Adie at 94 Nicolson's Street; Stark's *Directory* 1805, 1806; *Edinburgh Post Office Directory* 1807-1809. The apparent change of location from 94 to 96 Nicolson Street in about 1807 seems to be the result of a re-numbering.

38. Bryden, *op. cit.* (1), 24-25. As Bryden points out, Miller's entry as a burgess was probably not voluntary but the result of the Dean of Guild's pressure on tradesmen who traded without tickets. Because his father had just been created a Burgess Miller could claim his ticket without further payment or qualification and did not have to appeal to an apprenticeship completed under an Edinburgh freeman: Miller senior may have remained a journeyman, and if Yeaman was free, this will have been from the Canongate Hammermen.

39. See H. Woolf, *The Transits of Venus* (Princeton, New Jersey, 1959).

40. See above, reference 3.

41. See above, reference 3.

42. Emerson, *op. cit.* (15), 174: Emerson provides incorrect dates for Lind, confusing him with his namesake, physician to the Royal Naval Hospital, Haslar.

43. Harold W. Thompson (ed.), *The Anecdotes and Egotisms of Henry Mackenzie* (London, 1927), 108-109. Alemoor's house, known in full as Hawkhill Villa, was designed for him by John Adam and completed in 1757 (James Simpson, 'Lord Alemoor's Villa at Hawkhill', *Bulletin of the Scottish Georgian Society* 1 (1972), 2-9). The observatory is marked on John Ainslie's 1775 large-scale map 'The Counties of Fife and Kinross, with the Rivers of Forth and Tay Survey'd & Engraved By John Ainslie', where it is represented as a canopied structure reminiscent of the portable observatories designed by the engineer John Smeaton. Nevil Maskelyne, the Astronomer Royal, supplied details of Smeaton's design to Lind (Note by Maskelyne to J. Lind, 'An Account of the late Transit of Venus, observed at Hawkhill, near Edinburgh', *Philosophical Transactions* 59 (1769), 343). Other observations recorded at Hawkhill were of lunar eclipses in December 1769 and July 1776, of which the results were published in *ibid.* 59 (1769), 363-365, and the *Caledonian Mercury* of 3 August 1776. Hawkhill was later the home of Alexander Bryson, brother of James Mackay Bryson (q.v.), and was demolished in 1971 by the Edinburgh City Council. For John Ainslie's links with Miller in the mid-1770s, see below, reference 66.

44. See above, reference 3.

45. Lind's account of the observations is Lind, *op. cit.* (43), 339-346. The observations, together with the instruments used, are tabulated in Woolf, *op. cit.* (39), 186. Meteorological observations were recorded at Hawkhill from 1764: the

original observations made in the second phase of this work, 1769-1775, by James Hoy, clerk to Lord Alemoor, are in the Royal Society of Edinburgh, MS Met 5.1.

46. Thompson, *op. cit.* (43).

47. Science Museum, London, inventory 1906-71, engraved 'John Miller Edinburgh' on the hour circle: the telescope matches the description in Lind's account, *op. cit.* (43), but this appears to have led to the assumption that this was Lind's own instrument by Dollond. The stands of Lind's Dollond telescope and Mackenzie's by Ramsden(?) are described in David Gavine, 'James Stewart Mackenzie (1719-1800) and the Bute MSS', *Journal of the History of Astronomy* 5 (1974), 212. The telescope stand acquired for Edinburgh University does not survive: it is described as item C.6 'Equatorial Stand for Tellescope by Miller' in the 1833 MS 'Catalogue of Apparatus forming the public property of the Class of Natural Philosophy' (Royal Museum of Scotland).

48. *Philosophical Transactions* 59 (1769), 344-348.

49. '... Map of the Three Lothians ... [by] Andrew & Mostyn Armstrong ... engraved by Thos Kitchin, Hydrographer to his Majesty ... MDCCLXXIII'

50. NMS T1988.25.

51. Mount Stuart, Marquis of Bute's Papers, N.R.A.(S.) survey 631, Box 1768-1771: letters of Lind to Loudoun, 23 June and 15 August 1769.

52. Described and illustrated in A. Ewing, *The Synopsis of Practical Mathematics* (Edinburgh, 1771), v-vi; surviving examples are to be found in the Royal Museum of Scotland (NMS T1974.L12); the Museum of the History of Science at Oxford, and the University Museum, Utrecht.

53. T.C. C[ooper], 'James Lind', in S. Lee (ed.), *Dictionary of National Biography* XXXIII (London, 1893), 273.

54. Y. O'Donaghue, *William Roy 1726-1790 Pioneer of the Ordnance Survey* (London, 1977), 40.

55. James Lind, 'Description and Use of a Portable Wind Gage', *Philosophical Transactions* 65 (1775), 353-365; the instrument is discussed by W.E.K. Middleton, *Invention of the Meteorological Instruments* (Baltimore, Maryland, 1969), 193.

56. An example of 'Lind's Wind Gage' was in the Natural Philosophy demonstration apparatus at Edinburgh in 1833 as item B.51 (see reference 47 above). The earliest surviving example by an Edinburgh maker is one by Alexander Adie, once owned by the Scottish Meteorological Society, and now in the Royal Museum of Scotland (NMS T1983.119).

57. Miller's instrument is described in [D. Brewster], 'Anemoscope', in *Edinburgh Encylopaedia* (Edinburgh, [1808]-1830), II, 77. Miller was not, however, given the prestigious commission for the 'wind dial' for Robert Adam's Register House, Edinburgh: this, and the clock, were supplied by Joseph Vulliamy of London (S.R.O., SRO 4/70, account dated 18 September 1790).

58. A domestic barometer by Miller is in the Royal Museum of Scotland (NMS T1972.86), and another was offered for sale by Christie's, 8 December 1976, Lot 104. Besides Miller's mountain barometers, discussed elsewhere (see reference 129 below), he also produced a siphon barometer, examples of which are in the Science Museum, London (inventory 1921-565) and the Natural Philosophy collection at the University of Aberdeen: the division of the scales of these two instruments are discussed by John Reid in a forthcoming article in *Annals of Science*.

59. An example of a garden sundial by Miller (provenance unknown) is in the Royal Museum of Scotland (NMS T1979.19); another, purchased by the Duke of Buccleuch, is at Bowhill House, Selkirk, and several others have been recorded.

60. A conventional 10-sided wooden-barrelled telescope engraved 'J. Miller Edinr.' on the draw tube, is stamped 'SPENCER BROWNING & RUST / LONDON' on the barrel (in a private collection). For Spencer, Browning & Rust acting as Adie agents, see below, reference 140.

61. A reflecting telescope by John Gibson of Kelso, with a stand inscribed 'J. Miller 1790', was offered at Christie's South Kensington, 14 April 1988, Lot 145.

62. Adams, *op. cit.* (9).

63. Alexander Keith, 'Description of a Mercurial Level', *Transactions of the Royal Society of Edinburgh* 2 (1790), 16 and plate. The Miller level in the Royal Museum of Scotland is NMS T1980.124. A surviving account in Miller's hand to Robert Hay of Drumelzier, Peeblesshire, 27 November 1789, is for fitting a level to a theodolite and repairing the needle and sights (Hay of Duns, N.R.A.(S.) survey 2720, bundle 522: we are grateful to Mr and Mrs A.D. Hay for permission to cite this). It would have been a simple theodolite, with plain sights, similar to the John Miller theodolite offered at Sotheby's, 11 June 1985, Lot 209: this latter instrument had remained in service until at least 1835 because it had been fitted with a telescopic sight and altitude scale engraved 'Adie & Son, Edinburgh'.

64. J.F. Erskine, *General View of the Agriculture of the County of Clackmannan ... drawn up for the Consultation of the Board of Agricultural and Internal Improvement* (Edinburgh, 1795), 80.

65. *Caledonian Mercury*, 31 December 1793: '... all kinds of circular instruments for Astronomical Observations and surveying; divided with the utmost accuracy by a dividing engine, constructed agreeable to Mr. Ramsden's of London' (quoted in Bryden, *op. cit.* (1), 37 n.191).

66. Bryden, *op. cit.* (1), 21; examples of the 1793 pocket globe are in the Royal Museum of Scotland (NMS T1964.5), and were offered for sale by Sotheby's, 28 February 1984, Lot 129 and Christie's South Kensington, 30 June 1988, Lot 2. The episode is discussed in A.D.C. Simpson, 'Globe Production in Scotland in the Period 1770-1830', *Der Globusfreund* Nos. 35-37 (1987), 21-32.

67. Edinburgh City Archives, Council Records (Town Council Minutes), 16 April 1777.

68. Edinburgh City Archives, Macleod's Bundles, College: bundle 11, shelf 36, bay C, letter from Robison to Town Council, 25 August 1779. The pump is listed as item B.30 in the 1833 catalogue (see above, reference 47).

69. Item A.78 in the 1833 catalogue (see above, reference 47). J.T. Desaguliers, *A Course of Experimental Philosophy* (London, 1734), I, 465-6.

70. Item A.77 in the 1833 catalogue (see above, reference 47). A copy of a MS inventory of 1789 identifies 30 items, including this piece and the cometarium, as having been covered the account from Miller paid from the Stewart bequest (Royal Museum of Scotland).

71. Smith, *op. cit.* (2), 262.

72. Williamson's *Directory* 1783.

73. Advertisement in the *Caledonian Mercury*, 25 April 1789, quoted in Bryden, *op. cit.* (1), 31 n.158.

74. Advertisement in the *Edinburgh Herald*, 1 January 1790, quoted in Smith, *op. cit.* (2), 262.

75. Grant, *op. cit.* (2), 5.

76. S.R.O., GD 76/461/1, f.1. In a family tree of c.1880, John Adie is described as partner with Donaldson of the *Edinburgh Evening Courant* newspaper, whose half share was sold by John Miller (and see reference 81 below); we are grateful to Captain A.H. Swann for this information.

78. S.R.O., GD 76/464/1, f.1. This second marriage may have been that of Elizabeth Adie to Walter Pringle, 28 February 1797 (Grant, *op. cit.* (2), 5), in which case Miller may have become Adie's guardian after John Adie's death rather than after his wife's death.

79. S.R.O., GD 76/464/1, f.1.

80. David Herd (1732-1810) held a small position in an Edinburgh lawyer's office, and is important as a collector, recorder and editor of Scottish ballads, which he published as *Ancient and Modern Scottish Songs, Heroic Ballads &c. ...* (Edinburgh, 1776) (T. B[ayne], 'David Herd', in L. Stephen and S. Lee (eds.), *Dictionary of National Biography* XXVI (London, 1891), 236-237).

81. S.R.O., GD 76/464/1, ff. 1, 4. Adie may have paid lecture and tuition fees from a private income derived from the sale of his father's estate. After his brother's death his uncle decided that his young ward's health was too delicate to risk the Fife property falling to the Crown if he died while still in his minority, and he accordingly sold the property (S.R.O., GD 76/476, f.1).

82. Post Office and other Edinburgh directories, 1804-1822. The date of the move from 94 to 96 Nicolson Street has not been determined because the entries in the directories for 1805/6 and 1806/7 (Stark's) give no number.

83. Alexander Adie married Marion Ritchie on 22 November 1804; Marion (baptised Mary) was born on 16 September 1787, daughter of John Ritchie and Janet Sibbald, married 15 April 1782 (I.G.I., British Isles, Scotland, fiche F0385, May 1988, p.42; *ibid.*, fiche F0418, May 1988, p.15,228; Grant, *op. cit.* (2), 654).

84. *Edinburgh Post Office Directory* 1807.

85. S.R.O., RHP 38141/1, 'Plan of Cannaan 1802 / Thos Johnston Decr 1802'. The house is marked as 'Mr Edies Property' on Robert Kirkwood's 'large' plan of Edinburgh, 1817. It is now known as Canaan House and is part of the Astley Ainslie Hospital, Lothian Health Board.

86. See above, reference 81.

87. The example at the Royal Museum of Scotland (NMS T1925.57) is dateable from the partnership's address, given as 86 South Bridge, which they had quit by the time *Denovan & Co.'s Edinburgh and Leith Directory from July 1804 to July 1805* (Edinburgh, 1804) had been published.

88. The barometer and box sextant in the Royal Museum of Scotland are respectively NMS T1980.321 and T1967.110. For the description of the barometer, see below, reference 129.

89. NMS T1980.125.

90. Strathclyde Regional Archives, T-KF 6/3, pp.229-230. Register of William Kyle, Land Surveyor, Glasgow, letter of Kyle to Alexander Adie, 17 April 1817, quoted in Bryden, *op. cit.* (1), 13.

91. Theodolite in the Royal Museum of Scotland (NMS T1958.64). For another example of a long-lasting instrument see reference 63 above.

92. T.C., 'Improvement in the Air-Pump', *Glasgow Mechanics' Magazine* 1 (1824), 226.

93. St. Andrews, Hay Fleming Reference Library, MS by D. Smith, 'Inventory of the Apparatus of the Nat. Philosophy Class of the United College May 1847'; entry reads 'Air pump 1812 Miller & Adie £31-10.'

94. NMS T1980.221 and T1980.272.

95. *Caledonial Mercury* 16 March, 1808.

96. *Edinburgh Evening Courant*, 8 August 1818. The kaleidoscope 'in addition to the elegance of its own patterns, adds so much to the natural beauty of the Flower Garden'.

97. See A.D. Morrison-Low, 'Brewster and Scientific Instruments', in A.D. Morrison-Low and J.R.R. Christie (eds.), *'Martyr of Science': Sir David Brewster 1781-1868* (Edinburgh, 1984), 58-65.

98. *Scots Magazine, op. cit.* (31). The name of the partnership was not used exclusively in this period: the 1818 advertisement cited above (reference 96) begins 'A Adie, Optician, 15 Nicolson Street, begs to inform ...'

99. [David Brewster], 'Hydrodynamics' in *Edinburgh Encyclopaedia* (Edinburgh, [1808]-1830), XI, 440 section 12: an example in the Royal Museum of Scotland is referred to in reference 87 above.

100. *Scots Magazine, op. cit.* (31), 239.

101. S.R.O., RH4/21/2, letter from David Brewster to James Veitch, 22 November 1798: 'I arrived in Edinburgh on Thursday evening pretty late, and had not time to call at Miller's before the Carrier left the town, however, I called on him about the beginning of the week, when he said that the prices of object glasses forty-two inches long were in proportion to their goodness, & that they generally cost about Five Guineas ...'; however, the following year Brewster visited an unnamed optician in Edinburgh who had failed to make achromatic refractors (RH4/21/10, 16 August 1799). This entire correspondence has been microfilmed and can be consulted at the Scottish Record Office, RH/4/21, and we are grateful to W.H. Veitch Esq. for permission to quote from the letters.

102. D. Brewster, *A Treatise on New Philosophical Instruments* (Edinburgh, 1813), 59 n.

103. *Ibid.*, 90 n.

104. [D. Brewster], 'On Different Circular Micrometers', *Edinburgh Journal of Science* 1 (1824), 179. The instrument is described in more detail, but not credited to Adie in Brewster, *op. cit.*, (102), 48.

105. D. Brewster, *A Treatise on Optics* (London, 1831), 337.

106. W.H. Brock, 'Brewster as a Scientific Journalist', in Morrison-Low and Christie, *op. cit.* (97), 37-42.

107. S.R.O., GD 76/476, f.1.

108. V.A. Eyles, 'Sir James Hall', in C.C. Gillispie (ed.), *Dictionary of Scientific Biography* VI (New York, 1972), 53-56; idem., 'Sir James Hall Bt. (1761-1832)', *Endeavour* 20 (1961), 210-216; idem., 'The Evolution of a Chemist: Sir James Hall ...', *Annals of Science* 19 (1963), 153-182.

109. S.R.O., GD 76/476, ff. 6-7. Sir James Hall's papers form part of the Dunglass Muniments (S.R.O., GD 206) and include diaries; a scrutiny of these may reveal more of Adie's role in the experiments.

110. *Caledonian Mercury*, 16 March 1808.

111. John Leslie, *A Short Account of Experiments and Instruments depending on the Relations of Air to Heat and Moisture* (Edinburgh, 1813), 178, 180-181. Surviving examples at the Royal Museum of Scotland are noted in R.G.W. Anderson, *The Playfair Collection: the Teaching of Chemistry at the University of Edinburgh 1730-1858* (Edinburgh, 1978), 90 n.10, and at Teyler's Museum in Haarlem by G. L'E. Turner, *Martinus van Marum ... Volume IV. Van Marum's Scientific Instruments in teyler's Museum* (Haarlem, 1973), 263.

112. John Leslie, *Description of Instruments designed for Extending and Improving Meteorological Observations* (Edinburgh, 1820), 48.

113. *Ibid.*, 48.

114. Undated newspaper report headed 'Curious and Interesting Experiment', in National Library of Scotland, MS vol. X223 d.i., '[Alexander] Adie's collection of newspaper cuttings etc mainly relating to Edinburgh 1792-1848', 42. (Hereafter cited as 'Newspaper Collection'.) The similarity of a passage in the report predicting a wide application of the 'prodigious powers of refrigeration' to the equivalent section in Leslie's work of 1813 (*op. cit.*, (111), 155-157), suggests Leslie wrote the report himself.

115. See reference 93 above.

116. D.A. Stevenson, *The World's Lighthouses before 1820* (Oxford, 1959), 295.

117. See reference 90 above.

118. Adie hoped to solve the problem of corrosion by caking the two plates thickly in a resin, apparently based on material found in Egyptian mummies, and by hermetically sealing the coins and papers in glass bottles; it was thus hoped that 'evidence of this magnificent undertaking will be handed down to the latest posterity' (Report of the grand Masonic Procession, September 1815, *Weekly Journal*, 27 September 1815, 327).

119. James Hunter, 'Description of the Odometer, exhibited in January 1821 ...', *Transactions of the Highland Society of Scotland* 6 (1824), 600-604; 'Account of an Improvement on the "Odometer" ', *Edinburgh Journal of Science* 3 (1825), 44-46.

120. The odometer was available from James Howden, a watchmaker with substantial premises at 9 South Bridge (on Howden, see Smith *op. cit.* (2), 194-198). Howden was, like Adie and Hunter, a Fellow of the Royal Society of Edinburgh, and he was a neighbour of Adie's at Canaan. The device may initially have been made by Adie under contract, and it was subsequently marketed by Adie: one signed by Adie & Son is at the Royal Museum of Scotland (NMS T1969.27) and another (also post-1835) signed by Richard Adie of Liverpool has been recorded.

121. S.R.O., GD 76/464/1. f.5.

122. Adie's original records for the period 1801 to 1804-5 are at the Royal Society of Edinburgh, MS Met 5.6.

123. J.D. Forbes, 'On the Climate of Edinburgh for Fifty-six years, from 1795 to 1850, deduced principally from Mr Adie's Observations', *Transactions of the Royal Society of Edinburgh* 22 (1861), 331, 333. Professor John Playfair started his register in Windmill Street, off George Square, in 1794 (*ibid.*, 330).

124. *Ibid.*, 333.

125. S.R.O., GD 76/464/1, f.6.

126. Leslie, *op. cit.* (111), 90.

127. A. Adie, English Patent 4323, 23 December 1818. The Specification is printed in *Repertory of Arts, Manufactures and Agriculture* 2nd series, 35 (1819), 257-261 with 'Observations by the Patentee' (*ibid.*, 261-265).

128. NMS T1980.210 in the Frank collection, which is signed 'Adie & Son' and therefore was made after 1835. An earlier instrument, signed 'Miller & Adie' is T1980.321.

129. A. Anderson, 'Barometer', in *Edinburgh Encyclopaedia* (Edinburgh, [1808]-1830), III, 290. Brewster's *Encyclopaedia* was issued in part form, and this section was published in September 1810. The instrument is also discussed in the article 'New Self-adjusting Portable Barometer invented by Mr John Condie', *Scots Mechanics Magazine* 1 (1825), 71-73.

130. *Ibid.*, 72.

131. A. Adie, 'Description of two New Philosophical Instruments', *Memoirs of the Wernerian Natural History Society* 3 (1821), 483-498. This is a slightly later version dated 14 December 1820 of the text of the Specification and the 'Observations', *op. cit.* (127); it includes a separate account of the 'New Hygrometer' and Adie's note referring to Hooke's experiment (but omitting a testimony noted in reference 135 below).

132. *Ibid.*, 488, 492-494.

133. *Ibid.*, 489-490.

134. *Ibid.*, 490-491.

135. *Ibid.*, 491-492. A graphic illustration of the four hours' start over the barometer in predicting a storm with a sympiesometer on board H.M.S. *Nimrod* in November 1818, is given by Adie ('Observations', *op. cit.* (127), 264-265: the passage is omitted in Adie's 'Description', *op. cit.* (131)).

136. Newspaper report headed 'New Barometer', Edinburgh, 10 July 1817, in 'Newspaper Collection', *op. cit.* (114), 42.

137. *Scots Magazine* new series, 4 (1819), 259-260.

138. Adie, 'Observations', *op. cit.* (127), 265. Examples have been noted: one offered for sale at Sotheby's, 11 June 1985, Lot 71; a second, signed 'Thomas Jones, 62 Charing Cross, No. 29' was offered for sale by Phillips, 17 July 1985, Lot 13.

139. Sotheby's, 10 December 1981, Lot 245 and Phillips, 10 December 1986, Lot 5, now in the Royal Museum of Scotland, NMS T1987.30. For the importance of Greenock in the early 19th century see Chapter 18 'Nautical Instrument Makers'. For William Heron, see Chapter 22 'Whyte, Thomson & Co.'.

140. NMS T1967.99. As late as 1854, William Spencer Browning was advertising 'Adie's Sympiesometers' for sale at £3 3s 0d (Marwood's *Maritime and Commercial Advertiser* (Newcastle, 1854), 5; we are grateful to Michael and Diana Crawforth for this reference).

141. An 11½ inch octant in the Royal Museum of Scotland (NMS T1982.57) is signed 'Adie. Edinburgh' and stamped 'SBR' on the divided scale.

142. This example, no. 1181, is the last one known from the signature 'Adie' to ante-date the formation of Adie & Son in 1835. The next recorded example, no. 1393, is signed 'Adie & Son, and therefore dates from 1835 or later (Science Museum, London, inventory 1918-14).

143. Bryden, *op. cit.* (1), 14 and n.17.

144. In the Royal Museum of Scotland, NMS T1984.38.

145. S.R.O., GD 76/464/1, f.6.

146. *Ibid.*

147. Quoted by Adie in 'Description', *op. cit.* (131), 491.

148. An example of a rival product was the portable compensating barometer patented by William Harris of London (an example was offered for sale by Christie's, 28 April 1982, Lot 4, numbered 258; another was offered for sale by Sotheby's, 28 October 1986, Lot 125, numbered 102). The Adie business continued to sell the sympiesometer through agents all over Scotland: Alexander Cameron of Dundee (q.v.) sold one numbered 1708 (offered for sale at Phillips, 20 July 1983, Lot 25); Duncan McGregor sold a few: one numbered 2203 (offered for sale at Christie's South Kensington, 20 August 1987, Lot 19), one numbered 2235 (offered for sale at Phillips Edinburgh, 9 December 1983, Lot 70), one numbered 2697 (offered for sale at Christie's South Kensington, 14 March 1985, Lot 12) and one numbered 3179 (in the Whipple Museum of the History of Science, inventory 2083). James White (q.v.) was also producing sympiesometers in about 1888, and 'improved' instruments by a number of other makers are recorded. Among later assessments of the sympiesometer came one from Jacob Swart, who compared Adie's instrument with that designed by Cummins, in J. Swart, 'Waarnemingen gedaan met zes gewone Adie's Sympiesometers, en een en twee Cummins Sympiesometers', *Verhandelingen en Berigten Betrekkelijk het Zeewezen en de Zeevaartkunde* 3 (1843), 613-626; and another from W. Matthews, 'On the Sympiesometer and Aneroid Barometer', in *The Alpine Journal: a Record of Mountain Adventure and Scientific Observation* 2 (1866), 397-404. We are grateful to Dr Anita McConnell for these references.

149. *Transactions of the Royal Society of Edinburgh* 9 (1823), 504.

150. Royal Society of Edinburgh, MS vol. 'Minutes of meetings of Physical and Literary Classes 1793-1824', 16 November 1818: this was presumably the date of his proposed election.

151. S.R.O., GD 76/464/10, 'Memoir of Alexander James Adie' which states that '... by the advise of his friend Sir David Brewster he gave [it] the name of sympiesometer'.

152. National Library of Scotland, Royal Scottish Society of Arts Archives, Dep. 230/1 Council Minute Book Vol. 1, 18 November 1822. The Society is discussed in an unpublished paper by A.D.C. Simpson, 'Brewster's Society of Arts and the Pantograph Dispute', read at the Second Greenwich Scientific Instrument Symposium, September 1982.

153. C.R. Goring and A. Pritchard, *Micrographia* (London, 1837), 139. Goring graduated M.D. in 1816, with a thesis 'De Apoplexia Sanguinea'.

154. *Ibid.*, 141-142. S.R.O., RH4/21/52, David Brewster to James Veitch, 7 August 1821, '... I have sought thro' all the opticians here and cannot get a glass of this kind, & they are such bunglers in the grinding of lenses that I could not trust to them for one on which so much depends.'

155. Brewster, *op. cit.* (102), 403; *idem., op. cit.* (105), 337.

156. R.H. Nuttall and A. Frank, 'Makers of Jewel Lenses in Scotland in the Early Nineteenth Century', *Annals of Science* 30 (1973), 407-416.

157. Brewster, *op. cit.* (105), 337-338 (unchanged in the 2nd edition of 1838); Blackie's work is mentioned in Brewster's article 'Microscope', in *Encyclopaedia Britannica* 8th edition (Edinburgh, 1853-60), XIV, 768, and presumably dates from after 1831. The dating of Adie's work is taken from the 'Memoir of Alexander James Adie', *op. cit.* (151). Nuttall and Frank have speculated that Blackie may have made the optics of the Adie & Son jewel lens microsope in the Frank collection, now in the Royal Museum of Scotland (NMS T1979.47; described in R.H. Nuttall, *Microscopes from the Frank Collection 1800-1860* (Jersey, 1979), 35) and that pressure of work would have made it cheaper and easier for Adie to contract out lens work to Blackie who worked at home (*op. cit.* (156), 416).

158. See Nuttall and Frank, *op. cit* (156), 412.

159. G. L'E. Turner, 'The Rise and Fall of the Jewel Microscope', *Microscopy* 31 (1968), 85-94.

160. Nuttall and Frank, *op. cit.* (156), 412-413, corrected in A.D. Morrison-Low, 'Scientific Apparatus associated with Sir David Brewster,' in Morrison-Low and Christie, *op. cit.* (97), 92 item 19. (Although J.R. Hutchison's father, the optician and former Adie apprentice George Hutchison, is understood to have acquired material from the Adie business in the 1880's (Bryden, *op. cit.* (1), 33), the items may have been acquired in the course of J.R. Hutchison's activities as a collector).

161. Nuttall and Frank, *op. cit.* (156), 413. However, the second manuscript, written in pencil and apparently in Adie's hand, lists four types of lens each ground to focal lengths of $1/10$ and $1/50$ inch. Of these, only the pale garnet lens of $1/50$ inch appears in both lists, but the figures given against it (whether or not they are prices) do not correspond. There are no $1/10$ inch focus lenses in the group.

162. *Ibid.*, 408.

163. Turner, *op. cit.* (159), 90.

164. NMS T1982.90.

165. A.D. Morrison-Low, 'The Origins of the Polarising Microscope: Sir David Brewster *versus* William Nicol', paper read at the Second Greenwich Scientific Instrument Symposium, September 1982.

166. *Ibid.*; NMS T1933.65. A solar microscope, signed 'Adie, Edinburgh' is in the Museum of the History of Science, Oxford (inventory 1954.11).

167. 'Notice respecting Mr Cuthbert's Elliptic Metals for Reflecting Microscopes: communicated by a Correspondent [D. Brewster]', *Edinburgh Journal of Science* new series, 1 (1830), 321-2. The Adie microscope is not a retailed Amici item, as initially proposed by Nuttall and Frank, *op. cit.* (156), 414 n.35.

168. J.T. Stock and D.J. Bryden, 'A Robinson Balance by Adie & Son of Edinburgh', *Technology and Culture* 13 (1972), 44-54.

169. NMS T1979.47 (described in Nuttall, *op. cit.* (157), 35).

170. *Ibid.* Three other examples of this form of instrument have been recorded, although none has jewel lenses: one is in

the Wellcome Collection, Science Museum, London, inventory A56356, and is signed 'Adie Edinburgh'; another, 'with sliding stage on an oval base', was offered for sale by Phillips, 14 May 1974, Lot 79, signed 'Adie & Son'; the third was offered for sale by Trevor Philip & Sons, and described in their Catalogue [1] (March 1984), item 13, signed 'Adie & Son EDINBURGH'.

171. Forbes, *op. cit.* (123), 330-331.

172. Alexander Adie, 'Account of some of the rare Atmospherical Phenomena observed in 1824', *Edinburgh Journal of Science* 3 (1823), 49-59. 'On the Mean Temperature of Bombay, deduced from observations made in 1827, &c.', *ibid.* 10 (1829), 17-22. Compare with MS 'Register of the Pluviometer at Bombay 1827 Chowputty' [daily, June - October] (*Newspaper Collection, op. cit.* (114), 52). Adie evidently had contacts in India, for his son John and daughter Elizabeth both married into the Barron family of Aberdeen and Edinburgh, one of whose members lived in Bombay.

173. 'Newspaper Collection', *op. cit.* (114), 91. Adie did not fill in the column for electrometer readings, though all the standard instruments are recorded.

174. David Brewster, 'Results of the Thermometrical Observations made at Leith Fort, every Hour of the Day and Night during the whole of the Years 1824 and 1825', *Transactions of the Royal Society of Edinburgh* 10 (1826), 364; 'Report on the Hourly Meteorological Register kept at Leith Fort in the Years 1826 and 1827', *ibid.* 24 (1866), 351-362. The results for October 1827 were vitiated by false readings supplied by the negligent N.C.O.s who had volunteered for the 'liberally' paid thermometer reading duty (*ibid.*, 358n). The two MS volumes in which the readings were recorded are in the Royal Museum of Scotland (NMS T1981.93).

175. 'Quantity of Rain near Hexham', *Edinburgh Philosophical Journal* 12 (1825), 407.

176. Adam Anderson, 'New Corrections for the Effects of Humidity on the Formula for Measuring Heights by the Barometer', *ibid.* 12 (1825), 248-260; 13 (1825), 224-240.

177. M. Miller, 'Register of the Weather at Corfu ... 1821', *Memoirs of the Wernerian Natural History Society* 5 (1824), 90.

178. Leslie, *op. cit.* (112), 48.

179. John Leslie, 'Enumeration of the instrument requisite for Meteorological Observations', *Edinburgh New Philosophical Journal* 2 (1826-7), 141-145.

180. W.T. Hay Craft, 'On the Specific Heat of Gases', *Transactions of the Royal Society of Edinburgh* 10 (1826), 200.

181. St. Andrews University Library, Forbes Papers, correspondence of D. Brewster to J.D. Forbes, 1829/29, dated 22 May 1829: 'Mr Adie has I know some new information on these subjects' (various forms of thermometer) and 1829/33, dated 19 June 1829: 'Mr Adie has sent me something about the Sympiesometer. I think you should converse with him on the subject, as he thinks there is some error in your observations.'

182. J.D. Forbes, 'Discussions of One Year's Observations of Thermometers sunk to Different Depths in Different Localities in the Neighbourhood of Edinburgh', *Proceedings of the Royal Society of Edinburgh* 1 (1832-44), 223-224. Also discussed in H.A. and M.T. Brück, *The Peripatetic Astronomer: the Life of Charles Piazzi Smyth* (Bristol, 1988), 144-147: the remains of this set of thermometers, which were broken in 1876, are in the Royal Museum of Scotland (NMS T1986.L2.121). Parts of the replacement set of 1879, also made by Adie & Son, are also in the Royal Museum of Scotland (NMS T1966.52, T1975.15). Forbes discussed the construction of these thermometers some years later with Sir George Biddell Airy, the Astronomer Royal; 'As touching the Long Thermometers I had these Sets made for different situations, each 3, 6, 12, & 21 *French* feet long; with such thermometers to place inside the cases which enclose the scales above ground in order to correct for Temp. of Expans. column of Alcohol. They were made by Adie here & placed by him; a delicate & difficult operation - & cost £50. ... The graduation was performed by suspending the thermometer in a tall staircase *during winter*, with the Bulbs in along with 3 Standards. The staircase was artificially heated & after the whole had remained for a night the readings were made. They were read at four different Temperatures. The length of 1° on the 12 and 24 ft. instruments was from 1 to 2 inches & small in the others on account of the increased Ranges...'(St. Andrews University Library, Forbes Papers, Letterbook III, pp. 715-717, J.D.Forbes to G.B. Airy, 17 July 1845).

183. David Brewster, 'Report respecting the Two Series of Hourly Meteorological Observations kept at Inverness and Kingussie', *Report of the Tenth Meeting of the British Association ... 1840* (London, 1841), 349.

184. John Caldecott, 'Observations on the Temperature of the Earth in India', *Proceedings of the Royal Society of Edinburgh* 1 (1832-44), 432.

185. Adie moved to the new and fashionable Regent Terrace development in 1831 where his neighbours were the natural philosopher George Lees of the School of Arts and the Military Academy and George Baird, the Principal of the University. Canaan Cottage was leased to the merchant James Sinclair until 1838.

186. Forbes, *op. cit.* (123), 331.

187. Report of 'Proceedings of British Association Meeting at Edinburgh in September 1834', *Edinburgh New Philosophical Journal* 17 (1834), 420-421. Subsequently, short abstracts of Adie's weather register were printed with those of other observers in *ibid.* in April 1835, July 1836 and April 1838 (all for Regent Terrace); April 1840, April 1841 and April 1845 (all for Canaan).

188. Forbes, *op. cit.* (123), 327-333. Graphs drawn by Forbes to accompany his discussion of the reduction of Adie's measurements are preserved in the Royal Museum of Scotland (NMS T1984.80). The original MS records are at the Royal Society of Edinburgh, MS Met 5.13 a-c. Forbes reported that Adie's original rain gauge at Merchant Court had been broken up, but that its successor at Canaan Cottage was still standing in March 1859, according to Adie's son A.J. Adie (*ibid.*, 340).

189. Reports of the Royal Society of Edinburgh's proceedings in *Edinburgh Philosophical Journal* 10 (1823-4), 170, 353.

190. Andrew Fyfe, 'Description of a Hydro-Pneumatic Lamp', *ibid.* 11 (1824), 341. Dr Fyfe claimed to have produced a cheap and good alternative design. Adie's version was described and illustrated in Alexander Tilloch (ed.), *The Mechanic's Oracle and Artisan's Laboratory and Workshop* (London, 1825), 93 and plate XII fig. 4.

191. Robert Stevenson, 'Account of the explosion of a Steam Boiler at Lochrin Distillery', *Edinburgh Philosophical Journal* 5 (1821), 147-154.

192. N.L.S., R.S.S.A. Archives, Acc. 4534, box 15. Lecture series June 1832; Dep 230/3, Minute Book vol. 2, p.6 (11 April 1838), p.111 (29 May 1839).

193. Bryden, *op. cit.* (1), 10. Adie was co-author with James Jardine and D. Murray of *Report to Adam Duff, Esq. His Majesty's Sheriff Depute of the County of Edinburgh, regarding the Weights and Measures heretofore in use in said County* (Edinburgh, 1826). This work is comparable with that done at the same time in the Borders by James Veitch (q.v.).

194. N.L.S., R.S.S.A. Archives, Dep 230/1, Minute Book vol. 1, *passim*. In this context, the most interesting of the items is the hydrometer, constructed by William Lunan, an Aberdeenshire instrument maker, probably still active at this date (cf. Bryden, *op. cit.* (1), 52). Adie exhibited and explained the instrument himself on 4 February 1829, and reported on it on 18 February 1829: this suggests that Adie was contacted by Lunan as a well-placed expert on the instrument, in order to promote his design.

195. *Transactions of the Royal Scottish Society of Arts*, 1-5 (1829-61), *passim*; N.L.S., R.S.S.A., Archives, Dep 230/3, lists at rear of Minute Book vol. 2.

196. The move to the New Town was announced in the *Scotsman*, 13 June 1828. The Nicolson Street premises were occupied for some time after this date, for John Adie addressed a letter from there dated 20 March 1829 (N.L.S., R.S.S.A. Archives, Acc 4534, box 11).

197. *Edinburgh Post Office Directory* 1822-1880. Bryden, *op. cit.* (1), 43.

198. MS family tree (Royal Museum of Scotland).

199. A.J. Adie was placed under the engineer James Jardine, (who was associated with Adie *senior* in the weights and measures work) on 10 April 1828, for ten years. (S.R.O., Henderson Papers, GD 76/457/72, letter of Alexander Adie to Alexander J. Adie, 7 July 1843). In 1838, A.J. Adie was appointed Resident Engineer on the Bolton and Preston railway, and he subsequently returned to Edinburgh to practise as a civil engineer, and lived in Linlithgow. The Henderson Papers largely consist of his correspondence, and are a prime source for biographical material concerning both him and his father, with whom he shared many interests, including instrument experimentation and meteorology. On the latter see a brief obituary notice on 21 July 1879 in *Journal of the Scottish Meteorological Society* new series, 5 (1880), 368. A notice on A.J. Adie, based on the Henderson Papers is in John Marshall, *A Biographical Dictionary of Railway Engineers* (Newton Abbot, 1978), 9. Adie married Louisa Sinclair and died aged 76 at his Linlithgow house on 3 April 1879 of 'cerebral apoplexy' (G.R.O.(S.), Register of Deaths 1879 (Linlithgow), 668/52).

200. 'Newspaper Collection', *op. cit.* (114), 38. Adie's presence on the island is unexplained.

201. William Galbraith, 'Barometric Measurement of the Height of Ben Lomond', *Edinburgh New Philosophical Journal* 6 (1828-29), 121-128. Thomas Henderson, Adie's future brother-in-law, and later the first Astronomer Royal for Scotland (H.A. Brück, *The Story of Astronomy in Edinburgh* (Edinburgh, 1983), 14-21), took a simultaneous reading at the Observatory on Calton Hill, Edinburgh.

202. William Galbraith, 'Barometric Observation', *Edinburgh New Philosophical Journal* 10 (1830-31), 46.

203. J.D. Forbes, 'On the Defects of the Sympiesometer, as applied to the Measurement of Heights', *Edinburgh Journal of Science* 10 (1829), 334-346; 'Memoir on Barometric Instruments ... Part 1 On the Defects of the Sympiesometer', *ibid.* new series, 4 (1834), 91-122. Forbes' own sympiesometers - a pocket-sized example, unsigned and unnumbered, and a portable version, described and illustrated in the later paper - are both now in the Royal Museum of Scotland (NMS T1984.53 and T1984.51 respectively). W.E.K. Middleton, *History of the Barometer* (Baltimore, Maryland, 1964), 379, describes Forbes' dissatisfaction with the instrument.

204. W. Galbraith, 'On the Magnetic Properties of the Rock on the Summit of Arthur's Seat', *Edinburgh New Philosophical Journal* 11 (1831), 287-288.

205. W. Galbraith, 'Barometric Measurement of the Height of Cheviot', *ibid.* 14 (1832-33), 70.

206. William Galbraith, *Barometric Tables for the Use of Engineers, Geologists, and Scientific Travellers* (Edinburgh, 1833).

207. N.L.S., R.S.S.A. Archives, Dep. 230/3, Minute Book vol. 3, p.354 (13 April 1836). W. Galbraith, 'A New Pocket-Box Circle', *Edinburgh New Philosophical Journal* 22 (1836-37), 229, where he refers to John Adie's 'able assistance' in constructing the circle, satisfactorily, although it might be simplified and improved in minor details. A mercurial thermometer by Adie [& Son] belonging to Galbraith was purchased along with an artificial horizon, two telescopes and a levelling stave, for the natural philosophy class at Edinburgh University in December 1850 (St. Andrews University Library, Forbes Papers, box IX, 2, Account book relating to purchase of instruments for Natural Philosophy class [of the University of Edinburgh] from Town Council funds, 1839-60, p.[17]). A rule made by J. Sisson of London and inscribed with Galbraith's name, was in the personal possession of J.D. Forbes, and is now in the Royal Museum of Scotland (NMS T1984.59).

208. John Leslie, 'Theory of Compression applied to Discover the Internal Constitution of Our Earth', *Edinburgh New Philosophical Journal* 6 (1828-29), 84. There is an example of this device, unsigned, in the Royal Museum of Scotland (NMS 1891.316). This was bought at the sale of the apparatus of Edward Sang (1805-1890), who had acquired some of Leslie's own equipment.

209. John Adie, 'Comparative Experiments on Different Dew Point Instruments; with a Description of one on an Improved Form', *Edinburgh Journal of Science* 1 (1829), 60-65.

210. John Adie, 'Account of a New Cistern for Barometers', *ibid.*, 338-340. A drawing of the improved barometer cistern was shown to the Society of Arts (N.L.S., R.S.S.A. Archives, Acc.4534, (1829)). Middleton notes that a design similar to Adie's was being developed in Russia at the same time (*op. cit.* (203), 229).

211. Edinburgh University Library, Special Collections Department, Gen. 1996 14/7, letter of J.D. Forbes to Thomas Jamieson Torrie, 31 August 1842. Forbes reiterated the fault of the column's tendency to separate when in transit on subsequent Alpine expeditions, in 1843 and 1846. Forbes used Adie thermometers to calculate heights by the boiling point of water (J.D. Forbes, 'Further Experiments and Remarks on the Measurements of Heights by the Boiling Point of Water', *Transactions of the Royal Society of Edinburgh* 21 (1837), 235-244); one of these is now in the Royal Museum of Scotland (NMS T1984.54)

212. John Adie, 'Notice of a new Method of Cutting and Working Glass by the Use of Turpentine', *Edinburgh New Philosophical Journal* 19 (1835), 215.

213. Middleton, *op. cit.* (203), 245, 262. Adie cleaned the tubes with spirit.

214. John Adie and Thomas Henderson, 'Observations made at the Observatory, Calton Hill, Edinburgh. 1. Of Moon-culminating Stars. 2. Of Occultations of Stars by the Moon', *Memoirs of the Royal Astronomical Society* 4 (1831), 189-192.

215. Dumfries Museum, MS Minute Book of the Dumfries and Maxwelltown Astronomical Society, 'Minute Book 1 [1835-1842]', entry for 24 February 1835. The date of the Adies' supply of a camera obscura to the Observatory is not known.

216. *Ibid.*, entries for 24 February, 19 March, 28 April 1835. For details of Thomas Morton's involvement, see Chapter 15.

217. C.P. Smyth, 'On the Stability of the Instruments of the Royal Observatory', *Proceedings of the Royal Society of Edinburgh* 3 (1850-57), 232.

218. *Exposition des Produits de l'Industrie de toutes les Nations. 1855. Catalogue Officiel* (Paris, 1855) 326, no. 388.

219. William Swan, 'On the Total Eclipse of the Sun on 28th July 1851 ... ', *Proceedings of the Royal Society of Edinburgh* 3 (1850-57), 73-76. John Adie, 'On the Total Eclipse of the Sun, observed at Goteburg in Sweden, 28 July 1851', *Edinburgh New Philosophical Journal* 51 (1851), 371-376.

220. John Adie, 'Description of the Marine Telescope', *ibid.* 49 (1850), 117-122. It is interesting to note that he was also still concerned with capillary instruments, as a paper of the same year shows: 'Experiments to discover the Cause of the Change which takes Place in the Standard Point of Thermometers', *ibid.*, 122-126.

221. NMS T1988.21; previously offered at Christie's Scotland, 27 & 28 April 1988, Lot 243.

222. W.K. Murray, *Description of an Astronomical Observatory* (Crieff, 1858); S. Korner, *Rambles round Crieff* (Edinburgh and Crieff, 1862), 58-61. We are grateful to Dr David Gavine for his advice.

223. John Adie, 'Description of an Instrument by which the Variation of the magnetic Needle can be determined with a greater degree of accuracy than has been attainable in Field Surveying', *Transactions of the Royal Scottish Society of Arts* 4 (1856), 138-141; this is followed by the Committee's Report, *ibid.*, 141-142. He was awarded the Society's silver medal in 1852 for the instrument (*ibid.*, Appendix, 124).

224. William Swan, 'On a Simple Variation Compass', *Edinburgh New Philosophical Journal* new series, 1 (1855), 78-83; *idem, Transactions of the Royal Scottish Society of Arts* 4 (1856), 313-318.

225. William Swan, 'On Errors caused by Imperfect Inversion of the Magnet in Observations of Magnetic Declination', *Transactions of the Royal Society of Edinburgh* 21 (1857), 353 and 349, Plate VII.

226. William Swan, 'On the Prismatic Spectra of the Flame of Compounds of Carbon and Hydrogen', *ibid.*, 419.

227. William Swan, 'On the Constitution of the Flame', *Transactions of the Royal Society of Edinburgh* 22 (1861), 24-25.

228. John Adie, 'Note on the Refractive and Dispersive Powers of the Humours of the Eye, determined by Experiment', *Proceedings of the Royal Society of Edinburgh* 2 (1844-50), 232. William Swan, 'On the Gradual Production of Luminous Impressions on the Eye, and other Phenomena of Vision', part 1, *ibid.*, 230-232. It is perhaps worth noting that Adie & Son supplied an optometer to the Natural Philosophy Classroom of Edinburgh University in 1849 (St. Andrews University Library, Forbes Papers box IX, 2: Account book relating to purchase of instruments for Natural Philosophy class [of the University of Edinburgh] from Town Council funds, 1839-60, p.[15]). The item is no longer extant.

229. J.D. Forbes, 'Note respecting the Dimensions and Refracting Power of the Eye', *Proceedings of the Royal Society of Edinburgh* 2 (1844-50), 251-256.

230. Alan Stevenson, 'Notice Relative to the Polyzonal Lenses belonging to the Commissioners of the Northern Lighthouses', *Edinburgh New Philosophical Journal* 18 (1834-35), 192; *idem, Proceedings of the Royal Society of Edinburgh* 1 (1832-44), 49-50.

231. N.L.S., R.S.S.A. Archives, Dep. 230/3, Minute Book vol. 2, p.15.

232. N.L.S., Dep. 230/4, Minute Book vol. 3, pp. 224, 307. Although in the latter two cases only 'Mr Adie' is mentioned, it is likely that John Adie is referred to. As convenor of a Society of Arts Committee in 1843 on the polymath Edward Sang's improved camera lucida, Adie was able to point out to Sang that his design incorporated no new lens arrangement that had not been described by Wollaston or Coddington in recent accounts (N.L.S., Acc. 4534, box 24, (copy) Adie to Sang, 30 December 1843).

233. N.L.S., Acc. 4534, box 24, paper 1014, delivered 12 June 1843. The paper was published as William Swan, 'On the Determination of the Index of Refraction by the Sextant, and also by means of an Instrument depending on a new Optical Method of ascertaining the Angles of Prisms', *Transactions of the Royal Scottish Society of Arts* 2 (1844), 287-298; *idem, Edinburgh New Philosophical Journal* 36 (1844), 102-113. In it, Swan acknowledged the use of prisms borrowed from William Nicol, John Adie and Alexander Bryson (q.v.). D. Thorburn Burns has pointed out the importance of this paper in the history of the development of the optical spectroscope, as Swan's prism arrangement predates the Littrow spectroscope of 1863 (D. Thorburn Burns, 'Towards a Definitive History of Optical Spectroscopy. Part II. Introduction of slits and collimator lens. Spectroscopes available before and just after Kirchoff and Bunsen's studies', *Journal of Analytical Atomic Spectrometry* 3 (1988), 285-291).

234. Adam Warden, 'Description of a Totally Reflecting Prism, ... or Prismatic Auriscope ...', *Transactions of the Royal Scottish Society of Arts* 2 (1844), 328. There is an example in the Wellcome Collection at the Science Museum, London, inventory A647346. An earlier camera lucida signed by Adie is in a private collection; another, signed by Adie & Son is in the Royal Museum of Scotland (NMS T1975.2).

235. John Adie, 'On the Advantages to be derived from the use of Metallic Reflectors for Sextants and other Reflecting Instruments: and on Methods of Directly Determining the Errors in Mirrors, and Sunshades used in Reflecting Instruments', *Transactions of the Royal Society of Edinburgh* 16 (1849), 61-66.

236. See, for instance, T.C. Smout, *A Century of the Scottish*

People 1830-1950 (Fontana edition, London, 1987), 40-46; J.A. Bennett, *The Divided Circle A History of Instruments for Astronomy Navigation and Surveying* (Oxford, 1987), 194-208.

237. Thomas Stevenson, 'Some Account of Levelling Instruments, with descriptions of one of an Improved Form', *Transactions of the Royal Scottish Society of Arts* 2 (1844), 307-313 and plate, Appendix, 82-3; David Stevenson, 'Description of Portable Levelling Instruments', *ibid.*, 314-315 and plate XVIII. Two examples of the latter pattern (one signed, the other unsigned) are in the Royal Museum of Scotland, NMS T1988.92 and T1988.93). The attribution is strengthened by the reference in the Society's Minute Book simply to 'Mr Adie', which by 1844 generally refers to John Adie (N.L.S., R.S.S.A. Archives, Dep. 230/4, Minute Book vol.3, p.433).

238. John Sang, 'Description of an Improved Apparatus for Levelling Small Theodolites', *Transactions of the Royal Scottish Society of Arts* 2 (1844), 306-307; the example is in the Royal Museum of Scotland (NMS T1971.16) and is divided into grades. It is further discussed at reference 265 below.

239. S.R.O., GD 76/464/4.

240. S.R.O., GD 76/457/63, Richard Adie to A.J. Adie, n.d. (c. 17 November 1841).

241. S.R.O., GD 76/464/4.

242. S.R.O., GD 76/457/59, Alexander Adie to A.J. Adie, 3 July 1840. Professor Leslie was also interested in soil, as a result of his hygrometrical experiments ('On the Absorbent Powers of different Earths', [William Nicholson's] *Journal of Natural Philosophy, Chemistry and the Arts* 4 (1801), 196-200).

243. S.R.O., GD 76/457/13, Janet Henderson to A.J. Adie, n.d.

244. S.R.O., GD 76/464/10, 'Memoir of Alexander James Adie'.

245. S.R.O., GD 76/457/24, Agnes Adie to A.J. Adie, 26 March 1839. Jane died 4 March 1839, Janet died in childbirth in 1842, and is mentioned in an obituary of Thomas Henderson (Philip Kelland, 'Biographical Notice of the Late Professor Henderson', *Proceedings of the Royal Society of Edinburgh* 2 (1844-50), 43).

246. S.R.O., GD 76/457, *passim*. E.g. 457/5, Agnes Adie to A.J. Adie, 24 April 1838, written when John was establishing himself, '... time and paper would fail to give you a catalogue of all his purchases'.

247. *Ibid.*

248. S.R.O., GD 76/457/80, John Adie to A.J. Adie, 1 May 1844. 457/49 & 50x, Janet Henderson to same, 9 April 1839. 457/28, Marion Adie to same, 16 August 1839.

249. 'Newspaper Collection', *op. cit.* (114), item 146, *News for the Millions, About the Great League Meeting in Edinburgh* No. 1 (15 January 1844). The Adies received a circular appeal for funds dated 12 January 1844 (*ibid.*, item 134).

250. *Ibid.*, items 136-151.

251. *Scotsman*, 7 January 1857. The report, entitled 'Melancholy Suicide of Mr John Adie', includes the gruesome detail.

252. Miller killed himself on the night of the 23-24 December 1856 (*Scotsman*, 7 January 1857). It is not known whether the two men knew each other personally, although they probably did by reputation.

253. G.R.O.(S.), Register of Deaths 1857 (Edinburgh, Heriot and Warriston), 685^7/11. Under Cause of Death is entered: 'His own hand by fire arms as certified by William Seller M.D. F.R.C.P.E. who saw deceased December 14 1856.'

254. S.R.O., GD 76/464/10, 'Memoir of Alexander James Adie'.

255. S.R.O., GD 76/464/10. G.R.O.(S.), Register of Deaths 1858 (Edinburgh, Newington and Grange), 685^2/482. Under Cause of Death is entered 'Decay of Nature had been failing for some weeks.'

256. Hon. Lord [Charles] Neaves, 'Opening Address', *Proceedings of the Royal Society of Edinburgh* 4 (1857-62), 226.

257. Dr Christison's opening address to the Royal Society of Edinburgh, 7 December 1857, *ibid.*, 6-7.

258. S.R.O, GD 76/457/28, Marion Adie to A.J. Adie, 16 August 1839.

259. J.D. Forbes's opening address to the Royal Society of Edinburgh, 1 December 1862, *Proceedings of the Royal Society of Edinburgh* 5 (1862-6), 20.

260. George Buchanan, 'Description and use of a Protracting Table', *Transactions of the Royal Scottish Society of Arts* 2 (1844), 177-178 and plate VII; Bryden, *op. cit.* (1), 37.

261. See Chapter 5, 'John and Thomas Dunn'.

262. Galbraith, *op. cit.* (207); Edward Sang, 'Account of an Improvement in the Construction of Wollaston's Goniometer', *Edinburgh New Philosophical Journal* 22 (1836-37), 213-219.

263. C.P. Smyth, 'On an Improved Form of Reflecting Instrument for use at Sea', *Report of the British Association ... 1852* pt.2 (London, 1853), 12-13.

264. This is the sole extant piece of equipment, and was bought in the 1880s from the Adie business by the Edinburgh maker George Hutchison (q.v.), and is now in the Royal Museum of Scotland (NMS T1967.148; Bryden, *op. cit.* (1), 37).

265. The Royal Museum of Scotland has a theodolite signed 'Adie & Son / Edinburgh' (NMS T1971.16), which is divided in grades. This is the same instrument discussed in reference 238 above, which has the levelling-head designed by John Sang.

266. E.U.L., Special Collections Department, MS AAF, Sir John Robison to an unnamed correspondent, endorsed by recipient, 8 July 1839.

267. In a private collection.

268. William Wallace, 'Account of the Invention of the Pantograph and a Description of the Eidograph', *Transactions of the Royal Society of Edinburgh* 13 (1836), 418-419. A full account of the controversial evolution of the rival versions of the pantograph in the 1820s is given in an unpublished paper by A.D.C. Simpson, see above reference 152. See also Chapter 5, 'John and Thomas Dunn', for their part in this episode.

269. W. Wallace, *Geometrical Theorems and Analytic Formulae with their Application to the Solution of Certain Geometrical Problems. And an Appendix, containing a Description of Two Copying Instruments* (Edinburgh, London and Cambridge, 1839), 136, 153. The accompanying plate illustrates the instrument, inscribed 'ADIE & SON, EDINBURGH' and 'W WALLACE / INV^T / 1839. / CHRORGRAPH'.

270. An account of this firm is given in [Anon.], 'British Engineers No. 12. Messrs. James Milne & Son, Engineers, Brassfounders, &c. Milton House Works, Edinburgh', *The Mercantile Age* 12 (1887), 354. On John Milne, his training in

Paris and his work for the N.L.B., see his obituary in D. Bruce Peebles, 'Address by the President', *Transactions of the Royal Scottish Society of Arts* 11 (1887), 391-393.

271. Example in private hands. In 1838 Alexander Adie sent A.J. Adie an eidograph, presumably for his engineering drawings for the Bolton and Preston Railway Company (S.R.O., GD 76/457/18 A. Adie to A.J. Adie, 25 November 1838).

272. Nuttall, *op. cit.* (157), 12, citing the order books of James Smith. Photocopies of these are held by the Department of Physical Sciences, Science Museum, London.

273. There is a group of four of these microscopes in the Royal Museum of Scotland, illustrating a number of combinations of the mechanical parts. The first is a drum microscope, with side pillar, unsigned, but in a case with a trade card for Adie & Son (NMS T1968.38); a similar example is also in the Museum, signed 'Adie & Son, / Edinburgh' (NMS T1982.133) and another, signed 'ADIE & SON / EDINBURGH' is in the Wellcome Collection at the Science Museum, London, inventory A601263. The second is a student microscope, with pillar support, on reversed claw foot, signed 'ADIE & SON / EDINBURGH' (NMS T1978.75). Thirdly, there is a microscope with semi Lister-limb and flat triangular base, signed 'ADIE & SON / EDINBURGH' (NMS T1980.22). And the fourth is a microscope with semi Lister-limb, on reversed claw foot, signed 'ADIE & SON / EDINBURGH' (NMS T1983.46); another example of this is in private hands and a third was offered for sale at Christie's South Kensington, 14 April 1988, Lot 235. Mary Somerville (1780-1872), the Jedburgh-born author famous for her books popularising science, owned a microscope sold by Adie & Son, which appears to be made from the base of a microscope similar to those discussed in reference 132, and a fairly straightforward drawtube (on loan to the Museum of the History of Science, Oxford, from Somerville College). Other microscopes, which have been noted but were not described fully enough to draw any conclusions from them, were one offered for sale by Christie's South Kensington, 10 August 1977, Lot 24, signed 'Adie & Son'; another with a Lister-limb, signed 'R. Adie', offered by Sotheby's Belgravia, 29 June 1977, Lot 226; a third, 'simple brass microscope' appears to have been unsigned, but the box had a trade card for Adie & Son, offered for sale at Sotheby's, 8 April 1974, Lot 104A.

274. S.R.O., GD 76/457/59, A. Adie to A.J. Adie, 3 July 1840.

275. S.R.O., RD 5/1066, Disposition by Alexander Adie, 10 December 1858; second codicil dated 25 April 1843.

276. G.R.O.(S.), Census of 1851 (Edinburgh, St. George's), 733/11, p.6. Ranken is described as 'Optician Foreman.'

277. Edinburgh City Archives, Guild Register, vol. 21 (1838-1854), 3 July 1849: 'George Ranken, Nephew of Christina Begbie residing in Northumberland Street Edinburgh enters Apprentice to Alexander Adie Optician and Mathematical Instrument maker for six years from the first day of January 1825. George Ranken, Optician Rose Street Compearing is made Burgess of this City in right of Alexander J. Adie Optician his master, and he paid his dues.'

278. G.R.O.(S.), Census of 1851 (Edinburgh, St. George's) 735/8, p.9. Entries in the 1841 census do not, unfortunately, record the size of the workshop.

279. S.R.O., SC 1/37/44 f.236v, Deed of Settlement by John Adie and his wife Elizabeth Barron, codicil dated 3 March 1854, witnessed by James Grassick clerk to Messrs. Adie & Son.

280. Forbes, *op. cit.* (123), 332.

281. G.R.O.(S.), Census of 1861 (Edinburgh, St. Andrews), 685²/60, p.11. Grassick is described as 'Optician's Cl[erk] Managing'.

282. *Edinburgh Post Office Directory* 1862-1865.

283. S.R.O., Register of Deeds RD 5/1797, p.386, Last Will of Richard Adie, 28 April 1870.

284. In 1840 Richard was quoting prices of microscopes and camera lucidas which he had in stock, for his brother Alexander (S.R.O., GD 76/457/47, Richard Adie to A.J. Adie, 24 March 1840). The origin of Adie's move to Liverpool is obscure; it may have been due to actual or anticipated friction if he remained in the Edinburgh shop. One possible connection with Liverpool was Dr T.S. Traill who lived there between 1803 to 1832, and was active in the scientific and literary society of the city, before returning to Edinburgh. He shared interests with Adie and was familiar with his meteorological instruments; Traill communicated Richard Adie's 'Description of the Hydrodynameter, a New Instrument for showing the Rate of Sailing of Ships and Velocity of Currents, Rivers, Tides, etc.' to the Royal Society of Edinburgh in 1838 (*Proceedings of the Royal Society of Edinburgh* 1 (1832-1844), 206-207). For Traill, see B.B. W[oodward], 'Thomas Stewart Traill', in S. Lee (ed.), *Dictionary of National Biography* LVII (London, 1899), 151.

285. List of Richard Adie's published papers in the *Royal Society Catalogue of Scientific Papers (1800-1863)* I (London, 1867), 20-21; *ibid.* (1864-1873) VII (London, 1877), 10.

286. N.L.S., R.S.S.A. Archives, Dep. 230/3, Minute Book, vol. 2, p. 359. The paper was published as 'Description of a new Anemometer ...', *Edinburgh New Philosophical Journal* 22 (1837), 309-313. In December 1837 an example of the instrument was exhibited in the Society courtesy of Robert Stevenson (probably its owner), and in January 1838 Edward Sang listed precautions to be taken in its use (N.L.S., Dep. 230/3, p. 441; Acc. 4534, box 19). Adie's medal of 1836 is in the Science Museum, London (inventory 1982-564).

287. *Transactions of the Royal Scottish Society of Arts* 2 (1844), Appendix, 95, 96, 112. Adie's statical windgauge of an earlier date is discussed in Dr Traill's article 'Physical Geography', in *Encyclopaedia Britannica* 8th edition, (Edinburgh, 1853-69), XVII, 569-647.

288. *Op. cit.* (285). Among items recorded with Richard Adie's signature are three pieces in the National Museums and Galleries on Merseyside: a double barrel vacuum pump (inventory 1976.484), night binoculars (inventory 1984.20), and a refracting telescope (inventory 1988.162). Our thanks to Martin Suggett for this information. Also recorded is a cometarium, once in the collection of the Royal Astronomical Society, which originally belonged to Sir James South, and is now in the Museum of the History of Science, Oxford. Inscribed 'John Taylor Invenit 1828. R. Adie Fecit Liverpool. 1835', it has the additional inscription 'Presented to the R.A.S. by Mrs Hannah Jackson Gwilt April 1880. This Cometarium was originally the property of Mrs Jackson Gwilt's friend, Sir JAMES SOUTH' (see H.D. Howse, 'The Royal Astronomical Society Instrument Collection: 1827-1985', *Quarterly Journal of the Royal Astronomical Society* 27 (1986), 219). Various barometers by Richard Adie have been recorded, including a

stick barometer signed 'R. Adie, optician, Bold Street, Liverpool' offered by Phillips Edinburgh, 25 July 1986, Lot 119; another, signed 'Adie Liverpool' offered by the same saleroom, 27 September 1985, Lot 84; a wheel barometer in the Royal Museum of Scotland (NMS T1983.98); and a sympiesometer, 'Patent Adie & Son Edinr. No. 1463. R. Adie. Liverpool' in the Musée d'art et d'histoire, Geneva (inventory 860, described in M. Archinard, *Baromètres* (Geneva, 1978), 21, 31).

289. S.R.O., GD 76/457/63, Richard Adie to A.J. Adie, n.d. (c.17 November 1841). Of his papers to the Liverpool Polytechnic Society, examples were 'On Imponderables' (*Liverpool Standard*, 18 January 1842) and 'On Ventilation' (*ibid.*, 15 February 1842).

290. St. Andrews University Library, Forbes Papers box IX, 2: Account book relating to purchase of instruments for the Natural Philosophy class [of the University of Edinburgh] from Town Council funds, 1839-60, p.[29]. This item is no longer extant.

291. N.L.S., R.S.S.A. Archives, Dep. 230/6, Minute Book vol. 5, p.337, 339. The convenor of the committee on the instrument was James Mackay Bryson (q.v.). R. Adie, 'Description of an hermetically sealed Barometer', *Journal of the Chemical Society* 13 (1860), 7-8; *Transactions of the Royal Scottish Society of Arts* 6 (1864), Appendix, 5. The silver medal awarded by the Royal Scottish Society of Arts is now in the Royal Museum of Scotland, NMS T1984.165.

292. Alexander Leslie, 'Description of an Improved Joint for a Levelling Staff', *Transactions of the Royal Scottish Society of Arts* 9 (1878), 402-405.

293. C. Piazzi Smyth, *Life and Work at the Great Pyramid* (Edinburgh, 1867), II, 167. His samples of the casing stones are now in the Royal Museum of Scotland. See also H.A.and M.T. Brück, *op. cit.* (182), 95-134.

294. For example, sympiesometers, which have the addresses of both the Edinburgh and London Adie firms engraved upon them: one, numbered 3137, is in a private collection; another, numbered 3262, is in the Royal Museum of Scotland, NMS T1984.38.

295. Trade card with a Continental microscope manufactured by Nachet of Paris, now in the Royal Museum of Scotland, NMS T1984.86. Other examples have been recorded.

296. G.R.O.(S.), Census of 1851 (Edinburgh, St. Andrews), 740/11, p. 18.

297. *Edinburgh Post Office Directory, passim*; Bryden, *op. cit.* (1), 50.

298. *Edinburgh Post Office Directory* 1874 and 1877.

299. NMS T1967.122. At a later date, Adie & Wedderburn advertised scale division to the trade (*International Exhibition of Industry, Science & Art, Edinburgh, 1886. The Official Catalogue* 4th edition (Edinburgh, 1886), 356); it may therefore be that division of this item was undertaken by the Adie business (see also reference 265).

300. NMS T1967.100.

301. Hutchison, *op. cit.* (6), 17.

302. Alexander Frazer, 'Improved Barometer exhibited and described to the Royal Scottish Society of Arts', *Transactions of the Royal Scottish Society of Arts* 9 (1878), Appendix, 78; 'Improved Barometer', *ibid.* 10 (1883), 384-387; 'On a New Thermometer', *ibid.*, 437; 'Description of a Ball and Socket Level', *ibid.*, Appendix, 103; 'On Graduation of Thermometers for Extreme Cold', *ibid.*, Appendix, 131; 'On an Improved Anemometer', *ibid.* 11 (1887), Appendix, 3; 'On the Improved Centering and Focusing Nose-piece for Microscope Objectives' *ibid.*, 345-347; 'On a Self-Centering Form of Shadbolt's Turntable for Ringing Microscope Specimens', *ibid.*, 347-349 (Frazer won a Brisbane Silver medal in 1886 for this paper); 'On Recent Improvements in Freezing and Other Microtomes' *ibid.*, Appendix, 101 (he gained a Hepburn Prize in 1886 for this paper; an example of Frazer's microtome is in the Whipple Museum of the History of Science and is described in O. Brown, *Catalogue 7 Microscopes* (Cambridge, 1986), item 394 (inventory 1421)); 'Proposed new Method of Graduating Barometers', *Transactions of the Royal Scottish Society of Arts* 12 (1891), Appendix, 6; 'On a Graphic method of Recording Weather Observations', *ibid.* 14 (1898), 25-27, 182-184; Frazer was elected a member of the Scottish Meteorological Society on 21 July 1880 (*Scottish Meteorological Journal* new series, 5 (1880), 323), before which he exhibited his anemometer (*ibid.* 6 (1882-83), 76-77).

303. E.U.L., Alphabetical List of Graduates of the University of Edinburgh from 1859 to 1888; Da 1870-1, First matriculation Book, 2 (1870-71), p.164.

304. *Scotsman*, 26 January 1881; G.R.O.(S.), Register of Deaths 1881 (Edinburgh, Cockpen), 676/6.

305. S.R.O., RD 5/1797, p.387.

306. G.R.O.(S.), Census of 1881 (Edinburgh, Newington), 685[5]/52, p.3. Thomas Weaderton (*sic*) is listed as 'Optician Master', aged forty-three.

307. Hutchison, *op. cit.* (6), 15, states that Adie took Wedderburn into partnership with him, although we have located no evidence for this.

308. *Edinburgh Post Office Directory* 1881-1913, *passim*.

309. *Transactions of the Royal Scottish Society of Arts* 11 (1887), Appendix, 38. Another optician, Alexander James Menzies, was elected the same day, and was perhaps connected with him; as he took over the firm after Wedderburn's death this is a probability (*Journal of the Scottish Meteorological Society* new series, 6 (1882-83), 3).

310. *Scotsman*, 10 August 1886. G.R.O.(S.), Register of Deaths 1886 (Edinburgh, St. Giles), 685[4]/786.

311. *Edinburgh Post Office Directory, passim*.

312. *Ibid.* Among Adie & Wedderburn items recorded, apart from the pieces in the Frank collection are: a dumpy level (NMS T1922.12), a measuring rod (NMS T1922.15), and a set of French curves (NMS 1958.52), all in the Royal Museum of Scotland; a brass universal equinoctial dial, offered by Sotheby's, 19 May 1983, Lot 131; a brass roller rule, offered by Christie's South Kensington, 12 November 1987, Lot 44; a level, offered by Phillips Edinburgh, 27 January 1984, Lot 76; another, offered by the same saleroom, 26 September 1986, Lot 45; and a travelling barometer, offered by Sotheby's, 1 May 1986, Lot 118.

313. *International Exhibition ... Edinburgh, 1886, op. cit.* (299), 356.

314. Donald Falconer Mackenzie was factor (estate manager) at Mortonhall, Liberton, near Edinburgh, between 1886 and

1910 (*Edinburgh Post Office Directory*, *passim*), and a member of the Highland and Agricultural Society of Scotland from 1884. His only recorded paper, 'The Identification of Timber, with a uniform series of Microphotographs', *Transactions of the Highland and Agricultural Society* 12 (1900), 183-224, shows an interest in forestry. He applied for a patent, 5546, on 27 March 1884 for 'Improvements in "Dendrometers" or instruments for taking or measuring the height or elevation and distance of objects' but this was abandoned or voided. A simplified, unnumbered version of the instrument is in the Royal Museum of Scotland (NMS T1967.171); and a letter from Mackenzie to Adie & Wedderburn describing the use of his timber measuring instrument, dated 3 December 1886, presumably led to the firm manufacturing the piece. Another example signed 'D.F. Mackenzie's Patent' and numbered '72' was offered for sale by Christie's South Kensington, 27 November 1986, Lot 238, and subsequently acquired by the Royal Museum of Scotland (NMS T1988.16).

315. Assuming that he was apprenticed to Adie & Son c.1852, aged 14.

316. Hutchison, *op. cit.* (6), 15-16, where he refers to him as Peter Mabon: however, the presumption must be that this was Alexander Fernie Mabon, mathematical instrument maker, who was recorded in Glasgow in 1905, and his business as Alexander Mabon & Sons, opticians, in 1910 (*Glasgow Post Office Directory*, *passim*). A trade catalogue issued by the firm, undated, is in the Whipple Museum of the History of Science, Cambridge; a miner's dial signed 'A. Mabon & Sons, Glasgow' was offered for sale by Sotheby's Belgravia, 9 May 1980, Lot 152; a level signed 'Alex. Mabon & Sons Glasgow' was offered by Christie's South Kensington, 11 July 1985, Lot 141, and another by the same saleroom, 10 October 1985, Lot 128.

317. Obituary of Thomas Haddow, *Edinburgh Journal of Science* 2 (1928), 185-186. A Fortin barometer, retailed by Haddow, was offered for sale by Phillips Edinburgh, 31 March, 1989 Lot 57.

318. 'ALEXANDER JAMES MENZIES, Optician, Edinburgh, has acquired the Business carried on at No 17 Hanover Street, Edinburgh by ADIE & WEDDERBURN, Opticians there, and he will collect all outstanding accounts due to the Firm. No other person or persons have any interest in the said Business. [signed] A.J. MENZIES' (*Edinburgh Gazette*, 1 February 1887, 11); 'THOMAS MEIN, Optician, 17 Hanover Street, Edinburgh, has acquired the Business formerly carried on in Edinburgh by ADIE & WEDDERBURN, Opticians there, and he will carry on business under that style. No other person has any right or interest in the said Business. [signed] CHARLES D. MENZIES Executor of the late ALEXANDER JAMES MENZIES, sole Partner of the Firm of ADIE & WEDDERBURN' (*ibid.*, 13 March 1888, 261).

319. *Edinburgh Post Office Directory* 1881-1920, *passim*.

320. *Ibid.* 1918-1949, *passim*. S.R.O., BT 2/10153, records of Defunct Companies; documents include the purchase agreement between Mortons and Richardson, Adie & Co., and the certificate of the firm's incorporation, 19 November 1918. There is a pocket aneroid barometer signed 'Richardson, Adie & Co., Edinburgh' in the Royal Museum of Scotland (NMS T1982.63); and a barograph with their signature was offered for sale by Phillips Edinburgh, 28 January 1983, Lot 93.

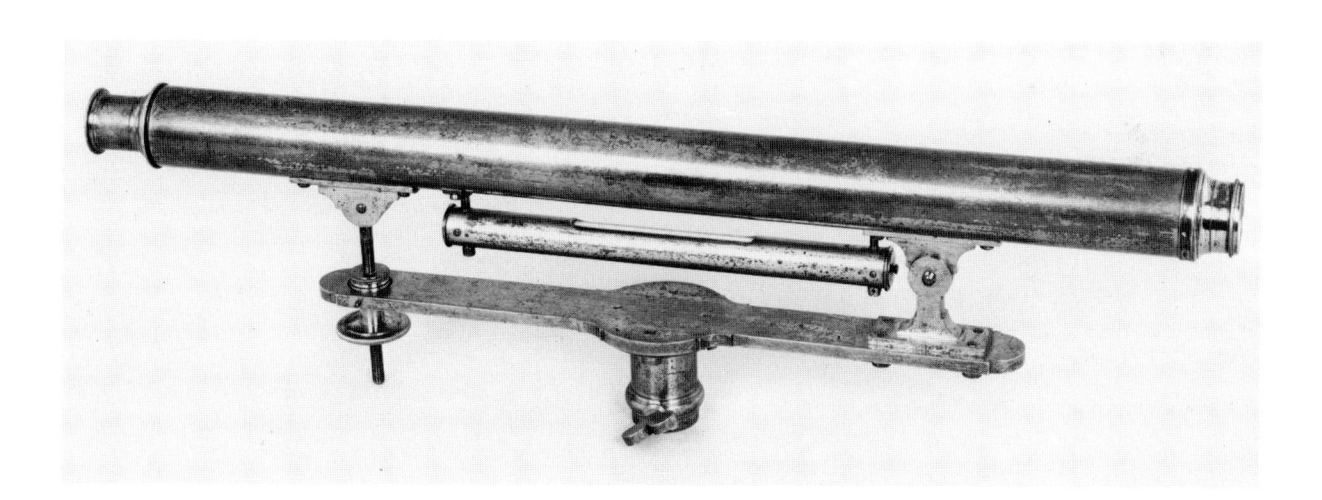

---

12. Level: J. Miller, Edinburgh, c.1790 (T1980.124)

18″ level in brass, with bubble tube suspended beneath the telescope, hinged at the objective end over a limb and with screw adjustment in altitude at the eye end; with push-fit focus, the eyepiece assembly incorporating an erecting system, and with single lens objective with sliding aperture. Engraved on the limb 'J=Miller Edinburgh'. Socket mount with clamp screw.

Telescope length: 450mm.
Tube diameter: 32mm.
Aperture: 13mm.

The long inner draw tube is fabricated frugally from a number of shorter components, and shows a varied quality of workmanship. A small patch has been expertly introduced into the base flange of the front pivot. The external eye stop may be a replacement for an earlier sliding cap. The graticule diaphragm is a later replacement. The rear mounting and adjustment screw for the bubble tube has been replaced and its attachment re-bushed: a corresponding slot has been cut in the draw-tube to accommodate the projection of this screw into the instrument. This has been used to preserve the orientation of the diaphragm wires in place of the engagement of one of the mounting screws of the telescope in an adjacent slot.

Provenance: Purchased from David Letham Ltd., Edinburgh, 1967.

13. Level: Miller & Adie, Edinburgh, c.1810 (T1980.125)

Compact 6½" level in brass, with bubble tube mounted alongside the telescope and carrying a calibrated ivory scale, single lens objective, and with the telescope and level bridged by a circular compass with 4-point compass rose and divided 0-90-0-90-0 by degrees, mounted on 2 pillars rising from the limb. Engraved on the silvered compass rose 'Miller & Adie Edinburgh.' The limb threaded for attachment to a levelling head.

Telescope length: 167mm.
Tube diameter: 22mm.
Aperture: 22mm.

Lacking the adjustment screws for setting the graticule wires (the graticule modified), and also the objective cover.

14. Microscope: Miller & Adie, Edinburgh, c.1810 (T1980.221)

Compound microscope, modified 'Martin's Universal' type, in brass on pillar stand and folding 3-toed tripod base. Engraved on the stage 'Miller / & / Adie / Edinburgh'. The fixed focus body with 2-component eyepiece assembly and field lens, screws into a bar-limb attached by an urn-headed screw to the top of a 2-part telescoping cylindrical pillar, connected by a rack and pinion, the lower end of the projecting rack sliding over the surface of the outer pillar, and the pillar terminating in a folding tripod. The inner pillar marked with the settings '1' - '5'. Four objectives (of 6), engraved '1', '2', '3' and scratched '6'. Cruciform stage with central aperture, mounted over a bar-limb extending from the top of the outer pillar. Concave mirror mounted over the fixed foot. With fitted case (restored) with accessories, including lieberkuhn, attached to a long sleeve which fits over the objective nosepiece, sprung aquatic stage, glass stage cell, fish plate, and stage condenser (glass missing).

Body length: 152mm.
Body diameter: 38mm.
Height: 330mm.
Maximum leg radius: 105mm.
Case size: 275 x 180 x 95mm.

Lacking the glass lenses in objectives 1 and 2. The single attachment of the rack, at its upper end, has broken and has been reinforced with a screw passing into the inner pillar. A plugged hole just above this attachment suggests that the rack had been previously relocated. The collar at the top of the outer pillar lacks its 4 fixing screws, and a partly obscured hole under the collar indicates further modification. The screw collar for the body tube has been repaired but has stretched and is now a poor fit. The lid of the case is a replacement.

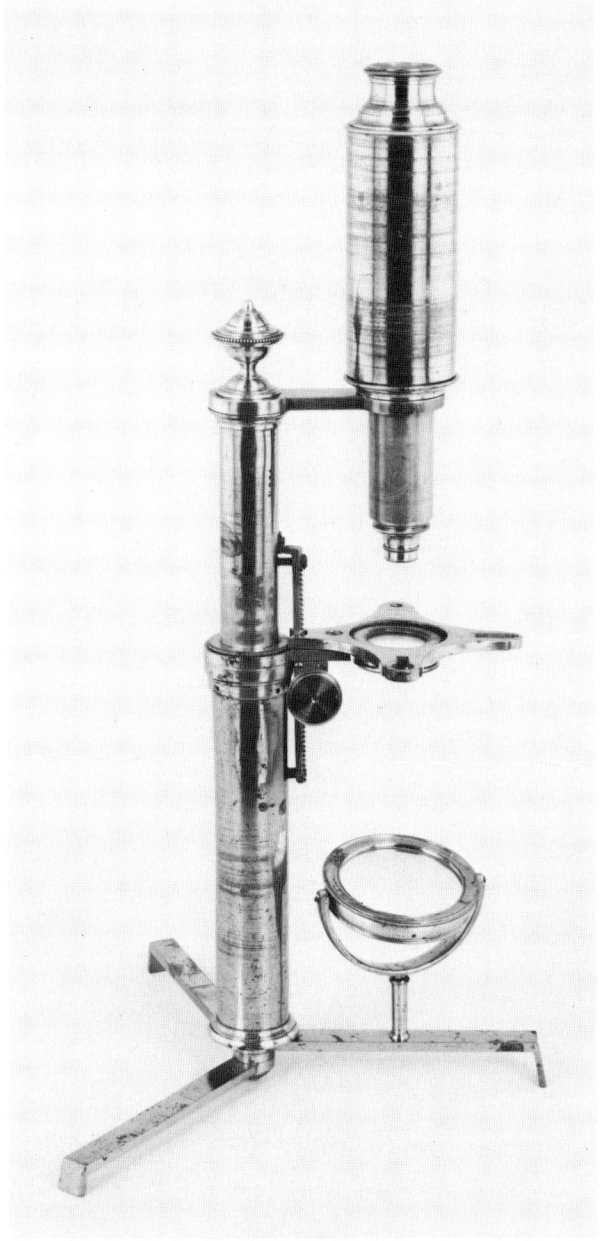

Provenance: Presumed to have been purchased from the private collector in Pinner, Middlesex, who corresponded with the Royal Scottish Museum about this instrument in April 1970. It is mentioned in R.H. Nuttall and A. Frank, 'Makers of Jewel Lenses in Scotland in the Early Nineteenth Century', *Annals of Science* 30 (1973), 414 n.35. Benjamin Martin's Universal Compound Microscope was first illustrated in his *Philosophia Britannica* 2nd edition (London, 1759) and is discussed by John Millburn, *Benjamin Martin, Author, Instrument-Maker, and 'Country Showman'* (Leyden, 1976), 114-116. An example of Martin's instrument in the Museum of the History of Science, Oxford, is discussed and illustrated in G. L'E. Turner, *Collecting Microscopes* (London, 1981), 60-61.

15. Reflecting telescope and stand: Miller & Adie, Edinburgh c.1815 (T1980.272)

2-foot focus Gregorian reflecting telescope, with speculum metal primary, the barrel in sheet iron, leather-covered, bound at the ends with brass, held in an iron trunion mount over an earlier wooden stand. This stand has a turntable revolving on 4 recessed brass wheels over a cross-braced mahogany tripod, and an inset ivory plate engraved 'Miller & Adie / Edinburgh.' Two wooden upstands with a connecting crosspiece have been reduced in height and their upper surfaces rounded to conform with 2 turned wooden sockets for the telescope trunnions.

Turntable height: 865mm.
Primary speculum diameter: 103mm.

The telescope presently mounted appears to be a mid-19th century instrument: this is now incomplete, lacking eyepieces, secondary speculum and the original focusing mechanism; it has been recovered in leather.

Provenance: Purchased from P. Couts Ltd., Edinburgh, c.1971.

16. Box sextant: Adie, Edinburgh, c.1825 (T1980.163)

3" diameter box sextant in brass with scale divided on silver [-5]-[160] by ½°, the index arm with clamp and tangent screw adjustment and vernier reading to 1 min. and a swinging magnifier. Engraved beneath the scale 'ADIE, Edinburgh'. Sliding aperture with pin-hole sight, but lacking the alternative telescopic sighting tube which would normally stow within the instrument. With 2 shades and key for adjustment of horizon glass. Screw-fit brass cover which reverses and attaches to the rear of the instrument to form a handle or support.

Scale radius: 45mm.
Diameter: 73mm.

Lacking sighting telescope. A clumsy replacement retaining bracket holds the index mirror in its frame. The cylindrical housing has been slightly deformed and one of the 3 attachment screws to the assembly plate is missing.

Provenance: Purchased at Christie's, 21 December 1971, Lot 30.

17. Sympiesometer: Adie, Edinburgh, c.1830 (T1980.209)

Two-liquid marine air barometer in hinged wooden carrying case. Silvered scale engraved 'PATENT / Adie / Edinburgh / N⁰ 1181.' Moveable scale on the right side of the plate sliding over a temperature compensation scale 25°-[122°]F divided to 0.2°. The moveable scale projecting as necessary through the lower edge of the case and graduated to read air pressure against the liquid level in inches of mercury or height in fathoms, 31"-27" divided to 0.2", and 0-[1650] fathoms divided to 2 fathoms. Mercury in glass thermometer graduated [-2⁰]-[117⁰]F divided to 0.2⁰. Sliding plug to lower barometer reservoir.

Case size: 50 x 35 x 540mm.

A wooden protective cover (with an air movement hole) enclosing the barometer hydrogen bulb and the thermometer bulb appears to be a more recent addition. Holes for the attachment of an earlier and smaller cover are visible at the rear of the case. The lower edge of the case trim, which would have been slotted to accommodate the sliding scale, is detached.

Provenance: Understood to have been purchased at auction in Paisley, c.1972. The sympiesometer, patented by Alexander Adie in 1818, is discussed in the text above.

18. Sundial: A. Adie, Edinburgh, c.1825 (T1981.35)

Circular horizontal pedestal sundial in brass, engraved at the North point of the dial 'A. Adie Edinburgh'. Engraved chapter ring IIII-XII-VIII divided to 1 min., central 8-point compass rose flanked by a 2-part calendar scale with 'equation of time' corrections given throughout the year in intervals of 1 min. The gnomon with a measured angle of 55°, pierced to give a plain inclined stile with an S-shaped support; attached to the dial plate by 4 screws from beneath. The intersection of the gnomon edges and the plate inscribed on 2 bronze plugs set into the plate. At the South side of the dial is engraved the armorial achievement of Agnew of Lochnaw.

Overall diameter: 398mm.
Gnomon height: 217mm.

Although the gnomon appears to be contemporary, the separation of the principal attachment holes in the plate has had to be reduced to accommodate it, and the central locating pin is not present.

The sundial was presumably commissioned from Alexander Adie between 1822 and 1835 by a member of the Agnew family whose seat is Lochnaw Castle, near Stranraer, Wigtonshire. The representation of the Agnew achievement is as follows: on an oval shield, set within a cartouche, the tinctures rendered with the Petra Sancta system: Argent, a chevron between two cinquefoils in chief Gules and a saltire couped in base Azure. The shield surrounded by the motto of the Order of the Baronets of Nova Scotia 'FAX MENTIS HONESTAE GLORIA' (Glory is the light of a noble mind), with the oval badge of the Order suspended beneath the shield (Argent a saltire Azure, en surtout, the Royal Arms of

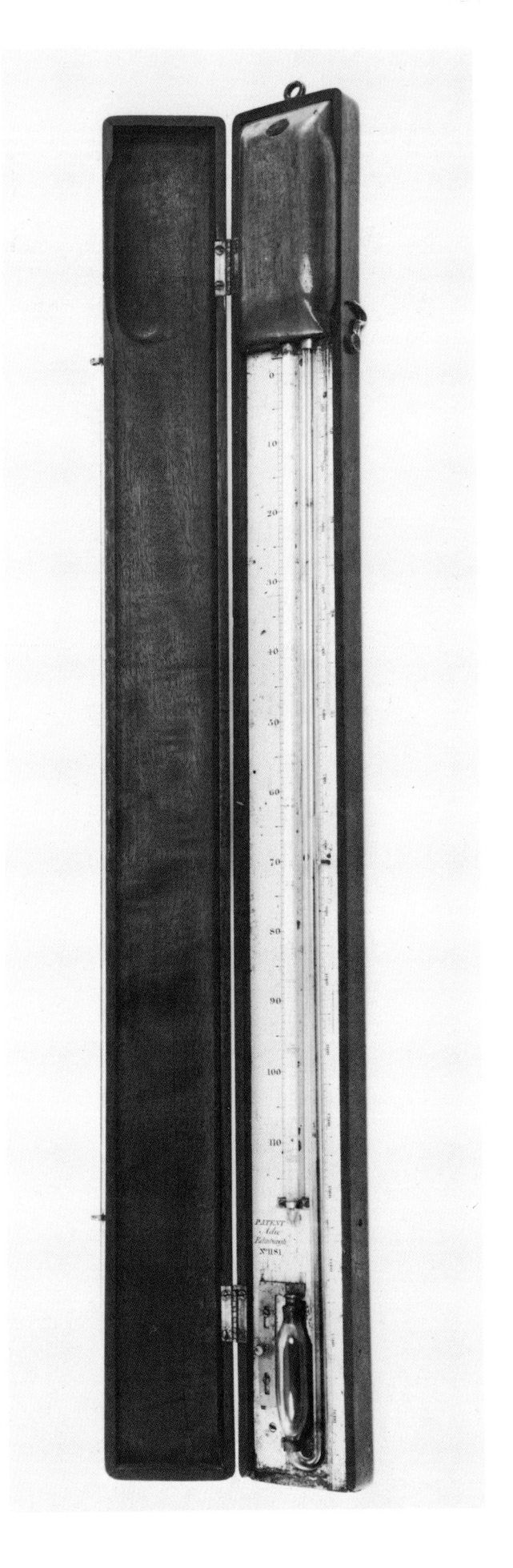

Scotland; the shield surmounted by an imperial crown). Above the cartouche a knight's helm with mantling Argent doubled Azure (should be Gules) and for crest an eagle issuant and regardant Proper. The supporters are two heraldic tigers Proper gorged with a coronet and chained Or, standing on an architectural compartment incorporating a ribbon bearing the motto 'CONSILIO NON IMPETU' (By counsel, not by force). We are grateful to Charles Burnett, Ross Herald, for his advice. It is most likely that the dial was made for Sir Andrew Agnew (1793-1849), 7th Bt., M.P. for Wigtonshire 1830-37. Agnew attended classes at Edinburgh University 1810-11, and spent subsequent years in the improvement of the castle and estate. John Hay (1758-1836), landscape gardener of Edinburgh, designed the formal garden in which the sundial was located. (T. McCrie, *Memoirs of Sir Andrew Agnew of Lochnaw, Bart.* (London and Edinburgh, 1850), 25; A.A. Tait, *The Landscape Garden in Scotland* (Edinburgh, 1980), 144, 255.)

19. Group of microscope objectives: unsigned, c.1825 (T1981.36)

Group of 6 high-powered objective lenses for simple microscopes in matching black-painted brass mounts to fit a threaded holder approximately 10mm. diameter. Four have garnet optics, a fifth is scratched on the lower painted surface '¹/₃₀ / Glass'. A sixth, which differs from the others in the form of its recessed upper surface, has a minute spherical lens mounted in a perforated lead foil. Associated with these is an unrelated and miscellaneous collection of optical components, including, however, one lens which has been identified as having been constructed from sapphire. A manuscript fragment in an unknown hand lists the characteristics of 6 lenses of garnet, sapphire and glass, which may be the lenses described above. Contained in a mid-18th century fishskin-covered instrument case with modified internal tray.

Garnet and glass lens mounts: 13-15mm. diameter.
Sapphire lens mount: 24mm. diameter.
Case size: 360 x 105 x 55mm.

The provenance of these items is discussed in Robert Nuttall and Arthur Frank, 'Jewel Lenses - a historical curiosity', *New Scientist* 53 (1972), 92-93, and also in *idem*, 'Makers of Jewel Lenses in Scotland in the Early Nineteenth Century', *Annals of Science* 30 (1973), 407-416. The second article includes a transcription of the manuscript, which lists 6 lenses by their optical material, focal length and another numerical characteristic. This last has been assumed by Nuttall and Frank to represent the prices of the individual lenses, but is not given in a conventional monetary form, but apparently in a decimal form: it may perhaps cover some other aspect such as a measure of the magnification or resolving power. They have also speculated that the lenses were unsold stock from the Adie business, but equally the material may have been sold and then subsequently acquired by J.R. Hutchison as part of his collection.

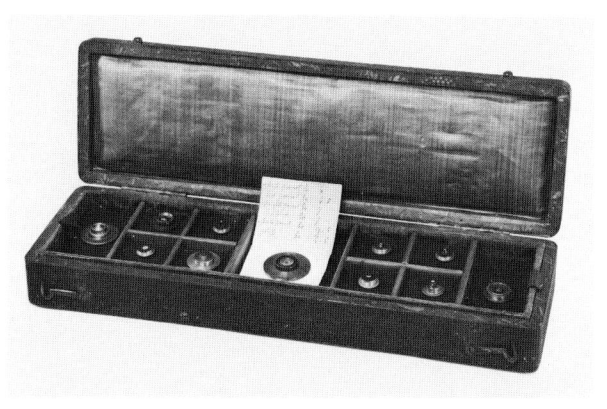

20. Level: unsigned, c.1860 (T1980.126)

10½" level in oxidised brass with bubble tube suspended beneath the telescope, rack and pinion focus, single lens objective, and objective cover. Socket mount for attachment to a levelling head. With fitted case containing trade label for Adie & Son, 50 Princes Street, Edinburgh.

Telescope length: 288mm.
Tube diameter: 29mm.
Aperture: 24mm (but 11mm. internal stop).
Case size: 300 x 130 x 65mm.

Top 23, left 21, right 22

21. Mining dial: Adie & Son, Edinburgh, c.1850 (T1980.148)

Glazed circular compass in brass with silvered dial, recessed circular level and 2 folding opposed slit and window sights. Engraved on the dial plate 'Adie & Son Edinburgh'. The dial with the 4 cardinal compass points, the scales divided in degrees 0-90-0-90-0 and in ½° 0-[360]. Flat 3-layer steel needle, blued at one end, on a relieved jewelled bearing. Socket mount with clamp screw, and geared azimuth rotation of the dial and sights from a knob beneath the dial, read to 1 min. against a silvered vernier within the dial.

Compass housing diameter: 171mm.

The baseplate has been bent at the extension piece for one of the sights. Sight wires detached.

22. Mining dial: Adie & Son, Edinburgh, c.1850 (T1980.149)

Glazed circular compass in brass with silvered dial, recessed circular level and 2 folding opposed slit and window sights. Engraved on the dial plate 'Adie & Son Edinburgh'. The dial with a 4-point compass rose, the scales divided in degrees 0-90-0-90-0 and [360]-0. Vertical-section steel needle on a relieved jewelled bearing. Socket mount with clamp screw, and azimuth rotation of the dial and sights, read to 2 mins. against a silvered vernier within the dial. With fitted case containing trade label for Adie & Son, 50 Princes Street, Edinburgh.

Compass housing diameter: 173mm.
Case size: 280 x 205 x 90mm.

Glass broken. The base of the case has been replaced.

23. Mining dial: Adie & Son, Edinburgh, c.1850 (T1980.150)

Glazed circular compass in brass with silvered dial, recessed circular level and 2 folding opposed slit and window sights. Engraved on the dial plate 'Adie & Son / Edinburgh'. The dial with the 4 cardinal compass points, the scale divided in degrees [0]-360 and 0-90-0-90-0. Flat 3-layer steel needle, blued at one end, on a relieved jewelled bearing. Socket mount with clamp screw.

Compass housing diameter: 172mm.

Six symmetrically-placed screws attach the compass ring to the baseplate: the base however also has 6 plugged holes displaced by approximately 15° about the same centre, indicating an aborted fabrication attempt before the baseplate had been cut. Lacking sight wires.

24, 25

24. Theodolite: Adie & Son, Edinburgh, c.1840 (T1980.156)

4″ theodolite in brass and bronze, the bevelled edge azimuth circle divided on brass 0-[360] by ½°, with clamp and tangent screw adjustment to the upper plate and vernier reading to 1 min. Engraved by the vernier 'Adie & Son / Edinburgh'. Crossed levels and silvered compass with the 4 cardinal point and divided [360]-0 by ½°: the needle with a relieved jewelled bearing. A-frames support the 1″ fixed-focus telescope with bubble tube above. Semi-circular altitude scale with rack edge divided on brass 50-0-70 by ½° with vernier reading to 1 min., and on the reverse 20-0-20 'Links to be Subt^d from Hypotenuse'. Socket mount. Objective cover.

Telescope length: 270mm.
Aperture: 25mm (but internal stop of 9mm.)
Azimuth scale diameter: 105mm.

Lacking the eye lens and the graticule wires.

25. Theodolite: Adie & Son, Edinburgh, c.1850 (T1980.157)

4″ theodolite in brass, the bevelled edge azimuth circle divided on brass 0-[360] by ½°, with clamp and rack and pinion adjustment to the upper plate and vernier reading to 1 min. Engraved by the vernier 'Adie & Son / EDINBURGH.' Crossed levels and silvered compass with the 4 cardinal points and divided [360]-0 by 1°, the needle with relieved jewelled bearing. A-frames support the fixed-focus telescope with 1″ doublet objective with bubble tube beneath. Semi-circular altitude scale with rack edge divided on brass 50-0-[65] by ½° with vernier reading to 1 min., and on the reverse 20-0-20 'Links to be Subt^d' Socket mount. With fitted case containing trade card for Adie & Son, 50 Princes Street, Edinburgh; packing instruction 'Obj. Glass' in MS at one end of the box.

Telescope length: 229mm.
Aperture: 25mm.
Azimuth scale diameter: 105mm.
Case size: 260 x 145 x 150mm.

Lacking objective cover. The clamp screw on the vernier plate is a replacement.

26. Theodolite: Adie & Son, Edinburgh, c.1880 (T1980.158)

5″ reversing transit railway theodolite in black-lacquered brass and bronze, the bevelled edge azimuth circle divided on silver [0]-360 by ½°, with clamp and tangent screw adjustment to the upper plate which covers the circle except at the verniers which are read to 1 min. with rotating magnifier. Crossed levels and silvered compass, engraved 'ADIE & SON. EDINBURGH.': the compass with the 4 cardinal points (transposed arrangement for reading bearings against the north point of the needle) and divided [360]-0 by 1°, the needle with a relieved jewelled bearing. A-frames support the telescope, with rack and pinion focus to the 1″ triplet objective, and objective ray shade. Integral parallel plate levelling head with axis collar clamp and tangent screw. With fitted case containing trade card for Adie & Son, 37 Hanover Street, Edinburgh, and another noting repair and adjustment by G. Hutchison & Sons, 18 Forrest Road, Edinburgh.

71

Telescope length: 250mm.
Aperture: 27mm.
Azimuth scale diameter: 130mm.
Case size: 180 x 180 x 480mm.

One bubble tube broken.

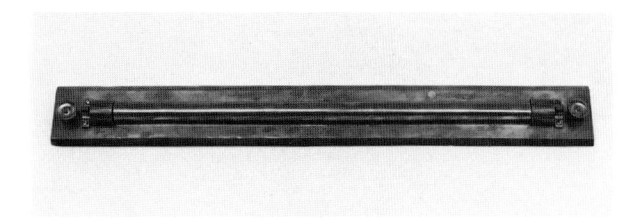

27. Parallel rule: Adie & Son, Edinburgh, c.1880 (T1980.169)

24″ rolling parallel rule in oxidised bronze (stripped), with chamfered edges and two lifting knobs. Engraved 'ADIE & SON, EDINBURGH.' and 'ROBERT BOATH.'

Length: 613mm.

The instrument's surfaces and edges are distressed.

Robert Boath has not been identified.

28. Eidograph: Adie & Son, Edinburgh, c.1860 (T1980.178)

30″ copying eidograph in brass, pivoted on a brass-cased lead-weighted base and with a lead counterweight. Engraved on the brass sleeve over the pivot 'Adie & Son Edinburgh.' A hollow-section central spar, graduated 80-0-80, passes through this sleeve, and at its extremities are 2 pulley wheels linked by an adjustable steel tape which maintains the parallelism of 2 hollow-section tracing and copying arms, similarly graduated 80-0-80, and passing through radially-mounted sleeves under the pulley wheels; the 3 scales set against indices in apertures in the sleeves and clamped. With fitted case, from which the trade card has been removed.

Length of spar between centres: 785mm.
Case size: 905 x 145 x 100mm.

Lacking pencil, tracing point and locating pins for the base.

Provenance: Purchased at Sotheby's, 15 October 1973, Lot 21. The eidograph, invented by William Wallace in 1821, is discussed in the text above.

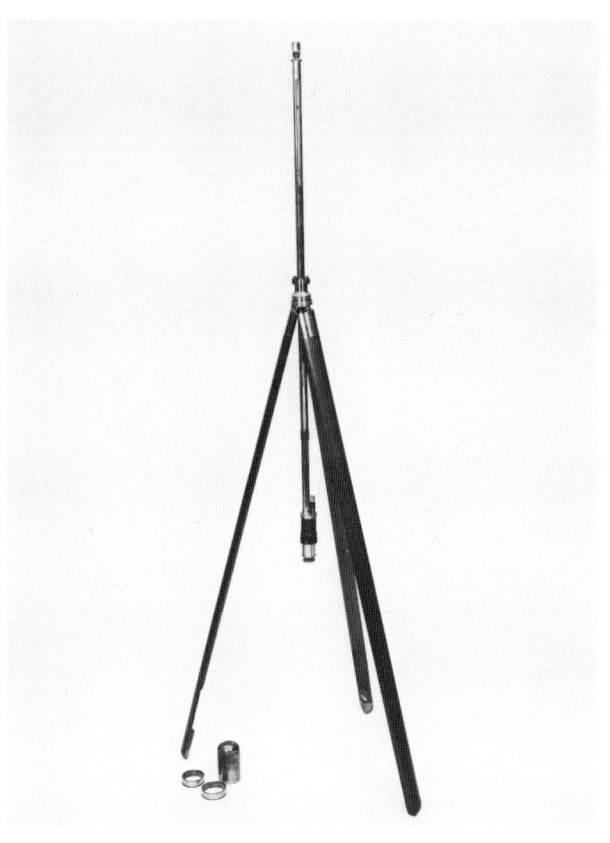

29. Mountain barometer: Adie & Son, Edinburgh, c.1835 (T1980.210)

Mercury mountain barometer in brass with folding wooden tripod which acts when closed as a carrying case. Assembled, the instrument is suspended at its mid-point by trunion screws resting in a ball and socket joint at the top of the tripod and steadied by 4 screws. Engraved at the top of the tube 'Adie & Son / Edinburgh'. The brass sleeve enclosing the glass barometer tube is pierced with 2 pairs of longitudinal slits with scales divided [18.1″]-25″ and [24.6″]-[32″], each with a sliding vernier reading to 0.001″ and adjusted by a micrometer screw at the top of the tube. The lower part of the cistern screws to raise the mercury level to the top of the tube for transit, and an adjusting screw at the base acts on an inner diaphragm to adjust the cistern atmospheric mercury level, the level being set against an ivory index in an open glass-sided projection over the cistern, closed with an ivory plug. Circular bubble level in the top of the tube casing. Two stalked magnifying glasses, for use in adjusting the mercury level and verniers, are stored in recesses in the tripod legs. A mercury in glass thermometer reading 5°-[119°]F and [-15°]-[48°]C is attached to the lower part of the instrument. In transit, the barometer is suspended by a further pair of trunion screws at the top of the tube, the suspension joint and adjustment protected by a brass screw cover, and the legs clasped by 2 brass rings.

Case size closed: 1500 x 45mm. diameter.

The thermometer is broken, and the outer brass sleeve for the cistern adjustment tube is lacking.

This instrument is a development from Edward Troughton's earlier mountain barometer. This pattern was first produced by Miller & Adie in or before 1810, and is discussed in the text above.

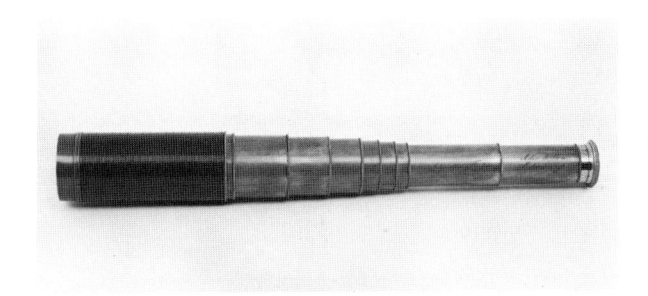

30. Refracting telescope: Adie & Son, Edinburgh, c.1860 (T1980.246)

1½" aperture 8-draw hand-held refracting telescope with doublet objective and 2-component eyepiece and erecting assemblies; objective ray shade of the same length as the outer tube and externally wound with dark grooved tape. Engraved on the first draw tube 'Adie & Son, / Edinburgh.' and on the second draw-tube (which contains the erecting optics) 'Adjusting Tube.' Components scratched with the assembly mark 'III' throughout. Closure on external eye stop.

Length closed: 148mm.
Ray shade diameter: 49mm.
Aperture: 40mm.

Lacking objective cover.

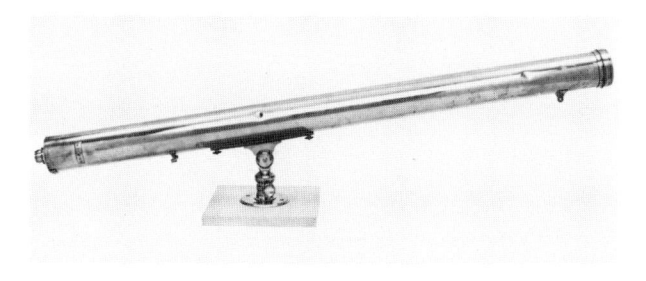

31. Refracting telescope: Adie & Son, Edinburgh, c.1860 (T1980.273)

3½" aperture 2-draw mounted refracting telescope in brass; with doublet objective but lacking other optics, and with rack and pinion focus (defective) to the second draw-tube. Originally fitted with a draw tube extension section, of which only the end ferrule remains, containing the image erecting assembly. Engraved on the barrel end plate 'ADIE & SON, EDINBURGH'. Mounted by a knuckle-joint with clamped azimuth motion to a flanged plate, for attachment to a wooden tripod base (lacking). Originally with extendible steady rods attached beneath the end plate, for which the stowage collars are present, and with a further attachment point towards the objective.

Length closed (less erecting assembly and eyepiece): 1,585mm.
Tube diameter: 98mm.
Aperture: 90mm.

Lacking eyepieces and image erecting unit, steady arms, focus knob, objective cover, and tripod support.

Detailed similarities in design and in the mounting of optical components suggest that this item is by the same manufacturer as items 81 (retailed by Gardner & Co.) and 116 (signed D. McGregor).

32. Barograph: Adie & Wedderburn, Edinburgh, c.1890 (T1980.213)

Recording drum barograph, in brass and steel, numbered '4419', with 9-part aneroid chamber, clockwork movement in recording drum, and with adjacent thermometer calibrated in Centigrade and Fahrenheit. The brass base plate mounted on a wooden base with drawer, and enclosed by a hinged glazed cover. An attached plaque inside the case is marked 'ADIE & WEDDERBURN / EDINBURGH'. With printed sheet 'Instructions for Self-Recording Barometer'.

Case size: 490 x 285 x 300mm.

Lacking ink pot.

Probably of London manufacture.

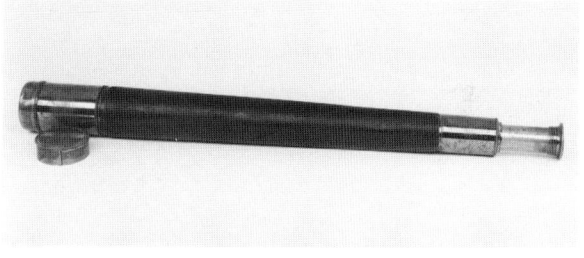

33. Refracting telescope: Adie & Wedderburn, Edinburgh, c.1890 (T1980.247)

2⅛" aperture single-draw hand-held refracting telescope with triplet objective and 2-component eyepiece and erecting assemblies (defective); in brass (with leather-lined draw slides), with objective ray shade and tapered leather-covered barrel. Engraved on the draw tube 'Adie & Wedderburn, / 17. Hanover St. / EDINBURGH.' Scratched assembly marks on draw tube components 'IX' or 'XI'. Objective cover, and closure on external eye stop.

Length closed: 641mm.
Maximum tube diameter: 62mm.
Aperture: 54mm.

Lacking the forward lens of the image erecting assembly. The leather cover has shrunk and the barrel is dented.

Probably by James Parkes & Son of Birmingham, cf. *Illustrated Trade Price List of Telescopes* (Birmingham, n.d.[1902]), 16, 'Captain's and Ship Telescopes, Best Make', no. 445.

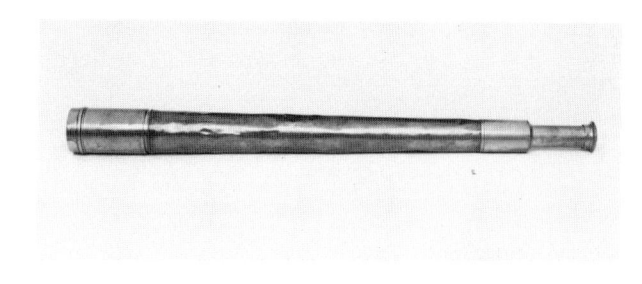

34. Refracting telescope: Adie & Wedderburn, Edinburgh, c.1890 (T1980.248)

2⅛″ aperture single-draw hand-held refracting telescope with doublet objective (damaged) and 2-component eyepiece and erecting assemblies; in brass (with leather-lined draw slide), with objective ray shade, the tapered barrel originally leather-covered. Engraved on the draw tube 'ADIE & WEDDERBURN /EDINBURGH' and 'BELL ROCK / LIGHTHOUSE'. Scratched assembly marks on draw tube components 'IV', on the objective mount 'II', and on the barrel under the end sleeve 'III'. Closure on external eye stop.

Length closed: 689mm.
Maximum tube diameter: 61mm.
Aperture: 53mm.

The objective is surface scratched and its edge is chipped; the objective mount is deformed. The barrel lacks its leather cover and is badly dented. Lacking the objective cover.

Probably by James Parkes & Son of Birmingham, cf. *Illustrated Trade Price List of Telescopes* (Birmingham, n.d.[1902]), 16, 'Captain's and Ship Telescopes, Best Make', no. 445. The Bell Rock Lighthouse, off the east coast of Scotland between Fife Ness and the Read Head of Angus, was built to mark the dangerous Inchcape Reef which claimed some 70 vessels in a storm in December 1799. Designed by Robert Stevenson (1772-1850), and constructed with enormous hardship and difficulty between 1807 and 1810, it still stands as a monument to its architect, who became engineer to the Northern Lighthouse Board (R.W. Munro, *Scottish Lighthouses* (Stornaway, 1979), 66-81; Keith Allerdyce and Evelyn M. Hood, *At Scotland's Edge: A Celebration of Two Hundred Years of the Lighthouse Service in Scotland and the Isle of Man* (Glasgow and London, 1986), 21-25, 118-119).

PATRICK ADIE

Patrick Adie was born in 1821, the fourth and youngest son of Alexander Adie.[1] At the customary age of fourteen he left the High School in Edinburgh and went for a short spell to gain mechanical experience in the workshops of Messrs. Milne & Son, gas engineers.[2] The following four years, apart from a six month break, were spent working under his father and brother in the shop in Princes Street. For part of 1842 he worked in Sir Thomas Makdougall Brisbane's magnetic and meteorological observatory at Makerstoun, near Kelso, under the management of John Allan Broun, then under John Welsh, who subsequently became Superintendent of Kew Observatory.[3] In July 1840 Alexander Adie reported that 'Patrick is now in the shop working[;] we shall soon see how he gets on.'[4] The answer was, very fast indeed, for after serving four years with Alexander and John, he sailed for London on 1 May 1844, at the age of twenty-three.[5]

On his arrival he set up as an optician and instrument maker, and was recorded at 1A Conduit Street, Regent Street in 1846, and at 14 Conduit Street in 1847. From 1848 to 1868 he was listed at 395 Strand, and between 1869 and 1873 he had two shops, one at 15 Pall Mall and the other at 29 Regent Street. In 1874 he gave up his Regent Street premises, but had a workshop at Broadway Works, Westminster.[6] By 1936, some considerable time after his death, the business was operating from 28 Medway Street, Horseferry Road; and by the time it made its final entry in the directories in 1942 it was as 'Patrick Adie Ltd.' at 45 Beaumont Road, London W.4. The following year the name was dropped.[7]

Patrick Adie's early independence can be attributed to a combination of mechanical precocity and an inherited capacity for hard work. As was the case with his great-uncle John Miller's move to London, the circumstances of his beginnings in 1844 remain obscure, but his father's wealth and connections with scientists and fellow-makers were probably helpful to him in setting up shop. Patrick later told his brother that 'at first his own hands work was worth 30/- per week',[8] and other evidence presented below gives an impression of success gathering momentum from these small beginnings. When his brother A.J. Adie was in London in April 1849 he reported that Patrick 'looks well, and not as if he were greatly troubled with the cares of the world'.[9]

At about this date Adie's talent in instrument work led to a renewal of the relationship with John Welsh, who moved from the privately-run Makerstoun to the government-sponsored Kew Observatory in 1850.[10] He was concerned with creating a standard for barometrical observation, and he seems to have co-operated with Adie in transforming by 1855 the then standard pattern of marine barometer into the so-called 'Kew type' which has formed the basis for marine instruments ever since.[11] It is interesting to note that Patrick's excellence lay in barometer design and construction, and if this can be attributed in part to training under his father and brother, he seems to have outstripped them both in his technical achievements in this specialisation. In one sense he was better placed than they to profit from important contacts and new developments in the field, but in another he was at a disadvantage as a young and relatively inexperienced maker in competition with long-established and prestigious instrument makers in the metropolis. His success is therefore all the more remarkable.

The first known improvement to the barometer which has been attributed to him, was his rack and pinion mechanism for adjusting the vernier scale on the Fortin barometer: this was perfected in about 1850 and was being copied in large numbers by London makers soon afterwards.[12] His association with Welsh was well established by summer 1852, for he constructed the meteorological instruments used in two balloon ascents by Welsh and a colleague in August and September 1852. These included a thermometer and dewpoint instrument and vacuum tubes for taking air samples at the greatest height, of 19,200 feet. Adie was on at least the first flight and was among the crew daguerreotyped by John Mayall as the basis for a woodblock in the *Illustrated London News*.[13]

Two years later Adie was at sea with Welsh in March 1854 testing three of his barometers, and again in May, this time with five.[14] At about this date he constructed a special adjustable-cistern barometer to be used as a standard for verifying marine and other fixed-cistern barometers at Kew Observatory.[15] But his glass blowing was not equal to that of Enrico Negretti, who produced the large-bore tube for the large standard at the Observatory in 1854; and, as noted above, it was Patrick's brother John whose method of cleaning the tube ensured eventual success.[16]

Adie's reputation as a barometer maker was therefore

MR. NICKLIN. MR. WELSH. MR. ADIE. MR. GREEN.
SCIENTIFIC BALLOON ASCENT FROM VAUXHALL GARDENS.—FROM A DAGUERREOTYPE BY MAYALL.

18 The first balloon ascent for scientific reasons, under the direction of the British Association for the Advancement of Science on 17 August 1852; woodblock engraving from the *Illustrated London News,* 4 September 1852

high by 1854, and in that year he was commissioned to produce fifty Fortin barometers for the United States Navy.[17] The indications are that Adie produced several types of barometer on a large scale and that these found their way to many parts of the world. The actual figures are not known, and the difficulty is increased by uncertainty both about dates and the registration system which Adie used to number his products. A fixed-cistern marine barometer, numbered 35, has been dated to 1855-56,[18] which places it after the order for fifty instruments of 1854, and anyhow seems *prima facie* too low a number for Adie to have reached after over ten years in business. A further puzzle is the Kew pattern barometer numbered 950 and dated to about 1903, and the Fortin barometer, number 1710, sent to America in 1878.[19] The explanation for these inconsistencies is either erroneous dating or more likely that Adie numbered different types of barometer in separate numerical sequences, not necessarily commencing with serial number '1'. Even so, in the absence of more details of individual series, his production run was clearly substantial. Some indication of the scale of his operation is given by the number of instruments Adie had under repair at 1 March 1862: fifty-eight barometers from the Board of Trade, compared with two at Elliott Brothers; and forty-eight from the Admiralty, compared with three at Casella. However, Negretti was completing an order for the Admiralty of forty-six barometers at this time, so that it is evident that these government contracts were the subject of fierce competition.[20] Whatever the explanation, Adie's specialisation in barometers kept him busy and perhaps necessitated the expansion of his premises in 1869. Large orders still came in: the American Navy's Signal Service imported ten large Fortin type barometers in 1878, of which number 1710 was one. Two of these instruments were so accurate as to be still in use as working standards in 1962.[21]

Patrick Adie was involved in the construction of other meteorological instruments besides barometers. In 1843 Francis Ronalds of Kew Observatory designed a rain and vapour gauge which was manufactured by Adie.[22] In 1851 John Welsh described a sliding rule for hygrometrical calculations to the British Association for the Advancement of Science, and concluded with the illuminating remark that 'it should be mentioned that the standard scales furnished to Mr Adie of London, the maker of the instrument, were divided at Kew by myself, with the aid of a dividing engine constructed by M. Perreaux of Paris'.[23] Another instrument was a cistern thermometer devised in 1871 by Sir Robert Christison Bt., the eminent toxicologist, to obtain deep water samples and temperature measurements simultaneously.[24]

Adie's success as a maker was already assured by the time he displayed seven examples of his barometers at the Special Loan Collection of Scientific Apparatus at the South Kensington Museum in 1876: one was of a type used in government meteorological observations and a second in the Medical Department of the War Office.[25] However, items from his workshop in the Museum's 1876 exhibition show that he produced a wider range of apparatus than the foregoing might suggest. In particular he specialised, like the Edinburgh business, in surveying instruments.

According to his obituarist he designed and supplied many of the instruments used by the engineers engaged in the great trigonometrical survey of India, and also in the construction of many railways, both in this country and abroad.[26] In fact, the only item by Adie which can with any certainty be given an Indian provenance is the tide gauge made along British Association recommendations, tested at Chatham, and subsequently brought out to India.[27] This is probably because Adie must have been acting as a subcontractor to one or other of the large instrument manufacturers who would have been supplying the Indian Service directly. The only trade catalogue by Adie traced so far, is an undated *Abridged Catalogue of Mathematical Instruments, &c. Manufactured by Adie, 395, Strand, (W.C.) London*. It was with other trade catalogues of other instrument makers, amongst papers belonging to the makers Thomas Cooke & Son of London and York, and this implies that either Cooke was acquiring material from Adie, or that Cooke was keeping an eye on his competitors.[28]

The catalogue can be dated from internal information to about 1865. Among the items offered for sale is the Kew pattern magnetograph, first described in 1859; Patrick Adie's patent theodolite level of 1859; his brother Richard Adie's new Hermetic barometer, first described in 1860; and his own patent telemeter of 1863; so the publication must be later than these dates. One item, unidentified so far, is described as 'Queen's College Galway Pattern [level], on 3 screw stage'. Adie appears to have had a strong connection with Queen's College, Galway, (now University College, Galway) one of the three university colleges founded by Act of Parliament in 1845, at which no religious tests were to apply.[29] In the Department of Civil Engineering there survive a number of instruments signed by Adie of London, among them three levels, each different from one another; also an alidade, a box sextant, a sextant, a transit theodolite, and an Everest theodolite.[30] Presumably Adie supplied these at the time when he was executing the Queen's College Galway level; the most likely candidate to have designed this was W.B. Blood, professor of civil engineering at Galway.[31]

Among Adie's twenty-one patents[32] was his theodolite level,[33] and he also produced a combination of the 'Y' pattern and 'Dumpy' levels,[34] and an unpatented 'Improved' level.[35] He also patented a new form of the tribrach stand for surveying instruments in 1876.[36] Another instrument, which was still not perfected and manufactured in 1876, although it had been provided

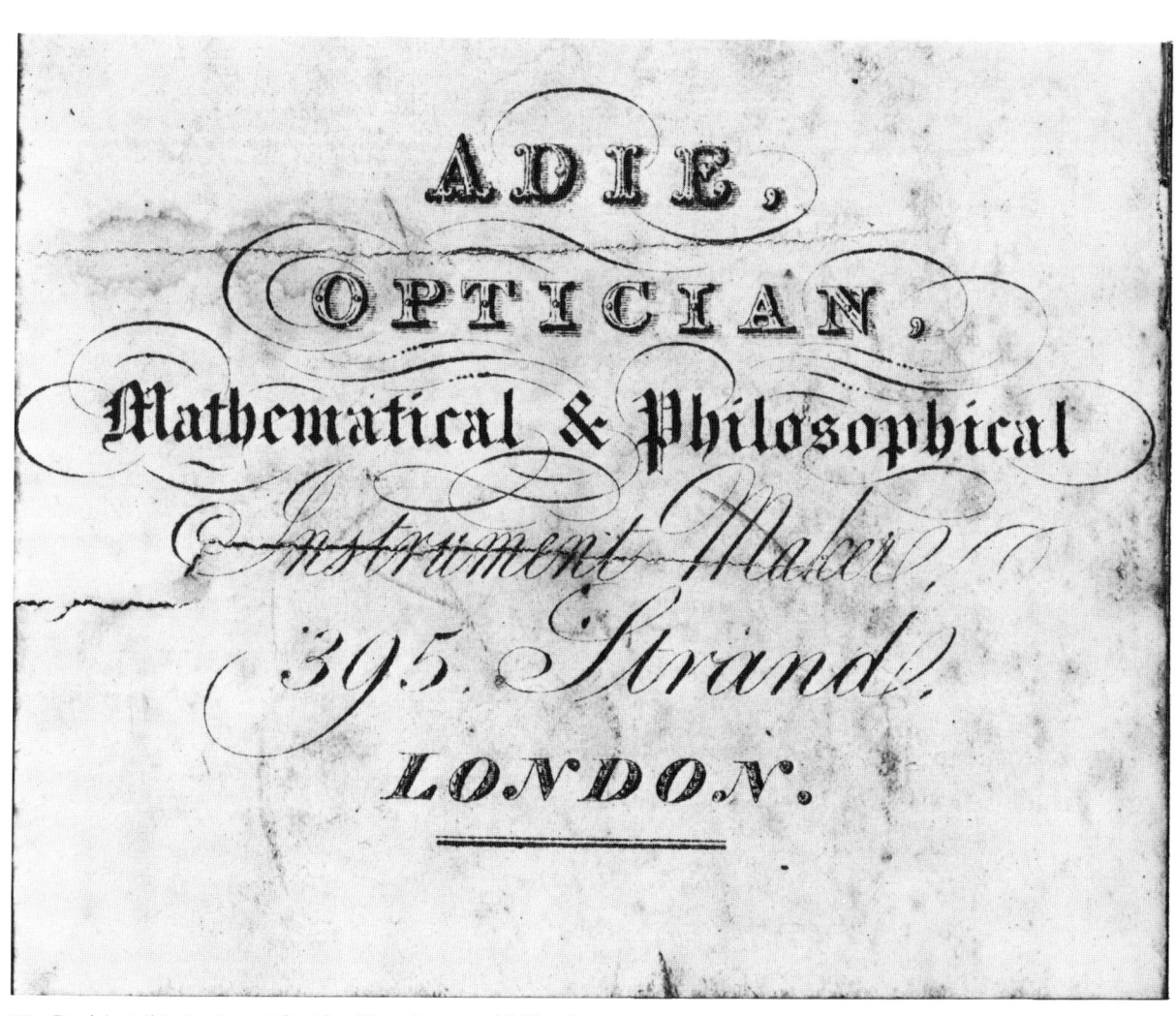

19 Patrick Adie's trade card for his address between 1848 and 1869.

20 Patrick Adie's trade card for his address from 1875. *Dundee Art Galleries and Museums*

21 Patrick Adie (1821-1886). *Trustees of the Science Museum, London*

with some form of patent protection, was his double telescope sextant.[37] Two further patents created his telemeter of 1863, a form of rangefinder;[38] this was the first English coincidence rangefinder[39] but in the long term proved to be insufficiently robust in the theatre of war.[40] He also obtained provisional protection for spectacles which enabled the wearer to 'see behind'.[41]

Not surprisingly Adie's inventiveness did not stop at more conventional instruments. He devised apparatus for ascertaining the tearing strain of cement and other substances for the Metropolitan Board of Works in London, and patented it in 1880.[42] He displayed it at the International Inventions exhibition in London in 1885 along with astronomical and surveying instruments, machinery for edging grass and velocipedes: the cement testing machine won a silver medal, and the instruments a bronze.[43] These were not his only exhibition medals. Although his obituarist claimed for him medals at the Crystal Palace in 1851, in fact he did not exhibit there, although he was awarded a bronze medal at the Paris exhibition in 1855.[44] At the 1862 exhibition he was also successful, gaining a medal for the 'ingenuity and excellence of construction of sextants, telemeter &c'.[45]

On one occasion his patenting took him to law, when one William Clark challenged his 1866 patent on the means and machinery for clipping horses and other animals: Clark felt that Adie had infringed another patent, to which Clark had obtained the rights, but both his claim and subsequent appeal were dismissed.[46]

Another group of instruments which Adie was supplying either with his own signature to clients (often institutions or governmental establishments), or to other wholesalers, were pieces of geomagnetic apparatus. In 1859 John Welsh described his photographic-recording magnetograph, (subsequently known as the Kew pattern) which was made by Adie and appears to have been reasonably popular; two surviving examples were used in Scotland and the United States respectively.[47] The same instrument was also offered for sale by J.J. Hicks, which implies that Adie was supplying at least geomagnetic apparatus through the instrument trade; and although difficult to prove, it is likely that he marketed his surveying equipment in a similar fashion.[48] In 1861, John Allan Broun, by now Director of the Trevandrum Observatory in India, compared magnetic instruments made by Adie with those produced by Thomas and Howard Grubb of Dublin.[49] At the Special Loan Exhibition at South Kensington in 1876 four magnetic instruments made by Adie - a magnetometer, a declination magnetometer, a vertical force magnetometer and automatic reading apparatus - were displayed by Dr H.C. Vogel, as the 'set of instruments made on the Kew Pattern, by P. Adie, for the Physical Observatory, at Potsdam'. Another instrument was supplied to the observatory at Nice, in the South of France; others were sold to Portugal.[50] Doubtless his interest in magnetic instrumentation dated, like that in meteorological equipment, from his youthful days at Makerstoun.

Patrick Adie appears to have been married twice; firstly to Kate Mitchell, by whom he had one son, Alexander James, about whom nothing further is known. In 1858 he married Clementina Hellaby, and they had four sons and three daughters; the sons all led brilliant undergraduate careers at Cambridge after their father's death; none of them married, and none went into the business.[51] Patrick Adie suffered from bronchitis and heart disease for the last ten years of his life,[52] for most of which time he lived at Worton Hall, Isleworth. However, after his death on 18 May 1886 his widow sold up and moved to Cambridge to be close to her sons. A family tradition names Patrick Adie's foreman Lloyd as the man who continued the business into the new century.[53]

REFERENCES

1. Obituary of Patrick Adie, *Minutes of Proceedings of the Institution of Civil Engineers* 86 (1886), 367-368; however, a family tree, in possession of one of Adie's descendants, gives his birth date as 26 June 1822. We are grateful to Captain A.H. Swann, R.N., for his help and encouragement with the genealogical details of the Adie family.

2. *Ibid.* An account of this firm is given in [Anon.], 'British Engineers No. 12. Messrs. James Milne & Son, Engineers, Brassfounders, &c., Milton House Works, Edinburgh', *The Mercantile Age* 12 (1887), 354.

3. *Ibid.*; 'During term-days, Mr Russell was assisted by Mr P. Adie of Edinburgh, Mr Hogg of Kelso, and myself. After the April term 1842, Mr Dods, teacher of Makerstoun parish school, replaced Mr Russell in the term observations; and after the term 1842, Mr Chisholm, teacher of Maxton parish school, replaced Mr Adie.' (John Allan Broun, 'Observations in Magnetism and Meteorology, made at Makerstoun in Scotland, in the Observatory of General Sir T.M. Brisbane, Bart., ... 1841 and 1842 ...', *Transactions of the Royal Society of Edinburgh* 17 (1845), xi). John Welsh arrived at Makerstoun in December 1842 (P.T. H[artog], 'John Welsh', in S. Lee (ed.), *Dictionary of National Biography* LX (London, 1899), 239-240). For a history of Kew Observatory, see the articles celebrating its bicentenary, published in *Meteorological Magazine* 98 (1969), 161-196. We are grateful to Dr Anita McConnell for this last reference.

4. Scottish Record Office, GD 76/457/59, A. Adie to A.J.Adie, 3 July 1840. Patrick is recorded aged fifteen the following year in the census record of the family at Canaan Cottage (General Register Office (Scotland), Census of 1841 (Edinburgh, Morningside), 685²/215A).

5. S.R.O., GD 76/437/80, John Adie to A.J. Adie, 1 May 1844.

6. *Kelly's London Post Office Directory* 1844-1941, *passim*. We are grateful to Dr Anita McConnell for this information.

7. *Ryland's Coal, Iron, Steel, Tinplate, Metal, Engineering, Foundry, Hardware and Allied Trades Directory ... 1936* 23rd edition (London, 1936); *Post Office London Directory* 1942. We are grateful to Ian Carter of the Science Museum Library, London for this information.

8. S.R.O., GD 76/462/2, A.J. Adie to his wife Louisa ('Loudie'), 4 April 1852.

9. S.R.O., GD 76/462/1, A.J. Adie to same, 22 April 1849.

10. *Meteorological Magazine*, *op. cit.* (3); obituary of John Welsh, *Proceedings of the Royal Society of London* 10 (1860), xxiv.

11. W.E.K. Middleton, *The History of the Barometer* (Baltimore, Maryland, 1969), 164.

12. *Ibid.*, 199-200.

13. *Illustrated London News*, 4 September 1852, 192.

14. Middleton, *op. cit.* (11), 165, citing *Report of the British Association ... 1854* (London, 1855), xxviii. The first voyage was a return trip to Leith, the second to Jersey. For Adie's description of his instrument, which he was willing to supply for £3 15s 6d in the London area (including the cost of packing and verification at Kew Observatory) see *ibid.*, xxx-xxxi, xli).

15. Middleton, *op. cit.* (11), 263, citing Welsh, 'Account of the Construction of a Standard Barometer, and Description of the Apparatus and Processes employed in the Verification of Barometers at Kew Observatory', *Philosophical Transactions* 146 (1856), 507-514.

16. *Ibid.*, 245.

17. *Ibid.*, 348. The author's doubts about whether the order reached America seem unfounded on the evidence he cites (*ibid.*, 348-349).

18. *Ibid.*, 164 n., 462: example in the Peabody Museum, Salem, Massachusetts (inventory M.10485).

19. *Ibid.*, 446 : example in the Meteorological Office, Bracknell. *Ibid.*, 350.

20. *Report of the Meteorological Department of the Board of Trade* (London, 1862), Appendix 4 'Return of Instruments', 304. We are grateful to Dr Anita McConnell for this reference.

21. Middleton, *op. cit.* (11), 350. The author cites other examples of the constancy of Adie's instruments, and their use in other locations e.g. Russia and Hamburg (*ibid.*, 211, 237).

22. Francis Ronalds, 'Report concerning the Observatory of the British Association, at Kew, from April the 1st, 1843, to July the 31st, 1844', *Report of the British Association ... 1844* (London, 1845), 128-129. An example of this instrument is in the Science Museum, London, inventory 1876-797; it was probably one of these which was examined at the Eastern Cemetery, Dundee in September 1867 and mentioned by J. Glaisher, *et al.*, 'Second Report of the Rainfall Committee', *Report of the British Association ... 1867* (London, 1868), 464-465.

23. John Welsh, 'Description of a Sliding-Rule for Hygrometrical Calculations', *Report of the British Association ... 1851* (London, 1852), 43.

24. Robert Christison, 'Observations on the Fresh Waters of Scotland', *Proceedings of the Royal Society of Edinburgh* 7 (1871), 570. An example of this instrument made by Adie and incorporating a thermometer by Negretti & Zambra is in the Science Museum, London, inventory 1913-163.

25. *Catalogue of the Special Loan Collection of Scientific Apparatus at the South Kensington Museum* 3rd edition (London, 1877), 677.

26. Obituary, *op. cit.* (1). The claim to have supplied instruments to the Survey of India is difficult to substantiate, for the *General Reports ...* tend to emphasise the primary instruments. However, to give an idea of the scale of the operation, it is perhaps worth quoting the following from the Mathematical Instrument Department: 'During the year 1880-81, the stock of instruments in the depot was increased as follows, *viz.*, about 7,540 instruments ... were obtained from England; 500 were purchased locally ... nearly 9,950 were manufactured in the workshop ...; nearly 5,970 instruments were received by interdepartmental exchange ... The number of instruments issued from the stock amounted to 20,158...' (J.T. Walker (ed.), *General Report of the Operations of the Survey of India comprising the Great Trigonometrical, the Topographical and the Revenue Surveys under the Government of India during 1880-81* (Dehra Dun, 1880), 52). Amongst these huge numbers of instruments, it is possible that items from Patrick Adie's establishment found their way to the Subcontinent.

27. Clements R. Markham, *A Memoir on the Indian Surveys* 2nd edition, (London, 1878), 318; a later tide gauge made by the firm was installed at Chelsea Bridge, London to collect data for the Chief Engineer's Department of the London County Council (now in the Science Museum, inventory 1980-559; described in A. McConnell, *Geophysics & Geomagnetism Catalogue of the Science Museum Collection* (London,1986), 24).

28. Vickers Collection, Vickers Instruments, York. The Adie catalogue was with the following: T. Cooke, *Telescope, Lathe and Clock catalogue* (1869); Richard C. Millar, *A Description of the Apomecometer* (1869); Thomas Ross, *Catalogue of Microscopes, Telescopes, Photographic Lenses and Optical Instruments* (1864); Powell & Lealand, *Catalogue of Optical Instruments* (1867); J.H. Dallmeyer, *Catalogue of Telescopes, Microscopes and Photographic Lenses* (1865); Troughton & Simms, *Catalogue of Instruments* (1864); and *American Tools from R. Lloyd, Birmingham agent for Darling, Brown and Sharpe, U.S.A.* (n.d.). We are grateful to Mrs Alison Breck, Archivist, for information about the catalogue.

29. Gearoid O Tuathaigh, *Ireland before the Famine 1789-1848* (Dublin, 1972), 193-194.

30. There is also an eidograph, and a station pointer, both signed by Adie, Edinburgh, and a miner's dial, signed by Adie & Son, Edinburgh. This is the largest 'collection' of instruments from the Adie business in Ireland, and we would suggest that Patrick Adie was retailing standard items from the Edinburgh firm when supplying specialist pieces from London. We are grateful to Dr Charles Mollan of the Royal Dublin Society for this information; and to Dr T. O'Connor of University College, Galway, for drawing the items to our attention.

31. He was the author of a single paper, written jointly with W.T. Doyne, 'An Investigation of the Strain upon Diagonals of Lattice Beams, with the Resulting Formulae', *Journal of the Franklin Institute of Philadelphia* 25 (1853), 224-228, 249-299.

32. Those which had nothing to do with instruments were as follows: UK patent 1747, 11 July 1861: apparatus in connection with railway carriage buffers for preventing damage in cases of railway collision; 2796, 30 October 1866: means and machinery for clipping horses and other animals; 1053, 27 March 1868: means and apparatus for shearing sheep and clipping horses and other animals; 3209, 6 December 1870: lamps; 60, 10 January 1871: lamps; 3397, 15 December 1871: oil lanterns; 421, 4 February 1875: clipping animals; 2640, 24 July 1875: clipping animals; 3714, 16 September 1879: making iron and steel; 2455, 17 June 1880: testing cements, etc.; 3185, 4 August 1880: edging grass and rapid cutting with shears or scissors; 5211, 13 December 1880: lighting mines; 3912, 16 August 1882 (with W. Simpson): strengthening and checking electric currents; 5569, 29 November 1883: velocipedes; 11,261, 14 August 1884: velocipedes; 1440, 1 February 1886: tempering steel. At the time of his death he was working on perfecting another idea for a patent: 'it consists in the employment of corrugated steel belting, in lieu of leather, which he believed would effect a large saving both in power and cost' (Obituary, *op. cit.* (1)).

33. *Catalogue ..., op. cit.* (25), 751. UK patent 326, 4 February 1859: apparatus for taking levels and measuring angles. An example of this instrument is in the Royal Museum of Scotland (NMS T1987.112); it was offered for sale at Christie's South Kensington, 4 June 1987, Lot 267. This instrument was described and illustrated in *The Builder* 52 (1887), 552-553; and further discussed in Arthur T. Walmisley, 'Modern Surveying Instruments', *Transactions of the Surveyors' Institution* 38 (1906), 136. We are grateful to Ian Carter, Science Museum Library, for this information.

34. *Catalogue ..., op. cit.* (25), 751.

35. An example of this, signed 'Adies Improved level No 40 / Manchester & Leeds Railway Comp^y / J. CASARTELLI MANCHESTER /ADIE LONDON' is in the collection of the Whipple Museum of the History of Science, Cambridge (inventory 2555), and was offered for sale at Sotheby's Belgravia,

30 November 1979, Lot 212, and again at the same saleroom, 9 May 1980, Lot 162; it is described in Olivia Brown, *Catalogue 1: Surveying* (Cambridge, 1982), item 148; and illustrated in G. L'E. Turner, *Nineteenth Century Scientific Instruments* (London, 1983), 254 and in J. A. Bennett, *The Divided Circle: A History of Instruments for Astronomy, Navigation and Surveying* (Oxford, 1987), 220.

36. UK patent 2697, 30 June 1876.

37. *Catalogue ..., op. cit.* (25), 756. This is described as 'patented by Mr Adie some years ago', but presumably this was either a temporary expediency, or the 'difficulties' mentioned in the catalogue entry proved insurmountable; full patent rights for this instrument were not obtained.

38. UK patent 357, 10 February 1860 and 608, 4 March 1863: means and apparatus for measuring angular and actual distances. The Adie telemeter, or rangefinder was extensively used by the Admiralty until superseded by the Barr & Stroud pattern. Examples are in the Science Museum, London, (inventory 1909-139, used on H.M.S. *Triton* between 1885 and 1904 during the survey of the east coast of England; and inventory 1948-367, which is unsigned but the case has a trade card for Adie). See also Patrick Adie, 'Adie's Telemeter', *United Service Institution Journal* 24 (1881), 230-233, which has some interesting discussion following the delivery of the paper. We are grateful to Dr Anita McConnell for locating this for us.

39. F.J. Cheshire, *The Modern Rangefinder* (London, 1917), 11.

40. Michael Moss and Iain Russell, *Range and Vision: The First Hundred Years of Barr & Stroud* (Edinburgh, 1988), 20.

41. UK patent 4365, 7 October 1881.

42. *Catalogue..., op. cit.* (25), 444. UK patent 2455, 17 June 1880.

43. *International Inventions Exhibition 1885 Official Catalogue* (London, 1885), 46, 86, 267; *Awards of the International Juries of the International Inventions Exhibition Division I Inventions* (London, 1885), 1. From 1883 until just before the Second World War, the cement testing machinery figured in the firm's description (*Kelly's London Post Office Directory, passim*).

44. Obituary, *op. cit.* (1); 'Class VIII Section 3 Optical Instruments and Apparatus of all Kinds used in measuring space ... ADIE PATRICK, 395 Strand London. Mathematical, optical and meteorological instruments. Bronze medal.' *Paris Universal Exhibition 1855. Catalogue of the Works Exhibited in the British Section of the Exhibition ...* (London, 1855), 51. His barometers were discussed by the jury: 'Médailles de 2e classe ...', in *Exposition Universelle de 1855. Rapports du Jury Mixte International ...* Tome I (Paris, 1856), 433.

45. *International Exhibition 1862 Reports by the Juries ...* (London, 1863), Class XIII, 98. Descriptions of the instruments which he exhibited - surveying instruments, barometers and a hygrometer - are also given (*ibid.*, 12, 34, 39). On at least three occasions his instruments were displayed at exhibitions sponsored by the Royal Meteorological Society: a station barometer in 1880 (*Quarterly Journal of the Royal Meteorological Society* 6 (1880), 162); a Fortin barometer used in 1857 by Palliser in the British North American Boundary Expedition (*ibid.* 9 (1883), 174); and a Fortin barometer, a Kew barometer, a Kew Marine barometer and a siphon barometer (*ibid.* 12 (1886), 200-203). All Portuguese meteorological stations had 'baromètres, mercure, système Adie' by 1875 (*Compte-rendu du II Congrès International des Sciences Géographiques II* (Paris, 1875), 274). Adie also exhibited instruments at the Third International Geographical Congress in Venice in 1881 (*Terzo Congresso Geografico Internazionale* (Venice, 1881), 213). We are grateful to Dr Anita McConnell for all these references.

46. *Law Reports: Appeal cases before the House of Lords ... vol.2 1876-7* (London, 1877), 315-343; 423-438. We are grateful to Malcolm Scott, advocate, for his advice about this reference.

47. Balfour Stewart, 'An Account of the Construction of Self-Recording Magnetograph at present in operation at the Kew Observatory of the British Association', *Report of the British Association ... 1859* (London, 1860), 200-220: 'The late Mr Welsh ... applied himself with much zeal to the task of constructing these magnetographs, and devised a plan which was transmitted to Mr Adie, optician, 395 Strand, who undertook to make the instruments. These were completed by Mr Adie in a satisfactory manner, and were in operation in July 1857.' An example of this instrument, used at Eskdalemuir Observatory until 1981, and with Adie's signature on it, is now in the Royal Museum of Scotland, NMS T1983.288. Another is in the Smithsonian Institution, Washington D.C., (inventory USC&GS 316525) and is illustrated and described, with its working history, in Robert P. Multhauf and Gregory Good, *A Brief History of Geomagnetism and A Catalog of the Collections of the National Museum of American History* (Washington, D.C., 1987), 29 and 50.

48. James J. Hicks, *Illustrated & Descriptive Wholesale Catalogue of Standard, Self-Recording and other Meteorological Instruments...* (London, n.d. [c.1880]).

49. John Allan Broun, 'The Bifilar Magnetometer, its Errors and Corrections, including the Determination of the Temperature Coefficient for the Bifilar employed in the Colonial Observatories', *Transactions of the Royal Society of Edinburgh* 22 (1861), 467-489.

50. *Catalogue ..., op. cit.* (25), 677; M. Perrotin, *Annales de L'Observatoire de Nice* I (Paris, 1899), 116: 'le magnétomètre enregistreur d'Adie'; the Portuguese instruments were displayed in Paris in 1875 (*Compte-rendu ..., op. cit.* (45), 274) and described in Fradesso da Silveira, *Relatorio do Servico do Observation do Infante Don Luis, no anno meteorologico 1870-71* (Lisbon, 1871). We are grateful to Dr Anita McConnel for these references.

51. Information from Captain A.H. Swann, R.N. The children were: Richard Haliburton, born about 1864, Fellow of St. John's College, Cambridge; Patrick, born 1868, Lloyd's insurance agent in Buenos Aires; Walter Sibbald, 1873-1956, Senior Wrangler and rowing blue, Indian Civil Service; Clement James Mellish, c.1875-1954, soliciter and housemaster at Eton College; Marion, born 1860, married Jason Gurney and had three children, who all died without issue; Louise Jane, c.1866-1954; and Elizabeth, 1862-1940, who became the second wife of William Laidlaw Purves and was grandmother to Captain Swann.

52. Obituary, *op. cit.* (1).

53. Information from Captain Swann.

35. Artificial horizon: Adie, London, c.1860 (T1980.165)

Mercury artificial horizon, Ordnance pattern, in black-painted brass and iron. Rectangular cast iron trough on 3 feet with covered corner and pouring hole; iron mercury bottle with plug and a pouring nozzle which doubles as a funnel. With fitted case containing trade card for Adie, 395 Strand, London.

Cover size: 167 x 93 x 112mm.
Case size: 195 x 150 x 130mm.

Although the mercury bottle is well restrained in the case, it is held in position in its mount by the side of the brass glazed cover: both sides of the cover have become bowed by blows from the bottle when the case has been jolted when on edge. The case has lost an inset name plate on the lid and most of a MS label inside the lid.

Artificial horizons were offered by Adie for from £2 2s to £5 in the only catalogue of his which has so far been recorded: *Abridged Catalogue of Mathematicial Instruments &c. manufactured by Adie, 395 Strand, (W.C.) London* (London, n.d.[c.1865]), 3. James Parkes & Son of Birmingham illustrate an artificial horizon described as the 'ordnance pattern': *Wholesale catalogue...* (Birmingham, n.d.[1867]), 26.

36. Current meter: Adie, London, c.1890 (T1980.206)

Recording water flow meter in brass, bronze and oxidised brass. Engraved on the frame 'ADIE. / LONDON.' and wheel components stamped with the assembly mark '6'. Driven by a 2-bladed vane on a steel arbor with a worm gear acting on the toothed perimeters of two divided dials. The rear wheel has one more tooth in its circumference than the forward wheel, which is pierced to expose the scale on the rear wheel; the slow differential rotation of the wheels is used to indicate the water flow in a fixed period. The forward dial divided in feet [0]-220; the rear wheel divided to read up to 10 miles, subdivided in miles and furlongs (of 220ft.). A screw-adjusted sprung linkage releases the wheels and brings the worm drive into engagement when a cord is pulled from the surface. Clamped mounting socket and large tail vane with engine spotting.

Vane size: 105mm.
Recording wheel diameter: 73mm.
Overall length: 325mm.

When acquired by Mr Frank the instrument had a fitted case which was not located in 1980.

The instrument is a development of Joseph Saxton's water current meter, which was first described in the Adelaide Gallery's *Magazine of Popular Science* 1 (1836), 108-112; Saxton's work is discussed by Arthur H. Frazier, *Water Current Meters in the Smithsonian Collections of the National Museum of History and Technology* (Washington, D.C., 1974), 51-55, where he suggests that Watkins & Hill of London, and their successors Elliott & Sons, both manufactured these instruments to Saxton's design. The mechanism predates Saxton's use: it was for example employed by Patrick Adie's father, Alexander, and James Howden in the commercial form of James Hunter's 1821 odometer, discussed in the text above, and was credited by David Brewster to Francis Wollaston (*Edinburgh Encyclopaedia* (Edinburgh, 1830), XV, 449). This example resembles no. 539, 'Current Meter, for use in small rivers and streams, to show the rate of flow of tide or number of gallons flowing from any reservoir or vessel... £4 4 0' in James J. Hicks, *Illustrated & Descriptive Wholesale Catalogue of Standard, Self-Recording, and other Meteorological Instruments ...* (London, n.d.[c.1880]), 114. Another London instrument maker, Louis Casella, also offered a similar instrument as no. 518, 'Current Meter, for showing the rate of flow of tide in any stream or river, and the amount in gallons per hour flowing off... £5 10 0': L. Casella, *An Illustrated and Descriptive Catalogue of Surveying, Philosophical, Mathematical, Optical, Photographic and Standard Meteorological Instruments ...* (London, 1871), 71-72. A closely similar example signed 'Elliot [*sic*] Bros., London' was offered by Tesseract, Catalog B (Fall 1982), item 62.

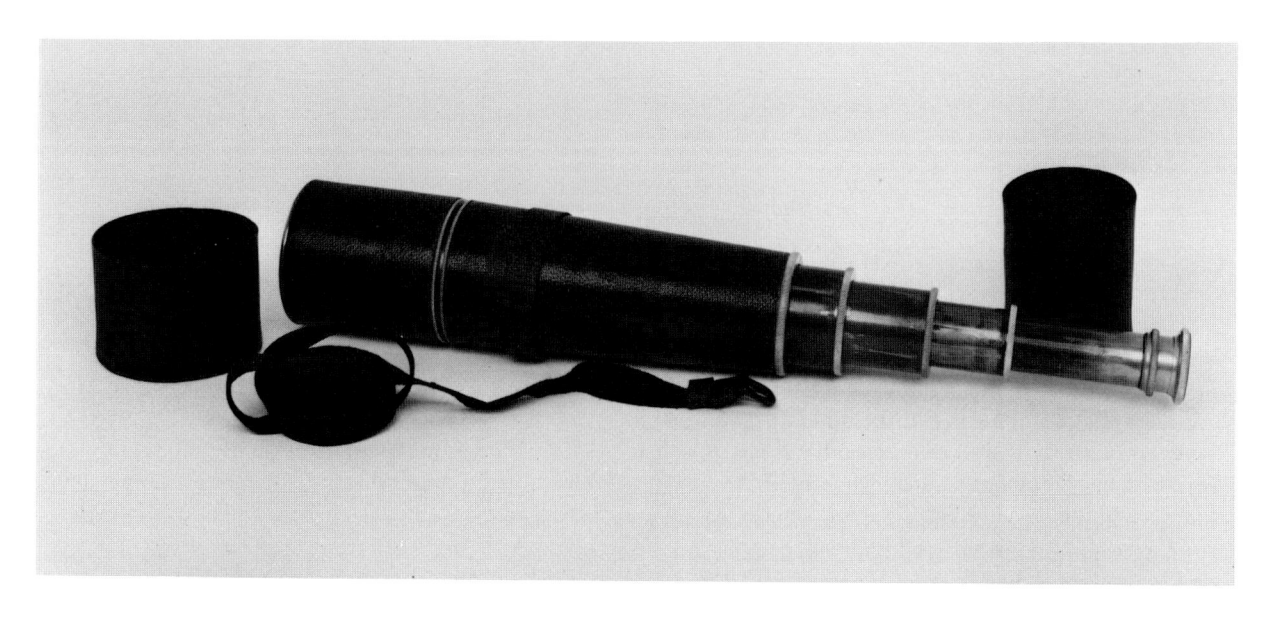

37. Refracting telescope: P. Adie, London, c.1920 (T1980.270)

3″ aperture 5-draw refracting telescope, intended for tripod use, with doublet objective and variable power eyepiece, comprising 2-component eyepiece and erecting assemblies of variable separation calibrated from 35 to 50 times; in oxidised brass with lubricated aluminium ferrules, and with leather-covered objective ray shade and tapered barrel. Engraved on the second draw tube 'P. Adie, / Broadway Works, / Westminster.' Scratched assembly mark 'II' on the second draw tube. Leather end-caps and strap.

Length closed: 398mm.
Maximum tube diameter: 95mm.
Aperture: 79mm.

When acquired by Mr Frank the telescope had a tripod which was not located in 1980.

GEORGE HUTCHISON

George Hutchison was an apprentice to the Adie business,[1] but with the ending of the family connections towards the end of the nineteenth century a number of workers left to set up on their own; Hutchison had done so by 1888,[2] when he first advertised in the street directory as an optician and 'manufacturer of surveyor's and architect's instruments'.[3] Between 1888 and 1907 his business was at 16 Teviot Place: in 1908 it moved round the corner to 18 Forrest Road, where it remained until at least 1967 when the firm disappeared from the street directories.[4] Between 1908 and 1916, Hutchison's shop was next door to the Edinburgh branch of W. Watson & Sons Ltd., the London microscope makers, who had established their premises at 16 Forrest Road in 1899, presumably to exploit the expanding student market of the adjacent Edinburgh Medical School.[5]

In 1895 Hutchison described himself as an 'ophthalmic optician and mathematical instrument maker'; in 1903 this changed to 'optician and mathematical surveyor, instrument maker', and two years later to 'ophthalmic optician, mathematical and surveyor's instrument maker'. In 1942 the firm became G. Hutchison & Sons, 'opticians and surveying instrument makers',[6] continued by the founder's two sons: 'George Hutchison practised as an ophthalmic optician, his brother John made and repaired surveying instruments'.[7] According to a trade card from this period, the firm advertised as 'DUMPY LEVEL MANUFACTURERS New and Second-hand Instruments always in stock'.[8] John R. Hutchison was a Fellow of the British Ophthalmic Association, and was involved in the Scottish Association; he was also interested in the history of scientific instruments[9] and to some extent a collector himself.[10]

The business and premises, together apparently with residual stock, were taken over in 1968 by the Glasgow-based optical firm Charles Frank Ltd.

22 George Hutchison's trade card for his address between 1888 and 1907

G. HUTCHISON,

Optician.

Manufacturer of Surveyor's and Architect's Instruments,

16 TEVIOT PLACE,

EDINBURGH.

> **DUMPY LEVEL MANUFACTURERS**
>
> New and Second-hand Instruments always in stock
>
> Send to us for Repairing and Adjusting
>
> **G. HUTCHISON & SONS, Opticians**
> 18 FORREST ROAD ——————— EDINBURGH, 1
> ESTABLISHED 1886   'PHONE 22000

23 G. Hutchison & Sons' trade card for their address from 1942

REFERENCES

1. John R. Hutchison, 'Notable Opticians of Auld Reekie', *Ninth Annual Conference [of] The Scottish Optical Practitioners ...* (Edinburgh, 1939), 17.

2. *Edinburgh Post Office Directory* 1888; D.J. Bryden, *Scottish Scientific Instrument-Makers 1600-1900* (Edinburgh, 1972), 50. Hutchison, *op. cit.* (1), 17 gives the date as 1886, but this has not been confirmed.

3. Trade card with an unsigned boxed level in the Royal Museum of Scotland, NMS T1978.97. There is another small level, signed 'G. Hutchison, Edinburgh' in a box with a trade label marked 'Repaired and adjusted by G. Hutchison ... June 1920' (NMS T1967.124); the Museum also has a compass marked 'HUTCHISON MAKER EDINBURGH' (NMS T1974.317).

4. *Edinburgh Post Office Directory* 1888-1967, *passim*.

5. *Ibid.* 1899-1916.

6. *Ibid. passim*.

7. Bryden, *op. cit.* (2), 27 n.132.

8. Trade card on a boxed level (the level signed 'Kelvin & James White Ltd'), in the Royal Museum of Scotland NMS T1976.80.

9. Although his comments must be treated with some caution where dates and details are concerned, these are of interest in preserving some of his father's recollections of the Adie firm.

10. Among items which have come to the Royal Museum of Scotland which were from Hutchison's collection are several by the Adie firm.

38. Microscope: W. Watson & Sons, London, c. 1905
(T1980.241)

Compound microscope, 'Edinburgh Students' Microscope 'H' model, in brass, bronze and oxidised brass, on inclining pillar and tripod base. Engraved 'W. WATSON & SONS / 313 High Holborn /LONDON / 8527' on the rear leg, and 'BRANCH / 16 FORREST ROAD, EDINBURGH' on a cross bar that restricts forward rotation of the instrument. Single-draw body, the draw tube divided in mm. to indicate the length of the body tube, and graduated in cms. 16-25, attached to the limb through a rack and pinion coarse focus adjustment and a spring-loaded slide advanced by an arm within the limb operated by a screw at the top of the pillar, divided 0-[10]. The pillar pivotted on trunions to the base, with the stage attached to the lower end of the pillar, and with a plane/concave mirror on a sleeve over a swinging tail under the stage. The mechanical stage with slow-motion adjustment in 2 directions and manual rotation of a rectangular stage plate with spring clips. Parachromatic condenser with iris diaphragm and filter carrier in a centering sub-stage mount with rack and pinion motion on a separate slide attached to the stage. With 4 objectives (1 Watson and 3 Leitz) and 2 eyepieces, and fitted with a rotating triple nosepiece by R. & J. Beck Ltd. With fitted case; a modern name plate in the door marked 'G. HUTCHISON & SONS / OPTICIANS / EDINBURGH' and '18 FORREST RD., (EST. 1886)'.

Body length (including Beck nosepiece): 154mm.
Body diameter: 38mm.
Height: 279mm.
Base size: 181 x 171mm.
Case size: 250 x 225 x 330mm.

Provenance: Understood to have been in the residual stock of G. Hutchison & Sons when the firm was acquired by Mr Frank in 1968. Watson's 'Edinburgh Student's Microscope', which incorporated the novel feature of a fine focus adjusting lever moving the whole body of the instrument, was produced in 1889 (*English Mechanic* 49 (1889), 471), and in a description later that year it was noted that it had been made 'on lines suggested by Dr. Edington, Lecturer on Bacteriology at Edinburgh University' (*Journal of the Royal Microscopical Society* 9 (1889), 802). Alexander Edington (d. 1928) was assistant to John Chiene, Professor of Surgery at Edinburgh, and was subsequently appointed to the Chair of Comparative Pathology at the Veterinary College, Edinburgh; he left Britain in 1891 to become Colonial Bacteriologist to the Cape Government (*Proceedings of the Royal Society of Edinburgh* 48 (1927-28), 227). Watsons later claimed that 'Dr. Henri Van Heurck, of Antwerp, the doyen of Continental microscopists, ... was so impressed with the design of the Edinburgh Student Microscope that he used it as the basis for the much more ambitious model which bears his name' (*Watson's Microscope Record* No. 41 (May 1937), 10). Van Heurck described Watson's instrument in 'La Microscope Anglo-Continental ou Microscope d'Etudiant de M. Watson and Sons', *Journal de Micrographie* 11 (1888), 314-318. The instrument was produced in a succession of versions designated by the letters A to H (of which only B, D, F and H had tripod as opposed to horseshoe stands), and the most complete version was the H which had a built-in mechanical stage (*Watson's Microscope Record* No. 41 (May 1937), 10). The H model is referred to in April 1902 in *Journal of the Royal Microscopical Society* 22 (1902), 267. Van Heurck's microscope was made by Watson to his specification, and the B version of this uses the same casting for the stand as the Frank instrument (*ibid.* 13 (1893), 92). The parachromatic condenser was described in 1900 (*ibid.* 20 (1900), 119-120). The 'Edinburgh H' enjoyed considerable success, and Watsons claimed for it in 1929 that it 'has had a more lasting popularity than any microscope of to-day and, we are assured, will outlive many of its present-day competitors' (*Watson's Microscope Record* No. 17 (May 1929), 24). W. Watson founded his business in 1837, and it became W. Watson & Son 30 years later. From 1882 until 1908, when it became a limited company, the name was W. Watson & Sons; the firm moved to 313 High Holborn, London in 1861 (*Watson Centenary 1837-1937* (London, 1937), vi).

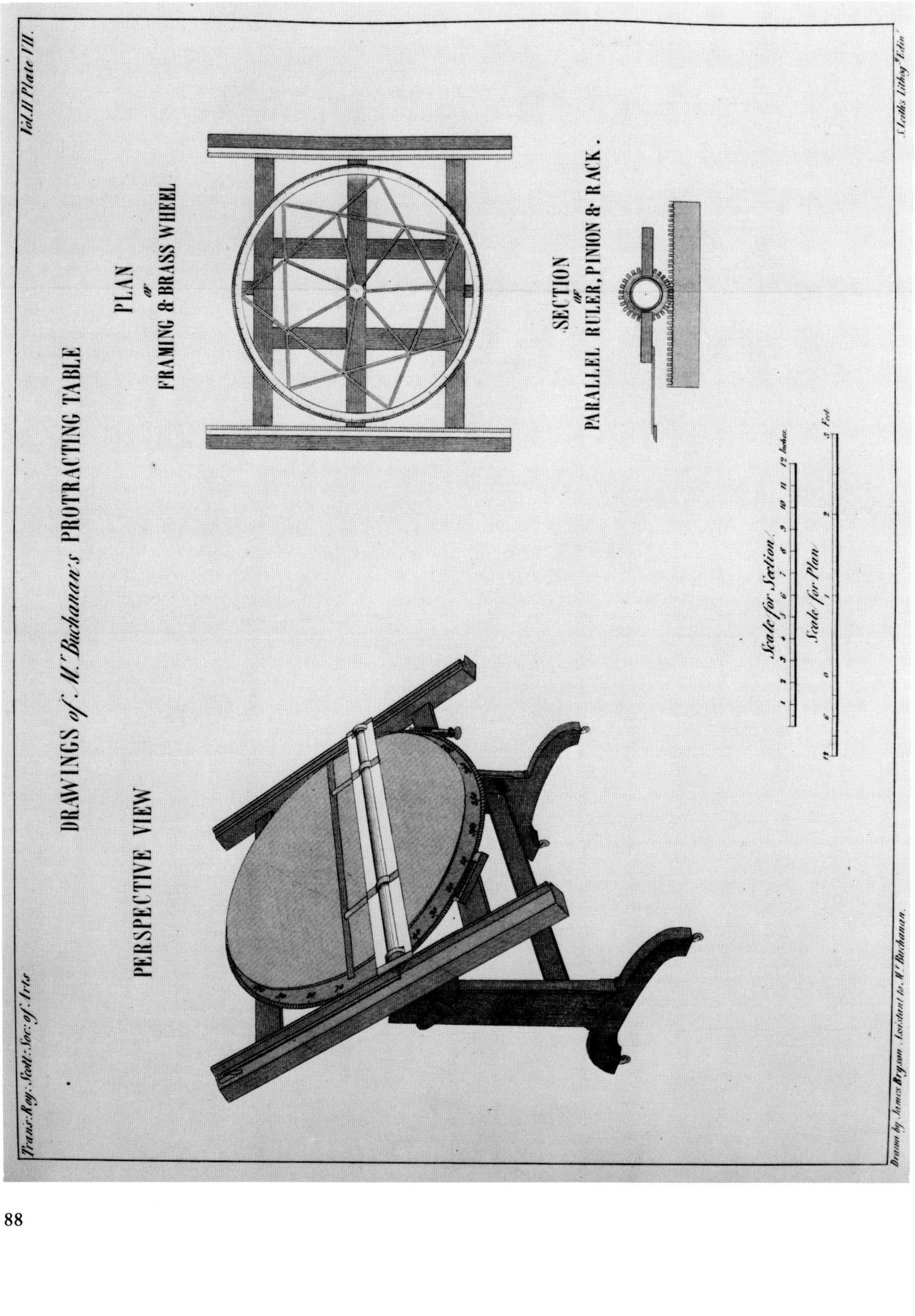
Trans: Roy: Scott: Soc: of Arts
Vol. II Plate VII.
DRAWINGS of Mr. Buchanan's PROTRACTING TABLE
PERSPECTIVE VIEW
PLAN
or
FRAMING & BRASS WHEEL
SECTION
of
PARALLEL RULER, PINION & RACK.
Scale for Section
Scale for Plan
Drawn by James Bryson Assistant to Mr. Buchanan.
S. Leith's Lithog. Edin.

5

JOHN AND THOMAS DUNN

John Dunn (c.1791-1841) and his brother Thomas Dunn (c.1803-1893) were the sons of Hamilton Dunn, an Edinburgh builder, and Elizabeth Purves,[1] and both practised as scientific instrument makers in the city during the second and third quarters of the nineteenth century. It is not known under whom John served his apprenticeship, but it is likely that he was placed with one of the few master opticians active in the 1800s (when he is likely to have begun serving) such as Miller and Adie (q.v.) or Peter Hill (q.v.). In his early twenties John began to attend the courses for the higher education of 'mechanics' offered by the Edinburgh School of Arts from its commencement in 1821. He was not a brilliant pupil, being listed eighteenth among the twenty-six most proficient students in the Mathematical Class during the School's second session 1822-3.[2] In 1825-6 he attended the Chemical Class, at the close of which he presented an ornamental clock to the teacher Dr Andrew Fyfe on behalf of his fellow pupils.[3]

By this date Dunn had already begun working independently, and first appeared in the Edinburgh street directory as an optician at 7 West Bow in 1824. Between 1825 and 1827 he was at 25 Thistle Street, between 1828 and 1831 at 52 Hanover Street, and from 1832 onwards at 50 Hanover Street.[4] In 1840 he opened a Glasgow branch at 157 Buchanan Street, which had moved to 28 Buchanan Street by the time of his death on 28 July 1841.[5]

John Dunn's output may appear slight, but his significance in the history of Scottish instrument making should not be measured by the relatively few extant signed pieces from his hand, for his high standard of workmanship, fertile mind and enthusiasm for science and scientific education earned him a position in Edinburgh scientific circles as a master optician second only to Alexander Adie and his son John (q.v.). Regarded simply as a maker, Dunn was inferior to them, and although he appears to have produced a broadly similar range of instruments, lacked the prestige which talent and patronage conferred on the Adies.

24 George Buchanan's protracting table, by John Dunn and Alexander Adie, 1827-42

Nonetheless, John Dunn was active as a maker of laboratory apparatus, surveying equipment and optical instruments, as is evinced by the items he displayed at an exhibition in Edinburgh in the winter of 1839-1840. These were a portable transit instrument, an altitude and azimuth instrument, a quadrant, a sextant, a theodolite and a thermometer.[6] While this group of pieces confirms Dunn's claim to be a maker of nautical as well as of mathematical and philosophical instruments,[7] it also underlines the important place which surveying instruments occupied in his workshop's production. This was in line with a general trend in Scotland, in which demand for all types of land-surveying instruments increased rapidly from the early nineteenth century. Dunn's share in one part of the market is represented in this collection by a theodolite, a widely used type of instrument. He also produced portable cases for surveyors, for instance containing a level, tripod and levelling rod, and several individual instruments by him are known.[8]

Dunn's repertoire covered experimental and demonstration apparatus, and not surprisingly his connection with the School of Arts seems to have been important in bringing him work in this line from 1826. However, he also executed work for experimenters connected with the Chemistry Class at the University of Edinburgh at about the same time, of which the only known extant example signed by Dunn is a pyrometer, built on a large scale to be visible to lecture audiences. It is undated, but may have been commissioned by Professor T.C. Hope, who is known to have ordered an 'Air Weighing Flask' from Dunn in about 1827.[9] A link between Dunn and the University around this time is further suggested by commissions from the newly-graduated chemist Thomas Graham: in November 1826 Dunn made 'some very minute weights' for him, and in September 1827 Graham spent his allowance on 'two very fine thermometers of great accuracy and beauty made by Dunn'.[10] Clearly Dunn's reputation was good within a few years of his commencement in business, and it is interesting to note how graduates who pursued their experiments turned to instrument makers to execute original designs, as with David Brewster and Alexander Adie (q.v.), or to purchase modest amounts of basic apparatus, in the case of Graham.

Dunn's subsequent recommendation of Thomas Graham to Patrick Chalmers the Edinburgh bookseller, who was setting up his new popular scientific monthly

Chalmers' Miscellany in about 1827, indicates the extent to which Dunn was already involved in scientific circles by this date. It also demonstrates the way in which the client on patron of an instrument maker could become his protégé.[11]

One of Dunn's particular concerns as a maker was to simplify instruments without impairing their performance, because he knew that 'the cost of apparatus prevented many gentlemen from engaging in philosophical pursuits'. According to his own account, among the first items he produced on this principle were air-pumps, which were normally very expensive. One example was made for Dr George Lees of the Edinburgh School of Arts in 1826, and led to a commission for another for the Chemical Class at London University in 1827.[12] It has been suggested that this was ordered by the first Professor of Chemistry there, Edward Turner, who had moved from Edinburgh to take up his chair in 1827 and presumably knew Dunn.[13] Some six years later, in about 1833 Dunn constructed a pump at the suggestion of an experimenter, Charles Chalmers, for whom he had previously made a working model of a bellows pump designed by Chalmers.[14] In addition, at about the time he made a pump for Dr Lees, Dunn built a double-barrelled pump for Henry Meikle, who praised its efficiency and fine manufacture.[15] With such a successful specialisation to his credit, it is not surprising that after Dunn had exhibited and described the Lees pump to the Society for the Encouragement of the Useful Arts on 19 December 1827,[16] he should have been proposed and elected a member in January 1828.[17] His subsequent involvement in the Society demonstrates very clearly his unceasing scientific activity in the 1820s and 1830s, and his work reflects, more directly than does that of the Adies, the impact of the Society's technical inventiveness and 'improving' values.

A striking example of the influence which these twin factors exercised on Dunn's work is another instrument in which he specialised, the drawing instrument known as the pantograph, used for copying, reducing or enlarging drawings and illustrations. Dunn's designs followed on from an original improvement to the pantograph, an instrument known as the apograph designed by a young Ayrshire manufacturer Andrew Smith, and the more sophisticated version called the eidograph, developed by William Wallace, Professor of Mathematics at Edinburgh University. Both versions were being promoted from 1821, and the subsequent dispute over priority of design and authorship of certain details was complicated when John Dunn executed in about 1828 what he claimed was an improved version of Wallace's instrument.[18] As noted above, one of Dunn's concerns was to simplify and reduce the cost of his instruments where possible, without harming their performance, so his knowledge of the eidograph not only led him to make technical changes to what was by then the standard form, but also to develop a second cheaper version for use where great accuracy was not required, for instance in reproducing maps in geography teaching at school. The potential market for this type of reproduction instrument was obviously large, and the first example of the latter type was used by a teacher at the newly-founded Edinburgh Academy in summer 1828.[19]

Dunn's pentagraph, or pantograph as it was alternatively known, would probably have been uncontroversial had it not been for the claims he advanced in support of his alleged improvements to Wallace's design. The eidograph had been re-engineered since 1821 and in its production version, as made by Alexander Adie (q.v.), it was demonstrated in the Society of Arts in June 1829.[20] Dunn, who was by then a member of the Society and had already produced his cheap workaday version, described his two types to it on 20 January 1830. While acknowledging the superiority of Wallace's design over previous ones, he criticised its inaccuracy which he remedied by replacing Wallace's chains and wheels with ingeniously pivotted linkage bars and conical bearings. This also eliminated the need to make adjustments prior to use, according to Dunn, and fulfilled his aim of 'delicacy, security and accuracy'.[21] His claims for the less accurate version were proportionately modest, but Wallace was incensed at Dunn's pretensions and vehemently denied that any improvements on his own design had been made.[22] He went further to attack Dunn's pentagraph as dangerously worse than the eidograph, but the Society's committee concluded that Dunn had merely altered Wallace's design, and had neither improved it nor produced a better instrument. This equable judgement was delivered after much postponement in 1836,[23] by which time Wallace had acted on Dunn's criticism and made improvements to the eidograph. Also of significance was Wallace's bitterness that Dunn was earning money from his (Wallace's) design, for the pentagraph seems to have found a market well into the mid 1840s, when more than 200 had been produced[24] and advertisements by Dunn's brother Thomas stressed the business's specialisation in the instrument.[25]

Another important instrument with which Dunn became closely involved through the Society of Arts was the electric telegraph as designed by the Edinburgh advocate and amateur scientist Mungo Ponton. In January 1838 Dunn read to the Society, in Ponton's absence, his description of an improved electric telegraph, and in June 1838 Ponton himself donated a model of the instrument to the Society's museum, of which Dunn was curator.[26] In December 1839 a working model of Ponton's 'Galvanic Telegraph' constructed by Dunn was exhibited in Edinburgh, accompanied by a description.[27] This description was reused in August and September 1840 for the exhibition at the Glasgow meeting of the British

25 John Dunn's trade card for his address between 1825 and 1827

26 John Dunn's trade card for his address between 1832 and 1842. *Trustees of the Science Museum, London*

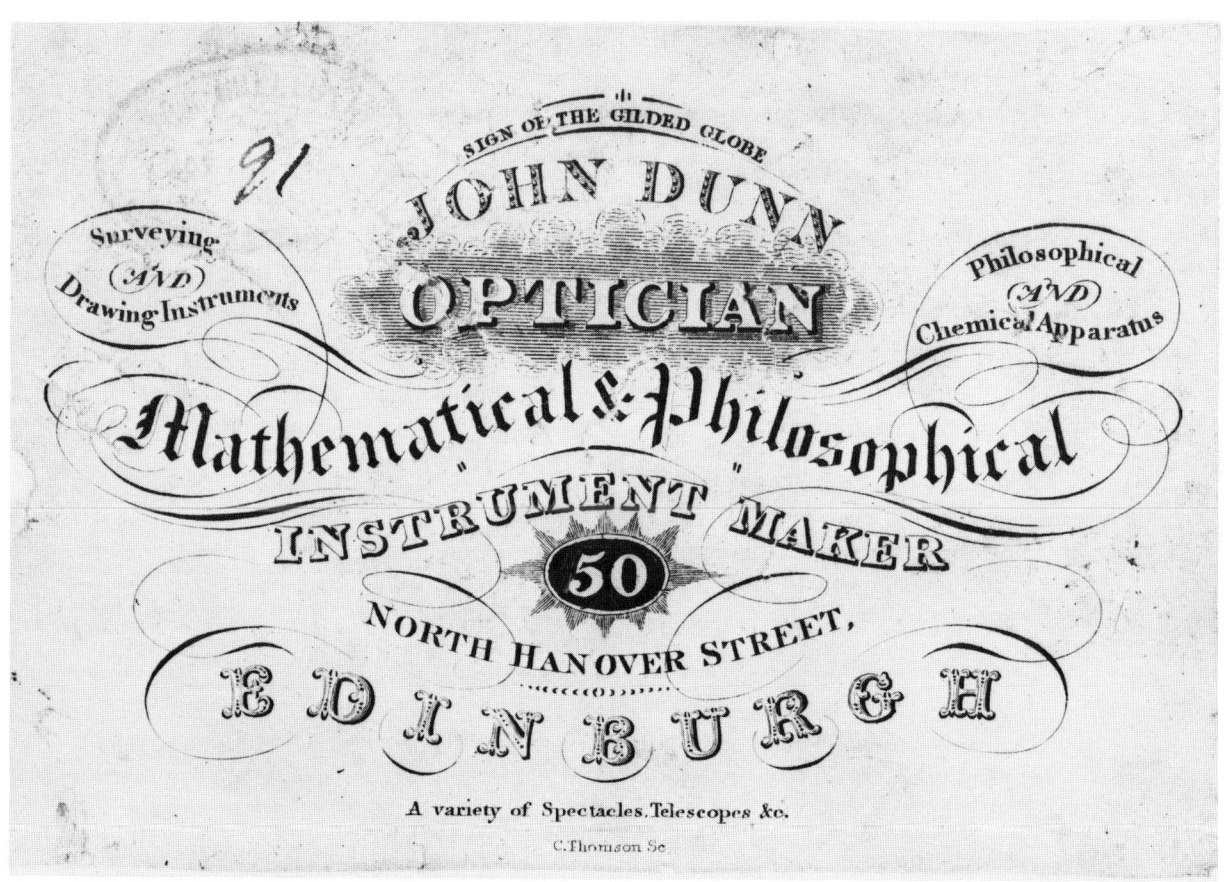

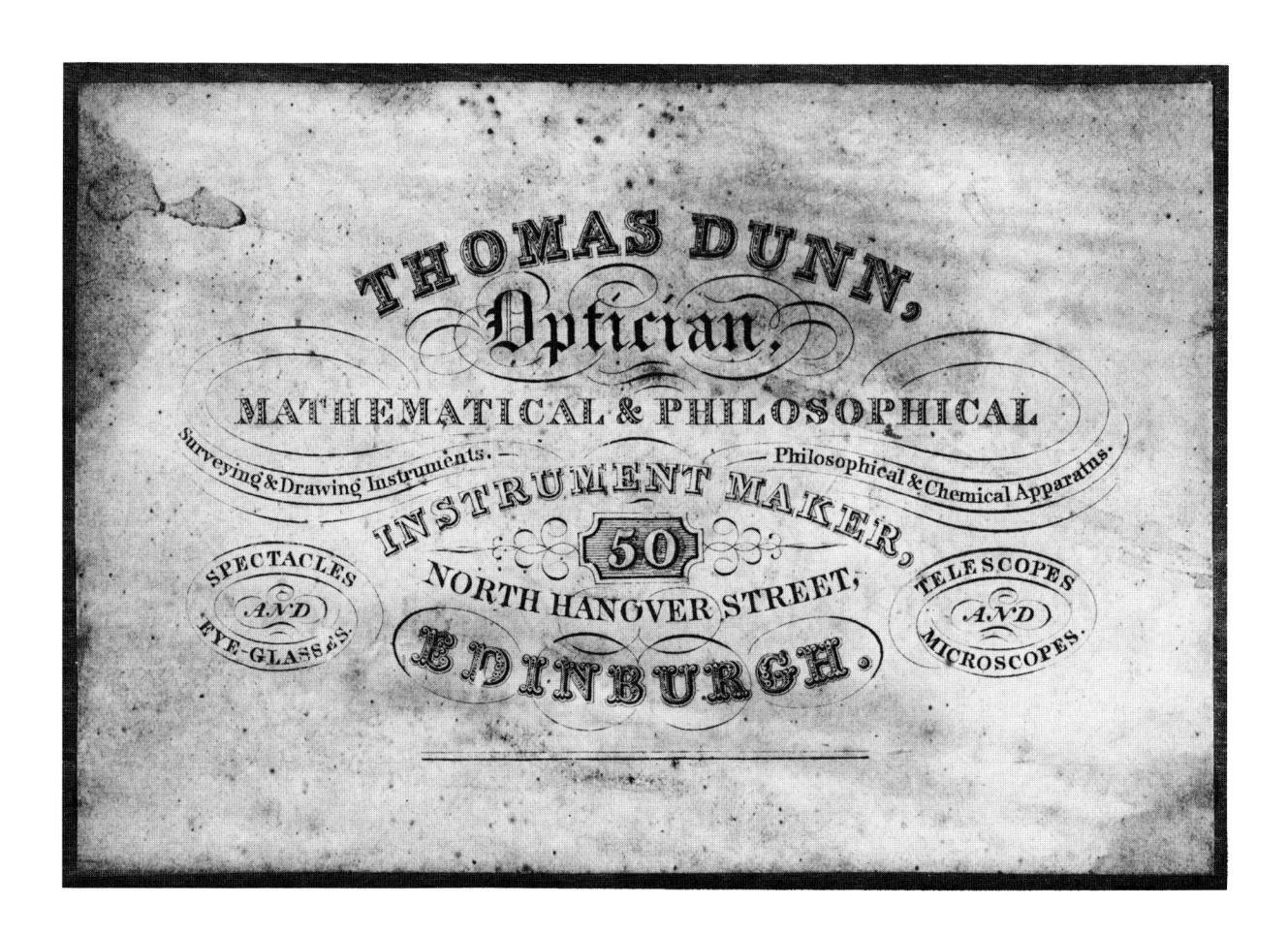

27 Thomas Dunn's trade card for his address between 1843 and 1866

Association, when the telegraph was again on show,[28] and Dunn gave a paper on it.[29]

Besides these specialisations John Dunn was active in other aspects of mechanics and physics, and his close co-operation with the designers of instruments illustrates the common bond of scientific zeal which linked the most talented master mechanics to both their professional and amateur patrons. For example in 1831 he and Edward Sang (one of the most brilliant members of the Society of Arts) gave a paper to the Society on the thermal expansion of marble, which had important implications for its use in clockmaking.[30] Other problems absorbed his attention, such as the 'Naphtha Lamp' on which he was experimenting in 1837,[31] an improvement to John Rutherford's thermometer in 1840, microscopic photography, and he was co-publisher of praxinoscope discs devised by Edward Sang.[32] Dunn's contributions to the Society's meetings extended far beyond the intermittent papers he delivered, for he not only served on the council 1831-3, before acting as Curator of the Museum from 1833 until his death, but he also sat on an impressive number of the committees appointed by the Society to evaluate and report on the papers delivered at its meetings. It says much for Dunn's mechanical ability and scientific knowledge that he was competent to examine papers in both the mechanical and chemical classes, which covered everything from an orrery to the extraction of gelatine from bones. He frequently sat alongside John Adie (q.v.) and Alexander Bryson, brother of James Mackay Bryson (q.v.) on the specialised committees.

While Dunn was one of the leading lights of the Society of Arts, he was also active in other fields of science and education. Along with fellow-alumni of the School of Arts he was instrumental in founding the Edinburgh School of Arts Friendly Society in 1828, whose committee he convened.[33] The Society's membership included subscribers to the Leith Mechanics Subscription Institution, and the Edinburgh Mechanics Subscription Library, founded in 1825 by members of the School of Arts to provide a wider range of literature than was available in the School's Library. Dunn, presumably one of its founders, was vice-president of the Edinburgh Mechanics Library in 1825 and its president 1826-7.[34]

Within a few years, however, Dunn's interests in scientific education changed, for he became involved in the running of the Edinburgh Philosophical Association, founded in 1832 to provide popular scientific instruction for the Edinburgh middle classes, who were neither catered for by the specialist Royal Society, nor by the Society of Arts, nor by the School of Arts (attended mainly by 'the operative classes').[35]

Dunn was a director for two seasons in the 1830s, and he was probably among those who in August 1835 pressed for a new code of laws both to stabilise its income and membership, and to obtain the legal status of an established scientific society.[36]

Dunn's preoccupation with the dissemination of science to a middle-class audience fits with his status as a leading master mechanic, which gave him a respectable entrée into their society at the same time as forming professional ties with the scientific community. It is therefore interesting to find him on the committee of the 'The Reform Association for the New Town of Edinburgh', (chaired by Sir James Gibson-Craig), a body designed to stimulate the exercise of the franchise by Edinburgh electors, and which numbered among its members the professional men with whom Dunn rubbed shoulders in the Philosophical Association and the Society of Arts.[37] His political views can thus be compared to the reformist attitudes of the Adie family.

The last work which John Dunn is known to have executed was the construction of the main part of a protracting table designed by George Buchanan, an Edinburgh engineer, by summer 1841.[38] It has been suggested that the completion of the work by Adie & Son (q.v.), using their dividing machine to measure out the circle, shows that Dunn did not possess a dividing machine of his own and had to contract out to the Adies who did.[39] Given his respected position as a maker it seems improbable that he lacked this piece of apparatus, but even if it was the case, the main reason for his non-completion of the protracting table was his death in Glasgow on 28 July 1841.[40] He was declared bankrupt, and his estate sequestrated on 14 August,[41] and on 30 October an advertisement appeared in the local press stating that 'the whole stock in trade belonging to the deceased [was] to be sold off without reserve, at 25 per cent Reduction'.[42]

His brother Thomas managed to recover the business by discontinuing the Glasgow branch and he then ran the Edinburgh shop himself. He had begun in the trade in about 1819, presumably as an apprentice, and had been his brother's principal assistant from about 1825.[43] He retained the premises at 50 Hanover Street until 1867 when he moved to 106 George Street, before apparently retiring about 1868,[44] despite going bankrupt himself in 1864.[45] He died aged ninety in 1893.[46] Advertisements for his business show that the stock which he offered in 1867 differed little from that which John was selling in 1840,[47] and he appears not to have been so conspicuously active either as an instrument maker or as an experimenter as his brother. However, in 1851 at the Great Exhibition he exhibited his own invention of an 'Electro Magnetic machine'.[48] He was a fellow of the Royal Scottish Society of Arts from 1844 and occasionally sat on its committees, particularly in the 1850s. Also, like John, he was a member of the School of Arts Friendly Society (which does not necessarily mean he attended the School) and of the Philosophical Institution. He is not to be confused with his contemporary namesake, a railway engineer and patentee from Manchester.

REFERENCES

1. General Register Office (Scotland), Register of Deaths 1893 (Edinburgh, St. Andrews), 685²/682.

2. *Second Report of the Directors of the School of Arts* (Edinburgh, 1823), 18.

3. *Glasgow Mechanics' Magazine* new edition, 5 (1832), 180; *Scots Magazine* new series, 18 (1826), 626.

4. *Edinburgh Post Office Directory* 1824-1832; *Gray's Directory of Edinburgh* 1832, contradicts the former in giving Dunn at his new address of 50 Hanover Street.

5. *Glasgow Post Office Directory* 1840. Notice of death, *Scotsman*, 31 July 1841.

6. *Catalogue of the Exhibition of Arts, Manufacturers and Practical Science, in the Assembly Rooms, George Street* [December 1839-January 1840] (Edinburgh, 1840), 25-26.

7. Advertisement in *Catalogue of Models and Manufactures at the Tenth Meeting of the British Association for the Advancement of Science* (Glasgow, 1840).

8. Besides the theodolite in the Frank Collection, NMS T1980.159, (which is signed 'Dunn, Edinburgh' and thus may have been made by either John or his brother Thomas), the Royal Museum of Scotland also has: a cased level set signed 'J. Dunn, Edinburgh', NMS T1982.175; a surveyor's graphometer signed 'J. Dunn, Edinburgh. Inv^t et Fecit', NMS T1986.6; and a surveyor's dial signed 'J. Dunn 50 Hanover Street Edinburgh', NMS T1950.5. A brass Everest theodolite with John Dunn's signature was offered for sale at Sotheby's, 23 June 1987, Lot 178.

9. R.G.W. Anderson, *The Playfair Collection and the Teaching of Chemistry at the University of Edinburgh 1713-1858* (Edinburgh, 1978), 40, 47, 82.

10. Thomas Graham to his sister Margaret, 20 November 1826, and Graham to his mother, 16 September 1827, in J.J. Coleman (ed.), 'The Life and Works of Thomas Graham ... Illustrated by 64 unpublished Letters', *Proceedings of the Glasgow Philosophical Society* 15 (1883-4), 267, 276. Graham became professor of chemistry at the Andersonian University, Glasgow in 1830, and from 1837 until 1855 was professor of chemistry at University College, London. His researches on the molecular diffusion of gases led him to formulate the law which bears his name.

11. *Ibid.*, 267, 269.

12. John Dunn, 'Description of an Improved Air Pump' (dated 19 December 1827), *Edinburgh New Philosophical Journal* 4 (1828), 382-386.

13. Anderson, *op. cit.* (9), 82-83, and n.11.

14. National Library of Scotland, Dep. 230/3, Royal Scottish Society of Arts Archives, Minute Book, vol. 2, p. 235, note of

John Dunn's 'Description of an Accurate and Cheap Air-Pump' (22 January 1834); *ibid.*, pp. 101-2, note of demonstration of transparent working model of bellows pump based on Chalmers' invention, on 28 December 1831.

15. Henry Meikle, 'Remarks and Experiments relating to Hygrometers and Evaporation', *Edinburgh New Philosophical Journal* 2 (1826), 28.

16. Dunn, *op. cit.* (12).

17. N.L.S., Dep. 230/1, R.S.S.A. Archives, Minute Book, vol. 1, entry for 23 January 1828. Given Sir David Brewster's pre-eminence within the Society and Edinburgh science as a whole, it is relevant to note an alleged connection between Brewster and Dunn; Dunn and his brother Thomas are supposed to have helped Brewster with 'much of the experimental work' on his inventions of the stereoscope, kaleidoscope and lighthouse projection apparatus (John R. Hutchison, 'Some Notable Opticians of Auld Reekie', *Ninth Annual Conference [of] The Scottish Association of Optical Practitioners* (Edinburgh, 1939), 17). We have located no evidence for this claim - as other facts given in this account are erroneous, the suggestion may perhaps be discounted.

18. A.D.C. Simpson, 'Brewster's Society of Arts and the Pantograph Dispute', read at the Second Greenwich Symposium on Scientific Instruments, September 1982.

19. N.L.S., R.S.S.A. Archives, Acc 4534, box 13, bundle 1830 (i), John Dunn, 'Description of an Improved Pentagraph' (17 December 1829).

20. Simpson, *op. cit.* (18).

21. Dunn, 'Improved Pentagraph', *op. cit.* (19).

22. Simpson, *op. cit.* (18).

23. N.L.S., R.S.S.A. Archives, Acc. 4534, box 13, bundle 1830 (i), 'Report of the Committee appointed to investigate Mr Dunn's "Improved Pentagraph"', (27 March 1836). The committee was convened by the brass founder John Milne (whose family business appears to have collaborated with Adie on the improvements to Wallace's eidograph), and included Edward Sang, and the engineers James Jardine and Robert Stevenson (all of whom were associated with the Adie business). However, Stevenson held a slightly more favourable view of Dunn's instrument than his colleagues (cf. Simpson, *op. cit.* (18)).

24. Simpson, *op. cit.* (18).

25. Eg. *Edinburgh Post Office Directory* 1844, 427; *ibid.* 1867, 51. Examples of these instruments at the Royal Museum of Scotland are NMS T1967.120 and NMS T1984.96 by John Dunn, and NMS T1983.49 and NMS T1948.X8 by Thomas Dunn.

26. N.L.S., R.S.S.A. Archives, Dep. 230/3, Minute Book, vol. 2, pp. 440-1, note of Mungo Ponton's 'Description of an Improved Electric Telegraph',(10 January 1838); Dep. 230/4, Minute Book, vol. 3, p. 20, description (20 June 1838).

27. *Catalogue, op. cit.* (6), 25-6. The model was placed in the museum after display to the Society on 15 January 1840 (N.L.S., R.S.S.A. Archives, Dep 230/4, Minute Book, vol. 3, p. 146).

28. David Murray, *Memories of the Old College of Glasgow* (Glasgow, 1927), 138-139.

29. *Report of the British Association ... 1840* (London, 1841), 213.

William Thomson (later Lord Kelvin) almost certainly saw the telegraph when he visited the exhibition, although when asked many years later whether it had influenced his own ideas on telegraph design, he could not recall having done so (Murray, *op. cit.* (28), 138-139).

30. John Dunn and Edward Sang, 'An Account of some Experiments made to determine the Thermal Expansion of Marble' (dated 30 March 1831), *Edinburgh New Philosophical Journal* 11 (1831), 66-71. Dunn's interest in the subject ties in with his construction of the pyrometer for the University of Edinburgh.

31. Dunn exhibited the lamp on 11 January 1837 (N.L.S., R.S.S.A. Archives, Dep. 230/3, Minute Book, vol. 2, pp. 382-383).

32. John Dunn, 'Description of an Improvement on the Mercurial Registering Thermometer of Rutherford' (dated 15 January 1840), *Edinburgh New Philosophical Journal* 29 (1840), 279-280. Dunn obtained an impression of a piece of cane using a solar microscope in 1839 (N.L.S., R.S.S.A. Archives, Dep. 230/4, Minute Book, vol. 3, p. 88, entry for 27 March 1839). The praxinoscope discs were described as 'recently ... invented upon the Continent, and introduced into Edinburgh by Mr Dun [*sic*], optician, and Messrs. Forrester and Nichol, lithographers' *Chambers' Historical Newspaper*, No 15, January 1834. Our thanks to Helen Smailes, Scottish National Portrait Gallery, for this reference. There are five discs in the Royal Museum of Scotland, marked 'As projected by Edward Sang, published by Forrester & Nichol, lithographers and John Dunn, optician, 50 Hanover Street' (NMS T1979.1).

33. *Rules of the Edinburgh School of Arts Friendly Society* (n.p., n.d. [Edinburgh, 1828]). Henry Cockburn, the eminent judge, and Robert Bryson, one of the School's founders, both sat on the Committee, and several instrument makers can be identified as members of the Society from successive *Report(s) upon the Affairs of the Edinburgh School of Arts Friendly Society*.

34. James Grant, *Old and New Edinburgh* (London, 1883), I, 291; *Laws and Catalogue of the Edinburgh Mechanics Subscription Library* 6th edition (Edinburgh, 1859).

35. The Association's original name was 'The Edinburgh Association for providing Instruction in Useful and Entertaining Science'; it later became 'The Edinburgh Philosophical Institution'.

36. *Fourth Report by the Directors of the Edinburgh Association ...* (Edinburgh, 1834), 8. *Fifth Report ...* (Edinburgh, 1835), 8. *Answers of the Directors of the Edinburgh Philosophical Association to the Protest of Mr William Fraser* (Edinburgh, 1836), 4-7. The changes of August 1835 which provoked the opposition of Fraser and others also centred on the role of the Association as an advisory body and clearing-house for science lectures throughout Scotland. The Association's role in the Scottish scientific community is examined in Steven Shapin, '"Nibbling at the teats of science" : Edinburgh at the diffusion of science in the 1830s', in Ian Inkster and Jack Morrell (eds.), *Metropolis and Province: Science in British Culture 1780-1850* (London, 1983), 151-178. Difficulties over the fee for lectures to the Association appear in letters between Dunn and George Combe, the phrenologist, in 1835. (N.L.S., MS 7234, ff. 189-190, MS 7386, ff. 382-383).

37. *New Town Reform Association. Rules and Regulations* (Edinburgh, 1839).

38. George Buchanan, 'Description and Uses of his

Protracting Table' (dated 21 February 1842), *Transactions of the Royal Scottish Society of Arts* 2 (1844), 176-9, and plate VII (read 14 March 1842).

39. D.J. Bryden, *Scottish Scientific Instrument-Makers 1600-1900* (Edinburgh, 1972), 37.

40. *Scotsman*, 31 July 1841. Dunn had evidently begun the brass circle, which he is unlikely to have done if he could not have divided it himself (Buchanan, *op. cit.* (38), 177).

41. Scottish Record Office, SC 39/17/6928, Sequestration Papers, John Dunn. This is marked 'No Process', and there are no documents relating to this case. However, notices appeared in the *Edinburgh Gazette*: the initial notice of sequestration (13 August 1841, 264); the announcement of the appointment of the Trustee and Commissioners (17 September 1841, 305); the announcement that a dividend would be paid on 12 April 1841 (1 March 1842, 95); the announcement that the dividend would be postponed (8 July 1842, 268); and the announcement of a second dividend to be paid on 31 March 1843 (3 February 1843, 41). The sequestration notice also appeared in the *Scotsman*, 14 August 1841.

42. *Scotsman*, 30 October 1841. By 15 January 1842 the 25% reduction was continuing, along with the statement that 'the Glasgow Branch of the Business being now closed, and the Stock removed to Edinburgh, the whole Extensive Assortment ... is now SELLING OFF at 50 Hanover Street'.

Advertisements of John Dunn's bankrupt stock continued to appear intermittently until August 1842.

43. Advertisement in the *Edinburgh Post Office Directory* 1844, 427.

44. *Ibid.* 1848-1868.

45. S.R.O., CS 319/1911/867, Sequestration Papers, Thomas Dunn. There is no Sederunt Book with these papers. Notices also appeared in the *Edinburgh Gazette*: the initial notice of sequestration (19 July 1864, 938); the announcement of the appointment of the Trustee and Commissioners (2 August 1864, 998); a postponement of accounts (9 December 1864, 1698); the first dividend to be paid on 19 May 1865, and a general meeting of Creditors called for 14 April to 'instruct the Trustee as to the disposal of the remaining assets belonging to the estate' (7 April 1865, 458); and the announcement of Thomas Dunn's Petition to be finally discharged without composition, which was granted (11 April 1865, 469).

46. G.R.O. (S.) Register of Deaths 1893 (Edinburgh, St. Andrews), 685²/682.

47. Advertisement in *Catalogue, op. cit.* (6); *Edinburgh Post Office Directory* 1867, 51.

48. *Catalogue of the Great Exhibition* (London, 1851), I, 472, no. 689A.

---

39. Theodolite: Dunn, Edinburgh, c.1850 (T1980.159)

5″ reversing theodolite in brass and bronze, the bevelled edge azimuth circle divided on silver [0]-360 by 1°, with clamp and tangent screw adjustment (defective) to the upper plate and verniers reading to 2 mins. Engraved by one vernier 'Dunn Edinburgh.' Crossed levels (replacements) and compass (defective), with 8 cardinal points and divided 0-90-0-90-0 by 1°, with relieved bearing. A-frames support the altitude semicircle and mounting limb; divided on silver [60]-0-[105] by 1° with clamp and tangent screw and vernier reading to 2 mins., and on the reverse 30-0-30 'Diff. of Hypo. & Base'. The telescope in Y-mounts over the limb, with ⅞″ aperture doublet objective, rack and pinion focus to the objective and draw tube focus to the graticule. Socket mount.

Telescope length: 255mm.
Aperture: 23mm.
Azimuth scale diameter: 120mm.

Lacking the level to the telescope, the compass needle, the glazing of the compass box, and the objective cover. The crossed levels are on modern mountings. The azimuth tangent screw is not of the correct size. The clamp plates of the Y-bearings are distorted and apparently lack restraining spacers. The instrument is in a distressed condition.

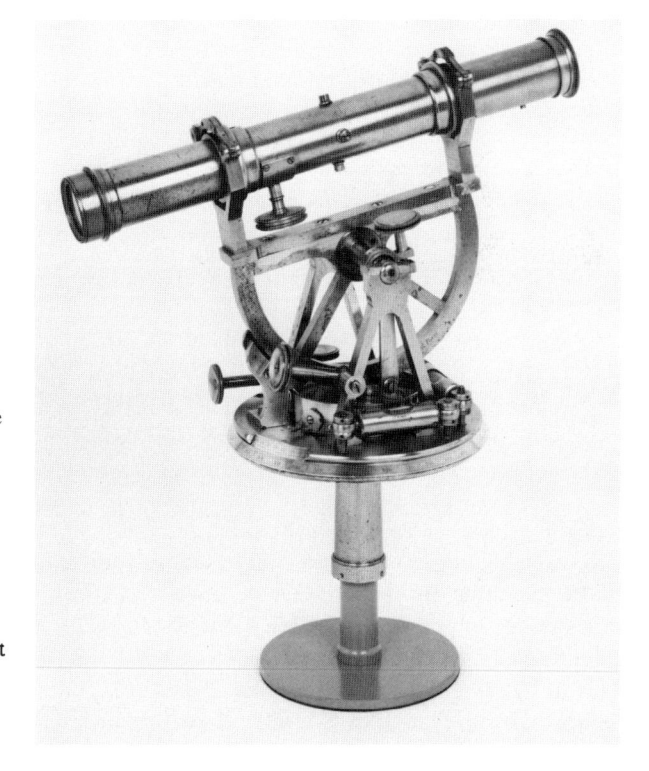

6

RETAILING INSTRUMENT MAKERS

The two Edinburgh businesses examined in this section were not alone in the nineteenth century Scottish capital in merely providing instruments rather than actually producing them. Indeed, by the end of the century, it is probable that most instruments sold in Edinburgh were manufactured south of the border, as was the case nearly everywhere in Scotland. This was the period during which radical changes occurred in both the techniques of manufacture and the organisation of that manufacture. This is not the place to discuss the general nature of that organisation, including for example, the changing reliance on 'outwork'. Suffice it to say that the changing patterns of labour management meant that the production of instruments (always a small scale activity when compared with the great industries which underwent transformation during this period, such as textiles, mining and metals) developed and blossomed from its original craft roots in the great industrial cities, among whose number Edinburgh did not figure.

Edinburgh's population almost doubled in the first quarter of the nineteenth century, and it was the centre of an area of extensive development, placed among areas of agrarian wealth and various subsidiary industrial sites. Lenman has described how:

> 'Edinburgh itself was the social, legal, medical and ecclesiastical centre of Scotland. Its own industries tended to be organised in smallish units and were rather old-fashioned, but there were plenty of them. Such craftsmen as skilled masons, cabinet-makers, jewellers and needleworkers abounded.'[1]

It is probably fair to say that a craft-based instrument making could be included in this classification. After this, the rate of population growth diminished: Edinburgh's population had doubled between 1801 and 1831 to 166,000 but by 1881 it stood at 295,000 (compared with Glasgow's 511,000) and in 1907 at 413,000 (when Glasgow's population was 761,000). Edinburgh's economy was securely based on small-scale trades providing services for a prosperous upper and middle class, and contrasted strongly with the situation in Glasgow.[2]

However, the real change came with the advent of the railways: the 1830s and 1840s saw tentative beginnings with short lines, essentially providing routes between the coal fields and the cities.[3] As had been said of the national railway system, 'not only did it unify the Scottish economy in an unprecedented fashion but it also linked that economy with England for the first time by land with fast, efficient and reliable transport.'[4] Now it was normally more economic to import parts for assembly and completed instruments from the South, especially from the two main centres of instrument making, London and Birmingham, than it was to make them locally.

By the early nineteenth century the British instrument trade, which had previously been monopolised by London, had to some extent moved its centre northwards to the recently industrialised English Midland cities of Birmingham and Sheffield. Even in eighteenth century London

> 'there were retailers with their own manufacturing workshops, who took apprentices and employed journeymen, but there were also retailers with minimal workshop facilities who turned to trade sources for their goods. These might come from another retailer who specialised in a particular line, or from a workshop, possibly of considerable extent, working solely for the trade. Some instruments, or their component parts, might come from an individual workman operating in his own house. Nevertheless it was the retailer who engraved his name on the finished article -whether or not he had any hand in its manufacture.'[5]

At the end of the century London lost its central position as instrument supplier to the provinces, partly through the demand for skilled mechanics needed for the construction and installation of machinery in the new factories of northern England, and partly through the expansion of educational and cultural activities such as those represented in the mechanics' institutes, the literary and philosophical societies, schools and universities, all of which helped create a new pattern of demand for instruments.[6] London's pre-eminence over the European market was also forfeited, mainly to German manufacturers, who were able to produce up-to-date designs more cheaply. In part this was due to the late advent of industrialisation on the Continent, but also because a closer relationship often existed between Continental workshops and scientists. The conventional working patterns which prevailed in London meant that although there were a number of instrument wholesalers based in the capital, others could be found in the major industrialised cities throughout the British Isles.[7]

Edinburgh instrument makers had been retailing instruments since the beginnings of the trade in Scotland.[8] By the early nineteenth century some items were appearing with signatures of firms which were only tenuously connected with instrument making. For instance, David Brewster's kaleidoscope was finally mass produced in both Birmingham and Sheffield in 1818-19 after the design was pirated during the cumbersome patenting procedure.[9] However, the list of Brewster-approved retailers included 'AT EDINBURGH ... Mr JOHN RUTHVEN'.[10] John Ruthven was, at different periods, a locksmith,[11] a printer,[12] and an inventor.[13] The reason why he was chosen by Brewster to retail the instrument remains obscure,[14] but probably relates more to his success as an entrepreneur because it is evident that he was not, strictly speaking, an instrument maker.

Similarly, the two firms discussed in this section were not makers of instruments. John Davis produced such a large and varied stock that it is difficult to believe that he, or his workshop, would have produced them all. Zenone & Butti were, by their own description, 'carvers & gilders', involved with furniture, looking glasses, and the needs of the 'dilettante' market of the Edinburgh professional classes. Other Edinburgh firms represented in this collection could be classed as retailers too. The Lennie firm (q.v.), for instance, does not appear to have produced its own instruments, and it is difficult to tell just what proportion of material signed by late nineteenth century businesses such as Turnbull & Co (q.v.) was actually produced by them.

Another example of an Edinburgh optician acting as a retailer was Alexander Reid, active from 1855 when he was at 66 Nicolson Street and trading from 31 Castle Street, Edinburgh between 1868 and 1901.[15] He was a member of the Royal Scottish Society of Arts from 1857, and died in December 1912.[16] He sold an unsigned reflecting telescope, which was possibly made by a wholesaler to the trade, such as George Dollond.[17] In an accompanying undated letter from Castle Street, to his customer George Thomas Clarke (1809-1898), Reid writes:

> 'On 31 Oct, I duly sent by Caledonian Rail[wa]y the Reflector, I hope you recd it in good order and that it gives you great satisfaction Mr Gaskin who made the Reflector Mirrors sold me & never made a finer pair; they have never been out of my hands since I bought them of him & they are as bright as the day I bought them, and the only [thing] I would use to clean them is a little spirit of wine...'

The 'Mr Gaskin' is possibly the George Augustus Samuel Gaskin who was apprenticed on 6 May 1802 to the instrument maker James Clarke, who in turn had been trained by Richard Rust, in the Grocers' Company of London.[18] He may well have been one of the considerable number of London outworkers who scraped a living doing specialised sub-contract work for the trade. It is only in occasional instances such as this that the unknown names of the actual makers - in this case the maker of the speculum mirrors of a telescope, rather than a complete instrument - ever come to light; and such examples must further erode our confidence in assigning a particular item to the craftsman whose name it bears.

REFERENCES

1. Bruce Lenman, *An Economic History of Modern Scotland* (London, 1977), 107.

2. Sydney and Olive Checkland, *Industry and Ethos: Scotland 1832-1914* (London, 1984), 36-38, 41.

3. *Ibid.*, 26. See also C.J.A. Robertson, *The Origins of the Scottish Railway System 1722-1844* (Edinburgh, 1983), 44-96.

4. Lenman, *op. cit.* (1), 166.

5. R.H. Nuttall, *Microscopes from the Frank Collection 1800-1860* (Jersey, 1979), 8.

6. *Ibid.*, 10-11.

7. As an illustration of this, see J. Steinhardt, *The Illustrated Guide to the Manufacturers, Engineers, and Merchants of England, Scotland, Ireland and Wales* (London, 1869). No supplier under the heading 'Opticians' comes from either Ireland or Wales. The following are listed: Frederick J. Braham (Birmingham); L. Braham & Co. (London); T. Cooke & Sons (London; manufactory, York); Fred. J. Cox (London); J. Crichton & Son (London); Cutts, Sutton & Son (Sheffield); Elliott Brothers (London); Field & Son (Birmingham); Thomas Froccatt (Sheffield); James How (London); H. Hughes (London); Joseph Hughes (London); John Jennings (Birmingham); Smith, Beck & Beck (London); John Spencer (Glasgow) (q.v.); James White (Glasgow) (q.v.); E.G. Wood (London).

8. D.J. Bryden, *Scottish Scientific Instrument-Makers 1600-1900* (Edinburgh, 1972), 3-6.

9. See A.D. Morrison-Low, 'Brewster and Scientific Instruments', in A.D. Morrison-Low and J.R.R. Christie (eds.), *'Martyr of Science': Sir David Brewster 1781-1868* (Edinburgh, 1984), 59-65.

10. *Ibid.*, 58; also illustrated by M.A. Crawforth, 'Evidence from Trade Cards for the Scientific Instrument Industry', *Annals of Science* 42 (1985), 544, fig. 65. There is an example of a Brewster patent kaleidoscope with Ruthven's signature in the Royal Museum of Scotland, NMS T1985.20.

11. A key marked 'I. RUTHVEN PATENT' has been noted in a private collection: Ruthven's patent has not been traced.

12. The anonymous author (probably David Brewster) of the

article 'Printing Presses', *Edinburgh Encyclopaedia* (Edinburgh, [1808]-1830), XVII, 166, devoted an entire column to Ruthven's printing press, which was protected by U.K. patent 3746, 1 November 1813. There is an example manufactured by Anderson, Leith Walk Foundry, Edinburgh in the Royal Museum of Scotland (NMS T1923.10).

13. See, for example, an article on a 'WATER PRESSURE ENGINE invented by John Ruthven' in the *Scotsman*, 11 June 1828. Ruthven patented another 8 inventions between 1818 and 1849.

14. M.M. Gordon, *The Home Life of Sir David Brewster* (Edinburgh, 1869), 96.

15. *Edinburgh Post Office Directory* 1855-1901.

16. National Library of Scotland, Royal Scottish Society of Arts Archives, Dep. 230, chronological Roll Book, death date inserted in manuscript. Reid was recorded as an apprentice optician aged 15 in 1841, in the Holyrood Abbey debtors' sanctuary (General Register Office (Scotland), Census of 1841 (Edinburgh, South Leith), 692^2/45.

17. Offered for sale by Christopher Sykes Antiques, Woburn, November 1981, as catalogue item S2880.

18. Joyce Brown, *Mathematical Instrument-Makers in the Grocers' Company 1688-1800* (London, 1979), 54.

JOHN DAVIS

John Davis first appeared in the street directory for 1836 as a mathematical and philosophical instrument maker at 64 Princes Street, Edinburgh, living at 32 Dundas Street. In 1839 the description 'optician' was added to his trade and in 1841 he moved to 78 Princes Street, remaining there for two years before disappearing from the directory by 1843.[1]

Little is known about the size of his workforce, although Thomas Davidson worked for Davis for 'about two years' from 1836 before setting up on his own.[2] Davidson was certainly capable of making apparatus, but whether Davis actually did so is unclear: of the items which survive with his signature, most, if not all, are stock pieces which could well have been bought in. The portable universal equatorial sundial in this collection[3] bears a remarkably close resemblance to examples bearing the name of W.& S. Jones of London, Dollond of London and Abraham & Co. of Liverpool and Glasgow.[4] Another example signed by Davis has been recorded.[5] Other instruments signed by Davis are a 6-inch Gregorian telescope,[6] a sympiesometer[7] and a 'brass surveyor's circumferentor'.[8]

On 5 August 1840 Davis's name appeared in the list of Scots bankrupts in the *Scotsman* newspaper, and he was examined before the bankruptcy courts the following month.[9] He managed to obtain a discharge on 11 November 1840.[10] However, he continued trading and advertisements for his stock appeared regularly in the *Scotsman*; this was explained, to some extent by one of these claiming a

> 'great sacrifice in the prices of optical, mathematical, and philosophical instruments, in consequence of retiring from the business. J. Davis, optician, 64 Princes Street, respectfully begs to inform the Nobility, Gentry and Public in General, that owing to the Shop possessed by him having been Sold to a Public Company, and that his time being limited, he is determined to SELL the WHOLE of his STOCK at GREATLY REDUCED PRICES, and that every Article purchased from this date will be at a Reduction of Twenty-five per Cent. The Stock is complete in all its branches, and such an opportunity may never occur again.'[11]

Just what the 'Public Company' comprised, was not further explained; possibly the premises had been sold over his head in order to allow him to obtain a discharge.

Davis appears to have been primarily an ophthalmic optician: frequent advertisements appeared proclaiming the virtues of 'his newly invented optometer' and 'his very superior Brazilian Pebble, a material retaining its natural frigid and achromatic qualities, thereby imparting refreshing coolness to the Eye, divested of the glare so offensive to the sight'.[12] 'SPECTACLES, SPECTACLES, SPECTACLES' began another of his puffs, which seemed to be directed against Benjamin Salom who advertised similar goods in the same newspaper.[13] He even resorted to doggerel:

> 'A Celebrated Optician, and DAVIS is his name,
> Has got for making SPECTACLES a fame,
> Or, Helps to Read, as we are told,
> Is writ upon his glaring Sign in Gold,
> And for all uses to be had from Glass,
> Is, by Readers, allowed to surpass.'[14]

However, he consistently advertised 'every description of Optical, mathematical and philosophical instrument',[15] and occasionally specified these. For instance:

> 'DEATH OF A GENTLEMAN
>
> In consequence of which J. Davis, Optician, 64 Princes Street, is empowered to offer by subscription, a splendid 3½ feet Brass Mounted ACHROMATIC TELESCOPE with three extra powers for Terrestrial and Celestial Purposes, the various motions by rackwork; also Pancratic Eye Piece, mounted on Brass Stand and Mahogany Case, complete in Twenty subscriptions at £1 1s each; the same cost, within six months, £42 ...'[16]

Telescopes, which Davis may well not have made himself, were the subject of another advertisement: 'No Tourist, Sportsman, or Seaside Visitor, should be without one' he claimed, continuing

> 'A Pocket Telescope, to show objects 9 miles off £0 18 0
> A Pocket Telescope, to show objects 12 miles off £1 10 0
> A Pocket Telescope, to show objects 16 miles off £2 2 0
> A Pocket Telescope, to show objects 20 miles off £3 3 0
> A Pocket Telescope, for Deer Stalking £4 4 0.'[17]

On 22 May 1841 Davis announced that he was 'removing from No 64 Princes Street, to No 78, Second

28 John Davis advertises his removal on the front page of the Scotsman, *5 June 1841. Trustees of the National Library of Scotland*

Shop West of Hanover Street, opposite the [Royal] Institution',[18] and on 29 May that he had 'removed from No 64 ...'[19] No reason for this was given but it probably had to do with the expiry of the lease of his original shop after his bankruptcy. By the beginning of 1842 Davis had decided to give up trading, possibly because the death of John Dunn (q.v.) and the subsequent sale at a reduced price of his sequestrated stock at 50 Hanover Street proved too competitive for Davis, just around the corner in Princes Street.[20] Davis also advertised 'a reduction of TWENTY-FIVE PER CENT. from the Original Price, in consequence of retiring from the Business' -something he had been saying since 1840 - on at least three occasions.[21] He was 'determined to SELL the WHOLE of his IMMENSE STOCK at GREATLY REDUCED PRICES ... The Stock is complete in all its Branches, and such an opportunity may never occur again.'[22] Despite this, he was declared bankrupt once more in November 1842,[23] and the hearing was held in January 1843.[24]

On this occasion, there is more evidence about his state of affairs. As with his previous sequestration the petition was brought by Davis, together with a 'Jacob Davis, wholesale Furrier and skin merchant, Bull Head Court, London, a creditor of the said John Davis to the amount required by law'. Unusually, there is an inventory of the stock-in-trade, and this reflects the material which Davis had been advertising: a very broad range, with an emphasis on spectacles and eyeglasses. The details of the workshop do not list tools, except a 'Pair of Bellows' '2 turning Lathes with 2 Vices - 1 pair of chops and tools of various descriptions [worth] £10 0s 0d' 'A Lot of Brass tubes' 'Brass Circle for dividing compasses [worth] 10s 6d' 'Brass wire' 'Binding iron wire': materials and tools more for repairs than for construction. In total, his stock-in-trade was valued at £524 0s 3d. He was discharged on 25 May 1843.[25]

There is nothing to indicate that he was any relation to the instrument maker of the same name who was working at a slightly later date in Cheltenham and Derby.

REFERENCES

1. *Edinburgh Post Office Directory* 1836-1843.

2. John Nichol, 'Reminiscences of Thomas Davidson, a Weaver Lad', *British Journal of Photography* 26 (1879), 391. For more about Davidson's instrument making career, especially his supplying photographic apparatus to Sir David Brewster and the pioneering Scottish photographers D.O. Hill and Robert Adamson, see A.D. Morrison-Low and J.R.R. Christie (eds.), *'Martyr of Science': Sir David Brewster 1781-1868* (Edinburgh, 1984), 94-95.

3. NMS T1980.166.

4. All in the Royal Museum of Scotland: NMS T1923.25, T1897.186 and T1980.167.

5. At the Ships of the Sea Museum, Savannah, Georgia, United States: R.S. and M.K. Webster, *An Index of Western Scientific Instrument Makers to 1850: C-F* (Winnetka, Illinois, 1971), 45.

6. [Anon.], *The Egestorff Collection: An Abridged Catalogue* (n.p., n.d., [Dublin,1983]), [5].

7. Noted in Nicholas Goodison, *English Barometers 1680-1860* 2nd edition (Woodbridge, 1977), 516. A stick barometer signed 'John Davies [*sic*] Optician Edinburgh' was offered for sale by Phillips Edinburgh, 25 July 1986, Lot 132; and a thermometer signed 'John Davis Edinburgh' was offered by the same saleroom, 24 February 1989, Lot 34.

8. Offered for sale at Phillips, 12 September 1978, Lot 51.

9. *Scotsman*, 5 August, 9 September 1840.

10. Scottish Record Office, CS 280/26/16, Sequestration Papers, John Davis: there is no Sederunt Book, only the Petition for Sequestration by John Davies (*sic*), dated 8 September 1838; and the Discharge, dated 11 November 1840. Notices appeared in the *Edinburgh Gazette*: the inital notice of sequestration (4 August 1840, 234); the announcement of the appointment of the Trustee and Commissioners, and that the Bankrupt had offered Creditors composition of six shillings per pound, this offer to be confirmed at a meeting called for on 30 September 1840 (8 September 1840, 270); the postponement of the meeting until 7 October (22 September 1840, 286).

11. This advertisement first appeared in the *Scotsman*, 2 January 1841; variations of it appeared on 6, 13, 27 January, 20, 27 February, 13 March and 24 April 1841.

12. *Ibid.*, 4, 25 November, 12 and 26 December 1840.

13. *Ibid.*, 20 January, 6 March, 15, 22, 29 May, 30 October, 13, 27 November and 15 December 1841.

14. *Ibid.*, 3 and 17 April 1841.

15. *Ibid.*

16. *Ibid.*, 2 December 1840.

17. *Ibid.*, 16, 26 June and 24 July 1841.

18. *Ibid.*, 22 May 1841.

19. *Ibid.*, 29 May 1841.

20. *Ibid.*, 30 October 1841.

21. *Ibid.*, 22 January, 18 June and 30 July 1842.

22. *Ibid.*, 5 March and 19 November 1842

23. *Ibid.*, 30 November 1842.

24. *Ibid.*, 7 January 1843.

25. S.R.O, CS 280/5/10, Sequestration Papers, John Davis: again, there is no Sederunt Book, and no list of Davis' debts. Again, notices appeared in the *Edinburgh Gazette*: the initial notice of sequestration (29 November 1842, 510); followed by the announcement of the appointment of the Trustee and the Commissioners, and that again, the Bankrupt had offered Creditors composition of six shillings and eightpence per pound, this offer to be confirmed at a meeting called for on 9 February 1843 (3 January 1843, 2).

---

40. Portable sundial: J. Davis, Edinburgh. c. 1840 (T1980.166)

Folding universal equinoctial dial in brass, with silvered scales. Engraved on the chapter ring 'J. Davis, Edinburgh,' and marked in ink on the reverse with the latitude '56 52'. A spring-loaded stile operating above or below the equator, pivotted in a chapter ring divided IIII-XII-VIII by 5 mins. on the upper and inner faces; hinged against an altitude arc graduated in latitude [84]-0 by 1° over a glazed compass dial incorporating crossed levels, and supported on 3 screw feet: the silvered dial with 8-point compass rose, divided [0]-[90]-[0]-[90]-[0] by 2°, with a needle on a relieved jewelled bearing. With a fitted leather-covered case.

Diameter: 109mm.

ZENONE & BUTTI

The partnership of John Zenone and Louis Joseph Butti appeared in the Edinburgh street directory for 1823 only, at 5 Calton Street, and may well have lasted for only a few months.[1] Both Zenone and Butti were of Italian extraction, and although their place of birth is not known, a recent study of Italian emigrants to Britain observed that those who traded in Britain as carvers and gilders and barometer makers 'represented a substantial proportion of the overall emigration until the 1860s, and people engaged in those occupations were, almost without exception, from the mountainous area around Como, in Lombardy'.[2] On the other hand, the same writer notes that the late nineteenth century Italian community in Edinburgh was 'almost monopolized by Italians from the Liri Valley', in particular from the commune of Picinisco, in Campania.[3] In fact, we know little about the origins of these individuals.

John Zenone, described as a 'carver and gilder and looking glass manufacturer' at 5 Calton Street, appeared alone in the Edinburgh street directory in 1825, following the appearance in previous years of Joseph Zenone, 'importer of French flowers and feather manufacturer' at 17 Calton Hill; Joseph moved to 8 Calton Street in 1825. The men were brothers, and shared their work to a greater or lesser extent.[4] In 1827 John Zenone added 7 Calton Street to his premises, and in 1830 John's address was given as 5 and 6 Calton Street and 77 Princes Street while Joseph's was at 7 Calton Street and 77 Princes Street: the two businesses, if they were separate, obviously shared premises. The Princes Street address vanished the following year, and in 1832 John was recorded at 9 Calton Street, with Joseph at 10 Calton Street. John had disappeared from the directory by 1833,[5] replaced by Molteni, Zerboni & Co., looking glass and picture frame manufacturers, at that address until 1838, while Joseph made a final appearance in the directory in 1841.[6]

Louis Joseph Butti was born in Italy, and after emigrating at an unknown date and becoming at some stage a naturalised British subject[7] he first appeared alone, after the termination of his partnership with Zenone, in the Edinburgh street directory of 1825 as a 'Looking glass manufacturer etc.' at 232 Cowgate, where he remained until the following year. Butti then vanished from the directories until 1836 when the entry 'Butti & Co. carvers and gilders (late A. Waterson & Co) 1 and 2 Ronaldson's Buildings' appeared for a single year. Butti reappeared at 2 Springfield Buildings, Leith, in 1849, and was joined by his son in 1853 when they opened premises at 14 Hanover Street. The following year James A. Butti, carver, gilder and looking glass manufacturer was given his own entry and in 1854 had exclusive use of the Hanover Street premises. L.G. Butti last appeared in the directory in 1867, but J.A. Butti continued, moving to 1 Queen Street in 1857, 7 Queen Street in 1868 and becoming J.A. Butti & Son in 1893. The firm became 'fine art collector and dealer in articles of vertu' in 1868 and by 1899 was described as 'fine art dealers'. It continued into the twentieth century.[8] Louis Joseph Butti, the founder of the business, married a Scotswoman named Elizabeth Mitchell, who predeceased him, and he himself died in 1868 at the age of seventy-four.[9]

It is unlikely that either Zenone or Butti constructed the telescope in this collection, and the signature is crudely executed.[10] No other instrument has been recorded with both names together, although a stick barometer,[11] two wheel barometers[12] and a telescope[13] by Zenone, and a wheel barometer[14] by Butti have been noted. A later London firm of carvers and gilders, Ciceri, Pini & Co, who claimed to be 'manufacturers of looking-glasses, barometers, thermometers and telescopes ... beg to announce that they have just succeeded to the business so long carried on in the above premises by MR JOHN ZONONE [*sic*], and latterly by MESSRS BATTISTESSA & CO', took over 8 and 9 Calton Street as their Edinburgh branch in 1842.[15] It is probable that Zenone's stock was similar to that of his successors, and that he sold telescopes and microscopes bought in from wholesalers.[16]

REFERENCES

1. *Edinburgh Post Office Directory* 1823. The announcement of the Dissolution of the Copartnery appeared the following year: 'Intimation is hereby given, that the Company lately carrying on Business, under the Firm of ZENONE & BUTTI, Carvers and Gilders, No 5, Calton Street, Edinburgh, was DISSOLVED by mutual consent, on the 20th day of July last ... [signed] JOHN ZENONE. LEWIS BUTTI.' (*Edinburgh Gazette*, 24 August 1824, 160).

2. Lucio Sponza, *Italian Immigrants in Nineteenth-Century Britain: Realities and Images* (Leicester, 1988), 32-33. Sponza then quotes Lady Morgan: 'These poor Comasques issue forth to every country in Europe ... to carry on a petty commerce, in which ingenuity is combined with great industry and frugality: these are they who are everywhere seen with barometers, looking glasses, coloured prints, gilt frames, and other works, which smack of the arts and ingenuity of their native country.' (Sydney Owenson, Lady Morgan, *Italy* (London, 1821), I, 320-321).

3. *Ibid.*

4. *Scotsman*, 30 January 1828; an advertisement states: 'PARIS GOLD AND SILVER FLOWERS. JOHN ZENONE, IMPORTER OF FRENCH FLOWERS, and FEATHER MANUFACTURER, has given up the above business to his brother JOSEPH, whom he recommends to his Customers, and assures them of being as well served, and in a more commodious situation, being in the front shop, next to the former place.'

5. 'Giovanni Zenone, glass mirror maker and gilder, married a Mackenzie but in 1834 sold out to a new firm ... and returned to Italy.' (David Keir (ed.), *The Third Statistical Account of Scotland: The City of Edinburgh* (Glasgow, 1966), 123).

6. *Edinburgh Post Office Directory* 1823-1842. The spelling of the name varies: 'Zenoni' 1823-29; 'Zenone' 1830-41.

7. General Register Office (Scotland), Census of 1861 (Edinburgh, St. Cuthbert's, 692²/55, p.3.

8. *Edinburgh Post Office Directory* 1823-1910. The spelling of the name is given as 'Buttie' 1824-26.

9. G.R.O.(S.), Register of Deaths 1868 (Edinburgh, South Leith), 692²/492.

10. NMS T1980.257.

11. Nicholas Goodison, *English Barometers 1680-1860* 2nd edition (Woodbridge, Suffolk, 1977), 335. Goodison misreads the signature as 'Lenone'. A similar piece, with signature given as 'Leoni Fecit', was offered for sale by Phillips, 22 June 1988, Lot 4.

12. Offered for sale by Sotheby's, 21 July 1983, Lot 151; and at Phillips Edinburgh, 28 October 1988, Lot 78.

13. Offered for sale by Phillips, 20 July 1983, Lot 138.

14. Goodison, *op. cit.* (11), 306.

15. *Scotsman*, 16 March 1842. Ciceri & Pini should not be confused with the later Edinburgh firm of carvers and gilders Ciceri & Co., who were unconnected, as an advertisement by P.D. Torre & Co. makes clear: 'P.D. Torre & Company, successors to Ciceri, Mantica & Torre, & Ciceri and Pini (established 1841 [*sic*]), carvers, gilders, mirror manufacturers, and plate-glass merchants, 81 Leith Street. No connection with Ciceri & Co., of Frederick Street, who commenced business in 1875. Caution -Interdict. - P.D. Torre & Co. v Ciceri & Co. The Court of Session after proof, have granted interdict against Ciceri & Co., with expenses. The Court thereby interdicted and prohibited Ciceri & Co. from holding out to the public that they are continuing the business formerly carried on at 81 Leith Street, Edinburgh, by Ciceri, Mantica, & Torre, which now belongs to P.D. Torre & Co.; or that Ciceri & Co. had any interest therein; and in particular interdicted and prohibited Ciceri & Co. from using or issuing in connection with their business any Invoice, Account, or Letter Headings, or other Written or Printed matter bearing any expression denoting or indicating that Ciceri & Co., or Joshua Ciceri ever carried on business at 81 Leith Street, or ever had any interest in the business carried on there, or that Ciceri & Co.'s Firm was established prior to August 1875; and also from having any expression to the said effect in any Advertisement by them. Date of Interdict, 10 March 1877.' (*Edinburgh Post Office Directory* 1878, 118).

16. Advertisements confirm that carving and gilding was his main line of business (*Scotsman*, 19 April and 20 December 1828).

---

41. Refracting telescope: Zenoni & Butti, Edinburgh, 1823 (T1980.257)

1½″ aperture 3-draw hand-held refracting telescope with doublet objective and 2-component eyepiece and erecting assemblies (but lacking the latter); in brass, with objective ray shade and wooden barrel. Engraved on the first draw tube 'Zenoni & Butti / Edinburgh./ Day or Night'. Objective cover with 1¼″ aperture stop.

Length closed: 274mm.
Tube diameter: 62mm.
Aperture: 39mm.

Lacking the image erecting assembly and the sliding closures to both the objective cover and external eye stop.

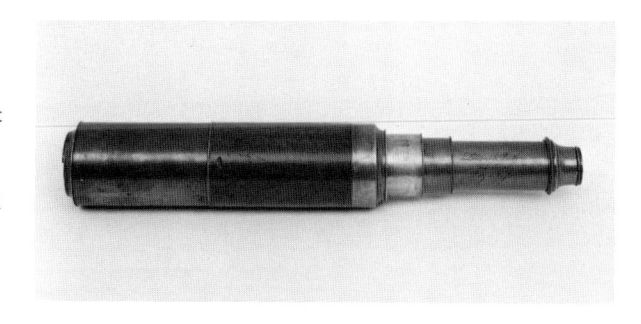

7
LEITH NAUTICAL INSTRUMENT MAKERS

Leith's geographical position, so close to the larger and more powerful capital city which greatly resented the port's attempts to win independent rights as a burgh, has meant that its history since the Reformation differs considerably from the other Scottish ports discussed in this work. Aberdeen, Dundee and Greenock were all able to establish and maintain foreign trade and commercial development, whereas Leith had to struggle for its existence as a port against the feudal stranglehold of its near neighbour Edinburgh.[1] Only in 1833 did Leith achieve its own short-lived municipal identity, which lasted until 1920;[2] but despite this stormy relationship with Edinburgh, Leith has been a place of considerable importance since the sixteenth century.

King James IV decided that Scotland should have a navy, and as there was insufficient deep water at Leith his shipyard was established a mile to the west, at Newhaven. The largest vessel built there was the *Great Michael* launched in 1511, two years before the King was killed at the Battle of Flodden where the idea of Scotland's navy died with him.[3] Despite being sacked by the English in 1544, occupied by a French army and besieged again during the Reformation struggles of 1560, Leith was nevertheless Scotland's chief port during the sixteenth century.

After the collapse of the shipbuilding, the people reverted to fishing for the herring and oyster which the Firth of Forth supported in abundance throughout the seventeenth and eighteenth centuries, although there were bad years when the herring failed to appear. Occasional whaling expeditions to the Greenland waters began in 1615, and the industry was put on a permanent basis with the founding of the Edinburgh Whale Fishing Company in 1750: by 1813 Leith was sending ten whalers to the Arctic, Britain's fourth whaling port after Hull's fifty-five ships, London's eighteen and Aberdeen's thirteen.[4] Whaling to Antarctica began at the turn of the century and continued until the 1950s. Shipbuilding continued in a small way so that by the end of the eighteenth century there were five businesses employing a total of 152 carpenters 'yet the facilities for shipbuilding were minimal - two dry docks and nothing else - owing to the extremely parsimonious attitude of the City of Edinburgh, which owned the Shore and Harbour of Leith'.[5] Eventually the harbour was developed and by the late nineteenth century, when the port had more control over its own affairs, there was a steady expansion in the volume of trade.

Leith's maritime outlook meant that ships' fittings were sold locally from a comparatively early date. Bryden discusses the case of Edward Buird, a ship's captain in Leith, who in February 1674 petitioned Edinburgh Town Council for permission to open a shop in Leith to sell ships' chandlery and nautical and mathematical instruments: 'it would appear ... that in importing navigational instruments Buird was supplying a need not otherwise met in Edinburgh.'[6] A similar licence, granted in 1681 to another Leith skipper, John Muir, specifically limited him to 'selling all manner of mathematical instruments for the use of seamen allenarly [only]'.[7] Another Leith ship's chandler selling nautical instruments was Thomas Mayo, who died in 1720, and the activities of these three, together with that of two Leith-based almanac makers, James Paterson (fl.1679-1693) and his nephew John Mann (fl.1693-1709), who also made and sold instruments, can be seen as 'a prelude to the existence of a local instrument-making trade'.[8] Over the next hundred years or so, the few instrument makers based in Leith tended to cater for the port exclusively with the selling and repair of nautical instruments: the close proximity of Edinburgh ensured that potential customers other than seamen would be catered for by opticians in the city.[9] As in the other Scottish ports during the late nineteenth century, there were also non-specialist firms retailing instruments bought in from elsewhere.[10]

REFERENCES

1. This struggle for autonomy is recounted by James Scott Marshall, *The Life and Times of Leith* (Edinburgh, 1986), 10-49.

2. *Ibid.*, 168-188.

3. *Ibid.*, 9-10.

4. David S. Henderson, *Fishing for the Whale* (Dundee, 1972), 11.

5. Marshall, *op. cit.* (1), 25.

6. D.J. Bryden, *Scottish Scientific Instrument-Makers 1600-1900* (Edinburgh, 1972), 4.

7. Marguerite Wood and Helen Armet (eds.), *Extracts from the Records of the Burgh of Edinburgh 1681-1689* (Edinburgh, 1954), 54.

8. Bryden, *op. cit.* (6), 6.

9. Bryden, *ibid.*, lists the following eighteenth and nineteenth century Leith-based instrument-makers: George Alexander, watch and compass maker, fl.1813-25 62 Shore, proprietor of British Patent No. 3646 of 1813 for suspending the card of the mariner's compass; George B. Brown, clock, watch and nautical instrument maker, 1836-37 38 Shore; Brown & Chalmers, clock, watch and nautical instrument makers, 1838-40 37 Bridge Street; John Dickman, chronometer, clock, watch and nautical instrument maker, 1794-96 Kirkgate, 1797-1813 Bernard Street, 1814-40 33 Shore; David Laird (q.v.); Peter Lyon, mathematical, nautical and optical instrument maker, 1784-88 Shore; Richard Millar, clock, watch and nautical instrument maker, 1845-51 45 Bridge Street, 1852-68 58 Bridge Street; Thomas Short, mathematical and optical instrument maker, fl.1748-68 Foot of the Broad Wynd (in 1768 he moved to London to run the business of his late brother, James Short (q.v.), returning to Edinburgh in 1776); David Stalker (q.v.); Robert Trotter, watch and compass maker, 1822-27 23 Couper Street, 1828-36 89 Kirkgate.

10. Examples of these are a sextant marked 'J. Grant Leith' offered for sale at Sotheby's, 21 October 1977, Lot 42, and another at Christie's South Kensington, 12 March 1981, Lot 122, a compass and another sextant marked 'J. Hogg Leith' offered for sale by Christie's South Kensington, 7 January 1982, Lot 159. A compass-adjusting kit by J. Hogg of Leith, is in the Royal Museum of Scotland, NMS T1974.100. Other instruments made in Leith, apart from those for use at sea appear to be unusual. The 18th century chemical glassware now in the Royal Museum of Scotland (NMS T1858.275.38-45), probably used by Joseph Black (1728-1799), professor of chemistry at the University of Edinburgh, may well have been manufactured at Leith (R.G.W. Anderson, *The Playfair Collection and the Teaching of Chemistry at the University of Edinburgh 1713-1858* (Edinburgh, 1978), 135-147). By 1700 the glassworks there employed about 120 skilled workers (S.G.E. Lythe and J. Butt, *An Economic History of Scotland 1100-1939* (Glasgow, 1975), 45) but heavy taxation had ended this profitable local industry by the close of the 19th century.

DAVID LAIRD

David White Laird's first appearance was in the Leith section of the 1834 street directory for Edinburgh, at 4 Bridge Steet, Leith, in premises previously occupied by a watchmaker, James Robertson.[1] Laird was also described as a watchmaker, but from 1836-40 his occupation was listed as a watchmaker and nautical instrument maker; and as a clock, watch and nautical instrument maker from 1841. In 1843 he took on additional premises at 58 Bridge Street, and the business had transferred to there by 1844. He was also listed between 1846 and 1850 at 59 Bridge Street, which presumably represents an expansion of the workshop, and the firm last appeared in the 1851 directory at the 58 Bridge Street address only.[2]

Laird was recorded in the Census return for 1841 as a reasonably young man with a growing family of four children.[3] At the next census, ten years later, his wife and most of the children had disappeared (although they may merely have been away on the day the census information was gathered). He was described in 1851 as a 'watchmaker, master employing 4 men',[4] and this may indicate that horological work was the firm's main source of business. The trade label with the sextant in the Frank collection[5] (which is of the late 1840s) boasts the supply of 'Sextants Quadrants & Telescopes, Sea Charts, Gunter's Scales, Ship & Pocket Compasses, Time Glasses &c.'. The sextant itself is unsigned, but other signed examples are known.[6] However, the same label also claims 'Gold & Silver Watches of every description, Watches, Musical Boxes & Jewellery repa[ired]' and, over a flamboyant representation of the Royal Arms 'Clockmaker to the Honourable the Pa[rliame]ntary Commissioners'.

It is possible that like other watchmakers elsewhere (for example, Alexander Cameron of Dundee (q.v.), or James Berry of Aberdeen (q.v.)), for Laird instruments were very much a sideline, the demand created by the location of the business in a port. Certainly the advertisement which appeared in the Post Office directory for 1847 would bear this out: the horological material and services were more prominently advertised, while nautical instruments were merely listed in a smaller type-face.[7]

On 16 April 1850, David Laird petitioned for sequestration, and was later discharged.[8] The business

29 David Laird's trade card for his address between 1843 and 1851

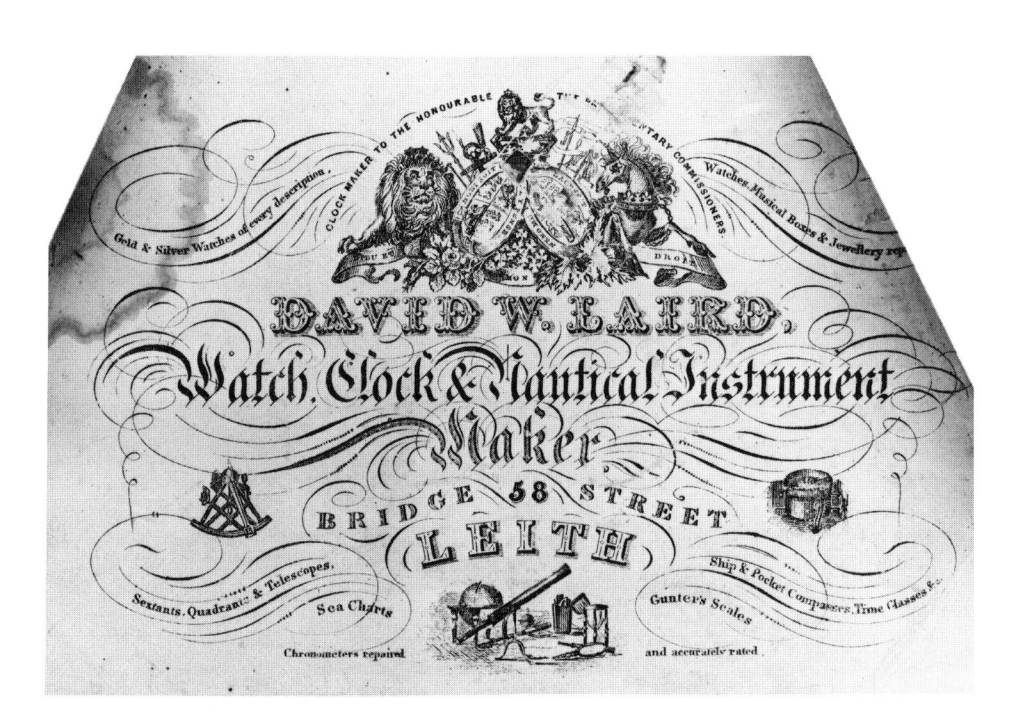

premises at 58 Bridge Street were taken over by Richard Millar, junior, in 1852, a clock and watchmaker from an Edinburgh horological family, who, together with his father had started in business in Leith in 1845,[9] and he continued to use Laird's trade card with the substitution of his own name.[10]

REFERENCES

1. *Edinburgh Post Office Directory* 1824-1834, *passim*. Robertson is recorded as working in Leith between 1818 and 1836, and having been admitted freeman clock and watchmaker of the Canongate Hammermen in 1818; in 1834 he moved to 4 West Nicolson Street, Edinburgh: John Smith, *Old Scottish Clockmakers 1453-1850* 2nd edition (Edinburgh, 1921), 319.

2. *Edinburgh Post Office Directory* 1834-1851, *passim*.

3. General Register Office (Scotland), Census of 1841 (Edinburgh, North Leith), 692¹/2, p.1. His age was given as 25, born in the county of Edinburgh.

4. G.R.O.(S.), Census of 1851 (Edinburgh, North Leith), 745/50, p.1. This time his age was given as 37, his birthplace as North Leith. Of the 4 children in the previous Census, only his daughter Mary remained, joined by the 4 year old Sarah.

5. NMS T1980.182.

6. For example, Dundee Art Galleries and Museums have an 11 inch radius octant, marked 'D.W. Laird Leith' (inventory 1974-409), and at Whitby Museum there is a 9⅞ inch radius octant similarly marked. An octant marked 'D.W. Laird, Leith' was offered for sale at Sotheby's, 20 May 1974, Lot 11. A single-draw refracting telescope marked 'D.W. Laird Leith' was offered for sale by Phillips Edinburgh, 31 July 1987, Lot 109.

7. *Edinburgh Post Office Directory* 1847, xlix. However, it has been pointed out that claims to be makers in advertisements, trade cards and trade directories were often untrue: M. A. Crawforth, 'Evidence from Trade Cards for the Scientific Instrument Industry', *Annals of Science* 42 (1985), 477.

8. Scottish Record Office, CS 279/1380, Sequestration Papers, David White Laird. There is no Sederunt Book, and the details of this sequestration are very sparse, merely the Petition for Sequestration, the Act of Warrant appointing the Trustee, and the Discharge. Notices appeared in the *Edinburgh Gazette*: the initial notice of sequestration (19 April 1850, 323); the announcement of the appointment of the Trustee and the Commissioners (21 May 1850, 410); a Petition for Discharge by Laird (20 September 1850, 796); an announcement of a dividend payment (1 November 1850, 912); three separate announcements that dividend payment was postponed (4 March 1851, 173; 4 July 1851, 522; 31 October 1851, 998); and the announcement that the second and final dividend would be paid on 17 April 1852 (2 March 1852, 188).

9. *Edinburgh Post Office Directory* 1845-1852. John Smith, *op. cit.* (1), 263-264 discusses Richard Millar senior.

10. An example is with an octant signed by Laird in Dundee Art Galleries and Museum (inventory 1974-409).

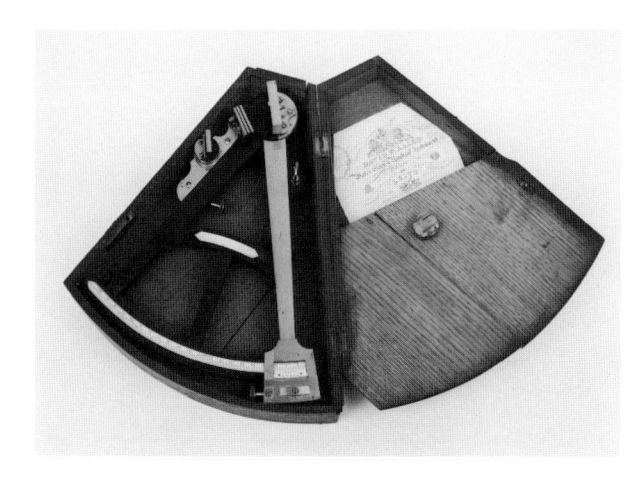

42. Octant: unsigned, c.1845 (T1980.182)

10" octant with ebony frame and brass fitments. Scale divided on inset ivory arc [-2° 40']-100 by 20 mins., the plain index arm with clamp, tangent screw and vernier reading to 1 min., and stamped with the divider's initials 'W. H'; the inset ivory name plate blank. The frame with single vertical strut and bowed horizontal strut; and provided with double pin-hole sights, 3 shades, adjustable horizon glass, provision for noteplate on reverse and pencil in vertical strut, and supported on feet. With fitted keystone-shaped stepped case containing trade card for David W. Laird, 58 Bridge Street, Leith.

Scale radius: 245mm.
Case size: 295 x 315 x 100mm.

Lacking the noteplate and pencil.

The initials W H probably indicate that the octant was wholesaled by William Harris of London. Harris was commissioned by Sir David Brewster to make a goniometer in 1809, and together they took out U.K. patent 3453, 21 May 1811 for a micrometer telescope (examples are in the Science Museum, London, inventory 1913-288, and the Whipple Museum of the History of Science, inventory 647). However, when Brewster's kaleidoscope was pirated during the patenting process during 1817, it was probably through Harris's indiscretion. Taylor gives his dates as fl.1800-1848 (E.G.R. Taylor, *The Mathematical Practitioners of Hanoverian England 1714-1840* (Cambridge, 1966), 365; A.D. Morrison-Low, 'Brewster and Scientific Instruments' in A.D. Morrison-Low and J.R.R. Christie, *'Martyr of Science': Sir David Brewster 1781-1868* (Edinburgh, 1984), 60-61; *idem.*, 'Scientific Apparatus associated with Sir David Brewster: An Illustrated Catalogue of the Bicentenary Display at the Royal Scottish Museum 21 November 1981-9 April 1982' in *ibid.*, 83-86).

DAVID STALKER

David Stalker was born in Leith in about 1830, the son of William Stalker, a master sailmaker, and Margaret Moodie his wife.[1] He set up in business as a watch and clock maker at 9 Commercial Street, Leith in 1855, taking over the additional premises at 10 Commercial Street in 1867, then moving in 1876 to 6 Commercial Street, from where the business operated until the twentieth century. In 1893 he was described for the first time in the street directory by the additional trade of 'optician'[2] although he had been trading in the instruments which were in demand in a port such as Leith for some time before that date. For instance, at the 1871 census he described himself as a 'watch, clock and nautical instrument maker employing 4 Men [and] 3 Boys',[3] and ten years later gave a similar description, 'employing 5 men'.[4] Unfortunately, the 1891 census gives no clues as to the size of his workshop at that date.

Stalker produced trade cards for each of his addresses: the one for 9 Commercial Street (1855-66) advertised sextants, quadrants (i.e. octants), compasses, telescopes, log and time glasses, barometers, thermometers, Adie's patent sympiesometers, and mathematical instruments.[5] The card features an engraved figure of an officer of the watch taking a reading with a sextant - an echo of the splendid and nearly life-size figure mounted as a shop-sign above Stalker's premises.[6] Stalker's trade card for 9 & 10 Commercial Street (1867-75) was redesigned and claimed that he was a 'manufacturer [of] sextants, quadrants ... telescopes, ship compasses, time glasses &c., Repairs Musical Boxes & Jewellery'.[7] For the final address at 6 Commercial Street (1876-1917) Stalker reverted to the design of his first trade card, with the additional information that he had been 'established 20 years'.[8]

In 1887 Stalker advertised in an almanac published by Duncan McGregor (q.v.) a whole variety of nautical

30 David Stalker's trade card for his address between 1867 and 1875

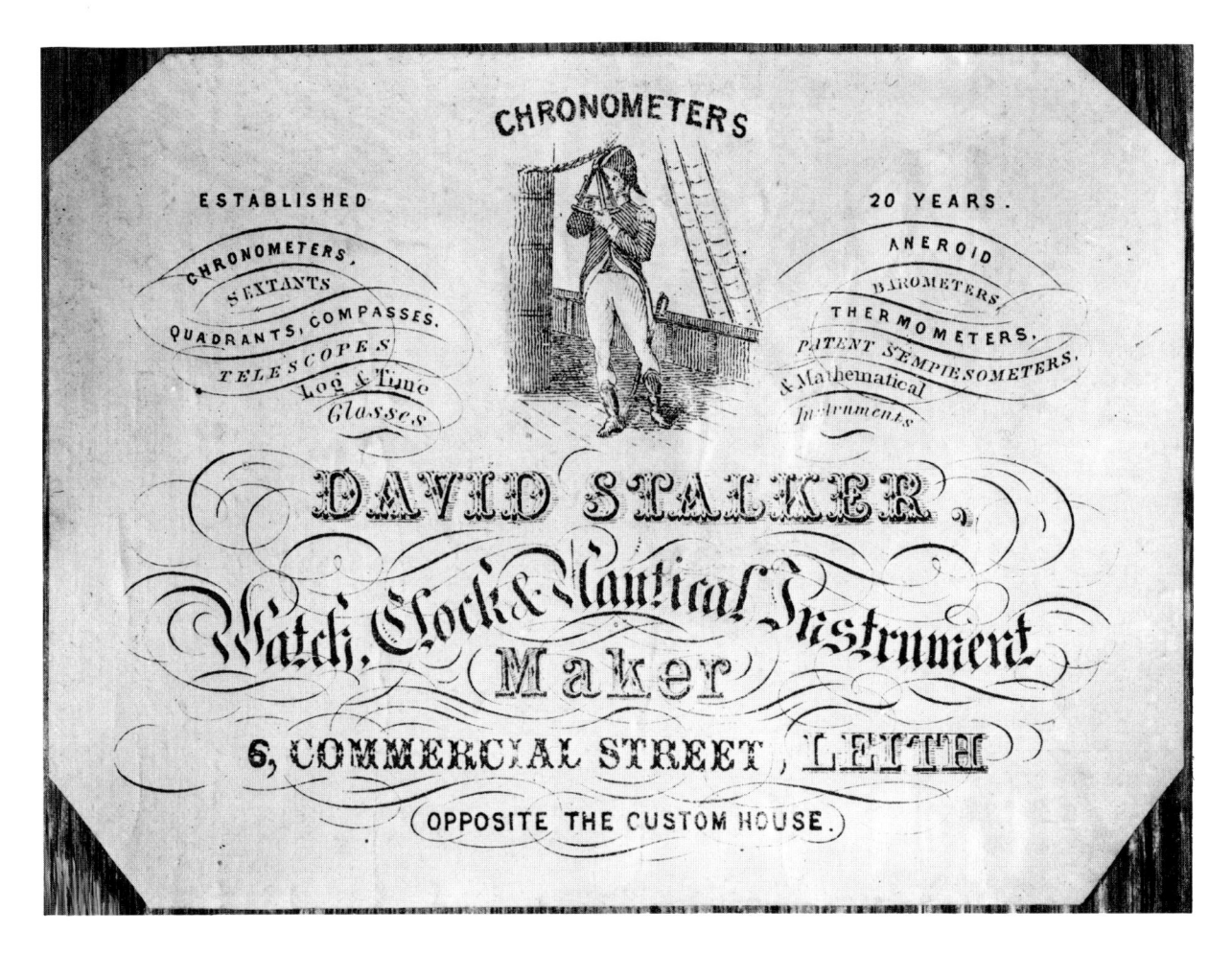

31 David Stalker's trade card for his address between 1876 and 1895

instruments: 'sextants & quadrants, telescopes, sea charts, ship & pocket compasses, aneroid barometers, Gunter's scales etc.'[9] At the same time he drew attention to a number of services he offered: 'Compasses adjusted on board iron vessels. Agent for Admiralty chart agents. Chronometers repaired & accurately rated.' It is possible that the people employed by Stalker were involved principally with these tasks rather than with the production or assembly of instruments. Certainly some instruments were merely retailed by Stalker; for instance, a combined marine barometer and improved sympiesometer with his name on it[10] had probably been wholesaled by J.J. Hicks of London. In other instances it is unclear whether he was the manufacturer, as for example is the case of a dry compass card bearing compass with the card made either by or for him, in a bowl signed by G. Whitbread of London.[12]

David Stalker died aged sixty-five on 11 May 1895.[13] He died intestate, leaving an estate worth £581 10s 4d.[14] He was survived by his wife Helen Archer, ten years his junior,[15] who appears then to have run the business herself in her husband's name, presumably until her own death in 1917 after which the business was incorporated into that of Turnbull & Co. (q.v.).[16]

REFERENCES

1. General Register Office (Scotland), Register of Deaths 1895 (Edinburgh, North Leith), 692¹/270.

2. *Edinburgh Post Office Directory* 1855-1893.

3. G.R.O.(S.), Census of 1871 (Leith, North Leith), 692¹/4, p.15

4. G.R.O.(S.), Census of 1881 (Leith, North Leith), 692¹/1, p.18

5. With an octant in the Leith Nautical College collection at the Royal Museum of Scotland, NMS T1987.151.

6. The figure is now in the collection of Edinburgh City Museums (inventory HH 3145/67), and is illustrated as the frontispiece in D.J. Bryden, *Scottish Scientific Instrument-Makers 1600-1900* (Edinburgh, 1972). The shops of both Berry & MacKay (q.v.) and P.A. Feathers (q.v.) were furnished with similar street signs. An example still survives in Duke Street Dublin, which once decorated the shop front of Richard Spear, mathematical instrument maker. Other examples have been recorded.

7. With a sextant in the Leith Nautical College collection at the Royal Museum of Scotland, NMS T1987.142.

8. With an octant signed 'STALKER LEITH' in the Royal Museum of Scotland, NMS T1962.48. A similar unsigned

instrument with David Stalker's trade label for 6 Commercial Street was offered for sale by Sotheby's, 23 October 1985, Lot 100.

9. [D. McGregor], *McGregor's Almanac and Tide Tables for 1888* (Glasgow, 1887), 237.

10. In a private collection.

11. James J. Hicks, *Illustrated & Descriptive Wholesale Catalogue of Standard, Self-Recording and other Meteorological Instruments ...* (London, n.d. [c.1880]), 25, item 32 figure 15, price £3 12s 6d.

12. Royal Museum of Scotland, Leith Nautical College collection, NMS T1987.156.

13. *Loc. cit.* (1). He died of a long-standing heart disease.

14. Scottish Record Office, SC 70/1/341, p.370, Inventory of the estate of David Stalker, optician, Commercial Street, Leith, recorded 2 July 1895.

15. G.R.O.(S), Census of 1891 (Leith, North Leith), 692¹/1, p.17.

16. *Edinburgh Post Office Directory* 1895-1917.

---

43. Octant: D. Stalker, Leith, c.1860 (T1980.181)

10″ octant with ebony frame and brass fitments. Scale divided on inset ivory arc [-3]-[108] by 20 mins., the reinforced index arm with clamp, tangent screw and vernier reading to ½ min., with swinging magnifier. Stamped on the inset ivory name plate 'D. STALKER LEITH'. The frame with two vertical struts and straight horizontal strut; provided with pin-hole sight, sighting tube and telescope with solar shade, 5 shades, adjustable horizon glass (with tilt adjustment key in the case), with handle at the rear and supported on feet. With fitted keystone-shaped case containing trade card for David Stalker, 9 Commercial Place, Leith.

Scale radius: 247mm.
Case size: 310 x 330 x 120mm.

Lacking the attachment screw for the sights and for one of the handle supports. The handle is distressed and one foot has sheared off.

W. CRAIG

Little is known about W. Craig, who described himself in the street directories between 1927 and 1933 as an optician and philosophical instrument maker. In 1927 he took over the premises at 6 Commercial Street from Turnbull & Co. (q.v.), although the business was still referred to in the trades section of the street directory as that of the former occupant, David Stalker (q.v.). From 1930, he gave his home address as 48 Clark Road, which is in the Trinity district of Edinburgh. In 1934 Craig described himself as an ophthalmic optician; the business disappeared from the directory in 1971.[1] In 1952, a retired scientific instrument maker named William Craig died in Glasgow, and this may have been the same man.[2] The telescope in this collection appears to be a standard mass-produced item which he retailed.[3]

REFERENCES

1. *Edinburgh Post Office Directory* 1927-1971, *passim*.

2. General Register Office (Scotland), Register of Deaths 1952 (Hillhead, Glasgow), 644[13]/429. Craig died on 4 April 1952, aged 67. The value of his estate was £363 2s 5d (Scottish Record Office, *Calendar of Confirmations* 1952, p. C149).

3. NMS T1980.265.

---

44. Refracting telescope: W.M. Craig, Edinburgh, c.1930 (T1980.265)

2¼" aperture single-draw hand-held refracting telescope, with doublet objective (damaged) and 2-component eyepiece and erecting assemblies; in brass, with objective ray shade and tapered leather-covered barrel. Engraved on the draw tube 'W.M. CRAIG / TRINITY / EDINBURGH' and 'MULL OF KINTYRE LIGHTHOUSE'. Scratched assembly mark 'V' on the barrel under the end sleeve.

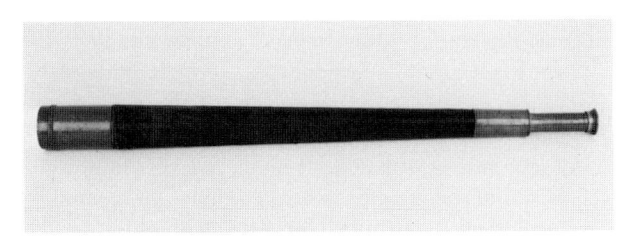

Length closed: 817mm.
Maximum tube diameter: 70mm.
Aperture: 60mm.

The objective mount is badly distorted and the lens severely chipped. The leather covering of the barrel has shrunk and been marked by an outer wound covering of string. The eyepiece assembly has parted and the external eye stop has no rear eye plate. Lacking objective cover. The instrument is distressed from heavy use.

Of the first 4 lighthouses built by the Northern Lighthouse Board, the most difficult site was that at the Mull of Kintyre, on the south-west coast of Scotland and within sight of the north of Ireland. Designed by Thomas Smith, father-in-law of Robert Stevenson, it was first lit in October 1788 (R.W. Munro, *Scottish Lighthouses* (Stornaway, 1979), 52-56; Keith Allardyce and Evelyn M. Hood, *At Scotland's Edge: A Celebration of Two Hundred Years of the Lighthouse Service in Scotland and the Isle of Man* (Glasgow and London), 17-18, 94-95).

8

THE BRYSON FAMILY

James Mackay Bryson (1824-1894) came from a family which held a position as one of the most important and prestigious watch and clockmaking business in Edinburgh for most of the nineteenth century.[1] His father Robert Bryson (1778-1852), who founded the family firm by 1810, not only distinguished himself at his trade by his mechanical ingenuity and scientific skill, but was also active outside his workshop in ways which rank him alongside Alexander Adie (q.v.) as a master craftsman and promoter of science. Robert Bryson's reputation as an horologist was such that he was included among the first twenty-four councillors of David Brewster's Society for the Promotion of the Useful Arts in Scotland, which was set up in 1821.[2] Although Bryson subsequently took little active part in the Society's meetings, his occasional communications reveal that not only was he an inventive improver of watches and clocks, but that he was also interested in scientific instruments which incorporated timepieces. For example, the 'ingenious mechanism' of his self-registering barometer, when exhibited to the Society of Arts in 1844, had already recorded 12,000 hourly readings, which Bryson duly tabulated.[3] This meticulous meteorological observation was very much the forte of scientifically minded mechanics in Edinburgh of whom the Brysons and the Adies were the leading examples.

So, while not actually practising as an instrument maker Robert Bryson made essays in the discipline, and was followed and then surpassed in this by his eldest son Alexander (1816-1866). Alexander Bryson trained as a watch and clockmaker in Musselburgh and London,[4] returning to Edinburgh by 1844 when he and his younger brother Robert (1819-1886) were presumably taken in partnership with their father, as the firm traded as Robert Bryson & Son(s) from this date.[5] His scientific interests spread wider than his father's, embracing not only horology and meteorology, but also photography, geology and the physical sciences as a whole, as can be judged from his contributions to the Society of Arts. From the time he joined in 1836 until his death in 1866 Alexander Bryson was one of its most conspicuously active members, serving it in many capacities, not least as a regular member of its committees and as an untiring recruiter of new Fellows.[6]

Several of his many communications to the Society concerned his father's horological and instrument work, in which he probably assisted him. Both men evidently constructed meteorological instruments[7] and as a result the authorship of pieces known only as by 'Bryson', is in doubt. For instance, the 'Metallic Thermometer' exhibited in 1839, was described as 'Invented by Mr Bryson';[8] so while the most likely attribution is to Robert Bryson, Alexander may also have been responsible for it. By that date he was well-established in the Society of Arts, and it is significant that in 1836 he was connected with microscope work for the Society by James Bell, an Edinburgh philosophical instrument maker who also advertised as a clock and machine maker.[9] Furthermore, on the Society's committees he frequently sat alongside John Adie (q.v.) and John Dunn (q.v.),[10] two of the city's most prominent instrument makers, and it is reasonable to suppose that the lively forum which the Society provided for the exchange of scientific ideas and techniques stimulated Alexander Bryson's considerable talent outside his daily trade. Thus the usually clear-cut distinction between the work of the horologist and the scientific instrument maker broke down in the case of the Brysons, a fact which has obvious implications for the mechanical education of Alexander's younger brother James Mackay Bryson.

Another important aspect of J.M. Bryson's upbringing was his father's role in the foundation of the Edinburgh School of Arts, which ultimately grew to become the Heriot-Watt University. According to Leonard Horner, its principal founder, it 'took its origin from an accidental conversation in the shop of Mr Bryson, Watchmaker, in March 1821'.[11] Much of the early recruitment was done through Bryson and his fellow master mechanics in related trades, and it was the further education of their apprentices and journeymen to which the scientific and mathematical classes were directed.[12] Alexander Bryson regularly attended the School for a number of years, in addition to the not uncommon practice of enrolling in the natural philosophy and chemistry classes at the University.[13]

Some years later, after initial schooling at the Southern Academy, J.M. Bryson began to attend the School of Arts,[14] although he is not listed among its prize-winning pupils. Presumably by this time he would have begun an apprenticeship, and he is therefore probably the James Bryson recorded in the Census of 1841 in Robert Bryson's house as an 'optician's apprentice'.[15] It

32 Robert Bryson (1778-1852), calotype by D. O. Hill and Robert Adamson. *National Galleries of Scotland, Edinburgh*

is not known under whom he served, but in view of his father's prestige it was probably with one of the city's best masters. The Adies clearly fell into this category, but surprisingly, there is scant suggestion that the Brysons had any dealings with them. Indeed, one of the few connections that can be found is that James Milne, the master brass founder, was closely associated with both businesses (probably supplying each with plate and cast pieces), and he joined Robert Bryson and Leonard Horner in the triumvirate which founded the School of Arts.

A further link was the presence of John Adie and Alexander Bryson in several committees of the Society of Arts, but the instrument maker John Dunn (q.v.) often formed a trio with them, and he too is a candidate to have been Bryson's master. The picture is not made much clearer by the fact that despite his failure to shine as a pupil in the school, J.M. Bryson was taken on as an assistant by George Buchanan, who taught mathematics there. The precise period of his employment is not known, although he was working for him by early 1842, since a drawing of Buchanan's 'Protracting Table' dated 14 March 1842 is signed 'Drawn by James Bryson, Assistant to Mr Buchanan'.[16] Buchanan was a consulting civil engineer who had been involved in metrological work with Alexander Adie in the 1820s, and the first of his protracting tables was begun by John Dunn in about 1841, although completed by Adie & Son in 1842.[17] Without anything more than circumstantial evidence, however, the question of Bryson's apprenticeship remains open.

By about 1843 the apprenticeship had ended, for at that

date Bryson went to Hamburg

> 'where for some years he studied and worked under Repsold, the distinguished German instrument maker. Thence he proceeded to Munich, and studied under the famous Mertz, of the firm of Mertz and Mahler, the construction of lenses for astronomical instruments, and other work of a like character.'[18]

In 1850, after seven years in Germany, he returned to Edinburgh and set up as an optician at 65 Princes Street, next door to his father and brothers' shop. He moved to 24 Princes Street in 1853, and subsequently was at No. 60 between 1855 and 1866; from 1867 until 1893 his address is given variously as 60 or 60A Princes Street.[19]

His business appears to have been successful, although comparatively little is known about the scope and quality of his instrument work, nor to what extent it was influenced by the unusual fact that he was trained abroad. As might be expected from a well-educated and finished craftsman with influential family connections, James Mackay Bryson was within a short period securing the kinds of commission which brought him prestige and respectability.

His earliest known works date from 1854 and form two related groups. First were several special commissions, the most interesting of which was the young physicist James Clerk Maxwell's order for 'improved tops'. These variations on the child's toy were made by Bryson in 1854 for Clerk Maxwell's experiments on colour, and were available with coloured papers made to his description from Bryson's shop.[20] Among his clients was the civil engineer Professor William Pole of University College, London.[21] Given this expertise in spinning tops, it is not surprising to find that Bryson supplied a gyroscope in 1855 for an Edinburgh scientist's paper on that instrument to the Society of Arts.[22]

The other early work was executed for the Natural Philosophy Department of the University of Edinburgh, under its professor, James David Forbes. In 1854 Bryson supplied and installed in the classroom a pendulum to demonstrate Foucault's experiment, and in 1858 adapted a brass hour circle for use with it. He also supplied a Leyden jar and stand in 1854, altering it in 1858, when he also renewed the tube of a water barometer.[23] In themselves they are unremarkable items, but they show that Bryson was well established by 1854, even if he was only one of several suppliers of Forbes' apparatus. Among the other Edinburgh businesses which sold instruments to Forbes, was Adie & Son (q.v.), and it may be that Bryson's connection with the professor, like theirs, was furthered through common interest in meteorology. It has already been noted that both Robert and Alexander Bryson were active in that field, and it may also have been important for J.M. Bryson that Alexander was an intimate friend of the great amateur scientist and patron Sir Thomas Makdougall Brisbane. There is, however, only limited evidence to show that J.M. Bryson was a maker of meteorological instruments.[24]

He made optical instruments, but probably only in a small way. For instance, by 1856 he was producing Gairdner's simple microscope, the invention of Dr William Gairdner, an Edinburgh physician.[25] The examples of microscopes in this collection show that not only did he manufacture instruments of his own, but that he also retailed continental models. This can partly be explained by his German training, and also by the fact that he seems to have worked with little assistance. He is said to have been the first person in Edinburgh to surface a cylindrical lens,[26] but given his training and the paucity of practised microscope makers in the city, this was not a particularly noteworthy achievement.

The same not wholly reliable source also states that Bryson 'is to be remembered because of his work with Nicol in producing the polarising prism'.[27] William Nicol, an elusive figure who was an important pioneer of optical techniques in the 1820s, resembled many scientists at this date in using private means to support his own experimental work, and although he called himself a lecturer in natural philosophy it is doubtful whether the occasional work he is known to have undertaken would have supported him unsupplemented. Nicol's achievement rests principally on the ingeniously simple use of two pieces of calcite as a double-refracting prism to polarise light. Since his method was first published in 1829[28] it can be assumed that the five year old James Mackay Bryson had nothing to do with Nicol's early work. His absence abroad and return shortly before Nicol's death in 1851 effectively rule out later collaboration.

J.M. Bryson's involvement in Nicol's optical work is thus a case of mistaken identity, and it is now clear that it was his brother Alexander who enjoyed an intimate association with Nicol. These two men were both deeply fascinated by geology, and seem to have struck a personal and scientific rapport before 1840.[29] Alexander Bryson was extremely active as a geologist,[30] and it was in his pursuit of a more exact understanding of rock structures that his experimental enthusiasm chimed with Nicol's. The latter had adopted a technique of slicing rock and mineral samples very thinly for examination by microscope, and Bryson followed suit in attempts to show the aqueous origin of granite.[31] More significant than his clinging to outmoded geological theory was the fact that Nicol bequeathed to him his entire collection of minerals, fossils, shells, books, philosophical and optical apparatus, together with his Edinburgh house and its contents.[32] Alexander Bryson's geological cabinet was thus of great interest to the rising generation of geologists who wished to pursue Nicol's investigations.[33]

The eminent sientist H.C. Sorby was led by examining the Nicol material in 1854 to the investigation of fluid cavities,[34] as was Bryson himself. Archibald Geikie, the geologist, who often inspected the cabinet in Bryson's shop as a youth, recalled the slices which 'Sandy' Bryson had made himself with his considerable dexterity as a manipulator'.[35] Some of Alexander's collection seems to have passed into the hands of James Mackay Bryson, who is known to have had a few Nicol's prisms towards the end of his life;[36] by that time they were probably old-fashioned curiosites in his shop.

Although Alexander Bryson made no significant contribution to geological science with any of his many papers, one instance of his indefatigable enthusiasm for 'geologising' in the field provides an example of his close working relationship with his brother James. This was his expedition to Iceland and the Faroe islands in 1862, undertaken with friends ostensibly to measure the internal temperature of the Great Geyser on Iceland. The jolly travelogue which Alexander later published reflects the spirit of Victorian scientific enquiry, and amid much topographical observation, reveals in the account of the actual observations that he used twelve large thermometers specially constructed by James with open tops to allow the mercury to escape in measurable quantities.[37] Indeed, J.M. Bryson appears to have specialised to some extent in high temperature work, for in about 1875 he supplied W.H. Chambers, a government safety officer, with a thermometer capable of registering 400° Fahrenheit for experiments on fireproofing.[38]

One reason for our scant knowledge of J.M. Bryson's work is that he was not as active in Edinburgh's various improving and scientific organisations as either his father or his brother Alexander. He was a member of the Society of Arts from 1857, sitting on its Council in sessions 1864-6, and 1887-9 and on its Prize Committee in 1865-6, 1880-1, and 1881-9. He also occasionally sat on committees appointed to evaluate contributions to its proceedings.[39] In July 1876 he described and exhibited Crookes' Radiometer to the Society. It was precisely the type of scientific and 'philosophical' novelty which opticians kept in their windows,[40] and Bryson probably sold it in numbers. On the whole, however, Bryson's absence from committees on which his specialist knowledge would presumably have been welcomed, and his failure to produce papers before the Society, tend to confirm 'the modesty and retiring disposition' which his obituarist noted.[41]

On the other hand, in common with almost everyone who had any scientific curiosity in mid-century, J.M. Bryson was interested in photography, although in this respect he was again over-shadowed by his brother.[42] He became a member of the Photographic Society of Scotland in 1858, and his specialised knowledge made him a member of one of its committees formed to evaluate landscape lenses for cameras currently on the market in 1859.[43] In 1864 he exhibited to the Edinburgh Photographic Society two 'photo-sculpture busts' of his brother Robert and his sister-in-law, presumably as their author. His keenness on the medium is evinced by the fact that he probably owned a stereograph and stereoscopes.[44] He sat on the Council of the Edinburgh Photographic Society in 1867. Following Professor Charles Piazzi Smyth's dramatic and successful use of a compact wet-plate camera to photograph the interior of the Great Pyramid in 1865, J.M. Bryson is said to have manufactured a camera based on Smyth's description.[45] Like other Edinburgh opticians who acted as photographic dealers, he is listed as a chemical dealer.[46]

The Piazzi Smyth-type camera, like the Gairdner simple microscope, is problematic in that although it appears to have been a particular specialisation, it is not known to have been produced in significant numbers. Indeed, there are indications that Bryson's workshop was never large, for the names of only three assistants have survived, and the state of his business at his death suggests that it was largely retail. The important name among the three is that of Angus Henderson, said to have worked for him and for Adie before setting up independently as an instrument maker in Edinburgh in 1862. The second is John Trotter, who allegedly served his apprenticeship with Bryson before moving to Glasgow, where he set up in business independently in 1890 after working briefly with George Mason & Co.[47] The third was Thomas Weir who witnessed Bryson's will in 1868 as his 'assistant',[48] but of whom nothing further is known.

J.M. Bryson's business seems to have been completely separate from the horological concern run by other members of his family; but he and his brothers appear to have had an interest in it. Founded by his father Robert by 1810, when it operated at the Mint, High Street, Edinburgh, the horological business 'removed ... to that commodious house, No 5 South Bridge, opposite to Hunter's Square' in 1815.[49] The firm moved to 66 Princes Street in 1840, and with Robert Bryson junior's death in 1886, the business and the Royal Appointment passed into the hands of Messrs. Hamilton & Inches.[50] Unfortunately, Robert Bryson junior died bankrupt, with the petition for the sequestration of his estates being brought by J.M. Bryson, who was owed £6000.[51] The Sederunt Book reveals that the deceased Alexander Bryson of Hawkhill, watchmaker, was owed £2300: and it was noted that

> 'Mr James Mackay Bryson the Bankrupt's brother claims to be a partner in the Firm under an alleged Contract dated 27th November & 7 December 1885 to the effect of practically extinguishing the Bankrupt's interest therein. The

33 Alexander Bryson (1816-1866), by an unknown photographer. *National Galleries of Scotland, Edinburgh*

Trustee is advised that this transaction is reducible and in that case about £4500 would be brought into the Estate but meantime no value can be put upon the Asset.'

The assets of the estate were valued at £797 12s, but it was apparent that the debts were entirely financial and not problems connected with the clockmaking business as such; Robert Bryson had arrears in shares and liabilities under cautionary obligations. However, the net effect was that the clock and watch making business was disposed of and Robert Bryson's estate was finally discharged in 1890.[52]

At the time of his own death in 1894 J.M. Bryson's estate was worth a gross total of £10,135, of which his stock in trade and other effects in his shop were worth £224 14s 6d.[53] Although such valuations are not necessarily reliable, the figure is low enough to suggest a modest business and this is supported by the sums of money owing to him, which fall into two groups, of which one was a list of instrument making firms with whom Bryson was evidently doing business. The smallness of the sums suggest small items supplied by Bryson for resale, or what is more likely residual credit with wholesale suppliers who provided the bulk of his stock: these firms included James White of Glasgow (q.v.), Parkes & Son (Birmingham), Yeates (Dublin) and Elliott Bros. (London).[54] The second list itemises the 234 customers who owed Bryson a total of £216 18s 7d, and makes it clear that his account-paying clientele consisted of the nobility and gentry, academics and professional men, a large body of respectable citizens, and a few institutions such as the North British Railway Co. and Fettes College.[55] There is no record of the cash paid by less worthy customers. Again, the smallness of the individual sums suggests that by 1894 Bryson was relying on ophthalmic work as his main

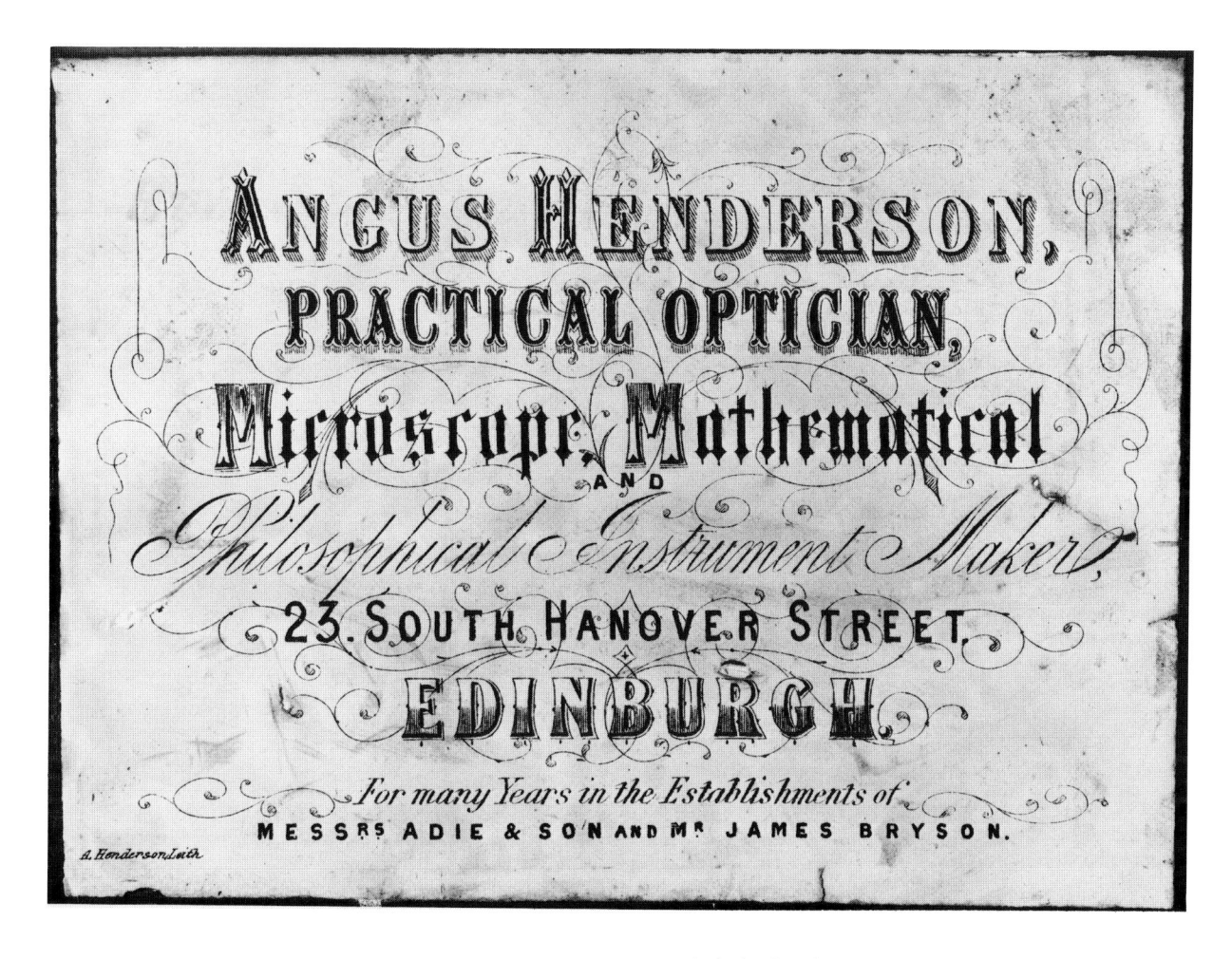

**ANGUS HENDERSON,
PRACTICAL OPTICIAN,
Microscope, Mathematical
AND
Philosophical Instrument Maker,
23 SOUTH HANOVER STREET,
EDINBURGH**

For many Years in the Establishments of
MESSRS ADIE & SON AND MR JAMES BRYSON.

34 Angus Henderson's trade card for his address between 1861 and 1868, advertising his previous connections with two of Edinburgh's foremost instrument making businesses

money earner. The balance of Bryson's estate was made up of insurance policies and a very mixed bag of shares in industrial, commercial and financial enterprises. The most important of these in the present context was his share of about £400 in the firm of Adie & Wedderburn (q.v.),[56] which he presumably acquired some time after 1881.

Like his elder brother Robert, J.M. Bryson served in the Edinburgh Merchant Company from 1889, first as convenor of Daniel Stewart's College, and later as a governor of the Merchant Maiden Hospital, a post which he retained until his death. He appears to have succeeded Robert junior, as Her Majesty's Clockmaker for Scotland, a position he resigned only a few months before his death,[57] on 6 January 1894.[58] He was sixty-nine. His obituary stated:

'His inclinations led him in the direction of the scientific side of his work more than towards the general business details, and brought him into relations with many of the prominent scientific men of this and the sister country. His reminiscences of the earlier stages of photography and many other modern discoveries and inventions, and of the distinguished men with whom he had been associated, would have been of great interest if he could have been induced to set them down for public perusal.'[59]

Unfortunately he did not record his 'fund of anecdote [and] his reminiscences of a past generation',[60] which his contemporary George Lowdon (q.v.) was persuaded to do, so we have little evidence of his own alleged scientific activity, nor that of his associates and patrons. Bryson was survived by his wife Mary, the daughter of David Dunn of Annet House, Skelmorlie, and their daughter and three sons. Although any of them were entitled to continue their father's business under the terms of his will, they did not do so, and the business came to an end.[61]

REFERENCES

1. For some account of the family business see John Smith, *Old Scottish Clockmakers 1453-1850* 2nd edition (Edinburgh, 1921), 69-74, where a foundation date of 1810 is claimed. This is borne out by the Edinburgh street directories, in which Robert Bryson appears for the first time as a clock and watch maker, with an address at the Mint, in the High Street, in 1810. However, Bryson clearly practiced as a clock maker before 1810. A mantel clock with an escapement triggered by a rolling ball, to the design subsequently patented in England by William Congreve (English patent 3164, 24 August 1808) and normally known as a 'Congreve clock', is signed and dated by Robert Bryson, Edinburgh, in 1804. (On Congreve's patent see [A.J. Turner], *William Congreve and his Clock* (London, 1972)). This clock is in the Royal Museum of Scotland (NMS T1972.127) and has a Bryson family provenance. Another dated example, of 1803, has been recorded (in a private collection, and also with a Bryson family provenance), indicating the some reliance can be placed on the dating (T.A.S. Drake, 'Congreve Clocks', *Antiquarian Horology* 1 (1953-56), 164). Bryson is likely to have completed an apprenticeship in the approximate period 1792-98 and therefore to have worked as a journeyman before setting up on his own. Similarities in the design of his clocks, notably his regulators, with the work of the prominent and innovative Edinburgh horologist Thomas Reid, suggests that Bryson may have trained under Reid. (One item to have been examined is a domestic regulator by Bryson in the Royal Museum of Scotland, NMS T1988.95, and we are indebted to Dr M. Dareau for suggesting this association). Bryson may have left at the formation of Reid's partnership (initially as Thomas Reid and Company) with William Auld in 1806. Auld had been apprenticed to Reid in the period 1793-99 but was in a favoured position as Reid's stepson (Smith, *ibid.*, 20, 312). Robert Bryson became a Burgess only in 1814, and a member of the Hammermen only in 1815 (Edinburgh City Archives, Burgess Roll, and Records of the Incorporation of Hammermen of Edinburgh).

2. National Library of Scotland, Royal Scottish Society of Arts Archives, Dep. 230/1, Minute Book, vol. [1], 1, entry for 9 July 1822.

3. Robert Bryson, 'Description of a new Self-Registering Barometer', *Transactions of the Royal Scottish Society of Arts* 3 (1851), Appendix, 6, 11-39. Alexander Bryson exhibited the barometer to the Society on 9 December 1844, but Robert Bryson himself had shown it to the Royal Society of Edinburgh (as one of its Fellows) on 2 January 1844: *Proceedings of the Royal Society of Edinburgh* 1 (1832-44), 450; *Transactions of the Royal Society of Edinburgh* 15 (1844), 503-505. Other examples of Bryson's horological designs can be found in the Society of Arts printed *Transactions*.

4. Obituary of Alexander Bryson, *Scotsman*, 10 December 1866.

5. *Ibid.*; Smith, *op. cit.* (1), 70. A fourth brother William Gillespie Bryson (born 1818) began as a banker's clerk, and became factor on the Seafield Estates at Cullen (General Record Office (Scotland), Census of 1841 (Edinburgh, St. Cuthberts), 685²/194, p.15; Report of funeral of Robert Bryson (junior), *Scotsman*, 25 March 1886).

6. *Transactions of the Royal Scottish Society of Arts*, *passim*; N.L.S., R.S.S.A. Archives, Dep. 230, Minute Books, *passim*.

7. Smith, *op. cit.* (1).

8. *Catalogue of the Exhibition of Arts Manufactures, and Practical Science, in the Assembly Rooms, George Street* December 1839-January 1840 (Edinburgh, 1840), 25.

9. Bryson communicated Bell's design for a microscope slide-rest to the Society (N.L.S., R.S.S.A. Archives, Dep. 230/3, Minute Book, vol. 2, entry for 27 April 1836); D.J. Bryden, *Scottish Scientific Instrument-Makers 1600-1900* (Edinburgh, 1972), 44.

10. N.L.S., R.S.S.A. Archives, Dep. 230, Minute Books, *passim*.

11. *First Report of the Directors of the School of Arts* (n.p., n.d. [Edinburgh, 1822]), 1.

12. *Ibid., passim*; *The School of Arts* (Edinburgh, 1828), 1, 6-7.

13. *Scotsman, op. cit.* (4).

14. Obituary of James Mackay Bryson, *Scotsman*, 11 January 1894. The Southern Academy was probably an episcopalian school, since the family appear to have been episcopalians (Report of funeral of Robert Bryson (J.M. Bryson's elder brother), *Scotsman*, 25 March 1886).

15. G.R.O. (S). Census of 1841 (Edinburgh, St. Cuthberts), 685²/194, p.15.

16. George Buchanan, 'Description and Uses of the Protracting Table', *Transactions of the Royal Scottish Society of Arts* 2 (1844), 176-179, and plate VII. The original drawing is preserved in the lecture series of the Society of Arts papers (N.L.S., R.S.S.A. Archives, Acc. 4534, box 23).

17. George Buchanan (c.1790-1852), was a civil engineer with a variety of interests; he was president of the Royal Scottish Society of Arts in 1847 (R[obert] H[arrison], 'George Buchanan', in L. Stephens (ed.), *Dictionary of National Biography* VII (London, 1886), 193-194). See Chapter 4 'The Adie Business' and Chapter 5 'John and Thomas Dunn'.

18. *Scotsman*, 11 January 1894. The standing of these particular firms in a European context has been assessed by A. Brachner, 'German Nineteenth-Century Scientific Instrument Makers', in P.R. de Clercq (ed.), *Nineteenth-Century Scientific Instruments and Their Makers* (Leiden, 1985), 117-157.

19. *Edinburgh Post Office Directory* 1850-1893, *passim*.

20. James Clerk Maxwell, 'Experiments on Colour', *Transactions of the Royal Society of Edinburgh* 21 (1857), 275 (read 19 March 1855). Maxwell's own example is described and illustrated in R.T.Gunther, *Early Science in Cambridge* (Oxford, 1937), 110. A later example, bought by Professor J.D. Forbes for the Natural Philosophy Class at the University of Edinburgh, came from the Aberdeen makers, Smith & Ramage, but has coloured papers marked 'Bryson, Edinburgh'; this is now in the Royal Museum of Scotland (NPM AU86).

21. Paul D. Sherman, *Colour Vision in the Nineteenth Century* (Bristol, 1981), 221 n.8.

22. Meeting of the Royal Scottish Society of Arts, 9 April 1855, communication 3: 'The instrument called the Gyroscope was exhibited in Action, and described by James Elliot, Esq.', Proceedings of the Royal Scottish Society of Arts in *Transactions of the Royal Scottish Society of Arts* 4 (1856), 197*-198*.

23. St. Andrews University Library, Forbes Papers Box IX, 2: Account book relating to purchase of instruments for Natural

Philosophy Class [of the University of Edinburgh] from Town Council funds, 1839-1860, items A.120, E.82 (1854 & 1858).

24. A fine domestic thermometer, 'Bryson's New Comparative Thermometer Arranged by A.K. Johnston. F.R.S.E.' is in the Royal Museum of Scotland (NMS T1856.38) dating from c.1835-1856 and may have been sold by Robert Bryson & Sons rather than by J. M. Bryson. Alexander Bryson's many friendships and prestigious connections are referred to in his 'Two Addresses delivered before the Royal Scottish Society of Arts' (reprinted from the *Transactions*, Edinburgh, 1861), *passim*, and his obituaries of Dr John Fleming and Sir Thomas Makdougall Brisbane (*Transactions of the Royal Society of Edinburgh* 22 (1861), 589-605, 655-680). Two further meteorological instruments, now in the Royal Museum of Scotland, should be mentioned. One is a copper three-inch floating-scale rain-gauge used at the Ben Nevis Meteorological Observatory, which was active between 1881 and 1904. The scale is stamped 'BRYSON EDIN^R' (NMS T1983.124). The second is a mercury maximum thermometer of the type devised by the geologist John Phillips in 1832, and is supported on a ceramic scale signed 'J.M. BRYSON, EDINBURGH' (NMS T1983.140). The thermometer was owned by the Scottish Meteorological Society which in the late nineteenth century consisted of a large proportion of prominent scientists, both professional and amateur. Either, or both, of these instruments might have been commissioned. Bryson also sold aneroid barometers (an example was offered for sale by Christie's South Kensington, 14 June 1987, Lot 29).

25. William Carpenter, *The Microscope* (London, 1856), 75 n. Examples of Gairdner's microscope are in the Science Museum, London (inventory 1917-15) and in the Wellcome Collection at the Science Museum (inventory A645015). We are grateful to Dr Jon Darius for this information.

26. John R. Hutchison, 'Some Notable Opticians of Auld Reekie', *The Ninth Annual Conference [of] The Scottish Association of Optical Practitioners* (Edinburgh, 1939), 17.

27. *Ibid.*

28. William Nicol, 'On a Method of so far increasing the Divergency of the Two Rays in Calcareous Spar that only one image may be seen at a time', *Edinburgh New Philosophical Journal* 6 (1829), 83-84. Nicol's role in optical instrumentation is discussed in A.D. Morrison-Low, 'The Origins of the Polarising Microscope: Sir David Brewster *versus* William Nicol', read at the Second Greenwich Symposium on Scientific Instruments, September 1982.

29. Alexander Bryson constructed a simple microscope for calculating the refractive indices of minerals, in a series of polarising experiments c.1840 (probably made with Nicol's help), which he described to the British Association in Glasgow that year: *Report of the British Association ... 1840* (London, 1841), 87-88. The instrument he used was presumably the 'Polarizing Microscope, with two of Nicol's prisms', exhibited in Edinburgh, the previous winter, 1839-40 (Catalogue, *op. cit.* (8), 25).

30. See obituaries of Bryson in the *Scotsman*, 10 December 1866, and in *Quarterly Journal of the Geological Society of London* 23 (1867), xlv.

31. Alexander Bryson, 'On the Aqueous Origin of Granite', *Proceedings of the Royal Society of Edinburgh* 4 (1857-62), 456-460 (read 29 April 1861).

32. Scottish Record Office, SC 70/4/17, pp. 957-9, Testament of William Nicol, dated 2 April 1842; *Scotsman*, 10 December 1866.

33. In addition to Nicol's collection, Bryson was bequeathed *viva voce* by 'my dear late lamented friend' Sir George Mackenzie, 'all his collection of minerals which he obtained in Iceland' (Alexander Bryson, 'Notes of a Trip to Iceland in 1862', *Scottish Guardian* 1 (1864), 118). Bryson's collection was sold to the British Museum in 1868.

34. Norman Higham, *A Very Scientific Gentleman* (Oxford, 1963), 38, 52-53.

35. Archibald Geikie, *The Founders of Geology* (London, 1897), 277; *idem.*, *A Long Life's Work* (London, 1924), 28.

36. Address by Sylvanus P. Thompson to an unnamed Optical Convention, 1905, cited in a letter of Joan M. Eyles to A.G. Bryson, 24 February 1952, copy in N.M.S. files. Archibald Gillespie Bryson was a son of J.M. Bryson and recorded his early memories of his father's shop, 'My father had a very few Nicol Prisms in his business as optician in Princes Street, and as a boy I was interested in the double image which the Iceland Spar showed': letter of A.G. Bryson to Miss Stevenson, 29 January 1952, copy in N.M.S. files. There is a prism made by William Nicol presented by Alexander Bryson in the Royal Museum of Scotland, NMS T1856.54, described in A.D. Morrison-Low, 'Scientific Apparatus associated with Sir David Brewster: an illustrated Catalogue', in A.D. Morrison-Low and J.R.R. Christie (eds.), *'Martyr of Science': Sir David Brewster 1781-1868* (Edinburgh, 1984, 86-87.

37. Bryson, *op. cit.* (33), 117. The account was published separately in Edinburgh in 1864 by R. Grant & Son.

38. W.H. Chambers, 'Fire and Damp Resisting Receptacle', *Transactions of the Royal Scottish Society of Arts* 9 (1878), 319 (read 12 April 1875). In 1886 Bryson exhibited 'a large mercurial barometer, indicating on a dial three feet diameter, and showing the rise or fall every morning at 9 a.m.' as well as 'scientific instruments, including optical, philosophical, and polarising apparatus' (*International Exhibition of Industry, Science & Art, Edinburgh, 1886. The Official Catalogue* 4th edition (Edinburgh, 1886), 126.

39. *Transactions of the Royal Scottish Society of Arts* 4-13 (1857-1894), *passim*.

40. *Ibid.*, 9 (1878), Appendix, 97; *ibid.*, 10 (1883), 341.

41. *Scotsman*, 11 January 1894.

42. Alexander Bryson claimed to have been the first, in Scotland at least, to have produced a daguerreotype portrait from life (*Scotsman*, 9 September 1840), which he showed to the British Association in 1840 along with his polarising microscope: Report of meeting of the Photographic Society of Scotland, *Journal of the Photographic Society of London* 5 (1859), 227; c.f. Bryson's obituary in the *British Journal of Photography* 13 (1866), 597. The presence of the pioneer photographer Thomas Davidson in his father's workshop c.1838, makes this claim very intriguing: John Nicol, 'Reminiscences of Thomas Davidson, a weaver lad', *ibid.* 26 (1879), 391.

43. Report of the meeting of the Society, *Journal of the Photographic Society of London* 5 (1858), 73; *ibid.*, 6 (1859), 226-227.

44. Report of meeting of the Edinburgh Photographic Society, *British Journal of Photography* 11 (1864), 428, 476.

45. D.B. Thomas, *The Science Museum Photography Collection*

(London, 1969), 12. Brian Coe, *Cameras* (London, 1978), 178, states that such a camera was being sold by Bryson c.1870. See also Larry Schaaf, 'Charles Piazzi Smyth's 1865 conquest of the Great Pyramid', *History of Photography* 3 (1979), 331-354 which refers to John Nicol, 'Photography in and about the Pyramids: how it was accomplished by Professor C. Piazzi Smyth', *British Journal of Photography* 13 (1866), 268-270. A recent discussion of Piazzi Smyth's life and work, including his photography, is by H.A. and M.T. Brück, *The Peripatetic Astronomer: the Life of Charles Piazzi Smyth* (Bristol, 1988).

46. E.g. *Edinburgh Post Office Directory* 1858.

47. Hutchison, *op. cit.* (26), 17 (however, this source is not entirely reliable); *Glasgow Post Office Directory, passim*.

48. S.R.O., SC 70/4/274, p. 116, Testamentary deed of James Mackay Bryson, dated 15 October 1868.

49. *Edinburgh Evening Courant*, 6 June 1815, quoted in Smith, *op. cit.* (1), 69. For the suggestion that Bryson operated from 1806 see reference 1 above.

50. *Ibid.*

51. S.R.O., CS 318/34/28, Sequestration Papers, Robert Bryson. There is a fairly full account of the process in the Sederunt Book. Notices also appeared in the *Edinburgh Gazette*: the initial notice of sequestration (30 November 1886, 1130); the announcement of the appointment of the Trustee and Commissioners (21 December 1886, 1198); a postponement of a dividend payment (8 April 1887, 325); a Creditors' meeting called for 30 May 'for the purpose of considering an offer for private sale of part of the heritable property belonging to the Bankrupt's Estate' (20 May 1887, 469); two new Commissioners appointed (24 June 1887, 589); another postponement of a dividend payment (16 August 1887, 764); first dividend payment on 2 January 1888 (15 November 1887, 1093); announcement of an equalising dividend payment (15 February 1889, 153); all funds recovered and expended; Trustee to seek discharge (17 May 1889, 443); and a final Creditors' meeting called for 12 September to consider discharging the Trustee (16 August 1889, 692).

52. S.R.O., CS 318/34/28, pp. 23-27.

53. S.R.O., SC 70/1/328, pp. 443-466, Inventory of estate of James Mackay Bryson, pp. 444, 459.

54. *Ibid.*, 446.

55. *Ibid.*, 447-455. Bryson's customers included the Duke of Argyll, the architect Hippolyte Blanc, and the Edinburgh professor of chemistry, Alexander Crum Brown.

56. *Ibid.*, 444.

57. *Scotsman*, 11 January 1894.

58. G.R.O.(S.), Register of Deaths 1894 (Edinburgh, St. Andrews), 685²/21. The cause of death was given as 'acute pulmonary congestion'.

59. *Scotsman*, 11 January 1894.

60. *Ibid.*

61. Bryson's surviving children seem to have been David Dunn Bryson, Robert Bryson, Archibald Gillespie Bryson and Mary Bryson. Two other daughters, Margaret Bannatyne Bryson and Jessie Gillespie Bryson, appear to have died young (S.R.O., SC 70/4/274, p. 116, J.M. Bryson's Testament).

---

45. Spectroscope: Bryson, Edinburgh, c.1890 (T1980.219)

Two-prism spectroscope, table model, in brass and oxidised brass, with one (only) 1¼ inch prism. The spectroscope platform engraved 'BRYSON EDINBURGH' and divided 0°-[127°] with a scale read against a vernier to 10 secs. for the position of the telescope, but with no separate rotation for the prism or collimator. The collimator and telescope mounted in sleeves with provision for limited adjustment in level and inclination. The spring-loaded slit adjusted by a screw; rack and pinion focus to the telescope and diaphragm wires. Supported on a brass pillar and tripod on a wooden base.

Path length: 660mm.
Table diameter: 150mm.
Height of prism table: 292mm.

Lacking the calibration prism from the slit, and one of the prisms from the optical train. The vernier support is distorted, and a repair has been made to a threaded joint in the telescope. Lacks protective cover for the slit. Fitted wooden case no longer present.

This instrument can be identified as John Browning's Analyst's Spectroscope, offered from before 1873 (J.N. Lockyer, *The Spectroscope and its Applications* 2nd edition (London, 1873), front endpaper); subsequent works reproduce the early block (e.g. R.A. Proctor, *The Spectroscope and its Work* (London, 1877), 26; John Browning, *Illustrated Catalogue of Spectroscopes, Spectrometers, Spectrum Apparatus, Induction Coils and X-ray Apparatus* (London, n.d. [c.1885]), [8]), but this instrument is of an improved form in that the earlier prism clamps have been replaced by screws acting from a circular platform mounted over the prisms.

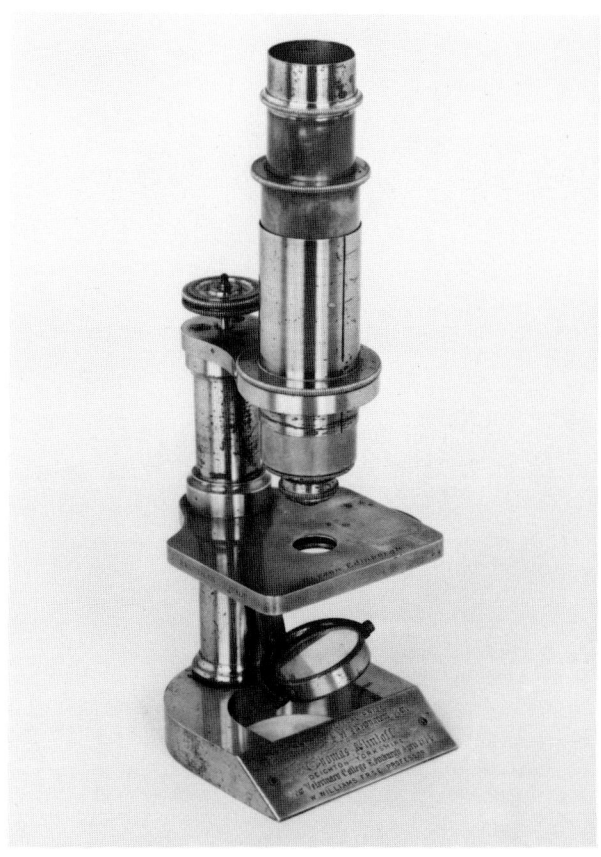

the stormy prelude to the inception of Edinburgh's New Veterinary College in 1873, documented in O.C. Bradley, *History of the Edinburgh Veterinary College* (Edinburgh, 1923), 64-69. The principal of the New College, William Williams, had been principal of the Edinburgh Veterinary College (now the Royal (Dick) School of Veterinary Studies, University of Edinburgh) from 1867 until his resignation after friction with the trustees in 1873, and he ran the new institution from 1873 until his death in 1900 (*Dictionary of Welsh Biography down to 1940* (Oxford, 1959), 1084). Between 1857 and 1866 Williams had pursued a successful career in Bradford, not 10 miles from Deighton, Yorkshire, the home of Thomas Flintoff. Flintoff matriculated at the Veterinary College in 1873 (*The Veterinarian* 46 (1873), 89) but appears to have left with Principal Williams along with the majority of the students (only 9 were left) and graduated from the New Veterinary College in 1875, joining the Royal Artillary as a veterinary officer. In 1878 he was transferred to the 8th Hussars, serving with them through the Afghan War. In 1900 he was gazetted to the 2nd Life Guards, was promoted to Lieutenant-Colonel in 1901, received the D.S.O. for his services in the Boer War and in 1904 retired. He died on 24 August 1907, at the age of 55 (*The Veterinary Record* 20 (1907), 155). Sir Frederick Wellington John Fitzwygram, 4th Baronet, served in the Crimea with the Inniskilling Dragoons and was Inspector-General of Cavalry between 1879 and 1884; a member of the Royal College of Veterinary Surgeons, he was their President from 1,875 to 1877, and published a number of works on horses (*Who's Who* (London, 1902), 468).

46. Microscope: Bryson, Edinburgh, 1874 (T1980.227)

Compound microscope, Continental model, in brass (the oxidised stage now stripped), on vertical pillar and flat horseshoe base. Engraved at the side of the stage 'R. WASSERLEIN BERLIN' and at the front of the stage 'Bryson Edinburgh'. A presentation plate attached to the chamfered forward edge of the base engraved 'SECOND PRACTICAL PRIZE / Presented by Sir F. W. Fitzwygram. Bart. / TO / Thomas Flintoff / DEIGHTON. YORKSHIRE. / New Veterinary College. Edinburgh. April 1874. / W. WILLIAMS. F.R.S.E. PROFESSOR.' Components scratched with assembly mark 'IIII' or stamped '4'. Single-draw body tube, in a draw sleeve for coarse focus, attached by a bar-limb to the pillar; fine focus adjusted by a screw at the top of the pillar. The rectangular stage plate attached to the pillar, and with a rotating 4-aperture diaphragm. Plane/concave mirror on a swinging tail beneath the stage.

Body length (less eyepiece): 104mm.
Body diameter: 29mm.
Height (less eyepiece): 188mm.
Base size: 71 x 79mm.

Lacks eyepiece and outer objective component. The fine focus mechanism is seized, and the bar-limb is bent. Lacking internal stop on the draw tube. The outer flange of the objective mount damaged.

The instrument was presumably manufactured or at least sold by R. Wasserlein of Berlin, and retailed by Bryson. Wasserlein is, however, not recorded in Alto Brachner, *Mit den Wellen des Lichts* (Munich, 1987) or *idem*, 'German Nineteenth-Century Scientific Instrument Makers', in P.R. de Clercq (ed.), *Nineteenth-Century Scientific Instruments and their Makers* (Leiden, 1985). Behind the inscription on this instrument lies

47. Microscope: Bryson, Edinburgh, c.1880 (T1980.228)

Compound student microscope, Continental model, in brass, bronze and oxidised brass, on an inclining pillar and black-painted flat horseshoe alloy base. Engraved on the body tube 'Mon. E. Hart. & Praz. / A. Praxmouski, suc.ʳ / Rue Bonaparte,, 1, / Paris' and on the front edge of the stage 'Bryson, Edinburgh'. Components stamped with the assembly mark '16' throughout. Single-draw body tube in a draw sleeve for coarse focus, attached by a bar-limb to the pillar; fine focus adjustment by a screw at the top of the pillar. Two objectives and two 2-component eyepiece assemblies, engraved '3' and '4'. The rectangular stage plate attached to the pillar, and with a rotating 5-aperture diaphragm. Compass joint, and concave mirror on a swinging tail beneath the stage. With fitted case (distressed) stamped '19442' and pasted in the lid a printed 'TABLEAU DES GROSSISSEMENTS OBTENUS' completed in MS to give the magnifications claimed for the two objectives (identified as of types 3 and 7) with eyepieces 3 and 4. Trade card in base of case for J.M. Bryson, Optician, Edinburgh.

Body length: 125mm.
Body diameter: 28mm.
Height: 242mm.
Base size: 74 x 115mm.
Case size: 280 x 155 x 105mm.

The fine focus mechanism is seized. Lacking one stage spring clip.

The dates for the Hartnack & Prazmowski partnership have been given as 1856-1876, and for Prazmowski alone 1876 - post-1881 (J. Payen, 'La Construction des Instruments

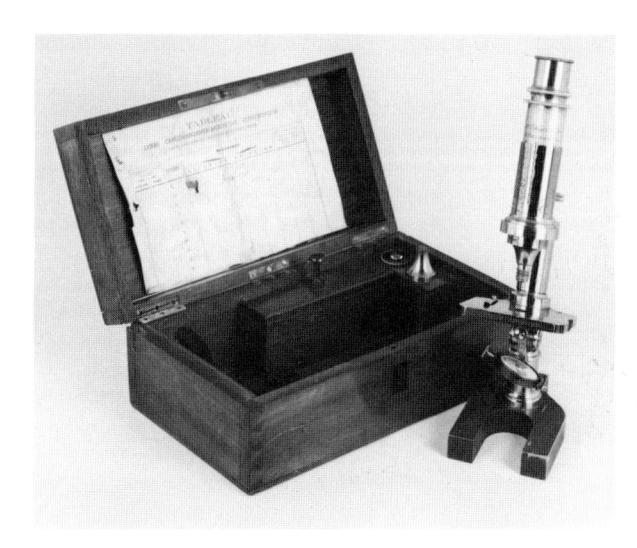

Scientifiques en France au XIXe Siècle', in P.R. de Clercq (ed.), *Nineteenth-Century Scientific Instruments and their Makers* (Leiden, 1985), 177; *idem*, 'Les Constructeurs d'Instruments Scientifiques en France au XIXe Siècle', *Archives Internationales d'Histoire des Sciences* 36 (1986), 130-131. At a slightly earlier date the London instrument maker James Swift was offering 'Swift's Improved Hartnack or Continental Form of Microscope' for which a ⅙″ objective was available 'fully equal to Hartnack's celebrated No. 7' (J. Swift, *Microscopes, ... and other Optical Instruments* (London, n.d. [c.1870]), 13). In 1970 Mr. Frank's collection contained an earlier Bryson-retailed Hartnack & Prazmowski microscope, suggesting a long-standing supply relationship (*The Frank Collection of Early Scientific Instruments, Interim List* (Glasgow, [1970]), item 171).

48. Microscope: Bryson, Edinburgh, c.1880 (T1980.229)

Compound Wenham binocular bar-limb microscope in brass and bronze, on an inclining pillar and flat 2-toed lead-weighted tripod base. Engraved on the body tube 'BRYSON, EDINBURGH.' and at the rear of the base 'BRYSON / EDINBURGH'. Single-draw body tube, with slide at the base of the body for moving the beam-splitting prism in and out of operation; attached to a bar-limb on a square-section slide within the pillar, raised by a rack and pinion for focus adjustment. Two objectives, with canisters engraved '1in / Bryson / Edinburgh' and '1', and '¼ / Bryson / Edinburgh' and '¼'; and one (only) 2-component eyepiece. The rectangular stage hinged between uprights rising from the base, with substage condenser, iris diaphragm and filter carrier in a push-fit sleeve; a concave mirror sliding on a pillar extension. With fitted case (distressed).

Body length: 215mm.
Body diameter: 30mm.
Height: 338mm.
Base size: 126 x 131mm.
Case size: 200 x 170 x 370mm.

Lacking second eyepiece. The orientation of the draw tubes is maintained by screws projecting through slots in the body tubes, but the screw on the main tube is lacking. The head of the screw attaching the limb to the pillar is damaged, and the focus pinion arbor is bent. Lead pads have been added under the base to improve stability. Lacking stage spring clips.

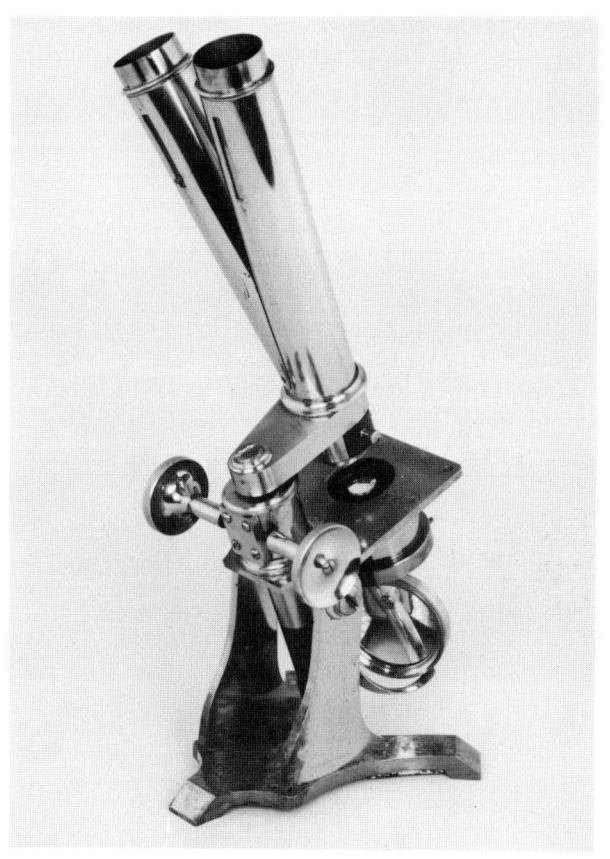

49. Microscope micrometer: J. Bryson, Edinburgh, c.1880 (T1980.242)

Ramsden screw micrometer eyepiece in brass. Engraved on the slide housing 'J. Bryson, / Edin.r' 2-component eyepiece assembly, with a fixed scale cut in brass at the edge of the field of view, divided to 0.1″ and subdivided to 0.02″, mounted in the eyelens focal plane. A spring-loaded sliding frame moves against this, advanced by a micrometer screw operated by an external threaded knob with silvered micrometer scale of 100 parts calibrated 0-[100], each division representing a movement of 0.0002″.

Eyepiece sleeve diameter: 33mm.
Size across micrometer: 110mm.

Lacking the fixed and moveable micrometer wires.

122

9

THE LENNIE BUSINESS

James Lennie was born in the parish of South Leith, Edinburgh, about 1817,[1] and is said to have begun in business as a jeweller and optician at a shop in the South Bridge, Edinburgh in 1835, and to have moved to Leith Street in the 1840s.[2] It is possible that the eighteen year old Lennie set up shop in 1835, but it seems more likely that initially he was employed. The earliest firm date for his independent business is 1840, when he appeared for the first time in the street directory as a jeweller and perfumer at 14 Leith Street. Subsequently in 1842 he advertised as a dealer in hardware, from 1843 to 1850 as a jeweller and optician, and from 1851 until 1856 as an optician only.[3]

James Lennie is recorded in the Census of 1851 as a master optician aged thirty-four, along with his wife Eliza aged thirty-three, their eldest son William, an apprentice optician aged thirteen, two other sons and three daughters.[4] William is likely to have been his father's apprentice, but after James Lennie died on 26 December 1854, aged about thirty-eight, no trace of him appears, and it fell to Mrs Eliza Lennie to continue her husband's business. This was valued at £400 in 1855,[5] and continued to trade as J. Lennie until 1857, when Eliza Lennie moved to 46 Princes Street, and set up as Mrs E. Lennie, Optician.[6] From 1862 until 1864 she advertised as 'optician and manufacturer of photographic apparatus', and between 1865 and 1872 as optician and jeweller, although the firm is said to have ceased trading in jewellery in about 1870.[7] Between 1873 and 1901 E. Lennie traded as optician and spectacle maker. From 1902 until 1906 the business was known as J. & J. Lennie; as Lennies from 1907 to 1922; as E. & J. Lennie, opticians, spectacle and eyeglass makers between 1923 and 1953; and as E. & J. Lennie, opticians and photographic dealers at 5 Castle Street between 1954 and 1959. In 1959, when Mr Jack Lennie was head of the firm, the photographic stock was disposed of in bulk, while the optical side of the business, along with the firm's records, were transferred to Turnbull & Co. (q.v.), of 56 George Street.[8]

In about 1872 the widowed Eliza Lennie married James Taylor, an optician from Haywood, England, who appeared under his own name at the 46 Princes Street premises from 1859 until 1887. He was probably responsible for the ophthalmic work which increasingly became the staple part of the business. Spectacles were prominently advertised in the street directory in 1867[9] and for 1873 the firm was specifically listed as spectacle makers. In this work Taylor was probably assisted at various times by at least three of his future stepchildren. In 1861 he was recorded as an optician (aged thirty-five), as were his stepsons John Lennie (aged twenty-three) and James Lennie (aged twenty-one); at the census ten years later a third stepson, Joseph Lennie, then aged twenty-three, was recorded as an optician.[10] This new generation of Lennies continued ophthalmic work after Taylor's apparent retiral in 1887, and in 1893 they were able to advertise:

> 'Many years of earnest attention and expertise in adapting the above [spectacles, etc.] to every condition of impaired vision. Oculists' prescriptions for Astigmatism, Cataract, Strabism, &c., carefully executed at moderate charges.'[11]

The idea that the Lennies were essentially retailers is borne out by advertisements for the firm's goods from the 1850s until the 1910s, and it is almost certain that the three refracting telescopes in this collection, which were supplied to the Northern Lighthouse Board, represent bought-in stock. Telescopes and other optical devices featured among goods advertised in 1893.[12] About forty years earlier, the stock ranged from cheap thermometers and chemical cabinets, to a wide selection of photographic apparatus and chemicals, spectacles, and microscopes priced between 1s and £14. Also available was the fashionable stereoscope, with a huge collection of views.[13] It is not surprising to find in 1863 that E. Lennie published a list of subjects of phantasmagoria and lantern slides which were for hire for the 1863-4 season, because such shows were then a fashionable and popular entertainment.[14] The firm also appears to have put considerable resources into the photographic side of its business at around this time, comparable to that of the Glasgow firm Lizars (q.v.).[15] Joseph Lennie seems to have taken a personal interest in photography, and was elected President of the Edinburgh Photographic Society for 1900.[16] In 1912 Lennie's issued, acting as their Edinburgh agent, the Glasgow maker Homan's catalogue of modern sundials, which were marketed as accurate timepieces rather than merely garden ornaments.[17]

The range of advertised stock in itself shows that the Lennies were not in any important sense specialists, except perhaps in the fitting of spectacles, and it is not even clear whether they offered the type of repair service that would have required them to maintain an instrument workshop. A considerable number of signed

35 Cover of an undated trade catalogue published by E. Lennie.
Trustees of the Science Museum, London

instruments which survive, including the cross-staff head in the Frank collection,[18] show characteristics of French manufacture of the period. The suspicion that much of the Lennie stock was being supplied by the major French wholesale houses seems borne out by the existence of items which have the pre-engraved signature incorrectly spelled. A notable example is a surveyor's barometric altimeter in the Royal Museum of Scotland signed 'LENNIE, / Optician / EDINBORO'.[19] The source of this phonetically signed material has not yet been identified.

REFERENCES

1. General Register Office (Scotland), Census of 1851 (Edinburgh, St. Andrews), 740/12, p. 14.

2. Article on E.& J. Lennie, quoting James Lennie's great grandson, Jack Lennie, head of the firm in 1959, *Edinburgh Evening News*, 9 February 1959. He was perhaps employed by James Bell, philosophical instrument maker of 54 South Bridge.

3. *Edinburgh Post Office Directory* 1835-1856. He may have been the James Lennie 'Toy Merchant and dealer in Fancy Goods in Edinburgh' for whom bankruptcy notices appeared in the *Edinburgh Gazette*: the initial notice of sequestration (7 April 1840, 104); and the announcement of the appointment of the Trustee and Commissioners, and that the Bankrupt had offered Creditors composition of 5 shillings per pound, this offer to be confirmed at a meeting called for on 8 June 1840 (12 May 1840, 143). Scottish Record Office, SC 39/17/6069, Sequestration Papers, James Lennie: these contain only the report from the Trustee, with the offer of composition. However, *Gray's Annual Directory* 1835, 1836 and 1837 give only a James Lennie at 207 High Street, with no trade; he does not appear at all in the *Edinburgh Post Office Directory* 1837 and 1838; however, in 1839 a Geo[rge] Lennie, 'toy warehouse' makes a single appearance at 9 North Bridge. Although there is no positive evidence in either the sequestration papers or the street directories to show that these two businesses were run by the same man, it is possible that Lennie began as a toy merchant, and after his bankruptcy, his business moved into jewellery and perfumery, only starting to sell optical goods at a later date.

4. G.R.O.(S.), Census of 1851 (Edinburgh, St. Andrews), 740/12, p. 14.

5. S.R.O., SC 70/1/90 Edinburgh Commissariot, Record of Inventories, pp. 646-650.

6. *Edinburgh Post Office Directory* 1857. However, she appeared in 1855 and 1856 under her own name in the classified section. She and Margaret Gardner (q.v.), widow of John Gardner junior are among the few women known to have been active in the Scottish trade.

7. *Ibid., passim; Edinburgh Evening News, op. cit.* (2).

8. *Ibid.* The records of the Lennie business were destroyed when Turnbull & Co. (S.C. Sorrell) moved to 3 Stafford Street (information from Mr Rattray, surviving partner of Sorrell, December 1981). The firm of McCall & Lennie, or Lennie & McCall, opticians, active between 1911 and 1951, was presumably connected with the Princes Street business.

9. *Edinburgh Post Office Directory* 1867, 49.

10. G.R.O.(S.), Census of 1861 (Edinburgh, St. Andrews), 685²/89, p. 5; Census of 1871 (Edinburgh, St. Andrews), 685²/89, p. 20.

11. *Scotsman*, 2 October 1893.

12. *Ibid.*

13. *Ibid.*, 7 January and 30 May 1857.

14. *Phantasmagoria and Dissolving Views produced by the Magic Lantern ... E. Lennie, optician ... Lists of Subjects for the Season 1863-64 ...* (Edinburgh, 1863), 4.

15. *Price-list of Apparatus and Chemical Preparations Used in the Art of Photography, Manufactured and Sold by E. Lennie, Optician, 46 Princes Street, Edinburgh* (Edinburgh, n.d.), 18pp., in the Science Museum Library, London.

16. J.C. Lennie, 'Photography as Applied to the Register of Sasines', *British Journal of Photography* 47 (1900), 695-696. On Lennie's presidency of the Edinburgh Photographic Society, see *Edinburgh Journal* 12 (1937), 41.

17. [Lennies], *Modern Sundials* (n.p., n.d. [Glasgow?, c.1912]), 11pp.

18. NMS T1980.164.

19. NMS T1987.52; described in Historical Technology, Catalog 130 (Spring 1987), item 164. A telescope signed 'Lennie Edinburg' was offered at Christie's South Kensington, 18 July 1985, Lot 173.

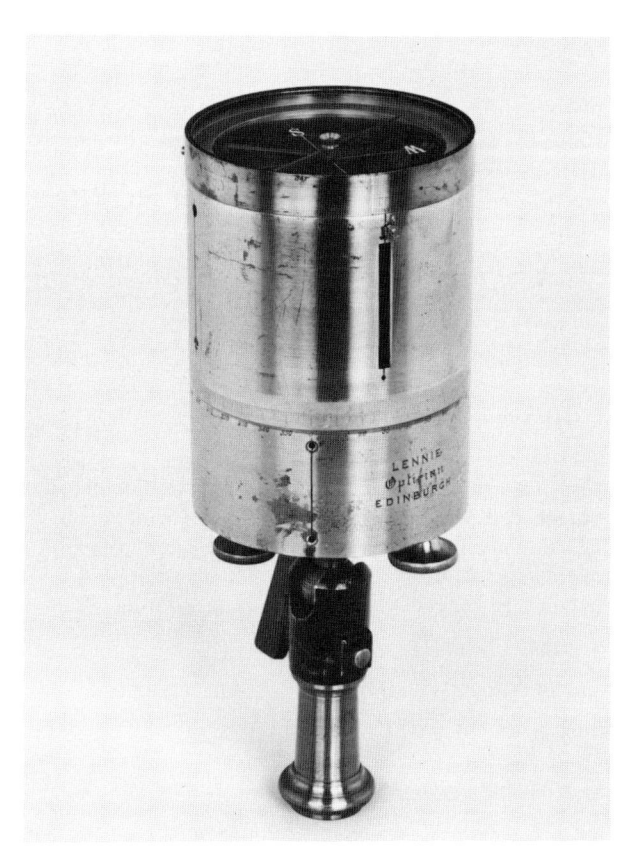

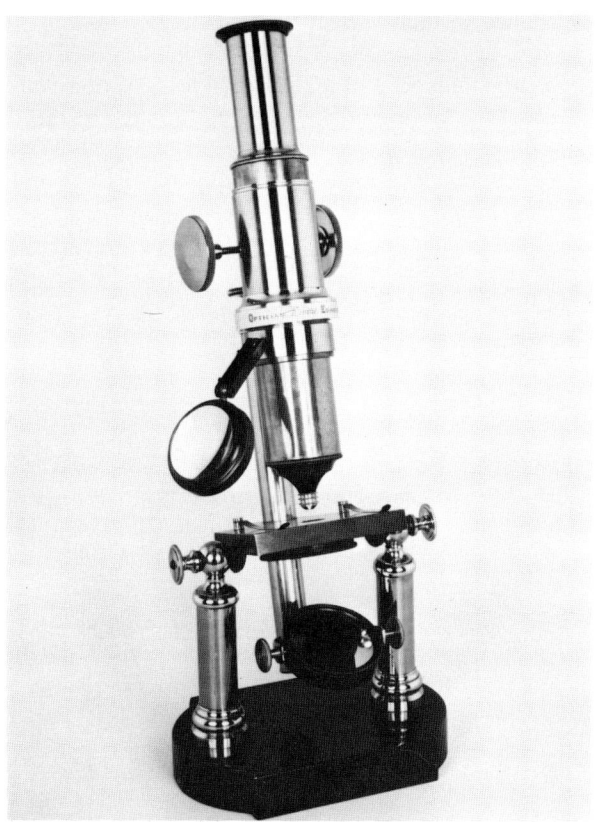

50. Adjustable cross head: Lennie, Edinburgh, c.1890 (T1980.164)

Cylindrical 2-part cross-staff head in brass and black-lacquered brass, with 2 orthogonal off-set slit and window sights, with clamp and geared rotation about a vertical axis over a similar fixed sight; the horizontal angle between these shown on a silvered azimuth scale divided [360]-0 by 1° with a vernier reading to 2 mins. Compass mounted above the off-set sights, with the 4 cardinal points, divided on a silvered scale 0-[360] by 1° and with a steel needle blued at one end on a relieved jewelled bearing. Engraved on the body 'LENNIE / Optician / EDINBURGH', and scratched on the milled clamp screw with the coded price 'DZ/-'. On a clamped ball and socket joint, cut to allow the instrument to be set horizontally, over a socket mount.

Diameter: 89mm.
Height of cylinder: 120mm.

The compass attachment screws are modern.

Probably of French manufacture.

51. Microscope: Lennie, Edinburgh, c.1880 (T1980.236)

Compound microscope in brass and oxidised brass, on an inclining pillar and black-enamelled cast iron base. Engraved around the collar supporting the body tube 'OPTICIAN, Lennie, EDINBURGH'. Stamped with the assembly mark '2' on the top of the pillar, and scratched with the price code '£D. TD/-' on one of the focus knobs and under the base. The body tube has a rack set into it, operated on by a focus pinion at the rear of a draw sleeve on a bar-limb. There is no provision for fine focus adjustment: the screw on the top of the pillar is purely decorative and engages only in the upper surface of the limb. Objective and 2-component eyepiece assembly. The pillar extends through the rectangular stage, which is mounted on trunnions between two turned pillars rising from the base. The stage with a rotating 4-aperture diaphragm, and spring clips. A plane/concave mirror at the base of the pillar. Condenser lens for illuminating opaque objects attached by a jointed arm to a sliding collar on the body tube mounting sleeve below the bar-limb. With fitted case.

Body length: 262mm.
Body diameter: 48mm.
Height: 447mm.
Base size: 180 x 123mm.
Case size: 205 x 140 x 475mm.

One focusing knob distorted.

Provenance: Purchased at auction. Probably of French manufacture. A very closely similar microscope, with identical base, pillar supports and stage, and differing principally in the detail of the body tube construction, is in the Collection of Historical Scientific Instruments at Harvard University (inventory 1225), engraved 'Microscope / Pasteur': we are grateful to Ebenezer Gay, Assistant Curator of the Harvard collection, for providing details. Another example of the same size, but with an oval base and circular stage, and engraved by Moreau of Paris, is in the Wellcome Museum, at the Science Museum, London (inventory A17828): we are grateful to Jane Insley for details of this. Two unsigned examples of a smaller size, one with a square base and the other with an oval base, which may be from the same workshop, are described and illustrated in H.R. Purtle, *et al.*, *The Billings Microscope Collection of the Medical Museum, Armed Forces Institute of Pathology* 2nd edition (Washington, D.C., 1974), 69-70, 80.

52. Microscope: unsigned, c.1860 (T1980.237)

Compound microscope, 'Martin's drum' type, in brass on a circular base. Body tube mounted in a push-fit collar which is moved within the outer cylindrical housing by a rack and pinion. The ends of the outer component of the collar distinguished by the scratched marks 'XII' and 'VI'. Objectives stamped '1' and '4' (of 3), and 2-component eyepiece assembly. The cylindrical housing cut to accept an axially mounted circular stage plate, and with 2 apertures below for illumination of the concave substage mirror. With fitted case containing trade card for E. Lennie, Optician, 46 Princes Street, Edinburgh.

Body length: 149mm.
Body diameter: 33mm.
Height: 234mm.
Cylindrical housing diameter: 48mm.
Case size: 255 x 90 x 80mm.

Lacking the lens in objective 1, and all accessories.

Benjamin Martin devised his 'Pocket Reflecting Microscope' in 1738, and illustrated it in his *Micrographia Nova: or a New Treatise on the Microscope, and Microscopic Objects* (Reading, 1742); this was the ancestor of what became known after Martin's death as 'Martin's drum microscope' and was most popular between 1820 and 1850. Another example, signed by Lennie, is illustrated in G. L'E. Turner, *Collecting Microscopes* (London, 1981), 45, and a third example, also signed by Lennie, is in the Collection of Historical Scientific Instruments at Harvard University (inventory 1281).

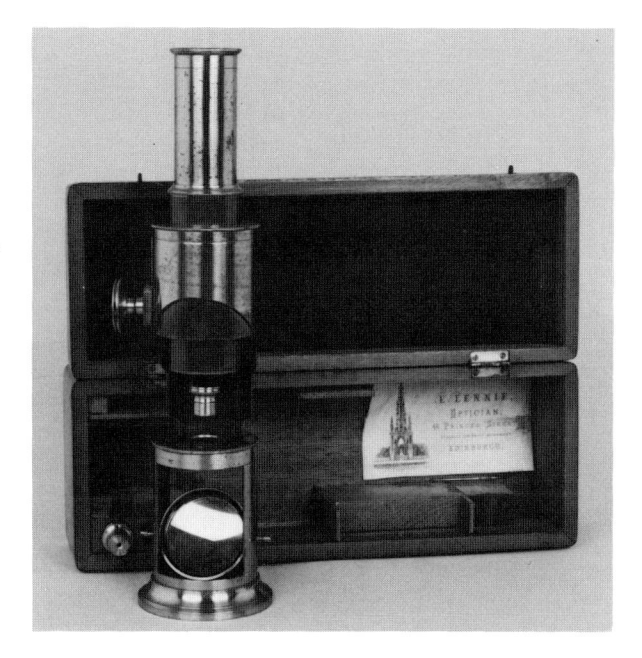

53. Refracting telescope: Lennie, Edinburgh, c.1850 (T1980.261)

1¼" aperture 3-draw hand-held refracting telescope with doublet objective and 2-component eyepiece and erecting assemblies; in brass, with wood-veneer covered barrel. Engraved on the first draw tube 'Lennie, Edinburgh'. Objective cover, and closure on external eye stop.

Length closed: 197mm.
Tube diameter: 4omm.
Aperture: 34mm.

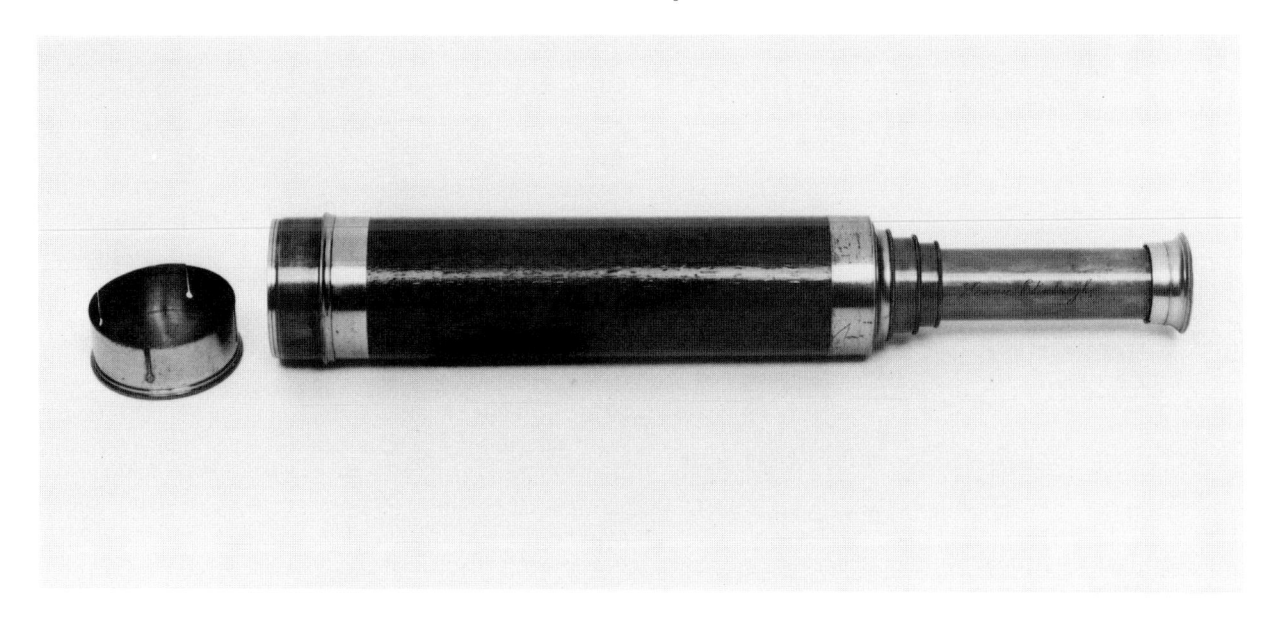

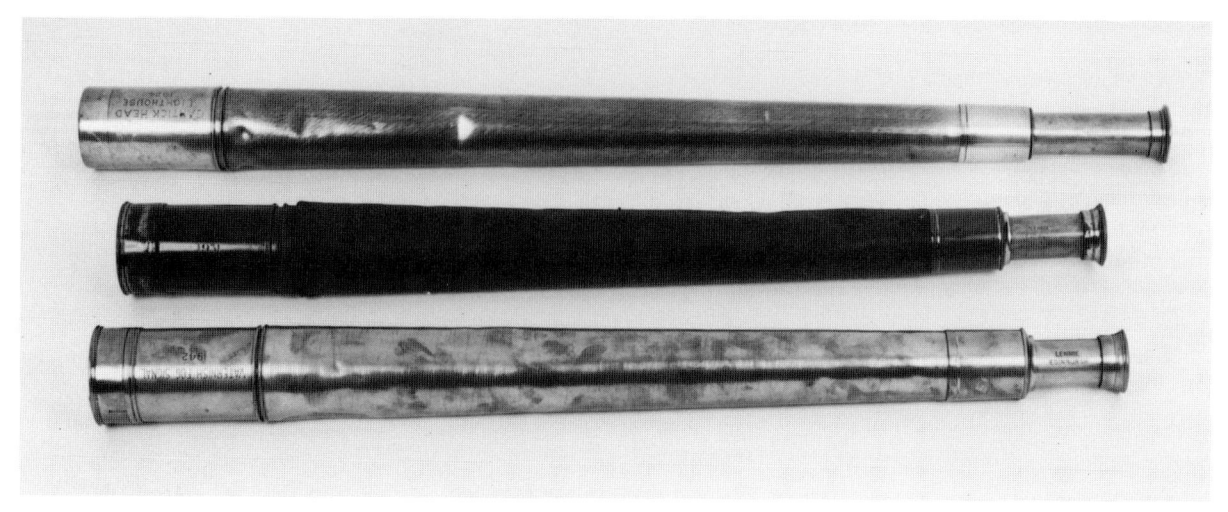

From the top, *54, 55, 56*

54. Refracting telescope: Lennie, Edinburgh, 1924 (T1980.262)

2″ aperture single-draw hand-held refracting telescope with doublet objective and 2-component eyepiece and erecting assemblies; in brass, with objective ray shade, the tapered barrel originally leather-covered. Engraved on the draw tube 'LENNIE / EDINBURGH' and on the ray shade 'CANTICK HEAD / LIGHTHOUSE / 1924'. Scratched assembly mark 'VII' on draw tube components, and 'II' on the barrel under the end sleeve.

Length closed: 734mm.
Maximum tube diameter: 61mm.
Aperture: 51mm.

The instrument is in a distressed condition with the longitudinal seam parting on the barrel and ray shade, and with dents in the draw tube and barrel, which has also lost its leather cover. The objective mount is distorted and the lens chipped at the edge. The ray shade has been affected by a corrosive liquid. Lacking objective cover, and the rear plate of the external eye stop.

Cantick Head Lighthouse, in the Orkney Islands was built as a manned lighthouse by the Northern Lighthouse Board in 1858, to designs by David Stevenson (1815-1886) (R.W. Munro, *Scottish Lighthouses* (Stornaway, 1979), 129-130; Keith Allardyce and Evelyn M. Hood, *At Scotland's Edge: A Celebration of Two Hundred Years of the Lighthouse Service in Scotland and the Isle of Man* (Glasgow and London, 1986), 47, 157).

55. Refracting telescope: Lennie, Edinburgh, 1931 (T1980.263)

2″ aperture single-draw hand-held refracting telescope with doublet objective (broken) and 2-component eyepiece and erecting assemblies; in brass, with objective ray shade and tapered leather-covered barrel. Engraved on the draw tube ' "LENNIE" / EDINBURGH' and 'POINT OF AYRE / LIGHTHOUSE / 1931'. Objective cover, and closure on external eye stop which is scratched with the assembly mark 'XXXII'.

Length closed: 645mm.
Maximum tube diameter: 61mm.
Aperture: 50mm.

The rear component of the objective is broken. The leather covering of the barrel is partly detached and shrunk at the objective end. The barrel has been distorted by dents and the draw tube is also distressed. The lining of the draw slide is lacking. The lacquer colour of the external eye stop does not match other lacquered parts.

Five major lighthouses were built by the Commissioners of the Northern Lights on the Isle of Man between 1818 and 1880. The Point of Ayre, at the northern tip, was designed by Robert Stevenson (1772-1850) and completed in 1818 (R.W. Munro, *Scottish Lighthouses* (Stornaway, 1979), 85; Keith Allardyce and Evelyn M. Hood, *At Scotland's Edge: A Celebration of Two Hundred Years of the Lighthouse Service in Scotland and the Isle of Man* (Glasgow and London, 1986), 27, 102-103).

56. Refracting telescope: Lennie, Edinburgh, 1942 (T1980.264)

2″ aperture single-draw hand-held refracting telescope with doublet objective and 2-component eyepiece (defective) and erecting assemblies; in brass, with objective ray shade, the tapered barrel originally leather-covered. Engraved on the draw tube 'LENNIE / EDINBURGH' and 'VATERNISH FOG SIGNAL / 1942'. Scratched with the assembly mark 'IIII' on the draw tube, 'VII' on the barrel and end sleeve, and 'VIII' on the draw slide. Objective cover, and closure on external eye stop which is scratched 'XXXIII'.

Length closed: 679mm.
Maximum tube diameter: 63mm.
Aperture: 50mm.

Lacking the leather covering of the barrel, which is badly dented. The field lens of the eyepiece is lacking as is the lining of the draw slide. The objective cover has soldered to it a nameplate engraved 'MᶜARTHURS HEAD / LIGHT-HOUSE', transferred from another instrument.

Vaternish Point on the north-west tip of the Isle of Skye, is now marked by a minor light; the fog signal is located at Neish Point, together with the lighthouse, on the westernmost headland of the island (Keith Allardyce and Evelyn M. Hood, *At Scotland's Edge: A Celebration of Two Hundred Years of the Lighthouse Service in Scotland and the Isle of Man* (Glasgow and London, 1986), 80).

57. Refracting telescope: E. Lennie, Edinburgh, c.1900 (T1980.277)

2¼" aperture 2-draw mounted refracting telescope in brass; with doublet objective recessed from the end of the barrel, with provision for 2-component eyepiece and erecting assemblies (lacking), and rack and pinion focus to the second draw tube. Engraved on the barrel end plate 'E. LENNIE / EDINBURGH'. Assembly mark 'III' scratched on the first draw tube and support tube for the erecting assembly.

Length closed (less eyepiece): 792mm.
Tube diameter: 59mm.
Aperture: 55mm.

Lacking eyepiece assembly and external eye stop, and also image erecting assembly (although one lens remains in the support tube).

Detailed similarities in design and in the mounting of optical components suggest that this item is by the same manufacturer as items 66 (Lowdon), 85 (Gardner & Co.), 93 (J. Brown) and 161 (Whyte, Thomson & Co.).

58. Refracting telescope: Lennie, Edinburgh, c.1920 (T1980.278)

4" aperture 3-draw mounted refracting telescope in brass; with doublet objective, rack and pinion focus to the third draw tube, and 1¼" finder telescope. Supplied with an eyepiece comprising a low-power 2-component eyepiece coupled with a 2-component erecting assembly, and presumably originally also with high-power astronomical eyepieces (now lacking). Engraved on the barrel end plate 'LENNIE / EDINBURGH'. Objective covers for telescope and finder. Pillar-mounted by a knuckle-joint with a clamped rack and pinion adjusted steady rod to the base of the pillar, and with free azimuth motion; attached to a folding wooden tripod with brass fittings, or to an alternative folding brass table tripod.

Length closed: 1,640mm.
Tube diameter: 103mm.
Aperture: 102mm.

Lacking astronomical eyepieces, and solar shade. Slots have been cut in the nut to the knuckle-joint on the pillar to facilitate tightening. The barrel of the telescope is spotted with corrosion marks over its upper surface, and there are patches elsewhere resulting from spillage or handling. The metal surface at the top of the pillar has been damaged by a wrench.

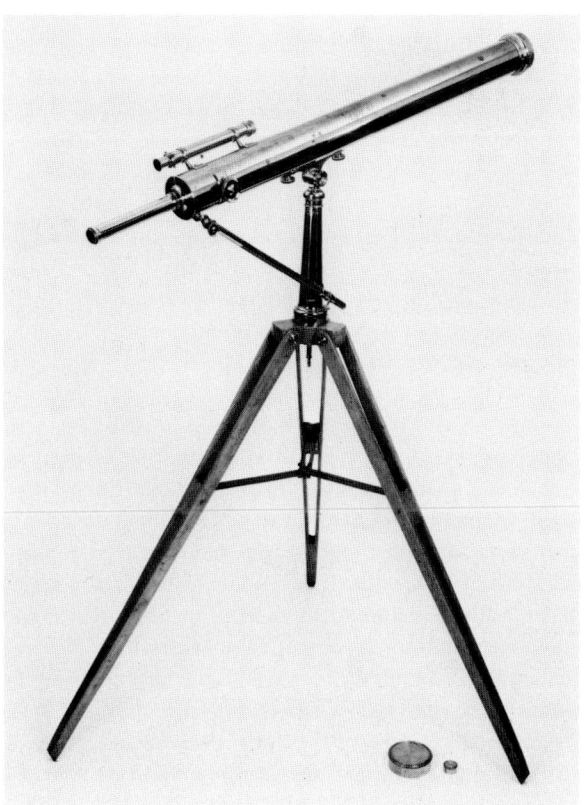

TURNBULL & CO.

John Miller Turnbull, optician, philosophical instrument maker, chemical dealer and photographic warehouseman, took over the premises at 14 Nicolson Square formerly occupied by James Gilchrist, photographer, in 1865. By 1881, and until 1884, he was at 19 South St. David Street, classified as an optician and philosophical instrument maker, moving to 6 Rose Street in 1885 where he remained until 1900. He finally disappeared from the street directories in 1904. Meanwhile, a second branch, Turnbull & Co., opened at 60 Princes Street as successors to James M. Bryson (q.v.) in 1894, when the firm was described as opticians and fishing tackle makers; the order of the description was reversed in 1896 and remained until 1940.[1] The fishing tackle side of the business proved to be the more important initially: at one time two girls were employed to make the tackle,[2] and there may have been some relationship between the proprietor of Turnbull & Co. and his contemporary namesake Robert Turnbull, fishing-rod and tackle manufacturer, of the Waltonian works at 10 and 12 Hanover Street.[3]

However, J.M. Turnbull was sufficiently interested in the optical side of his business to consider some of its practical problems. In 1886 he read before the Royal Scottish Society of Arts a paper on an improved sliding nosepiece and adaptor for the microscope.[4] For this, the Society awarded him a Hepburn Silver Medal.[5] That same year, Turnbull exhibited photographic apparatus at the Edinburgh International Exhibition of Industry, Science and Art, and advertised that he was the agent for the London optician W. Wray, 'manufacturer of telescope, microscope, and photographic lenses'.[6]

Turnbull had two sons, both of whom qualified as opticians. The elder, William Turnbull, first assisted his father,[7] then continued the business at 60 Princes Street until 1927 when he removed to 37 George Street. The firm remained there until 1956 when it moved across the road to 56 George Street. While at the first George Street address, Turnbull was joined by S.C. Sorrell, who took over the branch after Turnbull's death in 1939; and by J.W.A. Rattray, who eventually became Sorrell's partner and then successor.[8] In 1959 Turnbull & Co. (S.C. Sorrell), took over the optical side of the business of E.& J. Lennie (q.v.), 5 Castle Street,[9] and in 1976 the business moved to 12 Stafford Street, where it is now (in 1989), at present under the name of J.W.A. Rattray (Turnbull & Co.).[10]

J.M. Turnbull's second optician son, Frederick Brodie Turnbull (c.1886-1938) set up on his own in Leith in about 1914 as a compass adjuster, also making gun sights. By 1916 his business was based in Constitution Street when he sold out to a Mr Craig (q.v.). From then on he concentrated on optical work.[11] Turnbull & Co. had acquired a second branch at 181 Constitution Street, Leith, in 1913, and by 1916 Frederick Turnbull was running this, describing himself as F.I.O. (Fellow of the Institute of Opticians) or F.B.O.A. (Fellow of the British Optical Association) from 1934. In 1918 Turnbull & Co. took over 6 Commercial Street, Leith, formerly the premises of David Stalker (q.v.), and this shop with its distinctive shop sign[12] was also run by Frederick Turnbull. W. Craig (q.v.) took over at 6 Commercial Street in 1927, but Turnbull continued in business at 181 Constitution Street, Leith as an ophthalmic optician, and after his death in 1938[13] his daughter Miss Hilda B. Turnbull, F.B.O.A. ran the firm. Turnbull & Co. moved to 20 Great Junction Street, Leith, in 1982.[14]

During the early years of its existence, it is possible that Turnbull & Co. produced some of the instruments that they sold, the interest of the founder being an indication of their capability to do so. However, of the instruments known to have been sold by the business, it is probable that the pocket aneroid barometer in this collection[15] came from one of the large wholesalers in England such as Negretti & Zambra, J.J. Hicks or Louis Casella; and the same could be said about barographs retailed by the firm.[16] A folding wooden camera marked 'J.M. TURNBULL EDINBURGH' with Taylor, Taylor and Hobson lenses was probably made by Billcliff of Manchester.[17] Likewise, a brass sextant marked 'No 9593 Turnbull & Co Leith'[18] must have been from the numbered series of a major wholesaler, but without more evidence it is hard to say what proportion of items were bought in from outside. It would appear, however, from the trade description in the directory, that after the First World War the emphasis of the business moved away from fishing tackle with instruments as a sideline, towards the work of an ophthalmic optician, a role pursued by the successors to both branches today.

REFERENCES

1. *Edinburgh Post Office Directory* 1865-1940.

2. Information from Miss Hilda Turnbull, J.M. Turnbull's grand-daughter, in March 1982.

3. For example, *Edinburgh Post Office Directory* 1913.

4. J.M. Turnbull, 'On an Improved Sliding Nosepiece and Adaptor for the Microscope', *Transactions of the Royal Scottish Society of Arts* 11 (1887), 352-354; *Transactions of the Edinburgh Naturalists' Field Club* 1 (1886), 335-336; reported in the *Journal of the Royal Microscopical Society* 7 (1887), 293-296.

5. Report of the Prize Committee 1885-86, *Transactions of the Royal Scottish Society of Arts* 11 (1887), Appendix, 120.

6. *International Exhibition of Industry, Science & Art, Edinburgh, 1886. The Official Catalogue* 4th edition (Edinburgh, 1886), 196 and 336.

7. *Ref. cit.* (2).

8. Information from J.W.A. Rattray, December 1981.

9. *Edinburgh Evening News*, 9 February 1959.

10. *Edinburgh Post Office Directory* 1959-1974, *passim*; *Edinburgh and Lothians Telephone Directory* 1975-1988, *passim*.

11. *Ref. cit.* (2).

12. Now in the collection of Edinburgh City Museums and Art Galleries, inventory HH 3145/67, it is illustrated as the frontispiece in D.J. Bryden, *Scottish Scientific Instrument Makers 1600-1900* (Edinburgh, 1972).

13. Scottish Record Office, Calendar of Confirmations 1938, T58.

14. *Edinburgh and Lothians Telephone Directory* 1982.

15. NMS T1980.216.

16. Examples were offered for sale by Christie's South Kensington, 10 August 1977, Lot 54 and 10 July 1986, Lot 40, and by Sotheby's Belgravia, 12 July 1977, Lot 247.

17. In the Royal Museum of Scotland, NMS T1982.121.

18. Offered for sale by Sotheby's, 18 October 1971, Lot 64.

---

59. Pocket aneroid barometer: Turnbull & Co., Edinburgh, c.1930 (T1980.216)

Mining pocket aneroid barometer, temperature compensated, in a glazed brass case. Engraved on the silvered dial 'Compensated' and 'TURNBULL & C?. / 60, Princes S!. / EDINBURGH.' The pressure scale divided in inches of mercury, 23-31 by 0.05". An altitude scale divided in feet, 7000(altitude)-0-1000(below ground) by 50ft., can be adjusted to normalise the altitude by turning the knob at the suspension ring when used as a pressure-indicating altimeter. Adjusting point on the reverse.

Diameter: 48mm.

Lacking the index.

Although not identical, this instrument corresponds broadly to one of the many pocket aneroid barometers offered from an earlier date by the London wholesaler J.J. Hicks in his *Illustrated & Descriptive Wholesale Catalogue of Standard, Self-Recording, and other Meteorological Instruments...* (London, n.d. [c.1880]), 52: no. 199, 'Mining Aneroid Barometer, 2½-inch raised silver dial, scale of feet from 2000 below sea level to 6000 above, curved Thermometer, in morocco case. This portable Aneroid is specially adapted for engineers and surveyors, enabling them, without calculation, to ascertain heights of mountains and depths of mines £2.5.0.'

36 William Hume (c.1852-1924); possibly a self portrait.
W. W. Hume, Esq.

10

WILLIAM HUME

William Hume's first appearance in the Edinburgh street directory was in 1870, when he must have been about eighteen years old. He had no trade listed and his home address was given as 23 Lothian Road, where he remained until 1874. That year his trade was described for the first time as 'philosophical instrument maker and chemical dealer', with his shop separately listed at 16 South College Street, an address he kept until 1880 when he moved to 1 Lothian Street as a scientific instrument maker and chemical dealer. The business moved finally to 14 Lothian Street in 1907 where it was last recorded in 1921.[1]

It is not known from whom the young Hume learnt his trade. He was born in Glasgow about 1852, the eldest son of William Hume, a commercial flour traveller who came from Jedburgh, and his wife Elizabeth, originally from Peebles. None of the rest of his immediate family appear to have had any connection with instrument making.[2] It is not known either how large his workshop was, nor how many men he employed, but his business seems to have similarities to that run from a slightly earlier period by Kemp & Co. of Edinburgh.[3] For instance, Hume took an active interest in the development of specialist instrumentation; and just as Kemp & Co. helped devise philosophical apparatus for William Swan and Arthur Connell of the University of St. Andrews, Hume assisted at least two meteorologists in developing new apparatus. One was John Aitken (1839-1919), F.R.S.E., who devised the 'koniscope', which used an optical method to measure the density of atmospheric dust.[4] A number of these instruments, marked 'WILLIAM HUME MAKER EDINBURGH', are known.[5] Another instrument was the marine anemometer of William Galt Black (d.1909), an Edinburgh medical practitioner and Fellow of the Royal Meteorological Society, described by Black to the Royal Scottish Society of Arts in 1883.[6]

Hume's own connection with the Royal Scottish Society of Arts began on 28 July 1879, when he was elected a Fellow. In November 1880 he was on the committee which examined the optician Alexander Frazer's (q.v.) new form of thermometer.[7] Six years later, in May 1886 Hume lectured to the Society on 'Some Historical Demonstrations in Photography' in which he

> 'exhibited an extensive series of examples of the various processes in their historical order, and of the results. Among these he exhibited on the screen, the actual process of development of the latent image, and illustrated the general laws of the dispersion and absorption of light.'[8]

For his unpublished paper he received the special thanks of the Society, and was awarded a Keith Medal.[9] Hume continued to participate in the Society's work and served on a number of technical panels. In March 1887 he was on a committee which examined the optician William Forgan's paper 'On photo-micrography'.[10] He took part in the discussion in April 1888 on R. Norman Shaw's patent drainpipe and was appointed to the committee to which that paper was referred.[11] He was involved in a similar discussion about window apparatus in April 1889,[12] and in March 1890 he participated at a meeting about science and architecture.[13]

However, photography was clearly a subject which interested him deeply.[14] Like Kemp & Co. and Lennie (q.v.) of Edinburgh, Lizars (q.v.) of Glasgow and Feathers (q.v.) of Dundee, he produced an illustrated catalogue during the 1890s which was entirely devoted to 'photographic apparatus, chemical preparations and other requisites used in the practice of photography'.[15] But unlike them, his was more than a professional interest. In January 1895 he read a paper before the Edinburgh Photographic Society entitled 'Remarks on Photographic-colour printing' which reviewed the processes available (although he did not demonstrate them) and gave some idea of the technical problems involved:

> 'What is to be photographed in colour? Is it the strutting peacock with his tail spread? Is it the maiden of bashful fifteen? Or is it the widow of fifty? Let me tell you that none of these three, especially the two ladies, could bloom for six or seven hours in the sitting required for their colour portraits. Gentlemen, we have here a picture of a stuffed parrot, ... and we would require to have a stuffed widow of fifty before we could have her colour photograph.'[16]

He certainly took photographs himself, some of which were published.[17]

Although he retailed other makers' cameras,[18] he was interested in developing dark-room apparatus. In 1888 he designed his 'Cantilever enlarging apparatus', producing a booklet about it in 1890[19] and an improved version in 1906:

133

'Upon its introduction in the year 1888, Hume's Enlarging Apparatus was received with much favour and acceptance. It was built upon somewhat new lines, and it was introduced at a time when the Bromide process was beginning to be widely adopted for business and amateur enlarging.'[20]

Hume quotes the *Amateur Photographer*, of 26 December 1890: 'I have just purchased one of Hume's Cantilever Lanterns, and am more than satisfied with it ...'; and the photographic experimenter V.C. Driffield writing in the *British Journal of Photography* of 16 November 1894: 'The Lantern I am about to use is an excellent form of instrument made by Mr Hume, of Edinburgh, and known as the "Cantilever".'[21]

Hume also devised a projection microscope, which he demonstrated before the Edinburgh Field Naturalists' and Microscopical Society on 23 March 1887, when its 'clear definition ... [was] much appreciated by the members present'.[22] Hume was a member of the Society between 1883 and 1891.

In 1888 Hume retailed three different 'sets of apparatus for experimental physics, manufactured in Dresden by

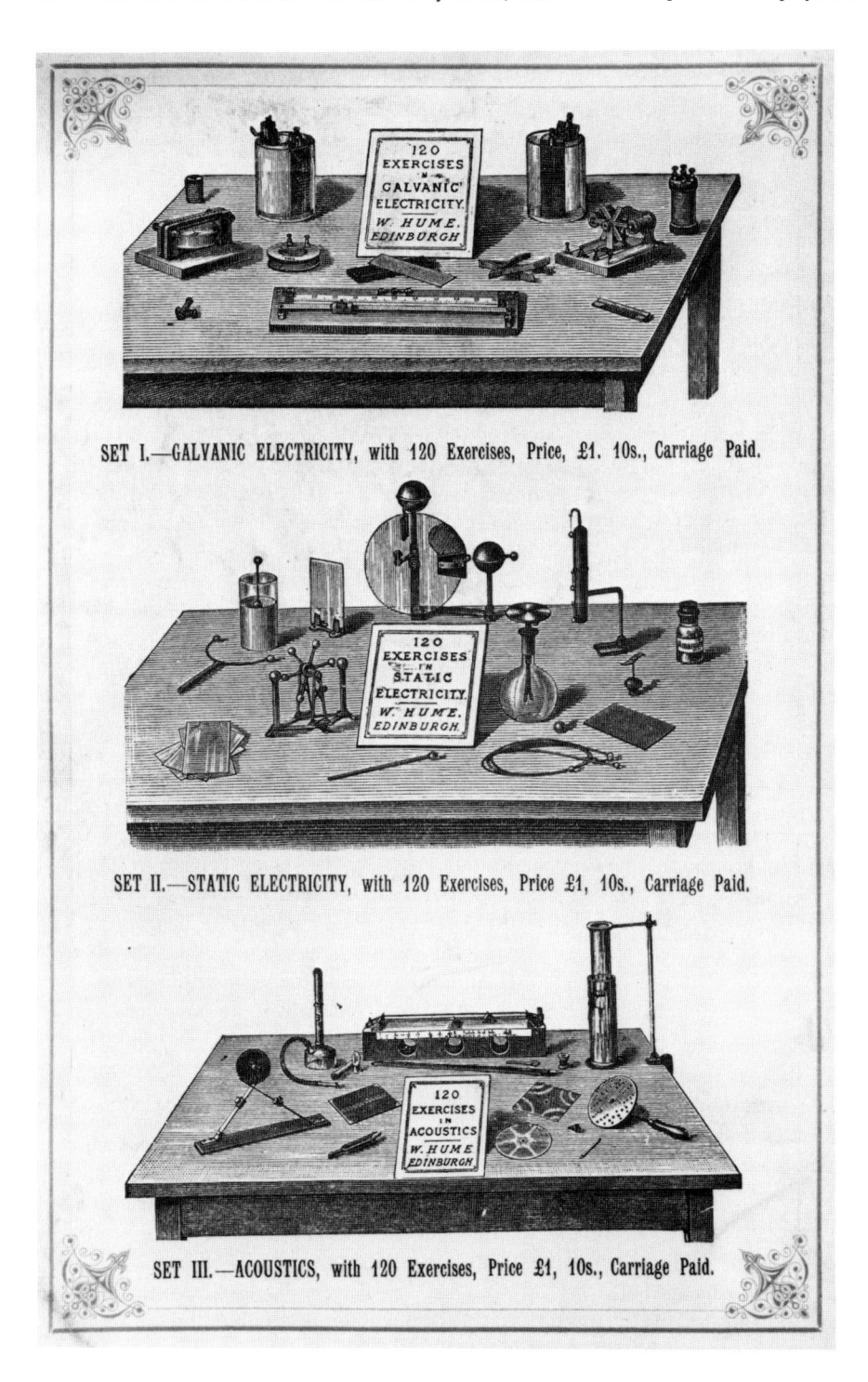

37 Three educational scientific sets retailed by Hume, 1888

134

Meiser & Mertig' and for each produced 'A Book of 120 Experimental Exercises and Problems, together with detailed descriptions of the nature and uses of the Apparatus ... furnished with the sets to facilitate the correct performance of each experiment by the student who is also recommended to study the ordinary Text-Books.'[23] The sets were Galvanic Electricity, Static Electricity, and Acoustics, and were priced at £1 10s each. The following year a fourth set, Optics, was produced, which retailed at £1 12s. These sets were particularly aimed at the educational market, and Hume observed:

> 'In an age when Science ministers to use in so many ways, it is desirable that all classes should acquire a knowledge of its main principles. Parents and friends will find these Sets to be admirably suited for giving as presents to boys and girls, as being likely to engage a portion of their leisure hours in pleasantly acquiring useful information, and in exercising and developing their powers of observation.'[24]

Hume exhibited at the Glasgow International Exhibition of 1888, when his business address was given as 1 Lothian Street and 1 and 2 West College Street, Edinburgh; the latter did not appear in the street directories and may well have been workshops. He advertised a 'workshop and skilled staff on the premises'[25] but with no indication of size or numbers. The range of material he displayed was considerable:

> 'Scientific Instruments and Apparatus for Research, Teaching, and Technical Purposes. Chemical Apparatus, Fine Balances, German and Bohemian Glass Goods. Apparatus in Porcelain, Stoneware, Metals and Wood, etc. Pure Nickle Basins and Crucibles. A Candle Balance for Gas Works. Physical Instruments and Appliances for Teaching Science. Students' Sets for Galvanism. Electricity, and Acoustics. Hand Dynamo, Electro-motor. Physiological Instruments. Revolving drums and Electro-magnetic Recording Pen, etc. Electro-medical Instruments for Physicians and Surgeons. Finest Microscopes and Accessories. Large new Laboratory Microtome for cutting Sections under Spirit. A new Projection Microscope for Lime-Light. Bacteriological Apparatus and Instruments. Photographic Apparatus. Pure Chemical Reagents.'[26]

It is difficult to gauge how much Hume made and what he merely retailed. The trade literature which survives is only unambiguous in places, as with the 'Thirty Shilling Sets' manufactured by Meiser & Mertig and a statement that 'microscopes for commercial and scientific purposes by Zeiss, Reichert, Hartnack, Beck, Swift, Pillischer, and other eminent Makers, [are] always in Stock for Choice'.[27] It is probable that the microscope apparatus in this collection was retailed by Hume.[28] On the other hand Hume would appear to have made the items such the 'Cantilever' which he described, and those such as the koniscope that he developed for clients like Aitken. The tone of his advertisements suggests that the major financial prizes to be won were the contracts for equipping the new analytic and research laboratories that were by now central to medical and industrial work, and the teaching laboratories that were of increasing importance in the school curriculum. Hume was in a position to supply:

> 'Laboratories completely Furnished with Chemicals, Apparatus, and Fittings for Public Analysts and Teachers, Agriculturists and Manure Merchants, Colleges and Schools, Photographers, Brewers, Health Officers, Oil Works, Paper Mills, and Chemical Works of every description.'[29]

No doubt the completion of the new Edinburgh Medical School in the early 1880s and the subsequent expansion of hospital and public health laboratories would have presented notable opportunities for the Hume business, and in these he was presumably vying with the major laboratory supply houses such as Griffins and Baird & Tatlock (q.v.). More important perhaps was the threat posed by his neighbour A.H. Baird whose expansion in the period after 1900 may well have been at the expense of Hume's business.

Between 1881 and 1891 Hume married Marion Watson, who was a year older than him, and who originally came from Glasgow; they had at least one son William, born about 1880, and she predeceased her husband.[30] The business ceased in 1920 when Hume was aged about sixty-eight, and on 5 January 1924 he was found dead by 'suffocation by lighting [i.e. coal] gas' by John Hume, his brother.[31]

REFERENCES

1. *Edinburgh Post Office Directory* 1870-1921.

2. General Register Office (Scotland), Census of 1871 (Edinburgh, St. Cuthbert's), 685¹/34, p.21. Hume's brothers John, aged 17, was a stationer's apprentice; Hall, aged 16, was a publisher's apprentice; while Scott and Elliott, 13 and 10 respectively, were 'scholars'.

3. For an account of that business see A.D. Morrison-Low, 'Kemp & Co: Laboratory Suppliers', in J.T. Stock and M.V. Orna (eds.), *The History and Preservation of Chemical Instrumentation* (Dordrecht, 1986), 163-186.

4. John Aitken, 'Portable Apparatus', *Proceedings of the Royal Society of Edinburgh* 16 (1890), 169-172. A recent assessment of

Aitken's work is by Josef Podzimek, 'One Hundredth Anniversary of John Aitken's Paper on Dust Particles in the Atmosphere', presented at the American Association for Aerosol Research 1988 Annual Meeting, 13 October 1988, Chapel Hill, North Carolina, publication forthcoming.

5. One example, in the Royal Museum of Scotland (NMS T1983.125) came from the Scottish Meteorological Society's collection, and was used at the meteorological observatory on top of Ben Nevis at some point between 1883 and its closure in 1904. Another, which belonged to Aitken, is in the Science Museum, London (inventory 1908-193).

6. William Galt Black, 'On a Wind-pressure Gauge for Ships at Sea', *Transactions of the Royal Scottish Society of Arts* 11 (1887), 47-49. An example of this instrument is in Bolton Museum and Art Gallery (inventory 1892-6), together with a book of meteorological observations made by Black in Bolton in 1833, both donated by Black's son. We are grateful to R.J. Bradbury, Bolton Museum and Art Gallery, for this information. Another example is in the Science Museum, London (inventory 1984-532). The Brisbane Prize medal awarded to Black for his invention is in the Wellcome Museum for the History of Medicine, Science Museum, London. We are grateful to David Wright for this information. Black was a Fellow of the Royal Meteorological Society, and his obituary reveals a life-long interest in the science (*Quarterly Journal of the Royal Meteorological Society* 36 (1910), 254).

7. *Transactions of the Royal Scottish Society of Arts* 10 (1880), Appendix, 98.

8. *Ibid.*, 11 (1887), Appendix, 103-104.

9. *Ibid.*, Appendix, 119.

10. *Ibid.*, 12 (1891), Appendix, 5.

11. *Ibid.*, Appendix, 40.

12. *Ibid.*, Appendix, 71.

13. *Ibid.*, Appendix, 101-102.

14. He advertised photographic apparatus in the annual *British Journal of Photography Almanac* frequently between 1888 and his final appearance in their 'Directory of the Photographic Trade' in 1921 (for example, *ibid.* (London, 1888), 675; *ibid.* (London, 1893), 142-143; *ibid.* (London, 1906), 1612-1613; *ibid.* (London, 1908), 1010-1011; *ibid.* (London, 1912), 1138-1139; *ibid.* (London, 1913), 1156-1157. We are grateful to John Ward of the Science Museum, London, and Sara Stevenson of the Scottish National Portrait Gallery, Edinburgh, for these references).

15. William Hume, *Illustrated Catalogue of Photographic Apparatus, Chemical Preparations, and Other Requisites used in the Practice of Photography* (Edinburgh, n.d.).

16. William Hume, 'Remarks on Photographic-colour Printing', *British Journal of Photography* 42 (1895), 25-28.

17. *Evening Dispatch*, 6 June 1923: 'Founder's Day at George Heriot's Hospital - 40 Years Ago'. We are grateful to W.W. Hume, grandson of William Hume, for this information.

18. For instance, in the Royal Museum of Scotland (NMS T1962.L14) there is an example retailed by Hume of the 'Instantograph' by the Birmingham maker J. Lancaster & Son.

19. William Hume, *The Cantilever Enlarging Apparatus for the Use of Photographers ... Designed ... by William Hume* (Edinburgh, n.d. [1890]).

20. William Hume, *Hume's Enlarging Apparatus, Formerly called the Cantilever Enlarging Apparatus for the Use of Photographers ... in making Enlargements upon Bromide and Chloride Papers ...* (Edinburgh, 1906).

21. *Ibid.*, quoting V.C. Driffield, 'The Principles involved in Enlarging', *British Journal of Photography* 41 (1894), 726.

22. William Hume, 'Hume's Projection Microscope', *Transactions of the Edinburgh Field Naturalists' and Microscopical Society* 2 (1887), 61-64.

23. William Hume, *Experimental Physics. I Electricity: Galvanic, with 120 Exercises. II Electricity: Static, with 120 Exercises. III Acoustics, with 120 Exercises* (Edinburgh, 1888), 2. There is an example of set II in the Royal Museum of Scotland (NMS T1960.44).

24. William Hume, *Experimental Physics. Part IV Optics with 120 Exercises* (Edinburgh, 1889), 2.

25. *Glasgow International Exhibition 1888 Catalogue* 2nd edition (Glasgow, 1888), 402.

26. *Ibid.*, 164.

27. *Ibid.*, 402.

28. NMS T1980.230 and T1980.243.

29. *Catalogue, op. cit.* (25), 402. A double-barrelled air-pump (NMS T1968.22) and a laboratory oil lamp (T1972.23) at the Royal Museum of Scotland come into this category.

30. G.R.O.(S.), Census of 1891 (Edinburgh, Newington), 685⁵/102, p.15.

31. G.R.O.(S.), Register of Deaths 1924 (Edinburgh, George Square), 685⁵/17, *Scotsman*, 7 January 1924.

60. Microscope: William Hume, Edinburgh, c.1900 (T1980.230)

Compound Wenham binocular microscope, in brass, bronze and oxidised brass, on an inclining pillar and bent claw-footed tripod base. Engraved on the rear of the base 'WM HUME / EDINBURGH'. Single-draw body tubes with linked rack and pinion adjustment, and slide at the base of the body for moving the beam-splitting prism in and out of operation; rack and pinion coarse focus and attached to the pillar through a spring-loaded slide, operated by a fine focus screw at the top of the pillar. Two unmarked objectives, the canisters engraved on their lids '1½ IN / Wm Hume / Edinburgh' and 'Wm Hume 3 IN / Edinburgh' (the latter with a printed label 'W. HUME, OPTICIAN, 1, LOTHIAN ST. EDINBURGH' and in MS '3" SVH. 25/-') and one additional unmarked objective; 2-component eyepieces. The stage pivotted between 2 uprights rising from the base, with the pillar mounted above, and a plane/concave mirror on a swinging tail sliding beneath; the rectangular stage of Nelson horseshoe form with an iris sub-stage diaphragm in a push-fit sleeve, and spring clips. With fitted case (distressed), stamped inside the base 'WM HUME EDINBURGH'.

Body length: 226mm.
Body diameter: 34mm.
Height: 357mm.
Base size: 126 x 135mm.
Case size: 200 x 150 x 390mm.

Concave face of the sub-stage mirror broken. Rack and pinion adjustment to the eyepieces partly seized.

The instrument bears some resemblance to the 'Improved Student's Microscope' of J. Swift & Sons of London described in *Journal of the Royal Microscopical Society* 11 (1891), 87-89, and may be based on this. E.M. Nelson's stage is described in *ibid.* 7 (1887), 292-293.

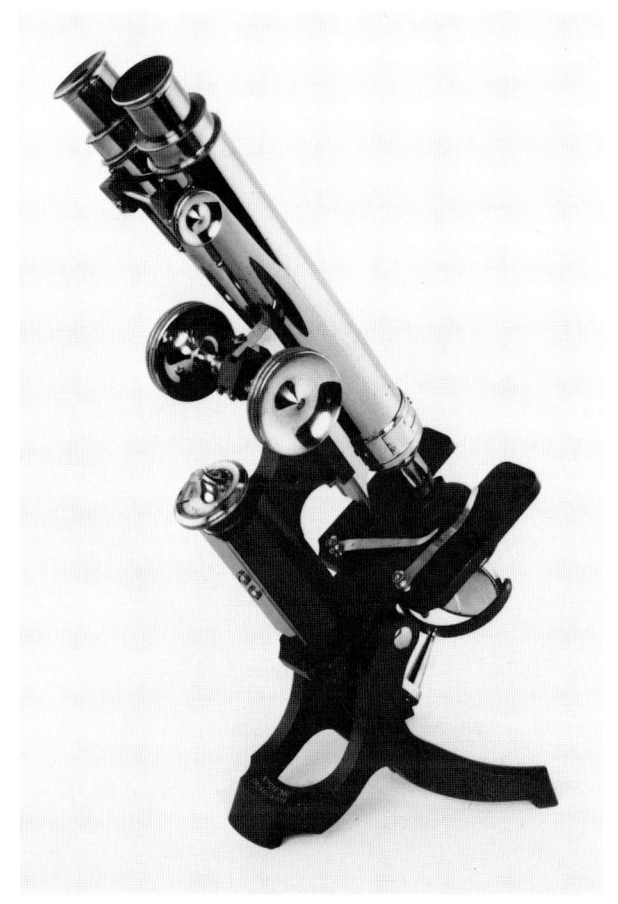

61. Microscope hot stage: William Hume, Edinburgh, c.1890 (T1980.243)

Water heated slide carrier, in brass, for use on a microscope stage. Engraved on one edge face 'Hote State / Wm Hume / Maker / Edinburgh'. The water circulation chamber pierced by a lined rectangular slot connecting the side faces, to take the microscope slide, and a lined cylindrical aperture connecting the upper and lower surfaces and intersecting the slot for the slide. The slide is held in place by sliding retention pieces at the side apertures. Two nipple connectors at opposite corners of one side face to take rubber tubing for water circulation, and a third for a thermometer. With velvet-lined leather-covered fitted case.

Chamber external size: 77 x 65 x 28mm.
Case size: 95 x 100 x 40mm.

11
DUNDEE NAUTICAL INSTRUMENT MAKERS

Instrument making in Dundee in the late nineteenth century was unusual, and not very lucrative work, if the statement of one of its practitioners is to believed. Writing in 1912, George Lowdon (q.v.) wrote 'Dundee was then less than a third of its present size, and for a few years I had a hard pull to live, besides having to teach Dundee what scientific instruments were, and how to use even such a simple apparatus as a thermometer.'[1] However, Dundee's shipping and commercial history can be traced back to the thirteenth century, and surviving records imply that this trade was by then already of long standing. Over the centuries, trading abroad and the domestic traffic grew so that the town's position on the Tay estuary could be exploited and a thriving port developed.[2] Seamen have always required navigational instruments from the times that vessels first left sight of land, and by the early nineteenth century it would appear that this demand had created a local market of nautical instrument suppliers.

In the late eighteenth and early nineteenth centuries Dundee began to participate in the whaling industry in the seas off Greenland, and with the end of the French Wars, Dundee's share in this boomed. Other east coast ports were involved in this profession, in decreasing order of importance the most prominent being Hull, Whitby, Peterhead, Aberdeen, Leith and Dundee. By 1845, Peterhead had taken over from Hull as the premier whaling port, and held this position for twenty years, losing it to Dundee by 1860 at about the time when sail gave way to steam. The steam-powered whaler brought a new prosperity to the industry, and in 1861 the Dundee fleet consisted of eight wooden steamers, which had the ability to pass through ice which would have stopped a sailing ship; many of them were built at Dundee. By the 1880s the success of whaling was waning, while the ships' owners had turned their vessels and crews to the more lucrative trade of catching seals in the Davis Straits. Whaling continued to decline towards the end of the nineteenth century, and with the outbreak of war in 1914 the Government commandeered the remaining ships from the Dundee fleet for the Hudson Bay Company. Many of these were destroyed carrying munitions to Russia and this was the effective end of Arctic whaling from the British Isles.[3] However, the vessels themselves were considered fine enough because of their special construction for exploration in Antarctica, the most famous of these being the Dundee-built *Discovery* launched in 1901 by Dundee Shipbuilders Co. Ltd., and built specifically as an expedition ship for Captain Robert Falcon Scott.[4]

Dundee's growth over the nineteenth century was due in particular to the jute industry over which the city gained a practical monopoly of a worldwide demand from the early stages in its developement. Textiles had always been an important local industry, feeding on the produce of an agricultural hinterland of part-time handloom weavers, with flax for linen the main staple.[5] However, large quantities of this had to be imported from Europe, and there was, for instance, a thriving trade between Dundee and Russia in flax: 'a goodly number of sailing brigs and schooners regularly trading to the Baltic'.[6] By the 1830s shortages and fluctuations in the price of raw flax led to a search for a continuous supply of steadily priced raw material, and that decade saw the Dundee textile industry's transition to jute.[7] 'The great jute-laden vessels from India form a yearly argosy far exceeding in value and importance those of Ragusa and Venice' wrote an observer in 1912.[8] Scotland's third largest city, Dundee had by 1851 a population of 79,000 rising to 140,000 in 1881 and 165,000 in 1911 (not including the growth of the suburbs, estimated at around 30,000 in 1911). In 1881 49% of the total employed population was involved in the textile industry, compared with 0.5% in Edinburgh (population 295,000), 13.6% in Glasgow (population 673,000) and 8.3% in Aberdeen (population 105,000). The figure for 1911 was only slightly less, at 48.2%.[9] Even after the disruption to trade caused by the First World War there was no depression in jute until after 1929.[10]

The healthily-growing port that Dundee became during the nineteenth century, ensured that there was a reasonable local demand for nautical instruments. It appears that about a dozen such businesses able to fulfil this need started up in the town, mostly located near the docks and supplying the practical instruments which would be required at sea.[11] However, as in the case of Aberdeen, as many of these businesses traded as 'clock, watch and nautical instrument makers' it is more than likely that 'many of them were primarily retailers whose sole claim to the "maker" lay in their undertaking repair work'.[12] There seems to have been little opportunity in the way of expanding a local 'dilettante' market with optical or philosophical instruments, and the lack of a university or major college until the end of the century meant few men of science were seeking out local instrument makers for special commissions. Those that there were, for instance the Rev. Thomas Dick (1774-1857), seemed happy to have made their own,[13] or in the case of the mathematician Sir James Ivory (1765-1842), to have had the opportunity of acquiring them elsewhere.[14]

REFERENCES

1. A.H. Millar, 'Other Pioneers of Invention: George Lowdon' in *James Bowman Lindsay and Other Pioneers of Invention* (Dundee, 1925), 85.

2. James Maclaren (ed.), *The History of Dundee ... A New Enlarged Edition of the Work, Published in 1847, by James Thomson* (Dundee, 1874), 273-280.

3. David S. Henderson, *Fishing for the Whale* (Dundee, 1972), *passim*.

4. Ann Savours, *The Voyages of the 'Discovery'* (London, forthcoming) gives an overview of the vessel's somewhat chequered career.

5. B.P. Lenman and E.E. Gauldrie, 'The Industrial History of the Dundee Region from the Eighteenth to the Early Twentieth Century' in S.J. Jones (ed.), *Dundee and District* (Dundee, 1968), 162-173.

6. J.M. Beatts, *Reminiscences of an Old Dundonian* (Dundee, 1882), 33.

7. Lenman and Gauldrie, *op. cit.* (5), 168.

8. J.H. Martin, 'Dundee: As It Is' in A.W. Paton and A.H. Millar (eds.), *British Association Handbook and Guide to Dundee and District* (Dundee, 1912), 29.

9. A.M. Carstairs, 'The Nature and Diversification of Employment in Dundee in the Twentieth Century' in Jones, *op. cit.* (5), 319-320.

10. S.G.E. Lythe and J. Butt, *An Economic History of Scotland 1100-1939* (London, 1975), 224.

11. These names are listed in D. J. Bryden, *Scottish Scientific Instrument-Makers 1600-1900* (Edinburgh, 1972), 43-59 and are as follows: Domenico Balerno, barometer-maker, 14 Yeaman Shore 1846-53; John Bon, chronometer, watch and nautical instrument maker, 17 Dock Street 1840, 25 Dock Street 1845, 26 Dock Street East 1846 continued by Mrs John Bon at 24 Dock Street East in 1850; Alexander Cameron (q.v.); James Clark, about 1800?; Peter Airth Feathers (q.v.); James Howie, optician and philosophical instrument maker, 76 Nethergate 1869; Peter Howie, optician, 52 Barrack Street 1871; John Kidd, nautical instrument maker, 6 East Dock Street 1853-56; Alexander King, nautical instrument maker, 37 High Street 1840; George Lowdon (q.v.); John Murray, compass maker, Dock Street 1834; Anthony Tarone, barometer maker, 37 Murraygate 1818-29, 32 Murraygate 1834, 12 Castle Street 1840, 90 Nethergate 1842, 12 Murraygate 1846; and James Taylor & Co., nautical instrument makers, King William Dock 1869-74.

12. *Ibid.*, 31.

13. Maclaren, *op. cit.* (2), 377.

14. *Ibid.*, 362.

ALEXANDER CAMERON

Alexander Cameron's main business seems to have been as a jeweller and cutler, and examples of his silverware are in various collections.[1] Cameron made his first appearance, described as a jeweller, clock and watchmaker and nautical instrument maker, in the Dundee street directory for 1818 in the High Street and was recorded there until 1824; after an absence from the directory for four years, he reappeared as a watchmaker at 120 Overgate in 1829 where he remained until 1848.[2]

In the *Edinburgh Evening Courant* of 6 January 1828 he advertised that he had 'fixed a Sidereal Clock, and procured a Transit instrument divided with N.P.D. [North Polar Distance] from a workman recommended by the Astronomer Royal at Greenwich Observatory, for rating chronometers and watches.'[3] George Lowdon (q.v.) commented in his autobiography that when he began in business in Dundee in about 1850,

> 'there were not at that time in town many who owned an astronomical telescope. One was owned by Mr Cameron, jeweller, whose shop was in a high building fronting to High Street, east side of Tindal's Wynd, since removed for the buildings of the Royal Bank of Scotland, and as Mr Cameron's house was in the top flat above his shop he was favourably situated for using his telescope to advantage ...'[4]

A number of watches and clocks by this maker have been noted,[5] including a report in 1832 that an Archangel business house had equipped two ships, commanded by officers of the Imperial Navy, at its own expence, to sail on a voyage of discovery 'to the great gulf of the Icy Sea ... to explore the entrance of the river Jenisky ... [They] purchased for the use of the discovery ships, the marine chronometer, No 1517, from Mr Alexander Cameron's chronometer depot, Dundee'.[6] In another later advertisement he mentioned that 'One of his Experimental Chronometers, sold for Forty Guineas to Captain Clerke of Dundee, has been twice round the World, and was in February 1835 at Sydney N.S.W., still found going at MEAN TIME.'[7]

Just how much of his horological work was his own is difficult to estimate, judging by his own 1828 advertisement:

> 'Alexander Cameron purposes being in the English market in a few weeks, and solicits orders for himself or his English connection, which may be executed direct from the different manufacturers. His friends in the south and north of Scotland will be waited upon at the usual times with a new set of patterns.'[8]

The same may be said of his nautical instruments which he 'sold and repaired'.[9] He described himself as a 'seller of watches and clocks' but a 'maker of chronometers'.[10] The octant in this collection[11] appears to have been made in London and merely retailed by Cameron. One other nautical instrument is known to have been sold by Cameron; it is a sympiesometer made by and signed 'Adie & Son Edinburgh' (q.v.) and further engraved 'Made for Alex[R] Cameron, 4 High St, Dundee' and numbered '1708'.[12] This example demonstrates that small businesses found it economic to buy in complete instruments for resale, as well as parts. It also shows the importance for larger firms of outlets near their markets, in this case a business based in the capital reaching customers in a port.

REFERENCES

1. Those in the Royal Museum of Scotland are as follows: three toddy ladles (inventory MEQ 60, MEQ 61, MEQ 251) and sugar tongs (MEQ 730). Dundee Art Galleries and Museums also have a number of pieces of silverware by Cameron: a pair of teaspoons, c.1810 (inventory 1986-87, 1 & 2) ; an electroplated fork made by Elkington (1986-91); a pair of dessert spoons, Newcastle 1828 (1986-84, 1 & 2); an egg spoon (1966-242); a whisky label (1962-743,1); a pair of candlesticks (1961-639,1 & 2); ten toddy ladles (1966-249,1-10); sugar tongs, Newcastle 1830 (1986-95); a cream boat (1980-114) and a table fork (1986-120). We are grateful to Miss Clara Young of Dundee Art Galleries and Museums for this information. Several items have passed through the auction houses: for example, Phillips Edinburgh, 22 October 1982, Lots 126, 130, 135, 136, 137, 138, 143, 146; 20 May 1983, Lots 118, 120, 121, 122, 123, 125; 18 May 1984, Lots 108, 110, 114, 116, 118, 119, 125; 19 October 1984, Lots 54, 55, 58; 24 May 1985, Lots 45, 46, 47, 57, 58, 59; 18 October 1985, Lots 47, 49, 52; 24 October 1986, Lots 47, 48, 49, 56, 57, 58, 60, 63, 202; 20 February 1987, Lots 110, 136, 137,143; 22 May 1987, Lots 62, 64, 72, 73; 20 May 1988, Lots 61-63, 68, 70-72.

2. *Dundee Post Office Directory* 1818-1848, *passim*. D.J. Bryden, *Scottish Scientific Instrument-Makers 1600-1800* (Edinburgh, 1972), 45. It has been supposed that Cameron was apprenticed in Edinburgh in the 1790s, on the basis of three items marked A.C., with Edinburgh assay marks and dated 1797-8 (two coconut cups, offered for sale by Sotheby's, 23 April 1970, Lot 183, one of which appears subsequently to have been offered by Phillips Edinburgh, 18 October 1985, Lot 46; and a silver salver of 1798, also offered by Phillips Edinburgh, 18 October 1985, Lot 49). However, we have no reason to believe that this is the same man, and no such apprenticeship has been found. Derek Graham of Phillips Edinburgh, has drawn our attention to Alexander Cameron, silversmith of Perth (J. Munday, *A History of Perth Silver* (Perth, 1980), [23]); it has been established that this maker was apprenticed as late as 1821 (information from Robin Roger, Perth Museum and Art Gallery, February 1989).

3. Quoted in John Smith, *Old Scottish Clockmakers* 2nd edition (Edinburgh, 1921), 75-76. No correspondence with John Pond up to 1835, has been located at the Royal Greenwich Observatory Archive, Herstmonceux. We are grateful to Adam Perkins, Archivist, for checking for references to Cameron.

4. 'George Lowdon, Optician and Scientist, Dundee. Sketch of his Reminiscences and Career', *Dundee Advertiser*, 3 and 6 February 1906.

5. A Duplex watch movement, marked 'Alex Cameron Dundee' is in the Royal Museum of Scotland (NMS T1931.234). A gold lever watch, with movement signed 'Alex Cameron Dundee', and case marked Chester, 1827, was offered for sale by Christie's, 8 April 1981, Lot 220. Another gold lever watch signed 'A Cameron Dundee 34697' in a case marked Chester, 1821, was offered for sale by Christie's, 24 July 1984, Lot 142. A mahogany longcase clock, with 13 inch dial signed 'Cameron Dundee', was offered for sale by Sotheby's, London, 9 June 1977, Lot 90. A quarter-striking regulator signed 'Alex Cameron Dundee' was offered for sale by Christie's, 30 October 1984, Lot 76.

6. *Scotsman*, 13 October 1832. Cameron is not known to have manufactured chronometers, and presumably this was an instrument supplied by a London chronometer maker, perhaps Charles Frodsham whose No. 1632 was purchased by the Admiralty in 1838 (V. Mercer, *The Frodshams* (Ticehurst, Sussex, 1981), 299).

7. *Ibid.*, 11 July, 9 September and 10 October 1840.

8. *Edinburgh Evening Courant*, 6 January 1828, quoted in Smith, *op. cit.* (5).

9. *Dundee Post Office Directory* 1818.

10. *Dundee Post Office Directory* 1845, 283-286.

11. NMS T1980.183.

12. Offered for sale by Phillips, 20 July 1983, Lot 25.

---

62. Octant: Cameron, Dundee, c.1820 (T1980.183)

12″ octant with ebony frame and brass fitments. Scale divided on an inset ivory arc [-2]-[99° 40′] by 20 mins., the reinforced index arm with clamp, tangent screw and vernier reading to 1 min., and scale divider's anchor stamp with indistinct initials. Engraved on the index arm 'Cameron, Dundee.'; the inset ivory name plate is blank. The frame with single vertical strut and bowed horizontal strut; and provided with double pin-hole sights for foresight and single pin-hole sight for backsight, both horizon glasses adjustable, single removeable group of 3 shades, noteplate on reverse and pencil holder in vertical strut, and supported on feet. With fitted keystone-shaped stepped case.

Scale radius: 296mm.
Case size: 335 x 375 x 100mm.

The frame has been split at the upper slot for the shades and repaired with a whitewood. Lacking the attachment washers for the sights, the forward horizon glass locking screw, the top plate of the index mirror frame, half of the index arm locking spring, and the pencil. The ivory scale is discoloured and becoming detached at the ends.

Provenance: Purchased at Christie's, 8 February 1966, Lot 2.

38 P. A. Feathers' shop in Dock Street, Dundee, c.1880.
A. I. Barry, Esq.

PETER AIRTH FEATHERS

According to an anonymous obituary, Peter Airth Feathers was born on 12 July 1821 in Dundee, son of James Feathers, a ship-owner and coal merchant, and his wife Helen Airth.[1] He began working as a watchmaker, served his apprenticeship in Dundee, then worked as a journeyman in Brechin, Leith and Glasgow, although it is not disclosed with whom he served. However, the firms in the seaports of Leith and Glasgow also made nautical instruments, and he was able to learn enough to be able to open his own business in this line in Dundee in 1840.[2] He first appeared in the Dundee street directory for 1842 as a watch and clock maker at 73 High Street. In 1845 he had moved to 10 Dock Street, where he was listed as a chronometer and nautical instrument maker, and remained there until 1853 when he moved to 26 Dock Street, an address he kept until 1868. From 1869 to 1873 he was at 40 Dock Street. The firm became P. A. Feathers & Son, opticians and chronometer makers, in 1874 and remained at 40 Dock Street for three years; between 1878 and 1909 it was listed at 43 Dock Street.[3]

His obituarist described the environment in which Feathers began business:

> 'at that time the Harbour presented a totally different appearance from that what it shows today. Though it was not developed to the same extent, it was perhaps a busier and livelier place. The river and docks were crowded with the smaller wooden ships which have almost entirely disappeared, and Mr Feathers' business brought him into contact with sailors of almost all nationalities. His nautical instruments made his name known in ports all over the world. He was familiar with all the changes that have taken place in shipping during these years. He saw wood giving place to iron and steel, sailing ships being replaced by steamers and had a fund of information regarding the shipmasters who

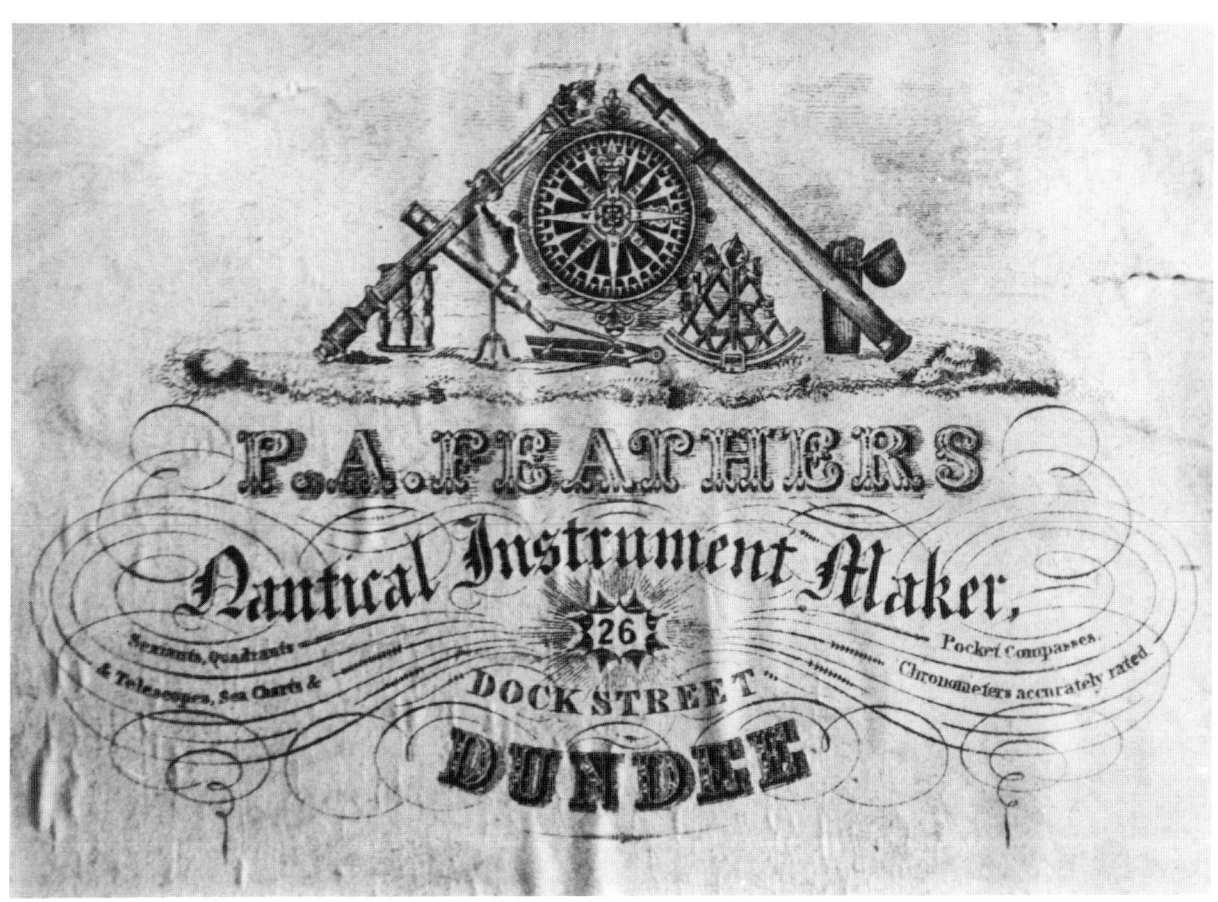

39 P. A. Feathers' trade card for his address between 1853 and 1868. Stirling Smith Art Gallery

143

visited the port from near and far. His work also brought him into immediate contact with the development of nautical appliances, and with all the inventions for facilitating navigation. Ever ready to take advantage of the progress of science relating to the equipment of ships, he continued to suit his business to the newest devices that were in demand.'[4]

Among the nautical instruments that have been preserved have been examples of octants - including one in this collection[5] - sextants[6] and telescopes,[7] while details of chronometers,[8] and other instruments have been noted.[9] It is not known how large his workshop was, but of his three sons James Feathers was working with his father from at least 1884 and Peter Feathers was a photographic stock dealer operating from his father's business premises from 1887 and moving to 6 Castle Street in 1891.[10] The firm produced a number of catalogues which were mainly photographic in content, and would appear to have been the work of the younger Peter Feathers: he continued to produce them after he had left Dock Street.[11]

Peter Airth Feathers died on 19 February 1904, aged eighty-four, at Somerville House, Broughty Ferry,[12] where he had lived since 1868. His estate was worth £2,305 12s 5d,[13] and in his will he left to his son and sole partner James the entire business of P. A. Feathers & Son. Peter Feathers was left £1,900 and the third son, John Murray Mitchell Feathers, a clerk, £1000.[14]

James Feathers ran P.A. Feathers & Son after his father's death until 1910, when he retired and the business disappeared from the street directories; Peter Feathers ran his business at Castle Street until 1920, when he too retired.[15] It is perhaps worth noting that the shop at 43 Dock Street had a shop sign similar to that once found at David Stalker's (q.v.) premises in Leith, that of a midshipman with an octant.[16] James Feathers lived in retirement at Barnhill, Broughty Ferry, until his death at the age of eighty-five on 6 April 1931.[17] Peter Feathers junior, who introduced cinematography to Dundee, owned and ran a cinema in Morgan Street between 1910 and 1923; he died, aged eighty-four, on 18 May 1943.[18]

REFERENCES

1. 'Mr Peter A. Feathers', *Dundee Yearbook* (Dundee, 1904), 55. The baptism is recorded in General Register Office (Scotland), OPR 282/15; we are grateful to Arthur I. Barry, a descendant of Feathers, for providing us with this and other information about the family.

2. *Dundee Yearbook*, op. cit. (1).

3. *Dundee Post Office Directory* 1840-1904, *passim*.; D.J. Bryden, *Scottish Scientific Instrument-Makers 1600-1900* (Edinburgh, 1972), 48.

4. *Dundee Yearbook*, op. cit. (1).

5. The Royal Museum of Scotland has a 7½ inch octant, marked 'P. A. Feathers & Son, Dundee' (NMS T1980.184). A 10 inch octant signed 'P. A. Feathers Dundee' is in the Museum of Science and Engineering, Newcastle upon Tyne, and a similar instrument is in Sunderland Museum and Art Gallery. A 9¾ inch ebony octant signed 'P.A. Feathers, Dundee' was offered for sale by Sotheby's, London, 10 March 1987, Lot 113; a 9½ inch octant signed 'P.A. Feathers & Son, Dundee' was offered for sale by Phillips, 8 May 1985, Lot 214, and a similar piece at the same saleroom, 17 July 1985, Lot 90; a 7 inch octant signed 'P.A. Feathers & Son, Dundee' was offered for sale by Phillips, 10 September 1986, Lot 191.

6. Dundee Art Galleries and Museums have three examples: a 9⅝ inch radius sextant marked 'P.A. Feathers Dundee' (inventory 1976-1182), an 8 inch brass frame sextant marked 'Feathers Dundee', (inventory 1965-77) and a similar example (inventory 1976-1587). The Smith Art Gallery and Museum, Stirling, has a sextant with the box labelled 'P. A. Feathers Nautical Instrument Maker Dock Street Dundee' (inventory STIGM; 1287). A 7¼ inch brass sextant, signed 'P. A. Feathers & Son, Dundee' was offered for sale at Phillips, 26 January 1983, Lot 108.

7. The Royal Museum of Scotland has a 3½ inch single draw refracting telescope marked 'Feathers Dundee' (NMS T1967.4). A similar instrument was offered for sale at Sotheby's Belgravia, 12 July 1977, Lot 217.

8. For instance, Sotheby's, 2 February 1976, offered as Lot 123, a two-day marine chronometer No 1087 by P.A. Feathers, 10 Dock Street, Dundee.

9. Dundee Art Galleries and Museums have a boat compass (inventory 1986-355). A sympiesometer with the signature recorded as 'D.A. Feathers, 10 Dog Street, Dundee' [*sic*] was offered for sale by Christie's South Kensington, 4 April 1984, Lot 283.

10. *Dundee Post Office Directory* 1884-1910, *passim*.

11. *Illustrated Catalogue of Apparatus, Chemicals and Every Requisite for Dry Plate Photography Supplied by P.A. Feathers & Son 43 Dock Street, Dundee* [Dundee, 1885]; *Season 1885-6. Illustrated List of New and Popular Lantern Slides for Sale or Lent on Hire by P.A. Feathers & Son 43 Dock Street, Dundee* [Dundee, 1886]; *Season 1889-90. Annual Catalogue of Lantern Slides kept in Stock by P.A. Feathers & Son, 43 Dock Street, Dundee ...* [Dundee, 1890]; *Season 1891-92. Catalogue of Lantern Slides and Lanterns kept in stock by Peter Feathers, 6 Castle Street, Dundee* [Dundee, 1892]; *1894-95. Illustrated Catalogue of Photographic Material Chemicals and Apparatus Supplied by Peter Feathers, 6 Castle Street, Dundee* [Dundee, 1895]; *1896-97. Illustrated Catalogue of Photographic Material Chemicals and Apparatus supplied by Peter Feathers, 6 Castle Street, Dundee* [Dundee, 1897]. All of these items are located in Dundee Public Library.

12. G.R.O. (S.) Register of Deaths 1904 (Forfarshire, Monifieth) 310/22. He died of apoplexy.

13. Scottish Record Office, SC 45/31/56, *Dundee Sheriff Court Record of Inventories*, 154-156.

14. S.R.O. *Calendar of Confirmations and Inventories* 1904, 189; S.R.O., SC 45/34/7, pp. 363-377. John M. Feathers subsequently became a flax merchant: he died on 30 November 1937 (*Dundee Courier and Advertiser*, 1 December 1937).

15. *Dundee Post Office Directory* 1904-1932, *passim*.

16. From a photograph of the exterior of the shop, c.1890, kindly supplied by Arthur I. Barry.

17. Information from James Feathers' grandson, Arthur I. Barry, May 1987.

18. 'Dundee Pioneer of Cinematography', *Dundee Courier and Advertiser*, 19 May 1943; 'A Century of Taking Pictures', *ibid.*, 12 April 1980.

---

63. Octant: P.A. Feathers & Son, Dundee, c.1900 (T1980.184)

8″ octant with plain cast frame, in oxidised bronze and brass (stripped). Scale divided on silver, [-5]-[125], the reinforced index arm with clamp, tangent screw and vernier reading to 15 secs. Engraved on the arc 'P.A. Feathers & Son, Dundee.' The frame with 7 shades and rear handle, supported on feet.

Scale radius: 189mm.

Lacking sight tubes and telescopes, the locking plate from the clamp, and the swinging magnifier. One of the handle supports has been replaced, and the sight tube attachment has been remounted.

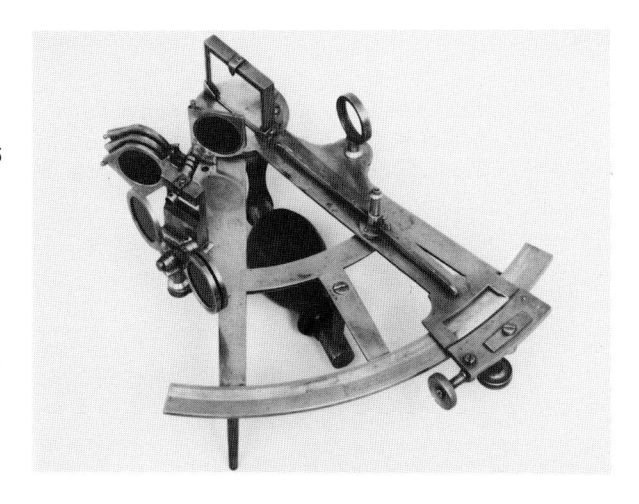

12

GEORGE LOWDON OF DUNDEE

George Lowdon was one of the few nineteenth century Scottish instrument makers to write his autobiography.[1] This opens with the remark: 'I was born in Kirk Entry, Wellgate, in 1825, and to show that I am, moreover, a Dundonian by descent, I can trace back my ancestors for 200 years, to a great-grandfather who had "a ship of his own", which traded between Dundee and Holland.'[2] His father, George Lowdon senior, was a grocer in Dundee but had also developed considerable mechanical skills and had a certain amount of equipment. His father's work clearly impressed young George and 'even at that time my father had successfully ground and polished lenses of different kinds, so naturally I took to what has proved the life-long making of scientific instruments'.[3]

40 George Lowdon (1825-1912)

However, at the age of fifteen, he went to work at a flax factory, where he remained for three years until depression in the textile trade forced him to leave. 'As a maker of scientific instruments I commenced business in Dundee in May, 1849, at the age of 24 years.'[4] He found it hard going at first:

> 'To show how one can live on little more than hope, I see that the total drawings (not income) I had at end of 1850, covering a year and a half's work of self and one man, amounted to £129 4s, but then a first-class workman's weekly wage was only fifteen shillings, and therewith he was content. I had then my workshop at top of Craig Pier, but in 1860 I removed it to special premises where the Caledonian Railway now stands, naming them "Railway Buildings." There I had a fine workshop along with a commodious sale

shop. This was to me a most unfortunate change, and what I had made before I lost, and more. After ten years I removed to 1 Union Street as a sale shop, and was there until it was sold. I was offered this shop and the adjoining one for £2800. Unfortunately, I could not buy it. It was sold some four years afterwards for over £6000, and I had to remove, first to the other side of Nethergate, then to Reform Street.'[5]

He was first listed in the street directory in 1850 as an optician and fishing rod maker at 25 Union Street, where he remained until 1861; in 1864 he was at Railway Buildings, South Union Street (or 1 Union Street); from 1876 until 1880 his address was 23 Nethergate; and from 1882 until 1904 he was at 65 Reform Street, after which he disappeared from the directory, although his son John Lowdon, electrician, was trading from the same address from 1887 until the mid-1930s.[6] Another son, Edward, became well known in the engineering world, establishing the firm of Lowdon Brothers in 1876, and becoming involved in traction engines, hydraulics, as well as installing electricity in large private houses throughout the country.[7] The tradition of being in the forefront of technology has continued with the firm's involvement in the North Sea Oil industry; and after its incorporation as a limited company in 1947 the firm celebrated its centenary in 1976.[8]

George Lowdon did not find an instant market when he began instrument-making, which perhaps accounts for his initial description as fishing-rod maker:

> 'There was when I commenced business no regular manufacturing optician in Dundee. One Balerno, who lived in Yeaman Shore, and another, Antony Tarone, who had his house in Murraygate, gave themselves out as "barometer and mirror makers". The former hawked his goods about the street, and usually had a barometer under one arm. Hawking was then in great vogue.'[9]

However, the young George Lowdon had a stroke of good fortune very early on in his new career when 'at the end of 1849 I got acquainted with that nobleman so well and favourably known to all Dundonians, George, Lord Kinnaird, and through him was introduced to many of the savants who were entertained by his Lordship at Rossie Priory'.[10] Amongst these was Sir David Brewster, scientist and polymath, 'who had at this period (1849) invented his stereoscope, and I got the making of the first one, and the sending of copies of it to many scientific men all over Europe'.[11] Lowdon claimed that he went on to patent an improvement in the device, to Brewster's disgust, but no trace of this can be found.[12]

He made other optical instruments, notably microscopes, of which a number have survived:

> 'In 1850-51 I made to order a monster microscope, said still to be the first and only one of its size ever made. It was intended to be shown in the great [sic] Exhibition of 1851, but was too late in being delivered. It stood four feet in height with a four-inch diameter tube and had every motion, vertical and horizontal that a fine instrument ought to have, besides having a photographic slide for Fox-Talbot paper to slip in the Huyghenian eye-piece. The glasses were ground by myself, and the instrument cost only £40. It was made to the order of Dr Wise, Tantalen Castle, Cork, and is now in the possession of the London Microscopical Society, and has been named by them "Jumbo", being contrasted with a very small one placed alongside and named "Tom Thumb". I have made a large number of microscopes, both monocular and binocular, at prices ranging from 60s to £60. This department constituted a large part of my business from 1860 to 1900, when the sale of microscopes fell away.'[13]

Clearly, Lowdon was making at least some of these instruments, although whether he also bought in parts and assembled them himself, or bought completed instruments for resale is a matter of conjecture. The microscope in this collection owes much stylistically to the microscopes of the great London maker, Andrew Ross, whom Lowdon knew:

> 'Mr Andrew Ross was then the principal maker of telescopes in London. I went twice to London to see him about glasses, etc. He supplied me with the curves for lenses of different classes. The lenses in the Harbour telescope were ground by me from curves supplied by Ross ... After this I made many telescopes, refracting and reflecting. I made a Newtonian telescope for Mr Robert Millar, teller in the Dundee Bank. The tube was 12 feet long, with a 12 inch mirror, and was erected for him between two walls in his garden ... '[14]

Lowdon also produced other optical and philosophical instruments: there are examples of refracting telescopes in this collection and elsewhere,[15] instruments such as theodolites[16] and sympiesometers[17] have been recorded, and he is known to have made meteorological apparatus.[18] It is certainly plausible that such a range of apparatus was being produced, because his workforce had increased until by 1861 he was employing four men and two boys.[19]

As for opticians such as Lizars (q.v.) in Glasgow and Hume (q.v.) in Edinburgh, the arrival of popular photography was something his business could exploit, and his autobiography mentions the arrival of the daguerreotype in Dundee[20] (with no indication of the date), but goes on to discuss the wet-plate method:

147

'Photography may be said to have commenced business when I commenced business. I was taking pictures by the Fox-Talbot process in 1849 (having failed to succeed with it in 1846), and since that period have taken pictures and made many cameras and given instruction to a large number of professionals and amateurs - notably to the late artist-photographer, Mr G.W. Wilson, of Aberdeen. I sold him a camera in 1853, and gave him his first lesson. The camera complete cost him £5 10s, from which out of his artistic hands sprang a considerable fortune.[21]

This connection with George Washington Wilson of Aberdeen, the photographer whose business career coincided with a revival in interest in the Scottish Highlands and the beginnings of tourism there, seems to exaggerate Lowdon's role. Although Wilson's change of profession from artist to phographer is relatively undocumented, it is known that he formed a partnership with the Aberdonian photographer, John Hay, junior, in late 1853; Hay's father and uncle, J.& J. Hay, were carvers and gilders who also had a shop in Market Street, Aberdeen, which sold various optical and philosophical instruments.[22] Lowdon continued:

> 'At this time time many artists were attracted by the aid which it was early seen photography was to lend to their art, and visited my shop for information. Amongst the most notable of these may be mentioned Sam Bough and John Cairne, for whom I took photographs of various scenes at the Harbour and in the outskirts of Dundee.'[23]

An early camera from Lowdon's workshop fitted with a London-made lens survives in this collection.[24] Lowdon's interest in the mechanics of photography led him to a further invention, which he described:

> 'I patented the first first fixed-focus folding pocket camera with a central shutter between the lenses, known as the "Eclipse". Of this camera a very great quantity was made and used all over the world. This camera was a new departure from the old forms, being capable of use by being simply held in the hand, and used in the streets for snap-shot work. It would be carried in the pocket and be ready for use in a few seconds. This was patented in 1885.'[25]

However, despite these optical and philosophical activities, it emerges from Lowdon's autobiography that his real interests and inventiveness lay in electricity and magnetism. He devoted larger sections of his time to magnetic communication on railways between guard and driver, magnetically detonated explosions of submerged rocks in the harbour, and above all to electric and wireless telegraphy and the introduction of electric lighting and the telephone to Dundee. He had a long association with the local inventor James Bowman Lindsay, who according to Lowdon 'had little or no ability as a mechanician',[26] a deficiency which Lowdon was able to remedy. He was 'for many years practically the only maker of electrical apparatus in the city... In the year 1876... the first dynamos in Scotland were brought by Mr Lowdon to light the works in building the first Tay Bridge.'[27] By concentrating in this area, electrical goods and electrical engineering proved the route to survival for the firm in the twentieth century.

George Lowdon married Frances Kenton, who predeceased him, and they had at least three sons and one daughter, the youngest of whom, Edward Joseph Bonar Lowdon, was his father's heir and successor when the old man died at the age of eighty-seven on 17 October 1912.[28] His personal estate was worth £106 11s 5d, and his heritable property less than £300,[29] which might account for the world-weary note on which his autobiography ends: 'Great at times has been the struggle.'[30]

REFERENCES

1. This appears in a number of forms. The fullest of these is 'George Lowdon Optician and Scientist, Dundee. Sketch of his Reminiscences and Career', *Dundee Advertiser*, 3 and 6 February 1906, which was also separately reprinted. An edited version was by A. H. Millar, 'Other Pioneers of Invention: George Lowdon', in *James Bowman Lindsay and Other Pioneers of Invention* (Dundee, 1925), 83-94. This appears to have formed the basis of an anonymous obituary, 'Death of Eminent Dundonian: Pioneer in Science' in the *Dundee Year Book* (Dundee, 1912), 68-69; and an article by Harry Ford, 'Born with Tools at his Fingertips', *Scots Magazine*, new series 121 (1984), 622-625.

2. Millar, *op. cit.* (1), 83.

3. *Ibid.*, 85. George Lowdon was initially in partnership with his father, but for how long is unclear as there is only his account to say that he began as an instrument maker in May 1849 (see below); the partnership ended in early 1850:

'Dundee, January 4, 1850. the Partnership carried on by the Subscribers, the Sole Partners, under the Firm of 'G. LOWDEN [*sic*] & SON', as Opticians and fancy Workers, at No. 25 Union Street, Dundee, was DISSOLVED this day by mutual consent, and the Subscriber, George Lowden, Senior, has no farther interest or concern in the Business to be carried on hereafter...' (*Edinburgh Gazette*, 11 January 1850, 33).

4. *Ibid.*

5. *Dundee Advertiser, op. cit.* (1).

6. *Dundee Post Office Directories* 1850-1936.

7. [Anon.], *Lowdon's Electrical Services 1876-1976* (Dundee, 1976).

8. *Ibid.*

9. *Dundee Advertiser, op. cit.* (1).

10. Millar, *op. cit.* (1), 86. The role of Rossie Priory in the early

history of photography in Scotland is discussed in A.D. Morrison-Low, 'Dr John and Robert Adamson: An Early Partnership in Scottish Photography', *Photographic Collector* 4 (1983), 202-203. The original article by Robert Graham, 'The Early History of Photography' is reproduced in *History of Photography* 8 (1984), 231-235, with calotypes of Rossie Priory.

11. Millar, *op. cit.* (1), 86. Brewster, Lowdon and the lenticular stereoscope are discussed in A.D. Morrison-Low and J.R.R. Christie (eds.), *'Martyr of Science': Sir David Brewster 1781-1868* (Edinburgh, 1984), 62, 98 and 99.

12. *Ibid.* No examples of Lowdon's stereoscope are known to have survived. However, he did advertise in the *Dundee Post Office Directory* 1858, xxxv, that he was the 'Sole licensed maker of the Patent Dioramic Stereoscope'.

13. Morrison-Low and Christie, *op. cit.* (11), 88-89. This instrument found its way with the collection of Sir Frank Crisp Bt., and was sold at Steven's Auction Rooms, London, on 21 January 1921 as Lot 705, when it was bought for the Wellcome Collection. It is now in the Wellcome Museum of the History of Medicine, Science Museum, London, where its accession number is A56559. It was described and illustrated in the *Journal of the Royal Microscopical Society* 2nd series, 2 (1882), 850-852. We are grateful to Jane Insley for this information. Other microscopes known to be by Lowdon are a compound achromatic instrument, which belonged to the electrician and lexicographer James Bowman Lindsay, in a private collection; a bar-limb binocular instrument is in the Royal Museum of Scotland, NMS T1980.239; and a Jackson-limb binocular instrument with many accessories was sold at Sotheby's, 19 May 1983, Lot 133.

14. *Dundee Advertiser*, *op. cit.* (1).

15. Royal Museum of Scotland, NMS T1980.260 and T1981.42; a refracting telescope is in Dundee Art Galleries and Museums, inventory 1974-560; another was offered for sale by Sotheby's Belgravia, 30 November 1979, Lot 242; a St. Ives Kromscope retailed by Lowdon is in the Royal Museum of Scotland, NMS T1936.67.

16. Offered for sale at Christie's South Kensington, 22 February 1978, lot 8; a surveyor's level described as marked 'Lowden Dundee' was offered for sale at the same rooms, 18 April 1985, Lot 100.

17. Offered for sale at Christie's South Kensington, 27 July 1983, Lot 287. This instrument is signed 'LOWDEN [*sic*]/ Dundee', and clearly demonstrates that Lowdon had a number of different ways of spelling his name: Sir David Brewster refers to him as 'Mr Loudon , optician, in Dundee' (David Brewster, *The Stereoscope: Its History, Theory, and Construction* (London, 1856), 29); Lowdon appeared as 'Lowden' in the Dundee Directory between 1850 and 1869, and his father as 'Loudon' in 1841.

18. A rain-gauge by Lowdon was recorded as instrument 234 in the survey by the Rainfall Committee of the British Association: *Report of the British Association ... 1867* (London, 1868), 448.

19. General Register Office (Scotland), Census of 1861 (Dundee), 282[1]/25, p.11.

20. Millar, *op. cit.* (1), 88.

21. *Ibid.*, 92-93

22. Roger Taylor, *George Washington Wilson Artist and Photographer 1823-93* (Aberdeen, 1981), 14-18. Barometers, probably retailed rather than made by the Hays, were offered for sale by Sotheby's, 21 July 1983, Lot 173 and 15 December 1983, Lot 135; by Phillips Edinburgh, 14 December 1984, Lot 45, 31 May 1985, Lot 111, and 30 August 1985, Lot 91.

23. Millar, *op. cit.* (1), 93. Some of Lowdon's photographs were exhibited at a photographic exhibition held in the Albert Institute Buildings, Dundee, from 10 February to 10 March 1882: [Dundee and East of Scotland Photographic Association] *Catalogue of Photographic Exhibition* (Dundee, 1882), 29, item 551 'Six Views in Dundee and Vicinity (from Wax Paper Negatives - No. 574), by George Lowdon, Optician' and item 552 'Nine Views in Dundee and Vicinity (from Wax Paper Negatives - No. 570), by George Lowdon, Optician'; *ibid.*, 30, items 570 and 574, which were the negatives for these prints. We are grateful to Sara Stevenson for bringing this information to our attention.

24. Frank Collection, NMS T1981.41; the lens is by Andrew Ross. A half-plate stereoscopic tailboard camera with a Lowden name-plate was offered for sale by Christie's South Kensington, 21 June 1984, Lot 48.

25. Millar, *op. cit.* (1), 93. U.K. patent 4102, 1 April 1885. The Royal Museum of Scotland has examples of this camera which was manufactured under license by J.F. Shew & Co., London, one a monocular (NMS T1984.77) and one a stereoscopic example (NMS T1984.78), both once owned by Professor George Forbes (1849-1936). F. Shew extended the original patent in U.K. Patent 11,394, 4 July 1891. Lowdon took out two other patents, both relating to camera structural improvements: 7121, 27 May 1886, and 5666, 17 April 1888.

26. *Ibid.*, 91. An electrical induction machine, signed 'Lowdon, Dundee' is in the Royal Museum of Scotland (NMS T1978.L15). Another electrical machine is in Dundee Art Galleries and Museums (inventory 1981-375).

27. Wm. C. Keay, 'Electrical Engineering in Dundee', in A.W. Paton and A.H. Millar (eds.), *British Association Handbook and Guide to Dundee and District* (Dundee, 1912), 307-308.

28. G.R.O.(S.), Register of Deaths 1912 (Dundee, St. Mary's), 282[2], 651.

29. Scottish Record Office, Dundee Register of Confirmations for 1912, SC 45/35/45, f.311.

30. Millar, *op. cit.* (1), 93.

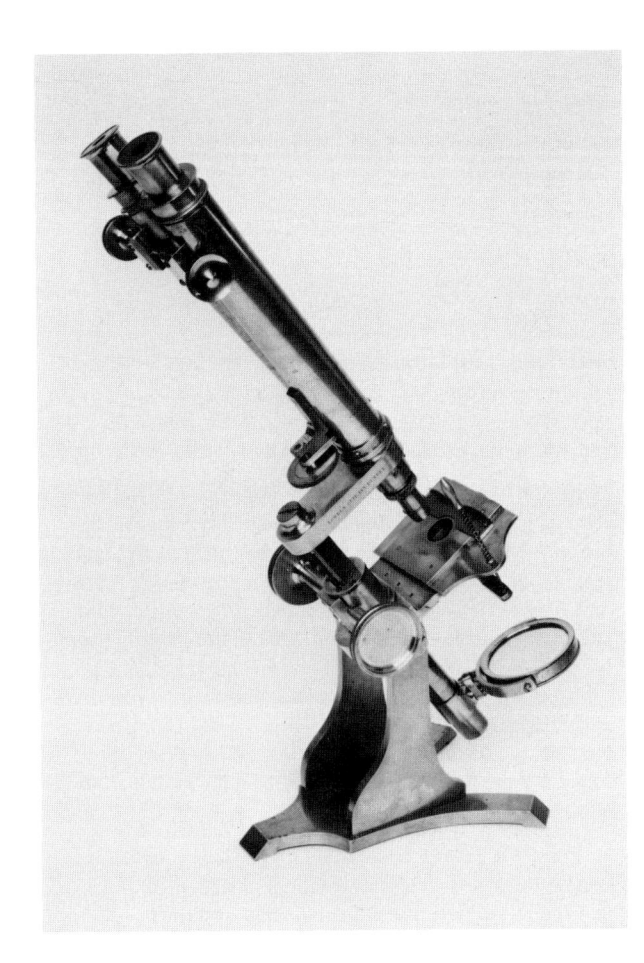

64. Microscope: Lowden, Dundee, c.1870 (T1980.231)

Compound Wenham binocular bar-limb microscope, in brass, bronze and oxidised brass (stripped), on an inclining pillar and flat 2-toed tripod base. Engraved on one side of the bar-limb 'LOWDEN, OPTICIAN, DUNDEE.' Single-draw body tubes with linked rack and pinion adjustment, and slide at the base of the body for moving the beam-splitting prism in and out of operation; the body attached to a bar-limb on a triangular-section slide within the pillar, raised by a pinion wheel for coarse focus adjustment (the reduced diameter pinion does not engage with a rack, but takes 3 turns of a fusee chain secured at top and bottom of the triangular pillar, which has not been drilled to take a rack); fine focus is by an external screw at the lower end of the body tube advancing a spring-loaded nosepiece (defective) with the objective. Objective and pair of eyepieces (engraved 'A'). The stage attached to a trunnion mount on the pillar, supported by two uprights rising from the base. Manual adjustment in two directions of a rectangular upper stage plate with slide mount, manipulated by a lever connecting the upper and lower stage at one side and with a slotted screw connection at the other. Rotating 4-aperture diaphragm, and plane/concave mirror on a swinging tail sliding on the pillar.

Body length: 275mm.
Body diameter: 33mm.
Height: 450mm.
Base size: 168 x 202mm.

The collar at the base of the body which provides alignment for the Wenham prism slide is split and deformed; the prism itself is chipped on two edges, and the objective centering screws which would pass through the collar are lacking. The screws attaching the off-set body tube and the twin racks to the eyepieces were replaced in the Museum in 1982. The sliding nosepiece operated by the fine focus is seized. Broken retaining screw on the plate which aligns the triangular-section pillar. Lacking the second slide holder. The plane mirror is a replacement.

There are detailed similarities with a binocular Jackson-limb instrument of the same size, signed on the base 'Lowdon / OPTICIAN / DUNDEE', offered at Sotheby's, 19 May 1983, Lot 133. The body tube construction, base, and sub-stage mirror assembly appeared to be identical, but the instrument had a complex mechanical stage and complete accessories including polariser and analyser.

65. Refracting telescope: Lowdon, Dundee, c.1850 (T1980.260)

1¾" aperture 2-draw hand-held refracting telescope with doublet objective and 2-component eyepiece and erecting assemblies; in brass (with leather-lined draw slides), with tapered leather-covered barrel. Engraved on the first draw tube 'Lowdon, / Dundee.' Scratched assembly mark 'II' on the barrel, draw tubes and draw slides. With velvet-lined carrying case and lid, but lacking the carrying strap.

Length closed: 164mm.
Maximum tube diameter: 50mm.
Aperture: 43mm.
Case size: 175 x 64mm. diameter.

Provenance: Purchased at Sotheby's, 15 October 1973, Lot 44.

66. Refracting telescope: Lowdon, Dundee, c.1890 (T1981.42)

3¼" aperture 3-draw mounted refracting telescope in brass; with doublet objective recessed from the end of the barrel and 2-component eyepiece and erecting assemblies in the first and second draw tubes respectively for variable power, and with rack and pinion focus (defective) to the third draw tube. Engraved on the barrel end plate 'LOWDON, / DUNDEE.' Closure on external end stop. Mounted by a knuckle-joint with clamped azimuth motion in a taper bearing on a folding wooden tripod with oxidised brass fittings.

Length closed: 1,182mm.
Tube diameter: 92mm.
Aperture: 85mm.

The knuckle-joint screw is distorted, and the clamp screw for the stand is lacking.

Detailed similarities in design and in the mounting of optical components suggest that this item is by the same manufacturer as items 57 (E. Lennie), 85 (Gardner & Co.), 93 (J. Brown) and 161 (Whyte, Thomson & Co.). It can also be compared with a similar telescope offered for sale by Sotheby's Belgravia, 30 November 1979, Lot 242.

67. Plate camera: Lowdon, Dundee, c.1855 (T1981.41)

Half-plate folding camera in wood with brass fittings and leather bellows. Sliding wooden lens mount with recessed name plate marked 'LOWDON / MAKER / DUNDEE'. Racked advance for bellows extension; removeable plate holder with ground-glass screen. All separable wooden components stamped '2'. Rack and pinion focused portrait lens designed for use with Waterhouse stops, engraved 'A. Ross, London / 6789'. The baseboard with threaded brass bush for tripod attachment.

Lens aperture: 42mm.
Size of wooden body closed: 200 x 210 x 75mm.

The objective lens is threaded to a brass mounting flange, attached by 4 screws. However, the sliding lens mount has previously had attached to it a flange secured by five evenly-spaced screws, indicating that the present lens is a replacement or that Lowdon fitted Ross lenses to complete cameras obtained from a specialist supplier. Lacking the photographic plate dark-slides.

13 ABERDEEN INSTRUMENT MAKERS

Aberdeen, Scotland's third largest centre of population after Glasgow and Edinburgh (with Leith), saw only sporadic local instrument making before the late nineteeth century. What occasional activity there was seems to have been connected either with scientific work at the universities, or with providing the local professional users of instruments with the tools of their trades, mostly surveying or navigational instruments, which were however not necessarily made in Aberdeen. By the late nineteenth century, as has been commented, 'many of these makers traded as clock, watch and nautical instrument-makers and it is suspected that many of them were primarily retailers whose sole claim to the "maker" lay in their undertaking repair work'.[1] The majority in Aberdeen seem to have had connections with watchmaking, and in the particular case of Berry and Mackay (q.v.), surviving business records show that this was the principal part of their trade. Ready access to wholesaling manufacturers in the South meant that most requirements could be catered for economically by small firms without the need to develop manufacturing capacity.

Special requirements for apparatus might be expected from the natural philosophy classes at Marischal and King's Colleges, where the subject was being taught from well before their amalgamation into the University of Aberdeen in 1860. However, much of the teaching and research apparatus used until at least the late eighteenth century was not constructed locally, but bought south of the border: for instance, prominent London makers such as Francis Hauksbee and George Hearne figure in the earliest list of physical apparatus at Aberdeen.[2] By the 1780s Patrick Copland, professor of natural philosophy at Marischal College from 1775 until his death in 1822, was able to call upon the assistance of John King for active participation in the creation of a collection of demonstration apparatus,[3] including the construction of an astronomical clock made to the design of James Ferguson,[4] who was also from the north-east. King's training was as a watchmaker, and it does appear that it was the practical horologists in this area who turned their hand, when required, to the needs of local scientists and such professional clients as and when necessary. For instance, other apparatus was made by another local clockmaker, Charles Lunan.[5] Later in the nineteenth century, Charles Smith and Charles Ramage made the prototype dynamic top for James Clerk Maxwell in 1857, during the short period that he was in Aberdeen.[6]

But demand for physical apparatus must have been small compared with that for navigational instruments, for Aberdeen was first and foremost a port. Little had been done to improve the harbour before the 1760s, when John Smeaton designed and constructed the North Pier and provided alterations which allowed the use of the harbour to be increased so substantially that by 1810 further civil engineering works could be paid for. This second stage was the work of Thomas Telford and John Gibb[7] and allowed for the growth of the whaling fleet. By 1813 Aberdeen had the largest fleet in Scotland, numbering some thirteen vessels, outnumbered only by the fifty-five from Hull and eighteen from London.[8] By 1845 the nearby town of Peterhead had overtaken Hull as Britain's leading whaling port until the rising success of Dundee led to its supremacy.[9] The growing herring fishing spread southwards down the East Coast from Caithness after 1815 with no single predominating centre, but Fraserburgh and Peterhead soon had over 200 boats each, with eight other local centres along the Moray Firth with between thirty and one hundred boats.[10] After a period of rapid growth from 1880 Aberdeen emerged as a principle port for herring and white fish by the end of the nineteenth century. In 1882 the first experiments in steam trawling led to a boom so that by 1913 the port had 218 steam trawlers based there. Daily auctions of the catch led to a growth in the associated trades of haddock curing and cooperage, while the increasing supply of fresh fish seemed unable to keep up with the demand. The advent of the railways meant that fish landed at Aberdeen could be sold at Billingsgate within twenty four hours. 'Between 1881 and 1911 Aberdeen was Scotland's most rapidly growing city, and fishing, with its ancillary branches, was the most rapidly growing of its industries.'[11]

Another major local industry, also connected with Aberdeen's proximity to the sea, was shipbuilding. The first Aberdeen steamboat, the *Velocity*, 256 tons, was built in 1821 and traded between Aberdeen and Leith. The first steamer built for the London run was constructed in 1827, the *Queen of Scotland*, of 530 tons. About 1839 Messrs. Hall acquired an international reputation for their design of fast sailing ships, the Aberdeen clippers, and although the Clyde with its local resources of coal and iron soon became the centre for shipbuilding, the Aberdeen yards continued to build large, fast and respected vessels.[12] Amongst these perhaps the most famous was the tea clipper

Thermopylae, launched in 1868 by Hood of Aberdeen, only ever beaten for speed by the Dumbarton-built *Cutty Sark*.[13]

As with Dundee, the growth of the particular industries which thrived in Aberdeen, ensured that there was a small but ready market for scientific instruments. However, as D.J. Bryden has commented, 'the active practitioner could and would acquire any necessary instruments outside Scotland; indeed, it is a recurring theme in the relationship between the Scottish instrument-makers and their professional customers that the latter often turned to the renowned London market even when the local artisan was capable of satisfying their demands'.[14] Thus, although Aberdeen boasted some fifteen firms during the nineteenth century which could supply such customers, more often than not their main business was horological rather than instrumental.[15]

REFERENCES

1. D.J. Bryden, *Scottish Scientific Instrument-Makers 1600-1900* (Edinburgh, 1972), 31.

2. G.J. Thorkelin, The Northern Traveller (1790), Letters 3 and 4. MS. in Det Kongelige Bibliotek, Copenhagen, published in W.T. Johnston (ed.), *Thorkelin and Scotland* (Edinburgh, 1982), 43-50

3. Bryden, *op. cit.* (1), 10 and n.51. King was employed by Copland between 1783 and 1790, seven years out of Copland's tenure of forty-seven, specifically to make models of machinery and kindred items. Copland himself must have made well over 200 items and hence most of the huge Copland collection was not bought from the South. For King, see R.M. Lawrance, 'Notes on the Old Clockmakers No. III', *Aberdeen Journal*, 30 March 1921 and John S. Reid, 'A Select Clock', *Antiquarian Horology* 13 (1981), 45-50. We are grateful to Dr John Reid for information about the Aberdeen collections.

4. Reid, *op. cit.* (3), 45-50.

5. A demonstration 'spring clock and movement for sound in vacuo' (inventory COP. 484) and proportional compasses (inventory COP. 262), both signed by C. Lunan are in the Natural Philosophy Collection at the University of Aberdeen. We are grateful to Dr John Reid for this information. Biographical information on Charles Lunan and his clockmaking relatives is given in R.M. Lawrance, 'Notes on the Old Clockmakers No. VI', *Aberdeen Journal*, 20 April 1921; also I.E. James, *The Goldsmiths of Aberdeen* (Aberdeen, 1981), 80. There is a globe friction electrical machine signed by William Lunan in the Royal Museum of Scotland (NMS T1902.29.7).

6. Bryden, *op. cit.* (1), 31 n.160; illustrated in R.T. Gunther, *Early Science in Cambridge* (Oxford, 1937), 91-92.

7. A.W. Skempton, *John Smeaton F.R.S.* (London, 1981), 201-204. For an overview of Aberdeen's maritime history see John R. Turner, *Scotland's North Sea Gateway: Aberdeen Harbour 1136-1986* (Aberdeen, 1986).

8. David S. Henderson, *Fishing for the Whale* (Dundee, 1971), 11.

9. *Ibid.*, 14-15.

10. Quoted in Malcolm Gray, *The Fishing Industries of Scotland 1790-1914: a Study in Regional Adaptation* (Oxford, 1978), 39.

11. *Ibid.*, 174. By 1913 there were 3,023 fishermen (*ibid.*, 177). The city's population in 1881 stood at 105,000 and in 1911 had grown to 164,000 (S.G.E. Lythe and J. Butt, *An Economic History of Scotland 1100-1939* (Glasgow, 1975), 245, Appendix II: Population of the principal towns).

12. W. Robbie, *Aberdeen: Its Traditions and History* (Aberdeen, 1893), 369-370.

13. Basil Lubbock, *The China Clippers* (Glasgow, 1914), *passim*; Turner, *op. cit.* (7), 81-84.

14. Bryden, *op. cit.* (1), 10.

15. Bryden, *ibid.*, 43-59 lists the following: Anthony Barazoni, optician, 23 Lodge Walk 1840; James Berry (q.v.); Francis Chickie, barometer maker, 127 Broad Street 1827, 127 Gallowgate 1828, 48 Broad Street 1829; William Duncan, mathematical, optical and philosophical instrument maker, 46 Dee Street 1841, 92 Union Street 1842-49; Alexander R. Easton, watchmaker, nautical and optical instrument maker 1856-81; John King, clock and philosophical instrument maker, fl.1780; William Lunan, watch, clock and philosophical instrument maker, 8 Castle Street 1824-25; John McKilliam, watch, clock and nautical instrument maker, 53 Marischal Street 1854-55; Peter McMillan (q.v.); Ogg & McMillan, William McMillan & Co, watch clock and nautical instrument makers, 53 Marischal Street 1844, 30 Regent Quay 1845-46, 28 Marischal Street 1847-48; John Murray, watch, clock and nautical instrument maker, 40 Quay 1824-27, 37 Quay 1828-35, 30 Quay 1836-41; John Ramage, Charles Smith, Smith & Ramage, John Grant, varying descriptions and addresses, 1806-65; J. Stevenson, watch and optical instrument maker, 8 School Hill 1887-1900+; John Stopani, barometer maker, 38 North Street 1824-25; 42 Queen Street 1827-40, 44 Queen Street 1841-43, 68 Broad Street 1844-50; Alexander Strachan, nautical and optical instrument maker, 50 Regent Quay 1864, 57 Wales Street 1867.

PETER McMILLAN

Peter McMillan was a versatile craftsman who specialised in clocks, watches and nautical instruments and who built up a flourishing business during the second quarter of the nineteenth century. He was born in Aberdeen in about 1802 and first appeared in business at 11 Guestrow in 1824, and was thereafter at Waterloo Quay in 1828, 43 Quay between 1829 and 1833, 49 Quay from 1834 to 1839, 52 Quay in 1840-1841, and finally at 45 Regent Quay from 1842 to 1851.[1]

Examples both of his horological and instrument work are now scarce, a fact which may reflect a modest output, but some have been recorded.[2] The sextant in this collection is therefore in a sense an isolated example of the type of instrument retailed by a competent but minor provincial maker, similar to the business of Alexander Cameron of Dundee (q.v.). McMillan habitually advertised himself as a watch, clock, compass and quadrant maker[3] but little else is known of his activities. In 1851, shortly before his death, he was employing two apprentices.[4]

It was common among smaller master clock and watchmakers to have only one or two assistants, and other contemporaneous evidence suggests that Peter McMillan falls into this category. When his stock in trade was sold in about 1851 it fetched £358 18s 10½d, which indicates a moderate-sized business. This is confirmed by the debts owed to him by his customers, which totalled just over £69. Of the twenty-three individuals and firms listed, at least seven were shipowners or masters of vessels, and four were shipbuilders; among the latter the large Aberdeen yards of Alexander Hall & Co. and Alexander Duthie & Co. were clearly major customers of McMillan, owing £30 and £19 19s 6d respectively. Other customers were the Blaikie Brothers, engineers, and a brassfounder, a tinsmith and a saddler, as well as larger commercial concerns.[5] Almost without exception they were Aberdeen customers, and the predominance of maritime connections suggests that the bulk of his business lay in that direction. An historian of clockmaking in Aberdeen noted from hearsay in 1921 that 'Peter was well known to the older citizens of Aberdeen as a very capable and conscientious tradesman'.[6] McMillan was married with one son, and died on 21 June 1851 aged about forty-nine.[7]

REFERENCES

1. *Aberdeen Post Office Directory* 1824-1851; *The Bon-Accord Directory* 1842-43; cf. D.J. Bryden, *Scottish Scientific Instrument-Makers 1600-1900* (Edinburgh, 1972), 53. John Smith, *Old Scottish Clockmakers*, 2nd edition (Edinburgh, 1921), 259, lists him at 45 Regent Quay in 1836.

2. R.M. Lawrance, 'Notes on the Old Clockmakers No. V', *Aberdeen Journal*, 13 April 1921, where he describes a fine grandfather clock by McMillan, then in private hands. He was apprenticed to John Gartly in 1821 and admitted to the Hammermen in 1829 (I.E. James, *The Goldsmiths of Aberdeen 1450-1850* (Aberdeen, 1981), 146.)

3. *Aberdeen Post Office Directory* 1848.

4. General Register Office (Scotland), Census of 1851 (Aberdeen, St. Clements), 133/2, p. 7. One was William McMillan, possibly a relative, but the other is not recorded (James, *op. cit.* (2), 147). Bryden records a William McMillan, watch, clock and nautical instrument maker, in partnership with William Ogg between 1844 and 1846, and continuing in business as William McMillan & Co. from 1847 to 1848; his relationship to Peter McMillan is not known (Bryden, *op. cit.* (1), 53 and 55).

5. Scottish Record Office, SC 1/36/55, ff. 561-565, Inventory of the personal estate of Peter McMillan.

6. Lawrance, *op. cit.* (2).

7. G.R.O.(S.), Census of 1851 (Aberdeen, St. Clements), 133/2, p. 7.

68. Sextant: P. McMillan, Aberdeen, c.1830 (T1980.189)

9″ sextant with ebony frame and brass fitments. Scale divided on an inset ivory arc [-4° 40′]-[137° 40′] by 20 mins., the reinforced index arm with clamp, tangent screw and vernier reading to ½ min., with swinging micrometer. Stamped on the inset ivory name plate 'P McMillan Aberdeen'. The frame with 2 vertical struts and straight horizontal strut; provided with sight ring for telescope, 6 shades, adjustable horizon glass, with handle at the rear and supported on feet.

Scale radius: 218mm.

Lacking sight tubes and telescopes. Two of the legs are detached and the lower surface of the frame damaged at these points. The ivory scale is discoloured and detached at the ends.

BERRY & MACKAY

The Aberdeen firm of Berry & Mackay was active between 1879 and its closure in 1975,[1] but the history of the business goes back further than this partnership. It was founded by James Berry (1808-1890) a native Aberdonian, third son of James Berry, shipmaster, and his wife, Christina Bisset.[2] He served his apprenticeship with William Spark of Craigiepark (1783-1870), watch and clock-maker in Marischal Street, Aberdeen. After this was completed he went to the nearby coastal port of Stonehaven, Kincardineshire, where he began a business as a watchmaker and jeweller, before moving back to Aberdeen in 1835.[3] There he was listed in the street directory for that year as a jeweller, watch and clockmaker at 52 Castle Street, a description which lasted until 1842. Between 1843 and 1850 he was described as a watch, clock and nautical instrument maker, and in 1851 had added chronometer maker to the list - still at the same address. In 1852 he had business premises at both 52 Castle Street and 53 Marischal Street, and between 1853 and 1856 was to be found at 88 Union street, described as an optician, chronometer and watch maker. In 1857 the firm became James Berry & Son, and traded under this name until 1865 at various addresses: 88 Union Street between 1857 and 1860, 29 Union Street in 1861, 29 St. Nicholas Street between 1862 and 1865, and with additional premises at 59½ Marischal Street in 1865.[4]

James Berry continued alone at the 59½ Marischal Street premises as a chronometer maker, nautical instrument maker and optician between 1866 and 1878. He was by all accounts a very energetic man, who

> 'had a strong mechanical genius, and had a penchant for astronomy. He studied the subject to good effect, and for a time he delivered numerous lectures throughout the country, illustrating them by means of the limelight. Having a fondness for astronomy, he was not unnaturally also interested in nautical matters, and rendered great assistance to the late Mr [Robert] Gray, mathematical teacher, who established the tide tables. Mr Berry's knowledge of nautical instruments led to his employment north and south, his journeys sometimes taking him to the Orkney Islands in order to adjust compasses.'[5]

He was admitted to the Hammermen on 27 February 1837, his essay being to make an eight-day clock.[6] He made the new turret-clock for Crathes Castle in 1845,[7] and a reflecting telescope made by him was shown at the Aberdeen Mechanics' Institution Exhibition of 1840.[8] So far only one nautical instrument made or retailed by James Berry has been noted.[9]

As with so many instrument makers of this period, it is difficult to judge how much of their stock they made themselves, and how much they bought in, either in parts or assembled. An advertisement dating from 1842 is worth quoting for this reason:

> 'MARINE CHRONOMETERS.
>
> JAMES BERRY, WATCH AND CLOCK-MAKER is happy to announce his having made arrangements for having always on hand a Supply of MARINE CHRONOMETERS, made by [Thomas] HEWITT, Maker to the Admiralty. Mr Hewitt's celebrity as a Chronometer Maker will be at once appreciated, when the fact is known, that out of twenty-three Chronometers sent by him, since 1839, to the Royal Observatory, Greenwich, for competition, no fewer than twenty were purchased by the Lords Commissioners of the Admiralty, for Government use, who were pleased to mark their approbation of the superior performance of the instruments by awarding premiums on the prices paid.'[10]

Only the best for Aberdonians; and, although Berry may have had the skills to undertake chronometer work, at this date there would have been no commercial incentive in doing other than purchasing from specialist chronometer makers in London, whose products in any case had a marketable cachet. The 1842 advertisement concluded:

> 'J.B. would embrace this opportunity of returning thanks for the liberal share of patronage he has received during the last twelve years, and which he hopes still to merit. He has always on hand a Superior Assortment of GOLD and SILVER WATCHES, on the most approved principles: also, SEXTANTS, QUADRANTS, TELESCOPES, COMPASSES, and every other article in the Nautical Department.'[11]

Fourteen years later, when his son George Allan Berry joined him in business, they claimed that their

> 'Watch department [is] second to no house in the North of Scotland. Their Stock has been extensively increased, and will be found to be

complete in Gold and Silver Watches; Marine Chronometers; Clocks and Time Pieces; Jewellery and Plate; Mathematical, Optical, Surveying, and Nautical Instruments.'[12]

With such a wide range of goods on offer, and with no reason to assume that James Berry had any particular interest in instrumentation, it seems likely that the nautical instruments were, like the other material in his stock, purchased from established wholesalers.[13]

The son who joined James Berry in business between 1857 and 1864 was his eldest son, George Allan Berry, born about 1833, who continued on his own account at 29 St. Nicholas Street from 1865 to 1870; he married in about 1858 Fanny Bristow from London, and moved to Inverurie and then subsequently to West Hartlepool.[14] When George Allan Berry first went into business with his father, an advertisement appeared in the local press stating that James Berry 'begs to intimate that he has been joined by his eldest son, who has just returned from London, where for the last two years, he has been acquiring a complete knowledge of Watch Manufacturing in its several branches'.[15] Another source claimed that he learned the trade from his father, being admitted as a member of the Hammermen Incorporation of Aberdeen on 25 January 1866, and on 5 February that year admitted a Burgess of Trade of the City of Aberdeen.[16]

The partnership with his father was dissolved in 1865,[17] and the reason for his ceasing to trade in Aberdeen is that he went bankrupt, petitioning for sequestration on 30 August 1867.[18] Under examination on 28 September 1867 he said:

> 'I commenced business in Aberdeen in September 1865. I had then no Capital. I was upwards of £300 in debt at that time which had been contracted before I came to Aberdeen in carrying on Business. I took over my father's business in Aberdeen. The stock having been valued over to me amounting to £1038. ... The goods were taken over at Invoice price. Two watchmakers being asked to state what deduction ought to be made therefrom. My father and I selected these watchmakers[.] 3/- in the £ was deducted from the Invoice price[.] I find now that I could have bought these goods much cheaper than what I bought them at. The principle part of these goods is still in stock. Of the valued price I have paid to my father £450 in cash. I have no account in my books between my father and me. The last payment I made to my father was one of £50 and interest at 5 per cent on the balance ...'[19]

In fact, George Allan Berry had a total deficit of £1167 19s 4d, of which £616 9s 8d was owed to his father.[20] He had lost £150 'on Watch Manufacturing during 1866 London business', and a further £300 debts on coming to Aberdeen.[21] Among the main creditors - there were fifty-two - were Messrs. Keyzor & Bendon, wholesale opticians, London; Messrs. Joseph Levi & Co., wholesale opticians, London; Messrs. Grimoldi & Co., barometer manufacturers, London; and A.&J. Smith, working jewellers, Aberdeen (q.v.).[22] After the estate was sold up, and the creditors paid two dividends, George Allan Berry was discharged on 18 March 1869.[23]

George Allan Berry first appeared in West Hartlepool directories in 1880, where he was described as a watchmaker, jeweller and optician at 63 Church Street until 1885, when he was joined by his son, George Francis Berry, who had been born about 1862 and apprenticed to a watchmaker.[24] He almost immediately branched out on his own account, with premises elsewhere in Hartlepool. However, by the turn of the century George Francis Berry had returned to the West Hartlepool business following his father's retirement and by 1906 he was described as a compass adjuster. By 1912, when he disappeared entirely from the directories,[25] he emigrated to Canada.[26]

In 1879 James Berry took into partnership Alexander Spence Mackay. He was a grandson of James T. Mackay, whose business as James T. Mackay & Co., working silversmiths, jewellers and opticians, first appeared at 104 Union Street, Aberdeen, in 1838; it was at 75 Broad Street in 1843, when it became James T. Mackay & Son, opticians, then at 30 St. Nicholas Street in 1853,[27] and the business had ceased by the time of James Mackay's death in 1887.[28] Alexander Spence Mackay was born about 1857 in Aberdeen, and in 1881 was living with his widowed mother, when he was described as a watchmaker.[29] In 1879, when he would have been aged about thirty-two, he was taken into partnership by James Berry, then aged about sixty-one, perhaps because none of the older man's four younger sons had shown any inclination to go into business with their father; indeed, after the experience of George Allan Berry, this is not surprising. After Mackay had been with the firm for some four or five years, James Berry retired in his favour,[30] and after Berry's death in 1890 at the advanced age of eighty-two,[31] Mackay continued as the sole partner in the firm bearing their joint names until his own death in London on 24 July 1914.[32]

Berry and Mackay retained the 59½ Marischal Street premises until 1880 when they removed to 65 Marischal Street. They remained there until the closure of the firm in 1975, when the proprietor was Walter Murray, who had taken over the business in the 1940s on his return to Aberdeen after being a prisoner of war.[33] He was presumably a relation, perhaps a son, of the Walter Murray, compass adjuster, mentioned in the 1914 inventory of Mackay's estate.[34]

Because intermittent records of the firm's transactions survive - and this in itself is surprising, since there is only one other example of a business archive of a nineteenth century Scottish instrument maker, that of

Kelvin & White (q.v.), whose business was far larger and more lucrative - some idea of the firm's trading patterns at certain dates may be gleaned. The entries date back to March 1854, the first recording 'one gold lever watch sold to Sheriff Watson, Dee Street' and 'one rosewood Sykes [sic] thermometer' (i.e. a thermometer, mounted on rosewood, for use with a Sikes hydrometer) to another customer.[35] An analysis of the stock sold by James Berry for the year 1855 can be summarised as mainly horological work. He sold seventy lever watches, four clocks, two American wall clocks and three chronometers (of which one, bought on 21 September by Baillie Adamson for the ship *Catherine Adamson*, was described as '1 best 2 day Cr. by Hewitt London £33').[36] His second most popular lines were the sale of cutlery, on forty-four instances, and jewellery, on twenty-eight instances (including, incidently, a 6s gold wedding ring sold to Mr James Mackay, the father or grandfather of his future partner, on 21 October 1854); and thirdly, eyeglasses (forty) and spectacles (thirty-four). Instruments as a group were perhaps the slowest to sell: twenty-seven telescopes were sold in 1855, with eighteen thermometers, thirteen cases of mathematical instruments, nine measuring tapes or rods or rules, eight compasses, eight sextants, six levels, six cameras (or their lenses), five microscopes, four quadrants, four dividers, four barometers, three dials and compasses, two plotting instruments, two lamps, two theodolites, a pair of globes, and a single sympiesometer, opera glass, botanical glass, reading glass, stereoscope, and magnet; ten repairs are also recorded.[37]

His customers were a cross-section of what D.J. Bryden has described[38] as the 'dilettante' and 'professional' market but did not seem to include the 'scientist', who would perhaps have gone to larger firms in either Edinburgh or Glasgow for special commissions. The presence of an ancient university with a long established natural philosophy class in Aberdeen, like that of St. Andrews, did not necessarily enhance the quality of the instrument making in those places, as it did in the larger centres of population. It would seem that firms needed the dilettante and professional markets in order to keep themselves viable first, before they could cater for specialist needs. Berry's dilettante customers came from the local nobility, gentry and professional class, ranging, for example, from the Marquess of Huntly, at Huntly Castle, who bought an eight-day spring American clock for £2 in August 1854, to James Brebner, advocate, who purchased a portable sympiesometer for £2 15s and a pocket dial and compass for 8s in February 1855.[39]

A far larger selection of customers came from professional users of instruments, such as navigators and surveyors.[40] For example, Messrs. Stephen & Forbes, shipbuilders of Peterhead, bought a marine barometer, a telescope, a binnacle top and brass compass, and other deck fittings in June 1854 for a total of £9 14s 4d. In August 1854, the owners of the brig *Governess* spent £20 on a two-day chronometer, and 19s on assorted instruments; a few days later the owners of the *Queen of England* spent £6 on repairs and the rating of two chronometers. Skelton Morrison, mate of the brig *Lion*, bought a new 'quadrant' (i.e. octant) for £2 15s, less 10s for an old one, on 16 April 1855, and it is interesting to note that this 'new lamps for old' practice appears to have been a regular one. For instance, on 27 July 1854, Lewis Robertson Esq., a land surveyor from Crathie, bought a best twelve-inch dumpy level with compass, silver ring, staff and rod, for fourteen guineas, and a best five-inch theodolite with latest improvements, boxed, for £26; the total of £40 14s was reduced by £6 discount, in exchange for an old level. Berry's sales of second-hand instruments are also recorded: a second-hand wooden sextant, for example, was sold in July 1855 for £3 7s 6d; a second-hand chronometer by Carter of London, no. 188, for £20 to Messrs. Alex Duthie & Co., shipowners, on 8 January 1855; and a £25 five-inch theodolite by Troughton & Simms, the famous London instrument makers, to a Mr Alexander Adam White on 17 May.

Among other professional customers was another land surveyor, James F. Beattie of Bonaccord Street, Aberdeen, who bought a 'four-fold Ivory Rule divided to order' for 12s 6d on 16 May 1854 - one of the few direct references to Berry's instrument making skills. Mr Brand, a railway contractor, bought a £1 compass on 22 May; Mr Reid, schoolmaster, of Silver Street, bought two globes, two cases of mathematical instruments, a telescope and a silver medal on 20 June, presumably as teaching aids or prizes for his pupils; a 'fine deer-stalking telescope' was sold to Mr Playfair, gunmaker, no doubt for resale, for £2 4s on 2 October 1854; and William Smith, architect, bought a case of instruments for £3 5s on 20 April 1855. More unusual are the sales of a 'best quadrant, double tangents' to Mr Dalyell, driver of the Peterhead coach, for £2 15s on 28 December 1854, and of a second-hand theodolite to Mr Marr, Inspector of the Poor at Udny, Aberdeenshire, for twelve guineas, on 5 April 1855.

However, horological activities predominated at this period, and indeed until the turn of the century. Some of his stock appears to have been made, or at least assembled, and numbered and signed by Berry himself; other movements were signed by London makers, or more local ones from elsewhere in Scotland, or Liverpool, or Dublin.[41] He was also selling Swiss watches. He was both buying and selling old and new watches, and one of his largest markets was to local watchmakers all over the north-east of Scotland.[42] On 22 December 1855 he sold a 'lever watch, compensated' to George A. Berry, London, his eldest son who was at that point serving his watchmaking apprenticeship in the South.[43] After the younger Berry had come north and then left his father's business, it appears that James Berry continued to supply him with stock, the composition of which was very similar to his own.

The chronometers which James Berry sold came from a variety of sources too. Apart from those bought from Hewitt of London, as noted above, other London-made chronometers (some bought new, others second-hand) were acquired during the 1850s from Shepherd, Brockbank & Atkins, Carter, Baker, McLaggan, Farquhar and Porthouse. The new 2-day chronometers were sold for £33, the second-hand pieces for around £20 each.[44] Twenty-five chronometers were sold in 1865, all apparently for use in local ships based at Aberdeen, Banff and other ports in the north-east.[45] The stock book for July 1876 shows a total value of £714 8s, with telescopes still popular, watches a main line and the usual miscellany of philosophical and nautical instruments which were to be found in 1855.[46] Later records, after Berry had retired from business, show a large number of shipowners as customers, especially the North of Scotland Steam Navigation Company.

By 1903 trading was largely nautical rather than watchmaking, and repair work made up a large element of business. By then, a significant portion of this was probably for the trawlers as Aberdeen had become a centre for the fishing fleet.[47]

In the surviving business records there is apparently no reference to the names of individual staff working for Berry, or Berry & Mackay, although further careful examination might reveal these. All that has been established is that in 1881 James Berry and Alexander Spence Mackay were employing two men.[48] In 1845 James Berry took on Arthur Beverley as an apprentice, but nothing further about him is known.[49] Aside from the octant in this collection,[50] the other surviving instruments with the 'Berry & Mackay' signature also appear to be nautical.[51]

James Berry led an active public career. After his admission to the Hammermen in 1837 he held various offices, including the convenership of the Trades for the year 1865-6. In 1849 he was elected to the Town Council, retiring in 1855. Then, for twenty-nine years he was uninvolved in municipal life, but after his retirement from business he once again was elected to the Town Council in 1884, until he lost his seat in 1888. He was also involved in education - apart from his own lecture tours, mentioned above - as a governor of Robert Gordon's College, and also one of the governors of the Education Trust. He was Commissioner to the Convention of Royal Burghs in 1887, and for several years one of the Town Council's representatives on the City Parochial Board, the Boys' and Girls' Hospitals and the Reformatories and Industrial Schools Board. At the Disruption of the Church of Scotland in 1843, James Berry left the Established Church, and became involved with the Free Church in Aberdeen. He was an elder of the congregation, and for ten years chosen representative elder for the Presbytery of Shetland. His obituarist described him as

> 'Cautious in forming opinions on public questions, his judgement generally proved sound, and his utterances always commanded respect, even from those who differed from him. A good speaker, his appearance in debate materially assisted the side on which he was ranged, for he took infinite pains in mastering his subjects.'[52]

He married Isabella Allan who survived him, and they had a family of five daughters and five sons. At his death four of his daughters had already died; the fifth had married the Rev. John Gall, Free Church minister at Rutherglen. Of his sons, George Allan Berry was in West Hartlepool, as already discussed; the second, James, was a produce merchant in Dundee; the third, William, was a doctor in Queenstown, South Africa, where he was several times elected mayor; the fourth, John Philip, was Free Church minister at Ceres, Fife; and the youngest, Andrew, was a pharmacist in Acton, London.[53] James Berry left an estate worth £1,173 6s 4d.[54]

Alexander Mackay also married and left a widow, Margaret Alexander, but apparently no children. Mackay had put his savings into various investments, such as local jute companies and shipping firms, and his total estate was worth £3,126 3s 7d.[55] He died in London on 24 July 1914 aged about fifty-seven, and his successor, as noted above, was Walter Murray who may already have been working with Berry & Mackay.

REFERENCES

1. 'Old firm closing down at Aberdeen', *Aberdeen Press and Journal*, 25 January 1975.

2. R.M. Lawrance, 'Notes on the Old Clockmakers No. IV', *Aberdeen Journal*, 6 April 1921.

3. *Ibid.*; 'Death of Councillor Berry', *Aberdeen Journal*, 20 September 1890. William Spark was in business at 27 Marischal Street in 1824, and by 1831 had moved to 29 St. Nicholas Street. He died, unmarried, in Aberdeen on 15 March 1870: R.M. Lawrance, 'Notes on the Old Clockmakers No. V', *Aberdeen Journal*, 13 April 1921.

4. *Aberdeen Post Office Directory* 1835-1865, *passim*; D.J. Bryden, *Scottish Scientific Instrument-Makers 1600-1900* (Edinburgh, 1972), 44. As in Glasgow at this time, street numbering in Aberdeen used fractions for addresses of parts of tenement buildings.

5. *Aberdeen Journal, op. cit.* (3). The Mr Gray mentioned in the account was the Rev. Robert A. Gray (d.1868), master of the Commercial and Mathematical School, Aberdeen c.1840-1850, who possessed a large library and instrument collection; he taught navigation and astronomy, and calculated astronomical information for Aberdeen Almanacks c.1856, and was the author of *Tide Tables for Aberdeen, London, Liverpool, Hull, Leith, Dundee, and Greenock, for 1848 ...* (Aberdeen, 1848); A.A. MacLaren, *Religion and Social Class: the Disruption Years in Aberdeen* (London, 1974), 236.

6. I.E. James, *The Goldsmiths of Aberdeen 1450-1850* (Aberdeen, 1981), 102.

7. R.M. Lawrance, 'Notes on the Old Clockmakers No. LI', *Aberdeen Journal*, 1 March 1922.

8. *Catalogue of the Aberdeen Mechanics' Institution Exhibition* (Aberdeen, 1840), 11.

9. A sextant by James Berry, Aberdeen, was offered for sale at Christie's South Kensington, 17 May 1978, Lot 8. A rosewood stick barometer with a signature recorded as 'T Berry Aberdeen' on the ivory register plates was offered for sale at Christie's Glasgow, 19 May 1982, Lot 36.

10. *Aberdeen Journal*, 23 March 1842.

11. *Ibid.*

12. James, *op. cit.* (6), 102.

13. The octant in this collection (NMS T1980.185), marked 'Berry & Mackay Aberdeen' must date from after 1879, and is clearly a retailed item. Further instruments are a brass sextant, marked on the arc 'Jas. Berry & Son Aberdeen', offered for sale at Sotheby's, 18 June 1986, Lot 144, and a similar item offered at the same saleroom, 28 October 1986, Lot 36.

14. *Aberdeen Post Office Directory* 1865-1870; R.M. Lawrance, 'Notes on the Old Clockmakers. No. LXII', *Aberdeen Journal*, 20 May 1922; Donald Whyte, 'Old Scottish Clockmakers: Review and Supplement', *Scottish Genealogist* 24 (1977), 39.

15. Quoted in James, *op. cit.* (6), 102.

16. Lawrance, *op. cit.* (14).

17. 'The Firm of JAMES BERRY & SON, Jewellers, Opticians, and Watch, Clock, and Nautical Instrument Makers, as carried on by the Subscribers at No 29 St. Nicholas Street and 59A Marischal Street, Aberdeen, was DISSOLVED the 16th day of September 1865 ... [signed] JAMES BERRY SNR. GEORGE ALLAN BERRY.' (*Edinburgh Gazette*, 27 October 1865, 1311).

18. Scottish Record Office, CS 318/12/53, Sequestration Papers, George Allan Berry. Notices appeared in the *Edinburgh Gazette*: the initial notice of sequestration (3 September 1865, 1030); the announcement of the appointment of the Trustee and Commissioners, of whom, James Berry senior was one (13 September 1867, 1072); the announcement that the first dividend would be paid on 2 March 1868 (12 january 1868, 69); the second and final dividend to be paid on 1 July 1868 (15 May 1868); and the application of the Trustee for discharge to be considered at a meeting to be called for 18 August 1868 (24 July, 1868, 933).

19. S.R.O, CS 318/12/53, 'Sederunt Book', pp. 31-32.

20. *Ibid.*, pp. 15, 17.

21. *Ibid.*, p.37.

22. *Ibid.*, pp. 90-99.

23. *Ibid., passim*

24. The Town Docks Museum, Hull, has a 10 inch ebony frame octant marked 'G. BERRY & SON WEST HARTLEPOOL' (inventory M4.462). We are grateful to Arthur Credland, Town Docks Museum, Hull, for this information. A 7½ inch radius brass sextant marked 'G. Berry, West Hartlepool', offered for sale by Sotheby's, 18 June 1986, Lot 152, might have been sold by either father or son.

25. Information from West Hartlepool directories 1880-1912, kindly supplied by H.S. Middleton, Curator, Hartlepool Museum Services.

26. Lawrance, *op. cit.* (14).

27. *Aberdeen Post Office Directory* 1842-1889; James, *op. cit.* (6), 126-127.

28. S.R.O., *Calendar of Confirmations and Inventories* 1887, 503.

29. General Register Office (Scotland), Census of 1881 (Aberdeen, Old Machar), 168²/4, p.24.

30. *Aberdeen Journal*, 1890, *op. cit.* (3). The *Edinburgh Gazette* gives an exact date: 'The Firm of BERRY & MACKAY, Chronometer and Nautical Instrument makers, and Opticians, No 65 Marischal Street, Aberdeen, of which the Subscribers were the sole Partners, was DISSOLVED by mutual consent on the 17th day of March 1884. Mr Berry retires from Business, and Mr Mackay is to continue business, under the same name of BERRY & MACKAY, for his sole behoof, in the same premises ... [signed] JAMES BERRY. ALEXR. S. MACKAY.' (1 April 1884, 270).

31. G.R.O.(S.), Register of Deaths 1890 (Aberdeen, Old Machar), 168²/799.

32. S.R.O., *Calendar of Confirmations and Inventories* 1914, 481; S.R.O., SC 1/36/170, Inventory of the Personal Estate of Alexander Spence Mackay, 31 December 1914, item 37.

33. *Aberdeen Press and Journal, op. cit.* (1).

34. S.R.O., SC 1/36/170.

35. Aberdeen Art Gallery and Museum, Records of Messrs. Berry and Mackay, marine opticians, chronometer and nautical instrument makers, 55-57 Marischal Street, Aberdeen: Day Book, March 1854 - December 1860, p. 1. See National Register of Archives (Scotland) Survey 518 for a list of these records.

36. *Ibid.*, p. 64.

37. *Ibid.*, pp. 33-77.

38. Bryden, *op. cit.* (4), 7-23.

39. Berry and Mackay, *loc. cit.* (35), pp. 17, 40.

40. The examples cited are all from *ibid.*, pp. 33-77.

41. *Ibid.*

42. *Ibid.*, pp. 257-325. Alexander Torry, Banchory, was clearly his main customer; for instance, between November 1857 and July 1858 Berry did almost £200 worth of business with him. (An account of Torry's career is given in R.M. Lawrance, 'Notes on the Old Clockmakers No. XXII', *Aberdeen Journal*, 10 August 1921.) Among others names were Alexander Sangster, Cruden (*ibid.*, 29 June 1921), James Walker, Ellon (*ibid.*, 13 July 1921), [Peter] Gordon, Huntly (*ibid.*, 11 April

1922), Petrie & Son, New Deer (*ibid.*, 5 October 1921) and John Grant, Fraserburgh (*ibid.*, 11 November 1921).

43. Berry and Mackay, *loc. cit.* (35), p. 71.

44. The examples cited are all from *ibid.*

45. Berry and Mackay, *loc. cit.* (35), volume containing accounts, etc.: December 1865 - July 1876, p. 289.

46. *Ibid.*

47. *Ibid.*

48. G.R.O.(S.), Census of 1881 (Aberdeen, Old Machar), 168²/36, p.55.

49. James, *op. cit.* (6), 147.

50. NMS T1980.185.

51. For instance, an octant, offered for sale by Christie's South Kensington, 31 March 1983, Lot 100; and a sextant, offered by the same saleroom, 22-23 September 1988, Lot 410.

52. *Aberdeen Journal, op. cit.* (3); [Anon.], 'Ex-Baillie Berry', *In Memoriam: An Obituary of Aberdeen and Vicinity for the Year 1890* (Aberdeen, 1891), 70-72.

53. *Ibid.*

54. S.R.O., SC 1/36/111, pp.507-513, Extract Inventory of Personal Estate of James Berry.

55. S.R.O., SC 1/36/170.

---

69. Octant: Berry & Mackay, Aberdeen, c.1890 (T1980.185)

8″ octant with plain cast frame, in oxidised bronze and brass. Scale divided on an inset ivory arc, [-5]-120, the reinforced index arm with clamp, tangent screw and vernier reading to 1 min. Engraved on the limb 'Berry & Mackay Aberdeen'. The frame with 7 shades and rear handle, and supported on feet.

Scale radius: 194mm.

Lacking sight tubes and telescopes, and one of the 3 feet.

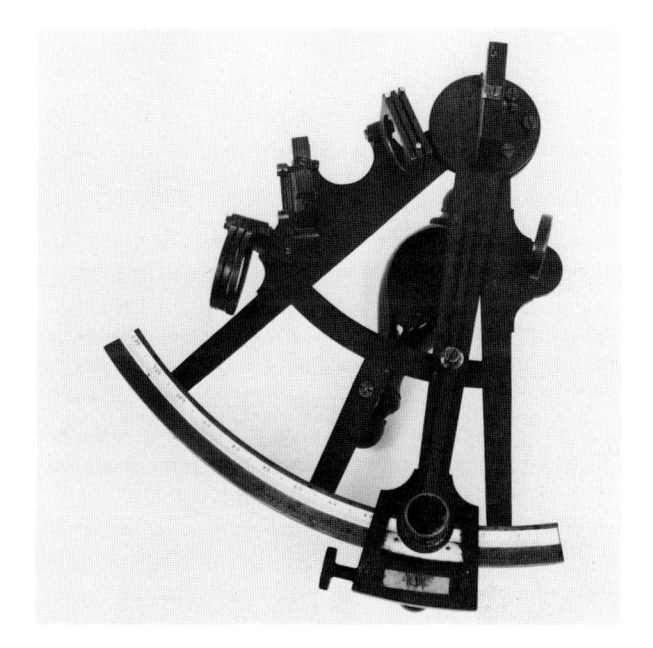

A. & J. SMITH

Alexander and John Smith were two brothers who established the firm of A. & J. Smith 'working jewellers' in 32 Netherkirkgate, Aberdeen, in 1867. The following year they moved to 25 St. Nicholas Street, expanding to include 23 St. Nicholas Street in 1876. By 1878 they were described as 'jewellers and watchmakers', and in 1879 moved their main premises to 113 Union Street, while the St. Nicholas Street branch was retained until 1919. By 1889 they were describing themselves as 'jewellers, watchmakers, and opticians', and by 1895 the Union Street branch had moved to 191 Union Street where the firm remained until it closed in 1965.[1]

Alexander Smith was the elder of the two brothers in this partnership, born in Aberdeen in about 1835 to Alexander Smith, a shoemaker, and his wife Isabella Donald.[2] He was admitted a member of the Aberdeen Incorporation of Hammermen on 5 November 1877, his essay being a gold marriage ring and the setting of a granite brooch; he was Deacon of the Hammermen in 1888 and 1889.[3] His younger brother John was born about 1844.[4] He was admitted a member of the Hammermen on 4 June 1880, his essay being similarly a gold marriage ring and a scarf pin, and the setting of a granite brooch.[5] Alexander married Catherine Nicol, and they had at least one son who survived infancy, Alexander George Nicol Smith,[6] while John married a widow, Ann Helm, née Brown, and they had at least two sons, Harold and Alfred.[7]

Little is known of their activities, or the range of their stock, but it is very likely, that, in common with other firms like Cameron of Dundee (q.v.) or Berry & Mackay of Aberdeen (q.v.) which were selling optical instruments as a sideline, such instruments as were sold would have been manufactured by wholesalers elsewhere. The instruments in this collection, two microscopes, would have been bought in from Birmingham for resale.[8] Other items which have been noted are a mahogany regulator clock,[9] a gold lever watch[10] and a brass sextant,[11] all of which would have been purchased from specialist makers to the trade, with the Smiths' name added as retailers. Without knowing more about the size of the firm and the type of stock it carried, there is no reason to suppose that they constructed any of their own instruments for sale.[12] Indeed, it is probable that this was merely a sideline to the jewellery business; for example, silverware sold by the firm has been noted.[13]

Alexander Smith, first of the brothers to die aged sixty-nine in 1904, had a share in the business of £6,341 6s 3d, out of a total estate valued at £6,938 19s 1d.[14] In his will dated 4 May 1882 he appointed various trustees; in a codicil dated 16 June 1898, he instructed that his son A.G. Nicol Smith was to take over his share of the business of A. & J. Smith watchmakers and jewellers 'of which I am a partner'.[15] John died in 1911, aged sixty-six, leaving his interest in the firm at £6,065 5s 7d, out of an estate valued at £8,941 9s 1d.[16] In his will he decreed that Alfred, his son, was to 'take over my share of business of A. & J. Smith ... of which I am a partner'.[17] The two cousins apparently ran the firm for over half a century, and shortly before its closure it employed twenty people. 'The firm had served a number of notable customers "from Balmoral downwards," said Mr Smith.'[18]

REFERENCES

1. *Aberdeen Post Office Directory* 1867-1965, *passim.*; I.E. James, *The Goldsmiths of Aberdeen 1450-1850* (Aberdeen, 1981), 136-137.

2. General Register Office (Scotland), Register of Deaths 1904 (Aberdeen, St. Machar), 168²/714.

3. James, *op. cit.* (1), 137.

4. G.R.O.(S.), Register of Deaths 1910 (Aberdeen, St. Machar), 168²/991.

5. James, *op. cit.* (1), 137.

6. Scottish Record Office, *Calendar of Confirmations and Inventories* 1904, 79.

7. *Ibid.* 1911, 623.

8. NMS T1980.239 and T1980.240.

9. Advertised in *Weltkunst* 50 (1980), 1328.

10. Movement signed 'A. & J. Smith Aberdeen No 4815', plain case (London 1877), offered for sale by Christie's, 10 March 1983, Lot 203.

11. Offered for sale by Sotheby's, 18 June 1986, Lot 151.

12. The announcement that 'an acetylene generator and lamp, designed for use in photo-micrography, is now made by Messrs. A. & J. Smith, of Aberdeen' (*Practical Photographer* 9 (1900), 92) may have meant that the item was assembled from bought-in parts. We are grateful to Sara Stevenson for this reference.

13. Offered for sale by Phillips Edinburgh, 18 May 1984, Lot 19.

14. S.R.O., SC 1/36/150, ff.320-1, Extract Inventory of the Personal Estate of Alexander Smith, 21 September 1904.

15. S.R.O., SC 1/37/125, ff.639-641, Extract Settlement and Codicils of Alexander Smith, 21 September 1904.

16. S.R.O., SC 1/36/163, f.306, Extract Inventory of the Personal Estate of John Smith, jeweller, 4 March 1911.

17. S.R.O., SC 1/37/132, ff.164-5, Extract Settlement of John Smith, 4 March 1911.

18. *Aberdeen Press and Journal*, 5 March 1965.

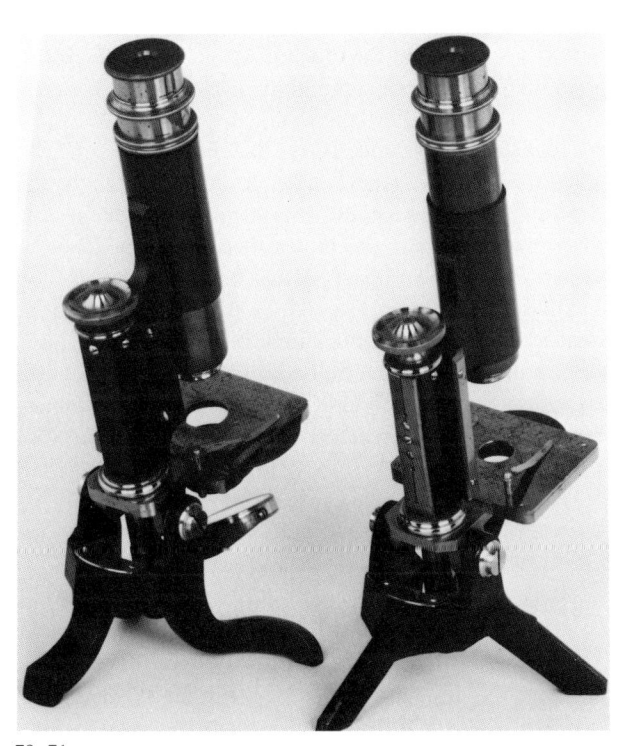

70, 71

70. Microscope: A. & J. Smith, Aberdeen, c.1890 (T1980.239)

Compound student microscope in brass and oxidised brass, on an inclining pillar and black-painted cast iron curved tripod base. Engraved on a brass plate screwed to the base 'A. & J. SMITH, / ABERDEEN.' Single-draw body tube (the draw tube with 3 settings marked), in a leather-lined draw sleeve for coarse focus, attached to the pillar through a spring-loaded slide, operated by a fine focus screw at the top of the pillar. 2-component eyepiece assembly. The stage hinged to the base, with the pillar mounted above, and a plane/concave mirror sliding on an extension beneath; the rectangular stage plate with an internal rotating 4-aperture diaphragm, swinging sub-stage condenser, and spring clips.

Body length: 153mm.
Body diameter: 30mm.
Height: 265mm.
Base size: 130 x 129mm.

Lacking objective.

71. Microscope: A. & J. Smith, Aberdeen, c.1890 (T1980.240)

Compound student microscope, 'The Worker' model of J. Parkes & Son, in brass and oxidised brass, on an inclining pillar and black-painted cast iron tripod base. Engraved on a brass plate screwed to the base 'A. & J. SMITH, / ABERDEEN.' Single-draw body tube (the draw tube with 3 settings marked), in a leather-lined draw sleeve for coarse focus, attached to the pillar through a spring-loaded slide, operated by a fine focus screw at the top of the pillar. 2-component eyepiece assembly. The stage hinged to the base, with the pillar mounted above, and a plane/concave mirror sliding on an extension beneath; the rectangular stage plate with an internal rotating 4-aperture diaphragm, and spring clips.

Body length: 153mm.
Body diameter: 31mm.
Height: 250mm.
Base size: 130 x 137mm.

Lacking objective.

The instrument can be identified as 'The Worker' microscope, offered for £3 10s to £3 15s in 1886 by J. Parkes & Son, *Wholesale Catalogue of Microscopes and Microscopic Apparatus* (Birmingham, n.d.[1886]), 9, no. 5019; and still available, with some modification, in 1903 (James Parkes & Son, *New Catalogue of Microscopes* (Birmingham, n.d.[1903]), 6). The instrument was being offered under the same name by Philip Harris & Co. of Birmingham in 1914, *Explanatory Price List of Physical Instruments* (Birmingham, 1914), 507. Parkes claimed that the Worker was 'designed to meet an ever increasing demand for a plain, serviceable instrument for students' use, of simple construction and sound workmanship, yet having high-class objectives - at a price sufficiently low to place it within the reach of all'.

14

THE GARDNERS OF GLASGOW

John Gardner (1734-1822), the son of John Gardner, gardener and maltman of Glasgow,[1] began his instrument making career working in Glasgow under James Watt, of steam engine fame, progressing to become his senior journeyman by 1769; his experience before joining Watt is unknown.[2] The Greenock-born James Watt had served a form of apprenticeship to an unknown Glasgow instrument maker, and with an introduction from Dr Robert Dick, Professor of Natural Philosophy at Glasgow, to James Short (q.v.) in London, he secured a period of training under the London mathematical instrument maker John Morgan. Returning to Glasgow in 1757, he set up as an instrument maker operating from the precincts of the College (i.e. Glasgow University). The business expanded with capital provided by Watt's partner of 1759-1765, the architect John Craig, and by 1764 Watt had a workforce of sixteen. In addition to providing specialist services to the University, Watt operated a retail business which from 1763 was prominently located in the Trongate, and in which a wide range of apparatus was on offer. His manufactured lines included instruments such as Hadley's quadrants (or octants), but he also constructed musical instruments.

Watt's energies were not devoted exclusively to instruments. In the mid-1760s he was preoccupied with the improvement of steam engines, leading eventually to the English patent for the separate condenser, taken out in 1769, and its industrial exploitation. After the death of his partner Craig in 1765 and the probable need to refund Craig's trustees, Watt turned to land surveying as another avenue of employment, and over succeeding years he became increasingly involved in civil engineering work, notably in canal surveying. During this period the instrument business continued, since the employment of journeymen is recorded until at least 1771. The running of the business and the shop may now have been to a larger extent in the hands of John Gardner, and it seems likely that he also assisted Watt in survey work because at a later date he also was involved in land surveying. The crisis in Watt's life precipitated by his wife's death in late 1773 led to his move to Birmingham in 1774, into closer industrial partnership with Matthew Boulton, but it is not known precisely when the instrument making business closed.

By late 1773 Gardner was operating independently, initially at the adjacent Candleriggs, by Bell's Wynd, as a mathematical, optical and philosophical instrument maker.[3] Bryden has discussed how Gardner used newspaper publicity to advertise his form of Bradford's statical balance 'for weighing and detecting frauds in light or counterfeit coin', capitalising on the current public awareness of the circulation of light and clipped gold coin, but also indicating his penetration of a market dominated by beam makers and specialist Lancashire coin scale makers.[4] Judging by the advertisements Gardner placed in the local press at this period, he had a wide and varied stock to offer his Glasgow customers, and the emphasis in two advertisements of 1773 of his 'continuation' in the business strongly suggests that he wished to be seen as Watt's successor.[5] His stock included at least one class of instrument in which he had undertaken development work, perhaps in conjunction with John Anderson, Dick's successor in the chair of natural philosophy:

> 'Barometers on the most improved principles, so as to admit of being carried to any distance with the greatest safety. Likewise Barometers, for measuring the height of hills, which have a peculiar adjustment to regulate the lower surface of the Mercury in one place, and have the mercury boiled in the tube. In this branch of business he has been favoured with particular methods and directions by some Gentlemen of this place of undoubted knowledge and skill in regard to this instrument.'[6]

Gardner was appointed assistant to the Glasgow land surveyor James Barry in June 1789, and on Barry's death succeeded him as surveyor in 1792.[7] A number of plans from his hand are extant.[8] Gardner's premises from 1773 onwards were at the corner of Bell's Wynd, and it was there that James Laurie joined him in 1792, possibly to ease the newly-appointed surveyor's workload. The year 1792 also marked the departure of James Sym, who became a Burgess and set up independently that year, and had previously served an apprenticeship with Gardner: another James Sym, perhaps his father, had been a fellow journeyman of Watt's with Gardner.[9] Laurie is not known to have practised as an instrument maker on his own either before 1792 or after the partnership ended in about 1798, but it seems likely either that he had trained under Gardner, or that he was a partner in an executive rather than an active sense.[10] Together they advertised an extensive range of items 'On the same terms with the Makers in London, WHOLESALE

GARDNER & CO.
OPTICIANS & MATHEMATICAL INSTRUMENT MAKERS,
21 BUCHANAN-STREET, GLASGOW,
SEVENTH SHOP FROM ARGYLE-STREET.
ESTABLISHED 1765.

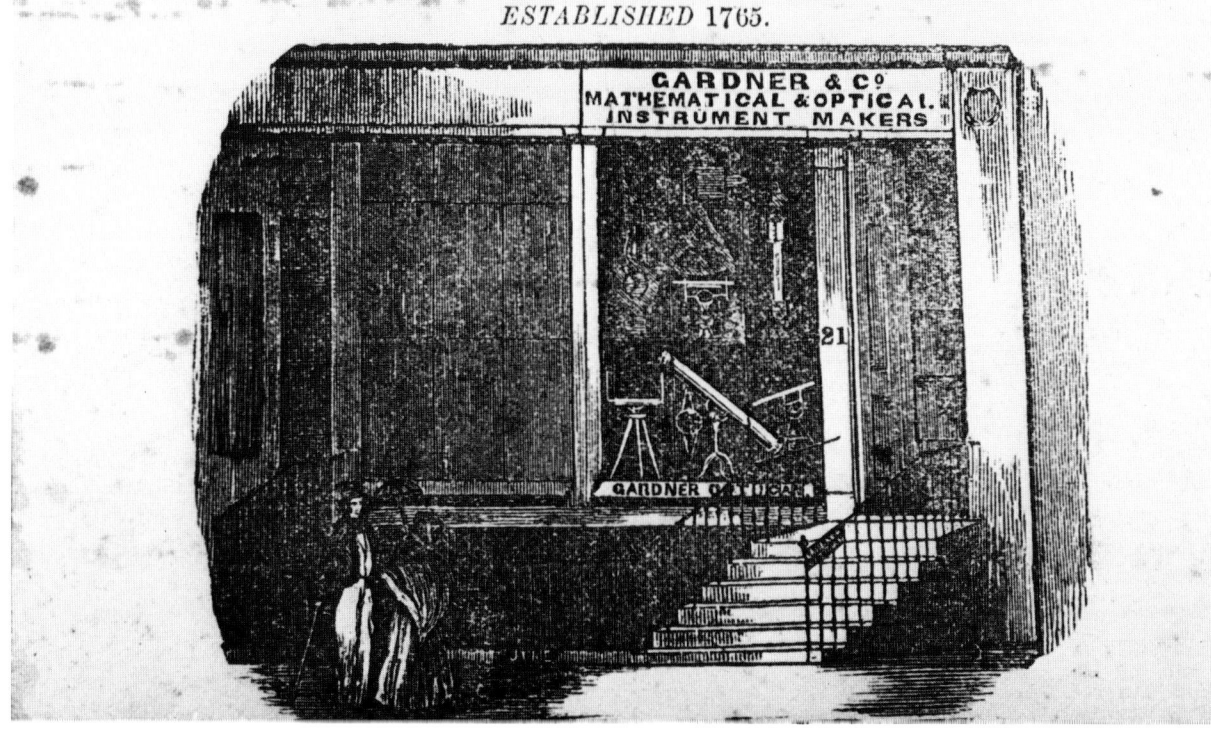

41 Gardner & Co.'s trade card for their address between 1839 and 1859. Whipple Museum of the History of Science, Cambridge

AND RETAIL',[11] but so far only one item from this period in the firm's history has been noted.[12] John Gardner had probably by this date taken his son John (1765-1818) into the firm, and it was as the partnership of J. & J. Gardner that they traded from 1799 until John junior's death in 1818.[13]

John Gardner junior died in the middle of the proceedings for sequestration, which was petitioned for by both Gardners, with the concurrence of Messrs. Middleton and Tennent, merchants, of Glasgow.[14] Unfortunately, not all the papers relating to the firm's bankruptcy survive, but it is apparent from those that do that the Gardners owned their shop on the corner of Bell's Wynd, and that they lived in or rented out the tenement above. Amongst their creditors was William Wood, 'merchant in Birmingham', for unspecified goods worth £78; William Fisher, pocket book maker in Glasgow, £11 11s 8d.; Thomas Jones, 'merchant in London' (presumably the scientific instrument maker of Charing Cross), £11 12s 6d. Despite the death of the junior partner, under the terms of the 1814 Bankruptcy (Scotland) Act, the sequestration continued. What is unknown is how quickly the matter was resolved and when the date of discharge was granted.[15]

As J. & J. Gardner the business had continued to sell a wide variety of instruments.[16] Subsequently, until about a year before the father's death in 1822 they were in partnership with Robert Jamieson as Gardner, Jamieson & Co., still trading at 43 Bell Street.[17] Between 1820 and 1828 a 'John Gardner Younger, Mathematical Instrument Maker' whose address is given as Ayton Court appeared in the directories; however, there seems to be no family link with the Gardners under discussion.[18] John Gardner senior died on 28 September 1822, on his eighty-eighth birthday,[19] and from 1823 until 1836 the business traded as M. Gardner & Co. (or & Sons); at 43 Bell Street until 1825, at 92 Bell Street from 1826 to 1831, and at 44 Glassford Street between 1832 and 1836.[20] The head of the firm would thus appear to have been Margaret Gardner (née Rankine), widow of John junior, and the sons to have been their children Thomas Rankine Gardner (c.1805-1884) and William Gardner (c.1809-1875).[21]

On 18 May 1832 M. Gardner & Sons, and each of the individual partners, petitioned for sequestration, in concurrence with one of their creditors Laurie & Hamilton, merchants, Glasgow.[22] At the first creditors' meeeting the 'Interim Factor' (appointed to manage the bankrupts' estate until the appointment of a 'Trustee') was authorised to pay the landlord of the Bell Street shop; this implies that after the previous bankruptcy the firm had been forced to sell its property.[23] It subsequently emerged that among the Gardners'

DRAINAGE LEVEL.

Registered for Mr. T. R. Gardner, *Optician,* 21 *Buchanan Street, Glasgow.*

(*From the* "Practical Mechanic's Journal," *February* 1, 1851.)

Mr. Gardner's "drainage level," is a simple and effective self-recording instrument, suitable for levelling drains, sewers, or roads, or for measuring the elevations and depressions of ground. Our engraving represents a complete side elevation of the instrument, one-fourth the real size.

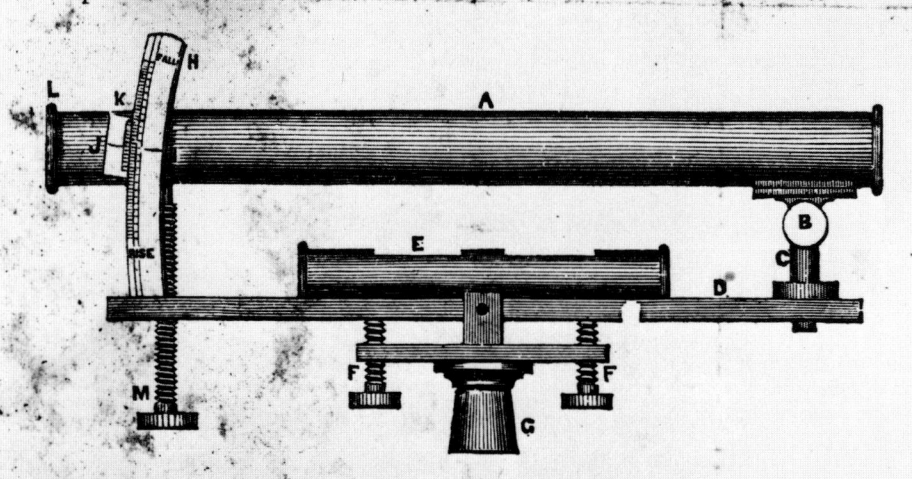

A, Is a telescope of the ordinary construction, hinged by a joint at B, to a stud-pin in the top of the short pillar, C, carried on the surface of the horizontal level plate, D. This plate has fitted to it a spirit-level, E, for adjustment to the horizontal position in the ordinary manner, by the inverted adjusting screws, F F, and the whole apparatus is fitted by means of a pivot or centre joint, G, to the usual tripod stand. H, Is a segmental graduated scale, set upright on one end of the level plate, D, the curve being struck from the centre of the joint, B. The zero of this scale is at I, in a line coincident with the horizontal axial line of the telescope; and an index, J, is carried on the side of the telescope on a vernier, K, for subdividing the graduations. The index figures of the graduations read upwards and downwards from zero, and are so proportioned as to show the amount of rise or fall per 100 feet of the horizontal distance, when the sight end, L, of the telescope is depressed or elevated, to bring the object into a line with the cross hairs by the adjusting screw, M. In using this drainage level, the plate, D, being first levelled by the spirit-level, the telescope is brought to bear upon the object by the screw, M, when the index on the telescope at once records the amount of rise or fall, in feet and inches, per 100 feet.

The graduations on the scale, H, are read as feet, subdivided to six inches; the index figures running from zero, in direct arithmetical progression, 1, 2, 3, &c., up and down, to the extent of the instrument's range. In the hands of even the most unlettered farm-servant, this little instrument will afford the most correct measurements, as the operator has only to level the plate, D, and bring his sight to bear on the object, when the elevation or depression is given at once. It will be a most useful contrivance for draining or road-making.

GARDNER & Co., OPTICIANS, 21 BUCHANAN STREET, GLASGOW.

42 Instructions for the use of Gardner's drainage level, 1851

43 Gardner & Co.'s trade card for their address between 1839 and 1859, but dating from after 1852 when they held the Royal Warrant as Optician to Queen Victoria in Glasgow

creditors (apart from £400 owed to the British Linen Co.) were Peter Firth & Co., Mathematical Instrument Makers, Sheffield, owed £302 16s 3d, and Spencer Browning & Rust, the London instrument wholesalers, £67 2s 8d. The Gardners' debts totalled £1171 0s 3d; their assets ('goods and tools on the premises') some £725.[24] At their examination, each partner said the same. William Gardner spoke first:

> 'The Company of which the declarant is a partner was formed in the month of August 1822, the partners consisting of the declarant, his brother Thomas Rankine Gardner and his mother Margaret Rankine or Gardner. Declares that there was no written contract of copartnery, but the parties were understood to be equally interested in the profits and losses of the Concern ...'[25]

A corrected list of debts owed by the firm emerged slightly later, and it appeared that amongst others, they owed Antoni Galletti (q.v.), £2 10s 1d; William Twaddle (*sic*),[26] £14 14s 6d; J.P. Cutts,[27] £30 1s 3d; and another local Italian glassworker Antoni Corti,[28] £3 10s 3d. On the other hand, they had unpaid accounts due from some very respectable customers, amongst whom numbered the Andersonian University, David Smith, a land surveyor and Port Glasgow Board of Health.[29] However, an advertisement was shortly placed in a local newspaper stating that:

> 'STOCK OF MATHEMATICAL AND OPTICAL INSTRUMENTS
>
> FOR SALE, AND SHOP TO LET
>
> To be Sold, by Private Bargain THE Whole STOCK, SHOP FURNITURE, UTENSILS, and TOOLS, belonging to the Concern of M. Gardner & Sons, for many years at No 92, Bell Street, now 44, Glassford Street, - consisting of Sextants, Octants, Quadrants, Azimuth Compasses, Telescopes, Visual Glasses and Spectacles, Microscopes, Nautical Instruments, Barometers, Electrifying Machines, Thermometers, Turning Tools, Burnishers, Gravers, Turning Lathes, &c. &c; together with almost every variety of article in their line of business.
>
> The Stock is in excellent condition; - the Shop is very neatly fitted up, - is in a central situation, and to any person wishing to follow this business, few such opportunities as the present of starting with advantage are to be met with.
>
> Sealed offers (marked Offer for Stock) will be received by Mr George Ord, 125, Virginia Street, up till the 10th December next, at ten o'clock forenoon, - if by post, the postage must be paid.
>
> An Inventory and Valuation of the Stock will, in the interim, lie at Mr Ord's office, for the inspection of intending Purchasers.
> Glasgow, 27th November 1832.'[30]

The first dividend was six shillings in the pound; the second and final dividend to creditors was 6d per pound.[31] However, the three partners were unable to petition for a discharge until 12 December 1840, by which time four fifths of the debts had been paid off, and there was still £324 11s 5¼d surplus.[32] During these eight years it appears that the business continued to operate, although the whole affair had been taken over, as the law required, by a Trustee who was personally

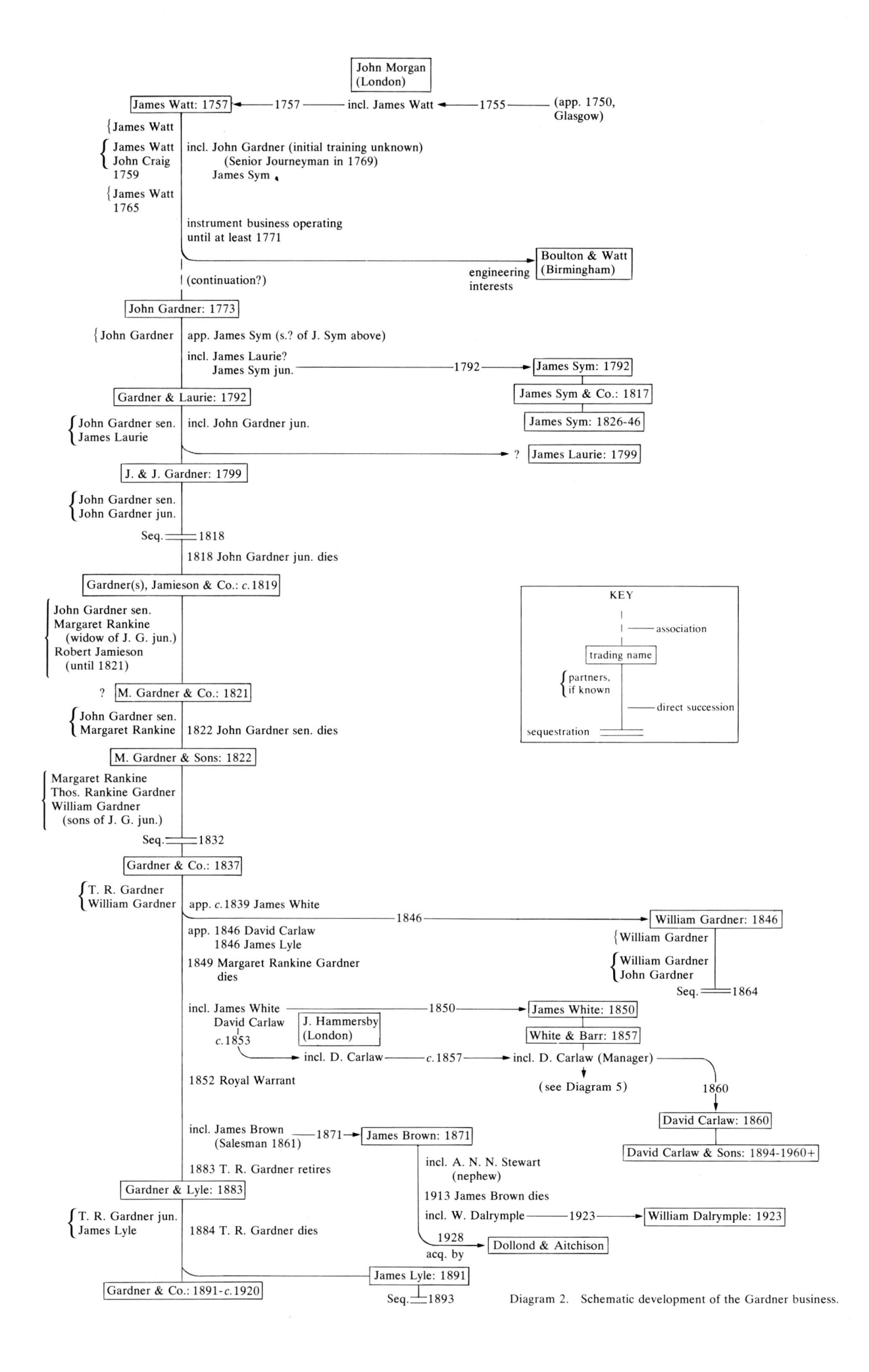

```
                          John Morgan
                           (London)
    James Watt: 1757 ◄──── 1757 ──── incl. James Watt ◄──── 1755 ──── (app. 1750,
   ⎰ James Watt                                                           Glasgow)
   ⎱ James Watt       incl. John Gardner (initial training unknown)
     John Craig           (Senior Journeyman in 1769)
     1759              James Sym
   ⎰ James Watt
   ⎱ 1765             instrument business operating
                      until at least 1771
                                                                    Boulton & Watt
                                                    engineering    (Birmingham)
                       (continuation?)              interests
     John Gardner: 1773
   ⎰ John Gardner     app. James Sym (s.? of J. Sym above)
   ⎱                  incl. James Laurie?
                            James Sym jun. ──────────── 1792 ────► James Sym: 1792
     Gardner & Laurie: 1792                                         James Sym & Co.: 1817
   ⎰ John Gardner sen. incl. John Gardner jun.                      James Sym: 1826-46
   ⎱ James Laurie
                                                              ? ──► James Laurie: 1799
     J. & J. Gardner: 1799
   ⎰ John Gardner sen.
   ⎱ John Gardner jun.
       Seq. ══ 1818
                      1818 John Gardner jun. dies
     Gardner(s), Jamieson & Co.: c.1819
   ⎧ John Gardner sen.                                    ┌─────────────────────────┐
   ⎪ Margaret Rankine                                     │          KEY            │
   ⎨  (widow of J. G. jun.)                               │           │             │
   ⎪ Robert Jamieson                                      │           │ ─── association │
   ⎩  (until 1821)                                        │      ┌─────────┐        │
     ? ─ M. Gardner & Co.: 1821                           │      │trading name│     │
   ⎰ John Gardner sen.                                    │      └─────────┘        │
   ⎱ Margaret Rankine  1822 John Gardner sen. dies        │  ⎰ partners, │          │
     M. Gardner & Sons: 1822                              │  ⎱ if known  │ ─── direct succession │
   ⎧ Margaret Rankine                                     │           │             │
   ⎪ Thos. Rankine Gardner                                │  sequestration ══       │
   ⎨ William Gardner                                      └─────────────────────────┘
   ⎩  (sons of J. G. jun.)
       Seq. ══ 1832
     Gardner & Co.: 1837
   ⎰ T. R. Gardner     app. c.1839 James White
   ⎱ William Gardner                                      ─── 1846 ────────────► William Gardner: 1846
                       app. 1846 David Carlaw                                    ⎰ William Gardner
                            1846 James Lyle                                      ⎱
                                                                                 ⎰ William Gardner
                       1849 Margaret Rankine Gardner                             ⎱ John Gardner
                            dies                                                    Seq. ══ 1864
                       incl. James White ──────── 1850 ────► James White: 1850
                             David Carlaw    ┌───────────┐   White & Barr: 1857
                             c.1853          │J. Hammersby│
                                             │ (London)  │
                                             └───────────┘
                                        ─── incl. D. Carlaw ── c.1857 ──► incl. D. Carlaw (Manager) ──┐
                       1852 Royal Warrant                                   (see Diagram 5)           │ 1860
                                                                                                      ▼
                                                                                           David Carlaw: 1860
                       incl. James Brown ── 1871 ──► James Brown: 1871
                            (Salesman 1861)                                                David Carlaw & Sons: 1894-1960+
                                                      incl. A. N. N. Stewart
                       1883 T. R. Gardner retires           (nephew)
     Gardner & Lyle: 1883                            1913 James Brown dies
   ⎰ T. R. Gardner jun.                              incl. W. Dalrymple ──── 1923 ────► William Dalrymple: 1923
   ⎱ James Lyle        1884 T. R. Gardner dies       1928
                                                     acq. by ► Dollond & Aitchison
                                       James Lyle: 1891
     Gardner & Co.: 1891-c.1920           Seq. ══ 1893           Diagram 2. Schematic development of the Gardner business.
```

ESTABLISHED A.D. 1765.

GARDNER & CO.,
OPTICIANS TO HER MAJESTY,
SPECTACLE MANUFACTURERS,
Mathematical and Philosophical Instrument Makers,
53 BUCHANAN STREET
(NEARLY OPPOSITE ARGYLE ARCADE),
GLASGOW,

Make and Sell

EYE GLASSES Set in Steel, Tortoise-shell, and Gold Frames.
SPECTACLES in Blued Steel and Gold Frames.
SPECTACLES with Neutral Tinted Glasses or Sun Shades, and for Preserving the Eyes from Dust or Wind.
OPERA GLASSES, RACE, FIELD, AND MARINE GLASSES.
TELESCOPES.
MICROSCOPES AND MICROSCOPIC OBJECTS.
STEREOSCOPES AND SLIDES.
MAGIC LANTERNS AND SLIDES, comprising Views of nearly all parts of the World,
POCKET MAGNIFIERS.
HAND MAGNIFYING GLASSES.
BOTANICAL MAGNIFYING GLASSES.
BAROMETERS.
ANEROID BAROMETERS.
POCKET ANEROID BAROMETERS.
POCKET ANEROID BAROMETERS, with Scale for Measuring the Height of Mountains.

DRAWING INSTRUMENTS.	SURVEYORS' LEVELS.
OPTIC SQUARES.	LEVELLING STAVES.
CROSS STAFF HEADS.	MINERS' COMPASSES.
THEODOLITES.	MEASURING CHAINS.

TAPE MEASURING LINES.
POCKET SPRING MEASURING LINES.
MAGNETO-ELECTRIC MACHINES.
GALVANIC BATTERIES, for Simultaneous Blasting.

And all other Articles in the Line.

44 Advertisement for Gardner & Co., 1876

responsible for it. At the creditors meeting on 12 June 1832:

> 'the meeting authorised the Trustee to continue the bankrupts in the possession of the shop and stock in the meantime security for their intromissions being found to his satisfaction and the delay for the present consideration as to the farther disposal of said stock.'[33]

They were formally discharged on 24 March 1841.[34]

In 1837 the family business began trading as Gardner & Co. at 44 Glassford Street moving to 21 Buchanan Street in 1839, to 53 Buchanan Street in 1860, and 53 St. Vincent Street in 1883.[35] The range of stock in 1840 shows that Gardners catered for both the non-specialist and the professional market. The standard items such as telescopes, microscopes, barometers, thermometers, sun dials and spectacles which the firm sold were the stock-in-trade of most optical and philosophical instrument makers. More specialised were measuring and drawing equipment, sextants and quadrants, Adie's sympiesometer, Allan's saccharometer and Sikes' hydrometer. The firm also repaired instruments.[36] It is difficult to estimate the extent of the Gardners' output in relation to stock bought wholesale, partly because of the difficulty of identifying the sources of instruments, and partly because the size of the workshop is not fully known. However, it is known that in 1861 T.R. Gardner employed three men and six boys, probably apprentices, and among them one salesman, James Brown (q.v.) who was to set up independently ten years later.[37]

Margaret Rankine Gardner had died aged sevety-seven in 1849.[38] By 1861 T.R. Gardner was probably the principal, if not the sole, partner in the firm, for his brother William had set up independently as an instrument maker in 1846.[39] William Gardner is recorded at 3 Royal Bank Place between 1846 and 1855, at 56 Gordon Street from 1856 until 1861 and finally at 134 Buchanan Street from 1862 until 1864, when he, too, petitioned for sequestration.[40] The individual partners in the firm were recorded as William Gardner and John Gardner, but the identity of the latter has still to be uncovered. Although the immediate cause of bankruptcy was a large debt of £132 15s 9d to a local wine merchant, among other creditors were William Wright, spectacle manufacturer of Wolverhampton, and Lejeune & Perken, importers of optical instruments, London. Others were Lowdon, optician, Dundee (q.v), 19s 6d; Casella, London, £14 10s 8d; Hugh Wilson, engraver, Glassford Street, 7s 1½d; John Spencer, optician, St. Enoch's Square (q.v.), £2 19s 4d; James White, optician, Buchanan Street (q.v.), £4 2s 3d; John Wrench, London, £16 13s 4d; J. Somalvico & Co., London, £10 18s 10d; Peter Firth & Co., opticians, Sheffield, £34 12s 4d; George Edward & Son, jewellers, Glasgow, £1 15s 6d; C.J. Chamberlain & Co., photographers, Chelsea, £1 15s 7d; Alexander Boyack, photographic goods merchant, Glasgow, £1 3s 5d.[41] However, the liabilities of the firm amounted to £423 3s 10d, while the assets came to £389 6s 5d; the Trustee wrote: 'the Trustee is of the opinion that the bankruptcy has arisen from innocent misfortunes and losses on business' and managed to pay off the debts, leaving a balance of £21. The partners were discharged in 1867.[42] Instruments with William Gardner's signature may well not have originated in his workshop, judging by the suppliers identified above.[43]

Shortly after William set up on his own, T.R. Gardner's prestige was increased when in 1852 he was appointed Optician to Queen Victoria in Glasgow.[44] However, his significance in the history of Scottish instrument making lies not only in the evident diversity and quality of his work,[45] but also in the fact that among his apprentices were two highly talented figures, James White (q.v.) and David Carlaw (q.v.). Both these men were to make important contributions to technological development in the latter half of the century, and credit for their mechanical brilliance must rest in part with T.R. Gardner.

At this time, the Victorians were becoming increasingly aware of the need to clean up the filth and squalor of their sprawling, unplanned, insanitary and disease-ridden cities, amongst whose number ever-growing Glasgow was included.[46] Perhaps the item for which Gardner is most famous is his drainage level, which was of a design registered as no. 2602, 28 December 1850: the level was sold in large numbers[47] and other retailers' signatures are known to have appeared on them.[48] He also appears to have been unusual among his fellow-makers in the west of Scotland insofar as he was an active meteorologist, and in this respect he resembles Alexander Adie of Edinburgh (q.v.). From at least the 1840s until about 1860 T.R. Gardner kept meteorological records at his house, Ibroxholm, near Glasgow, the results of which were used by contributors to the proceedings of the Philosophical Society of Glasgow.[49] He himself was a member of the Society. In 1846, the same year that David Carlaw was formally apprenticed to Gardner, James Lyle also joined his business, probably as an apprentice. He stayed with the firm and in 1883 became a partner with Thomas Rankine Gardner, junior; Thomas Rankine Gardner, senior took the opportunity to retire,[50] and the business traded as Gardner & Lyle until February 1891 when the partnership was dissolved.[51] Lyle subsequently set up under his own name as a manufacturing and consulting optician, with a range of additional stock very similar to that of Gardner & Co.[52] Lyle also went bankrupt in 1893, his sequestration petitioned by York & Son, photographic publishers, Nottinghill, London. He appears to have been discharged in 1895, but the case details are very sparse.[53]

T.R. Gardner senior died on 21 March 1884,[54] and it seems that T.R. Gardner junior remained in the firm, which after Lyle's departure resumed business as Gardner & Co. at 53 St. Vincent Street. In 1899 Gardner's moved to 36 & 40 West Nile Street continuing there until 1920 when they cease to appear in the Directory.[55]

REFERENCES

1. David C. Soutter, *The Gardner Family of Glasgow* (1976) and *Additional information relating to John Gardner, senior ...* (1977). Typescript. Copies in the Mitchell Library, Glasgow and Royal Museum of Scotland.

2. The discussion of Watt's business is drawn from J.P. Muirhead, *Life of James Watt* (London, 1859) and H.W. Dickinson, *James Watt, Craftsman & Engineer* (Cambridge, 1935).

3. Advertisement in *Glasgow Journal*, 9 September 1773.

4. *Ibid.* D.J. Bryden, *Scottish Scientific Instrument-Makers 1600-1900* (Edinburgh, 1972), 20. An example of the coin scale was

sold at Sotheby's, 16 June 1975, Lot 57, together with an accompanying pamphlet: *The Description and Use of an Instrument for Weighing and Detecting Frauds in Light or Counterfeit Gold Coin: Improved by John Gardner, Mathematical-instrument maker Candlerigs, opposite to Bell's Wynd, Glasgow ...* (Glasgow, 1773).

5. For instance: '... JOHN GARDNER, also Continues to MAKE and SELL, Theodolites of all kinds with their apparatus Surveyors and Coaliers Compass Boxes, Hadley's and Land Quadrants, Electrical Machines with all their Apparatus, Perspective Machines, and Camera Obscuras, Drawing Instruments, of all kinds, Telescopic Levels, Weavers Microscopes, With a great variety of other articles.' (*Glasgow Journal*, 9 September 1773); '... J. Gardner also continues to make and sell Theodolites, Compass boxes, drawing Instruments, Electrical Machines &c.' (*ibid.*, 11 November 1773); ' ... he has just now received from London, a fine assortment of Temple and Nose Spectacles ground in brass moulds, and a number of Day and Night Telescopes for sea. He also continues to make the following articles, viz. Theodolites of all kinds, with their apparatus. Circumferentors, being the principle instrument used by land-surveyors in the West Indies. Scots and English Land Measuring Chains, Hadley's quadrants of different sorts...' (*Glasgow Mercury*, 1 July 1779). A set of drawing instruments, signed 'J. Gardner, Glasgow' on the compass, was offered for sale at Sotheby's, 22 January 1973, Lot 120.

6. *Glasgow Mercury*, *op. cit.* (5). An example of John Gardner's domestic barometers is in the Royal Museum of Scotland, NMS T1980.69: instead of the normal words 'Very Dry' on the register plates, it has 'Serene'. Another was offered for sale by Sotheby's, 22 October 1987, Lot 118.

7. Robert Renwick (ed.), *Extracts from the Records of the Burgh of Glasgow* VIII (Glasgow, 1913), 303-304. Peter Eden (ed.), *Dictionary of Land Surveyors and Local Cartographers of Great Britain and Ireland 1550-1850: Supplement* (Folkestone, 1979), 414.

8. For instance, Scottish Record Office, RHP 16, Plan of the town of Glasgow's quarries, 1802; RHP 65/2, Plan for a new disposition of Stabtree and Fordneuck, being part of the estate of Barrowfield, 1815 (I.H. Adams, *Descriptive List of Plans in the Scottish Record Office* (Edinburgh, 1966), I, 86).

9. Bryden, *op. cit.* (4), 30.

10. However, a stick barometer signed 'James Laurie, Glasgow' has been recorded; private collection.

11. *Glasgow Courier*, 3 March 1792.

12. This is the octant in the Frank collection, NMS T1980.180.

13. *Glasgow Herald*, 28 August 1818; quoted in Soutter, *op. cit.* (1). Gardner was aged 53.

14. S.R.O., CS 235 SEQS. G 1/46, Sequestration Papers, J.& J. Gardner. Notices appeared in the *Edinburgh Gazette*: the initial notice of seqestration (7 August 1818, 212); the announcement of the appointment of the Trustee (25 September 1818, 249); and the announcement of the final dividend payment on 9 August 1819 (6 July 1819, 186). John Gardner senior had married Marion Tennent, eldest daughter of Hugh Tennent, on 2 October 1763 (*Parochial Records (Marriages) of the High Church of Glasgow (Glasgow Cathedral)*, quoted by Soutter, *op. cit.* (1)).

15. S.R.O., CS 235 SEQS. G 1/46: there is no Sederunt Book.

16. There is a plain theodolite in the Royal Museum of Scotland (NMS T1922.59), and a 9 inch brass protractor (NMS T1988.85), both signed 'J. & J. Gardner, Glasgow'. A sundial was recorded at Auchinleck House, Ayrshire by W.B. Stevenson, 'Sundials from Six Scottish Counties, near Glasgow', *Transactions of the Glasgow Archaeological Society* new series, 9 (1937-40), 286; another, signed 'J. & J. Gardner, Glasgow', dated 1811, and made for latitude 55° 56' was offered for sale by Phillips, 3 June 1987, Lot 111. A set of glass spirit beads was offered for sale by Christie's South Kensington, 9 February 1984, Lot 267; another set is in the Science Museum, London (inventory 1980-751); we are grateful to Dr Jon Darius for this information. Two stick barometers, signed 'J. & A. Gardner, Glasgow' (*sic*) were offered by the same saleroom, 10 July 1986, Lots 26 and 29.

17. *Glasgow Post Office Directory* 1819-1822. So far, only one item from this period of the firm's history has been recorded; a surveying compass, signed 'Gardner's Successors to J. & J. Gardner, 43 Bell St., Glasgow', offered for sale by Sotheby's, 22 April 1965, Lot 51. Notice of the Dissolution of the Copartnery was published: 'The Concern carried on here by the Subscribers under the Firm of GARDNERS (*sic*), JAMIESON & CO., Mathematical Instrument Makers, was this day DISSOLVED by mutual consent ... [signed] MARGT. RANKINE, THOS. R. GARDNER, ROBT. JAMIESON. Glasgow May 2, 1821.' (*Edinburgh Gazette*, 4 May 1821).

18. *Glasgow Post Office Directory* 1820-1828. A sundial, dated 1821 and signed by John Gardner, which may have been made by this man, was recorded at Pitcon, Dalry (Stevenson, *op. cit.* (16), 286).

19. *Glasgow Herald*, 7 October 1822; quoted in Soutter, *op. cit.* (1).

20. *Glasgow Post Office Directory* 1792-1836, *passim*. A stick barometer signed 'Gardner & Son, Glasgow' was offered for sale by Christie's, 20 June 1968, Lot 4; and a 2-draw refracting telescope signed 'M. Gardner & Sons, Glasgow' was offered for sale by Christie's Scotland, 3 June 1987, Lot 19.

21. Soutter, *op. cit.* (1).

22. S.R.O., CS 96/4215, Sederunt Book, M. Gardner & Sons. The partner representing Laurie & Hamilton was James Laurie, but there is no indication if there was any relationship between him and the James Laurie who was previously a partner in the firm. Notices appeared in the *Edinburgh Gazette*: the initial notice of sequestration (18 May 1832, 138); the announcement of the appointment of the Trustee and the Commissioners (26 June 1832, 183); a Creditors' meeting called for the bankrupts' offer of composition (9 October 1832, 292); their offer of 6s per pound as a Company, and 6d per pound as individuals (6 November 1832, 320 and 9 November 1832, 324); notice that accounts have been drawn up, and that the first dividend of 6s per pound would be paid on 20 May, and one of 9d per pound for Margaret Rankine or Gardner, as an individual (26 March 1833, 90); accounts available for inspection by Creditors until 21 January 1834, when the second and final dividend of 6d per pound would be paid (13 December 1833, 330).

23. S.R.O., CS 96/4215, Sederunt Book, p. 7.

24. *Ibid.*, p. 11.

25. *Ibid.*, pp.21 22.

26. Bryden, *op. cit.* (4), 35, discusses Twadell.

171

27. The Sheffield wholesalers are listed in directories between 1822 and 1876; see also R.H. Nuttall, *Microscopes from the Frank Collection 1800-1860* (Jersey, 1979), 48.

28. Antoni Corti appears at various addresses in the *Glasgow Post Office Directory* between 1837 and 1844. Barometers with his name were offered for sale by Phillips, 16 September 1981, Lot 9, and 11 November 1981, Lot 11; Phillips Edinburgh, 29 June 1984, Lot 70, and 25 September 1987, Lot 71; and Christie's South Kensington, 22 January 1988, Lot 12.

29. S.R.O., CS 96/4215, Sederunt Book, pp. 31-36. The Andersonian University was a forerunner of the University of Strathclyde.

30. *Ibid.*, p. 58. (*Glasgow Herald*, 30 November, 3 and 8 December 1832: the final appearance was followed by a second advertisement: 'TO OPTICIANS, &c THAT SHOP, No 92 BELL STREET, Occupied by the GARDNERS for upwards of 60 years as Opticians and Mathematical Instrument Makers. The Stock offered in the preceeding advertisement will be disposed of, and thus afford an excellent opportunity of succeeding to an established business with a select Stock. Pecuniary assistance will be afforded to an enterprising tenant. ... Apply to John Slater, 12 Wilson Street, and 17 West Nile Street.')

31. *Ibid.*, pp. 68, 78.

32. S.R.O., CS 236/G31/5, Sequestration Papers, M. Gardner & Sons, 'Petition for Discharge'. There was also a notice in the *Edinburgh Gazette*, 18 December 1840, 403.

33. *Ibid.*, 'Minute of Meeting of Creditors, 12 June 1832.'

34. *Ibid.*, 'Discharge'.

35. *Glasgow Post Office Directory* 1837-1883, *passim*. Apart from the instruments in the Frank collection, among other items recorded with the firm's signature are: a student microscope in the Royal Museum of Scotland (NMS T1983.58); another, offered for sale by Phillips Edinburgh, 31 January 1986, Lot 64; a miner's dial, offered for sale by Sotheby's Belgravia, 16 July 1982, Lot 92; another, offered by Sotheby's, 19 May 1983, Lot 113; a 1¾" refracting telescope, offered by Sotheby's Belgravia, 6 February 1981, Lot 105; a stick barometer, offered by Sotheby's Sussex, 9 June 1982, Lot 680; another, offered by Christie's, 7 October 1981, Lot 4; and a wheel barometer, offered by Christie's Scotland, 3 June 1987, Lot 56.

36. Advertisements in *Glasgow Post Office Directory* 1840/1.

37. General Record Office (Scotland), Census of 1861 (Lanark, Govan), 646¹/18, p.17.

38. Soutter, *op. cit.* (1).

39. 'DISSOLUTION OF COPARTNERY. The Copartnery of GARDNER & COMPANY, Opticians, Mathematical and Philosophical Instrument Makers, No 21, Buchanan Street, Glasgow, carried on by the Subscribers, was DISSOLVED on the 28th day of May last, by mutual consent. The Subscriber, Thomas Rankin Gardner, will continue to carry on the Business on his own account at No 21, Buchanan Street ... and the Subscriber, William Gardner, will continue the same line of Business also in his own account, at No 3, Royal Bank Place. [signed] T.R. GARDNER. WM. GARDNER.' (*Edinburgh Gazette*, 2 June 1846, 238).

40. S.R.O., CS 318/10/145, Sequestration Papers, William Gardner. Notices appeared in the *Edinburgh Gazette*: the initial notice of sequestration (11 October 1864, 1285); the announcement of the appointment of the Trustee and Commissioners (28 October 1864, 1347); the announcement of the first dividend to be paid 11 April 1865 (24 February 1865, 207); the announcement of a Creditors' meeting to be held on 19 May to consider discharging the Trustee (25 April 1865, 523); and the Bankrupts' petition for discharge (9 May 1865, 594).

41. *Ibid.*, pp. 13, 45, 46.

42. *Ibid.*, Sederunt Book no 2.

43. Amongst pieces with William Gardner's signature is his version of Gardner's drainage level (NMS T1978.100); and a circular bearing protractor, offered for sale at Christie's South Kensington, 12 December 1985, Lot 172.

44. Bryden, *op. cit.* (1), 49.

45. Among the items produced by the firm during these years were mining dials, microscopes, telescopes and the drawing protractor, the design of which was registered by Gardner, 4 May 1848.

46. 'The building of the cities was a characteristic Victorian achievement, impressive in scale but limited in vision, creating new opportunities but also providing massive new problems. Perhaps their outstanding feature was hidden from public view - their hidden network of pipes and drains and sewers, one of the biggest technical and social achievements of the age, a sanitary 'system' more comprehensive than the transport system.' Asa Briggs, *Victorian Cities* (Pelican edition, London, 1975), 16-17; see also, for the particular case of Scotland (Briggs deals only incidentally with Glasgow, although he states: 'It was Glasgow which was the real Victorian city in Scotland, with a rapidly growing proportion of Scotland's inhabitants living inside its boundaries. Victorian Scotland, indeed, concentrated population in Glasgow more dramatically than Victorian England concentrated population in London.' *ibid.*, 34), T.C. Smout, *A Century of the Scottish People 1830-1950* (Fontana edition, London, 1987), 32-57, which also gives a useful bibliography of primary and secondary sources.

47. These have serial numbers stamped on the back of the scale. Apart from the two in the Frank collection (NMS T1980.146 and T1980.147), there is another example in the Royal Museum of Scotland (NMS T1978.98) and one in the Whipple Museum of the History of Science, Cambridge (described in Olivia Brown, *Catalogue 1 Surveying* (Cambridge, 1982), item 151 (inventory 1450)), and another in the National Museums and Galleries on Merseyside (inventory 1967.316.2). Other examples have been offered for sale: Sotheby's, 19 May 1983, Lot 119; Christie's South Kensington, 17 October 1985, Lot 228; and Historical Technology, Catalog 104 (Spring 1972), item 93. The instrument was described in the *Practical Mechanic's Journal*, 1 February 1851.

48. An example with an engraved plate 'H.A. Thompson, Lewes' soldered over Gardner's name, and with trade cards relating to Thompson in its box, is in the Royal Museum of Scotland (NMS T1968.59). Another example sold by Thompson was offered by Historical Technology, Catalog 132 (Spring 1989), item 129.

49. Gardner was admitted a member of the Society 3 April 1850; *Proceedings of the Glasgow Philosophical Society* 3 (1848-55), 109. He had left it by c.1874.

50. 'The firm which has carried on business for many years as Opticians at 53 Buchanan Street, Glasgow, under the name of

GARDNER & COMPANY, and of which the Subscriber Thomas Rankine Gardner, Senior, was sole Partner, has been DISSOLVED as at and from 28th May 1883. The Business will be carried on by the Subscribers Thomas Rankine Gardner, Junior, and James Lyle at 53 St. Vincent Street, Glasgow, under the name of GARDNER & LYLE ... [signed] T.R. GARDNER. THOS. GARDNER. Junr. JAMES LYLE.' (*Edinburgh Gazette*, 13 July 1883, 528). Gardner & Lyle advertised their wares at the Glasgow International Exhibition of 1888, but did not exhibit (*International Exhibition Glasgow 1888: Official Catalogue* (Glasgow, 1888), 76).

51. The Notice of Dissolution of the Copartnery was published: 'The Firm of GARDNER & LYLE, Opticians, 53 St. Vincent Street, Glasgow, of which the Subscribers Thomas Rankine Gardner and James Lyle were the sole Partners, was of mutual consent DISSOLVED on 28th February 1891, by the retiral of the said James Lyle. The subscriber Thomas Rankine Gardner will continue Business at 53 St. Vincent Street, Glasgow, under the Firm of GARDNER & CO. ...' (*Edinburgh Gazette*, 19 May 1891, 536). Items recorded with a 'Gardner & Lyle' signature include a level, offered for sale at Sotheby's Belgravia, 30 November 1979, Lot 184; another, offered by Phillips Edinburgh, 28 October 1988, Lot 72; a monocular Society of Arts pattern microscope, offered by Christie's South Kensington, 13 November 1986, Lot 97; and a refracting telescope offered by Sotheby's, 16 November 1987, Lot 164.

52. [Stratten's] *Glasgow and Its Environs* (London, 1891), 159. Lyle was perhaps the same James Lyle, an engineer who developed a creosote oil-burning lamp for industrial use in 1883 (J.B.Hannay, 'On the Lucigen: a new Industrial Light', *Proceedings of the Glasgow Philosophical Society* 17 (1885-86), 393-402. A barometer with his signature was offered for sale by Phillips Edinburgh, 29 August 1986, Lot 113.

53. S.R.O., CS 318/38/173, Sequestration Papers, James Lyle. Notices appeared in the *Edinburgh Gazette*: an initial notice of the Petition for *Cessio Bonorum* presented by Perken, Son & Rayment, 99 Hatton Garden v James Lyle, summoned for 25 September 1893 (12 September 1893, 962); the notice of sequestration (15 September 1893, 975); and a Deed of arrangement providing for six shillings and eightpence per pound to be paid for in four equal installments, 'payable at three, six, nine and twelve months respectively ... and of the Sequestration being declared at an end' (27 October 1893, 1129).

54. Soutter, *op. cit.* (1).

55. *Glasgow Post Office Directory* 1890-1920, *passim*.

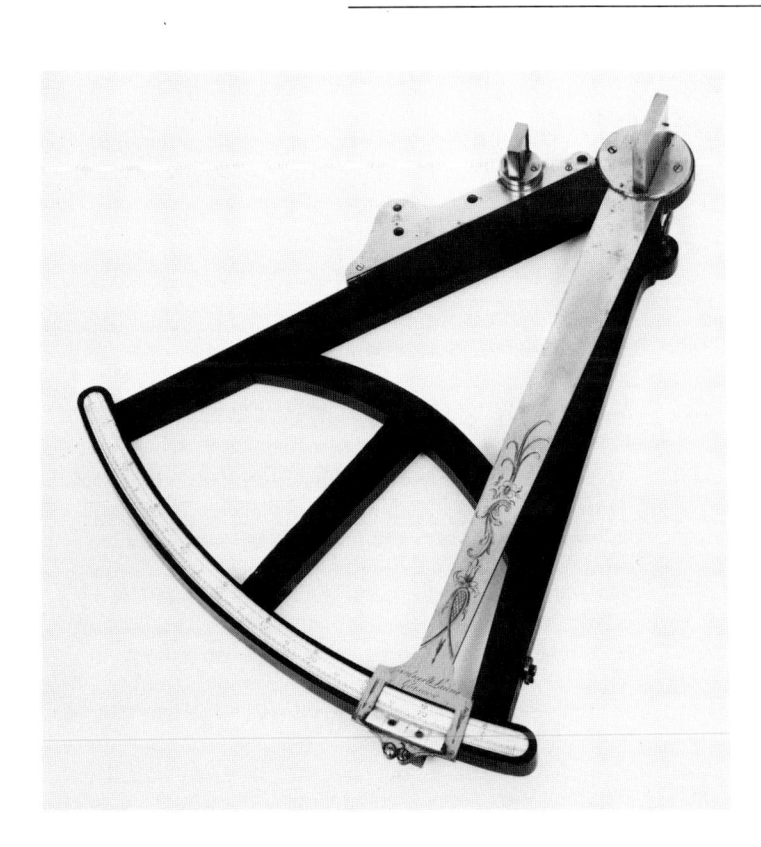

---

72. Octant: Gardner & Laurie, Glasgow, c.1795 (T1980.180)

16" octant with ebony frame and brass fitments. Scale divided on an inset ivory arc [-5]-[95] by 20 mins., with centre-zero vernier reading to 1 min. The index arm with foliate decoration and engraved 'Gardner & Laurie / Glasgow'. The frame with single vertical strut and bowed horizontal strut; and constructed for taking foresights and backsights, but with only the double pin-hole foresight and adjustable forward horizon glass present; noteplate on reverse and provision for pencil in vertical strut, and supported on feet.

Scale radius: 407mm.

Lacking the backsight, back horizon glass and adjustmemnt, both sets of shades, forward horizon glass clamp screw, 2 of the 3 feet, and the pencil. The frame has been repaired at the foresight.

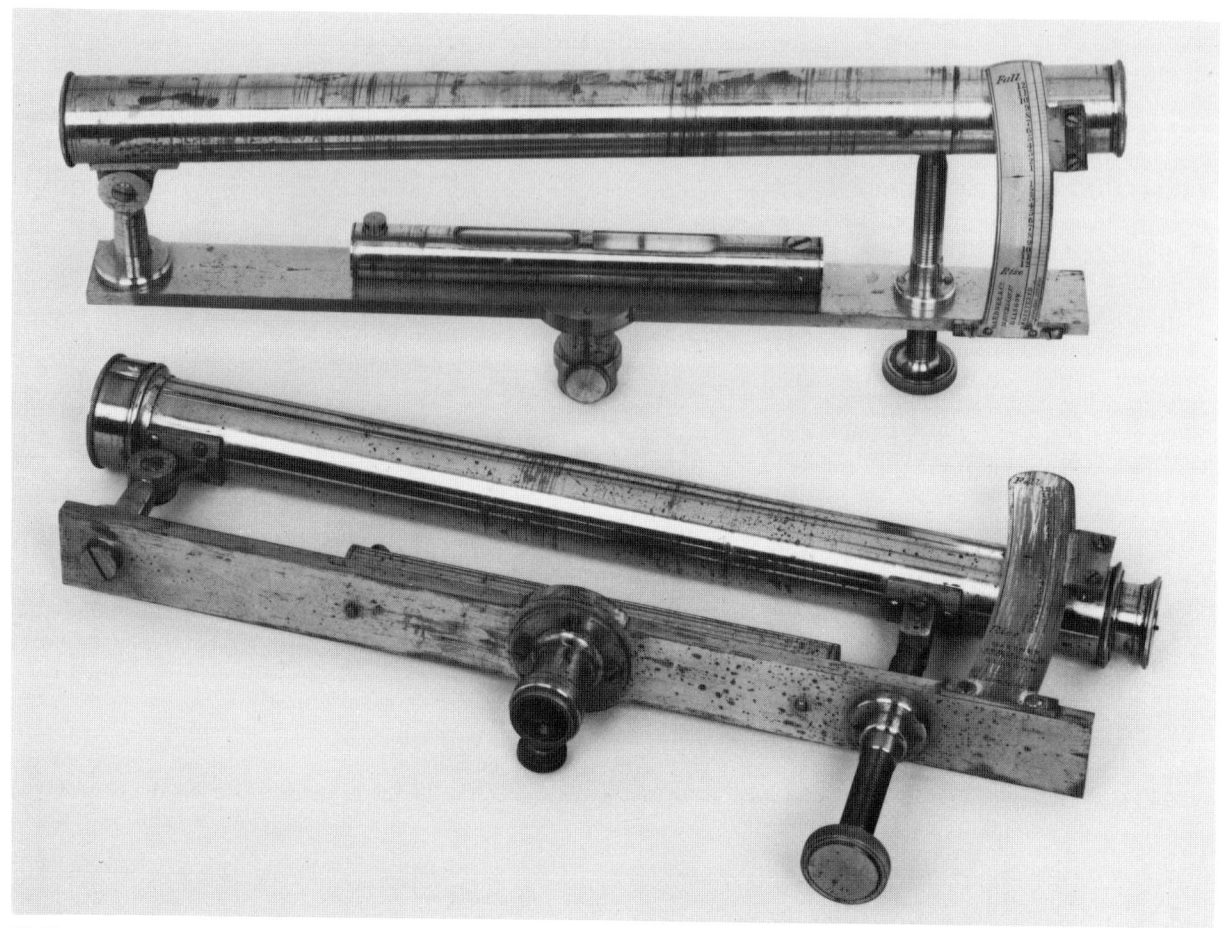

73, 74

73. Clinometer: Gardner & Co., Glasgow, c.1855 (T1980.146)

13½" drainage level in brass with telescopic sight hinged at the objective end over a limb with a bubble level, with screw adjustment at the eye end for altitude and read against a silvered arc calibrated for rise and fall of up to 12 feet per horizontal 100 feet, divided to ½ft. Engraved on the arc 'GARDNER & C^O / 21 BUCHANAN S^T / GLASGOW. / REGISTERED / N^O 2602 DEC. 28TH 1850.' and stamped on the reverse '85'. Single lens objective and draw tube focus, the orientation of the graticule maintained by a slot in the draw tube engaging on the end of the adjustment screw. The limb with a socket mount, stamped '2', for attachment to a levelling head. With fitted case containing trade card for Gardner & Co., 21 Buchanan Street, Glasgow.

Telescope length: 344mm.
Tube diameter: 29mm.
Case size: 357 x 170 x 70mm.

Lacking the graticule wires. When originally acquired by Mr. Frank the clinometer had a tripod (with integral levelling head) which was not located in 1980.

Provenance: Purchased privately in Chippenham, Wiltshire.
The device was registered in the Patent Office Design Registry.

74. Clinometer: Gardner, Glasgow, c.1860 (T1980.147)

13½" drainage level in brass with telescopic sight hinged at the objective end over a limb with a bubble level, with screw adjustment at the eye end for altitude and read against a silvered arc calibrated for rise and fall of up to 12 feet per horizontal 100 feet, divided to ½ft. Engraved on the arc 'GARDNER / 53 Buchanan S^t / GLASGOW / REGISTERED' and stamped on the reverse '213'. Doublet objective with objective cover; draw tube focus, the eyepiece assembly incorporating an image erecting system and external graticule, the orientation of the graticule maintained by a screw passing through a slot in a sleeve on the draw tube mounting flange. The limb with a socket mount for attachment to a levelling head. With fitted case containing trade card for Gardner & Co., 53 Buchanan Street, Glasgow.

Telescope length: 348mm.
Tube diameter: 29mm.
Aperture: 28mm.
Case size: 370 x 190 x 70mm.

The device was registered in the Patent Office Design Registry.

75. Level: Gardner & Co., Glasgow, c.1870 (T1980.127)

13″ dumpy level in oxidised brass and bronze (stripped), with bubble tube above the telescope, rack and pinion focus, doublet objective, and objective ray shade; the limb with a central circular level and attached to a parallel plate levelling head, with free azimuth motion. Engraved on the telescope tube 'GARDNER & C°. / 53 BUCHANAN STREET, / GLASGOW.' The base of the levelling head stamped '32'.

Telescope length: 328mm.
Tube diameter: 44mm.
Aperture: 40mm.

The glass of the bubble tube has been broken, and the mounts of the tube lack the four adjustment screws. The surface has been marked by mechanical abrasive cleaning.

76. Mining dial: Gardner & Co., Glasgow, c.1850 (T1980.151)

Glazed circular compass in brass with silvered dial, recessed circular level and 2 folding opposed slit and window sights. Engraved on the dial plate 'GARDNER & C°. OPTICIANS &c. / 21 BUCHANAN ST. GLASGOW.' The dial with the 4 cardinal compass points (transposed arrangement for reading bearings against the north point of the needle), the scale divided in degrees 360-[0] and [0]-[90]-[0]-[90]-[0]. Triangular-section steel needle on a relieved jewelled bearing. Socket mount with clamp screw, stamped 'B7'.

Compass housing diameter: 161mm.

Sight wires detached.

77. Mining dial: Gardner & Co., Glasgow, c.1850 (T1980.153)

Glazed circular compass in brass with silvered dial, recessed circular level and 2 folding opposed slit and window sights.

Top 76, left 78, right 77

Engraved on the dial plate 'GARDNER & C°., OPTICIANS &c. / 21 BUCHANAN ST. GLASGOW.' The dial with the 4 cardinal compass points (transposed arrangement for reading bearings against the north point of the needle), the scale divided in degrees 360-[0] and [0]-[90]-[0]-[90]-[0]. Triangular-section steel needle on relieved jewelled bearing.

Compass housing diameter: 160mm.

The socket mount is a later replacement. Sight wires detached.

78. Mining dial: Gardner & Co., Glasgow, c.1860 (T1980.152)

Glazed circular compass in brass with silvered dial, recessed circular level and 2 folding opposed slit and window sights. Engraved on the dial plate 'GARDNER & C°. / 53, BUCHANAN STREET, / GLASGOW.' The dial with the 4 cardinal compass points (transposed arrangement for reading bearings against the north point of the needle), the scale divided in degrees 360-[0] and [0]-[90]-[0]-[90]-[0]. Triangular-section steel needle on a relieved jewelled bearing. Socket mount with clamp screw, stamped '155'.

Compass housing diameter: 156mm.

The base plate has been bent at the extension pieces for both of the sights. Lacking sight wires. Glass broken.

79. Drawing protractor: Gardner & Co., Glasgow, c.1850 (T1980.173)

8" protractor in brass, with a reduced semi-circular arc divided on silver for 60° on either side of its centre line; pivoted at the centre of the arc, and clamped at its circumference, is a 14½" rule with 2 parallel drawing edges. Engraved at the pivot 'REGISTERED / 8th May 1848 / No 1444' and 'GARDNER & Co / 21 Buchanan St / GLASGOW.' The underside of the arc checked along a chord for locating against a raised horizontal edge on a drawing board and with an opposed locating guide. The inclination of the rule to this edge read against a silvered index to ½°, the arc calibrated 60-0-60, 30-90-30, 120-180/[90]-150. A brass tube mounted over the arm acts as a guide for a sliding cross-piece.

Arm radius: 481mm.
Scale radius: 95mm.

The form of the locating guide under the pivot has been modified and is probably not original. Only traces of the silvering on the arc now remain.

The device was registered in the Patent Office Design Registry.

80. Microscope: Gardner & Co., Glasgow, c.1870 (T1980.222)

Compound Jackson-limb microscope in brass, bronze and oxidised brass, on an inclining pillar and flat 2-toed tripod base. Engraved on the back of the base 'GARDNER & Co / Opticians to her Majesty / 53 BUCHANAN ST / GLASGOW.' Single-draw body tube with coarse focus rack and pinion adjustment (defective) to the limb; fine focus is by an external screw at the lower end of the body tube advancing a spring-loaded nosepiece with a greased push-fit internal slide for the objective. 3 objectives (2 perhaps by Swift) and 2 single-component eyepieces (the field lens remaining in the draw tube). The stage hinged between curved uprights rising from the base, with the pillar mounted above and a concave mirror on an extension beneath; the rectangular mechanical stage with slow-motion movement (defective) in two directions, rotating 4-aperture diaphragm and pair of slide holders. With fitted case.

Body length: 235mm.
Body diameter: 38mm.
Height: 405mm.
Base size: 172 x 196mm.
Case size: 445 x 225 x 200mm.

The pinion arbor for the coarse focus has sheared at one side so that one of the adjusting knobs is detached. The screw threads on the mechanical stage have stripped. Lacking one fixing screw on the support bracket for the fine focus adjustment.

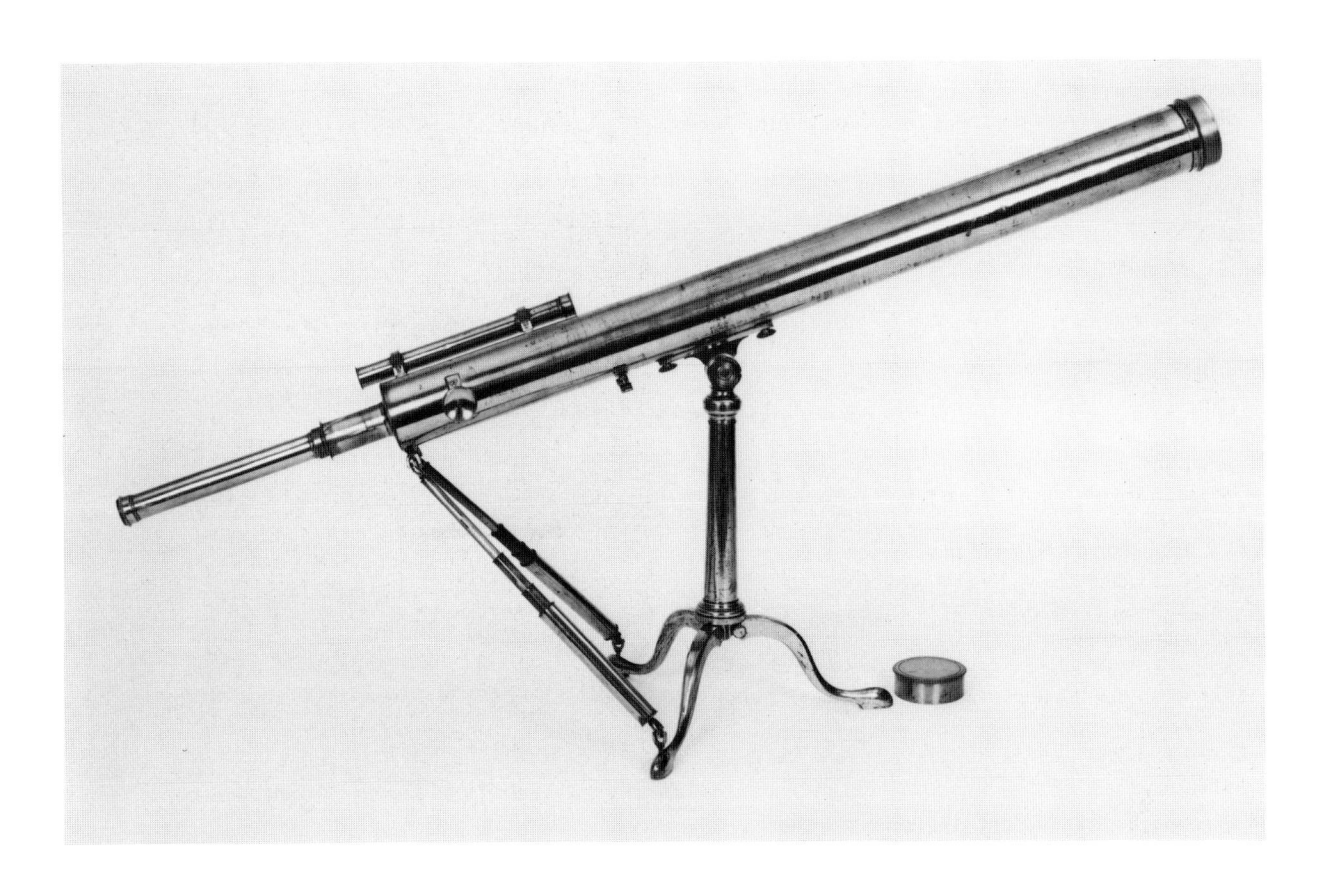

81. Refracting telescope: unsigned, c.1860 (T1980.274)

3¼″ aperture 2-draw mounted refracting telescope in brass with doublet objective, rack and pinion focus to the second draw tube and ⅞″ aperture finder telescope. Supplied with 2 low-power 2-component eyepieces, and 2 high-power compound eyepieces, the latter marked in MS '200' and '300', together with 2 solar shades; draw tube extension section with 2-component image erecting assembly for use on terrestrial objects. Objective covers for telescope and finder. Pillar-mounted on a folding brass table tripod (but with provision for mounting the pillar on a tall tripod) with two 5-part extendible steady rods to the tripod feet, which stow against the barrel when not in use. With fitted case containing trade card for Gardner & Co., 53 Buchanan Street, Glasgow.

Length closed (with erecting assembly): 1,365mm.
Tube diameter: 86mm.
Aperture: 81mm.
Case size: 1,555 x 255 x 145mm.

The mounting brackets for the finder telescope were not original and have been replaced. There are a number of fractures evident in the barrel wall.

Another example of this model (signed by Adie & Son, Edinburgh) was offered by Sotheby's, 22 October 1976, Lot 175. Detailed similarities in design and in the mounting of optical components suggest that this item is by the same manufacturer as items 31 (Adie & Son) and 116 (D. McGregor).

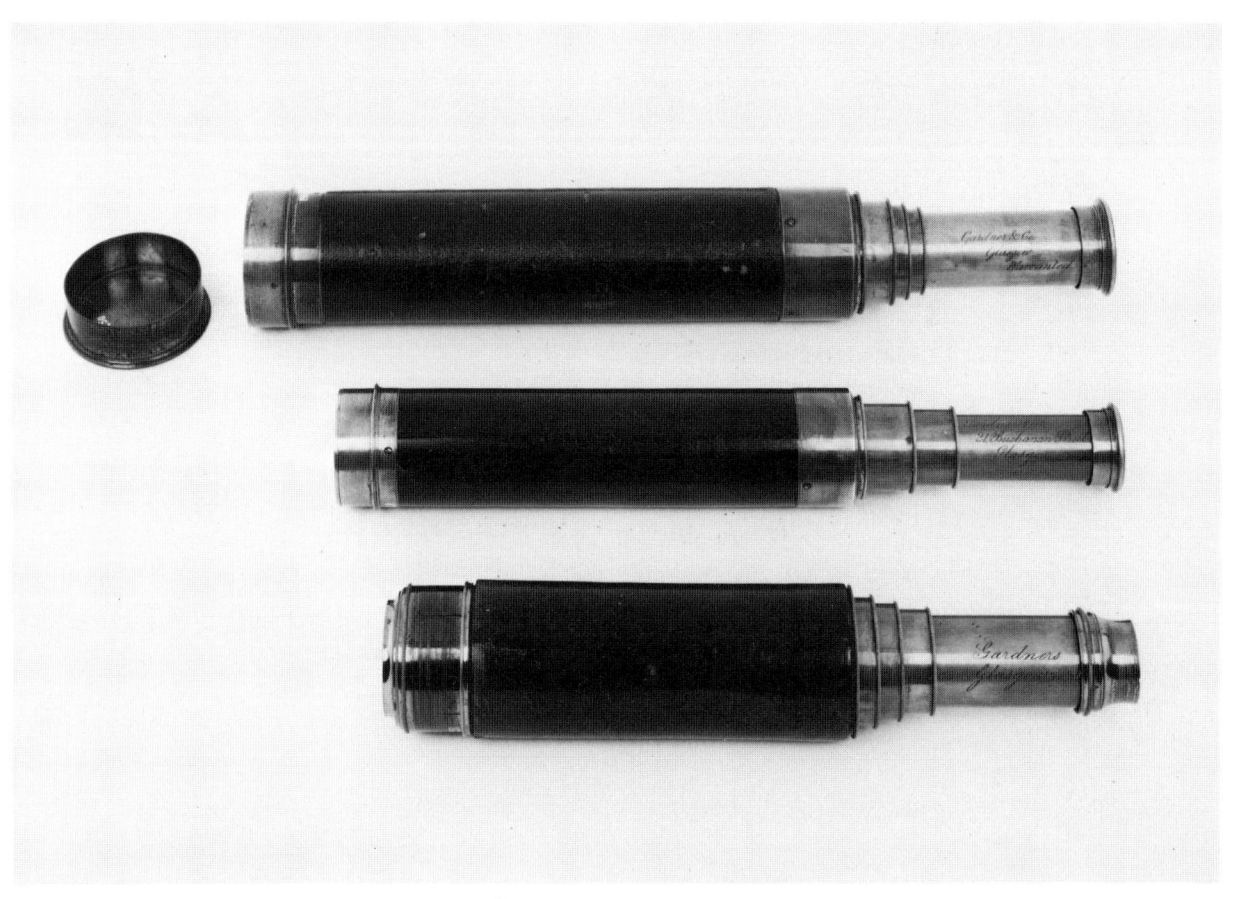

83, 82, 84

82. Refracting telescope: Gardner & Co., Glasgow, c.1850 (T1980.249)

1½" aperture 3-draw hand-held refracting telescope with doublet objective (defective) and 2-component eyepiece and erecting assemblies; in brass, with wooden barrel. Engraved on the first draw tube 'Gardner & C.º / Buchanan Street, / Glasgow.' Scratched assembly mark 'VI' on the 3 draw tubes and 2 of the draw slides and the external eye stop, which has a sliding closure.

Length closed: 240mm.
Tube diameter: 48mm.
Aperture: 40mm.

At acquisition the concave component of the objective was found to have been broken as a result of incorrect assembly in the mount. The wooden barrel has 2 longitudinal splits. Lacking objective cover.

83. Refracting telescope: Gardner & Co., Glasgow, c.1890 (T1980.250)

2" aperture 4-draw hand-held refracting telescope with doublet objective and 2-component eyepiece and erecting assemblies; in brass, with leather-covered barrel. Engraved on the first draw tube 'Gardner & Co., / Glasgow / Warrented'. Scratched assembly mark 'VII' on the draw tubes, draw slides and erecting assembly. Objective cover, and sliding closure on external eye stop.

Length closed: 305mm.
Tube diameter: 57mm.
Aperture: 51mm.

84. Refracting telescope: Gardners, Glasgow, c.1900 (T1980.251)

1½" aperture 4-draw hand-held refracting telescope with triplet objective and 2-component eyepiece and erecting assemblies; in brass, with leather-covered barrel. Engraved on the first draw tube 'Gardners / Glasgow'. The image erecting assembly forms a separate inner draw tube at the opposite end of the first draw tube to the eyepiece assembly, and is extracted by the draw slide when the first draw tube is withdrawn. Scratched assembly mark 'VI' on the draw tubes and draw slides, and 'I' on the erecting assembly. Objective cover (incomplete) scratched 'VIII'.

Length closed: 195mm.
Tube diameter: 53mm.
Aperture: 35mm.

Lacking the aperture plate of the external eye stop, the aperture and closure of the objective cover, and the stop on the third draw tube.

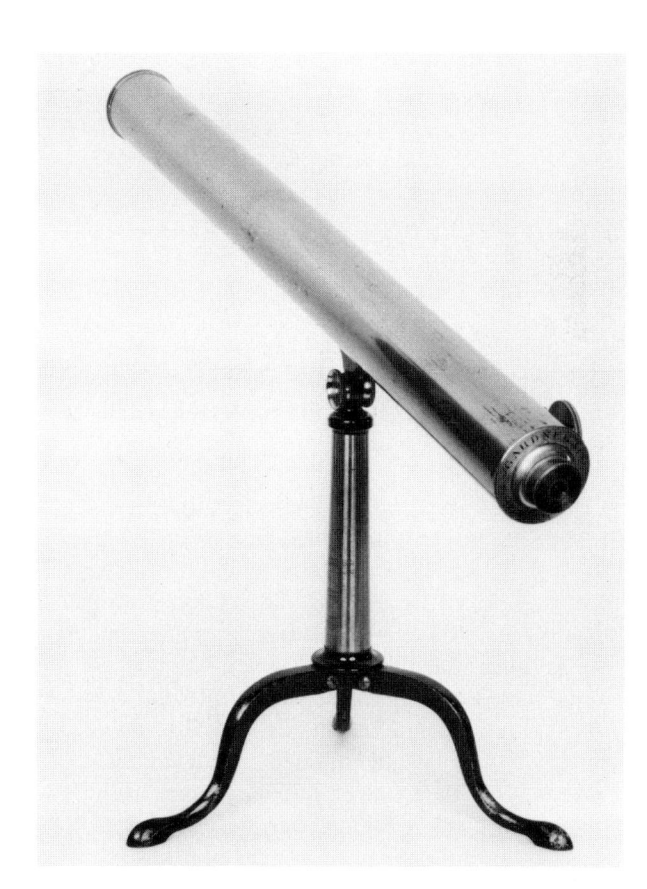

85. Refracting telescope: Gardner & Co., Glasgow (T1980.275)

3″ aperture 2-draw mounted refracting telescope in brass; with doublet objective recessed from the end of the barrel, and 2-component eyepiece and erecting assemblies, and rack and pinion focus to the second draw tube. Engraved on the barrel end plate 'GARDNER & Cº / GLASGOW.' and scratched with the assembly mark 'III' on the objective mount. Closure on external eye stop. Pillar-mounted by a knuckle-joint with free azimuth motion, and on a folding table tripod with black-painted iron legs, the tripod base and legs stamped 'VIII'.

Length closed: 983mm.
Tube diameter: 77mm.
Aperture: 72mm.

Another example of this size, also signed 'GARDNER & Co. / GLASGOW' was offered by Historical Technology, Catalog 105 (Fall 1972), item 202. Detailed similarities in design and in the mounting of optical components suggest that this item is by the same manufacturer as items 57 (E. Lennie), 66 (Lowdon), 93 (J. Brown) and 161 (Whyte, Thomson & Co.). This type of objective mount is found in James Parkes & Son's *Illustrated Trade Price List of Telescopes* (Birmingham, n.d. [1902]), 9, 'Astronomical Telescope on table tripod stand ... with 3 inch Object Glass', no.320.

DAVID CARLAW

David Carlaw (1832-1907), the son of Bernard Carlaw, a Glasgow warper, worked under T.R. Gardner (q.v.) from 1 January 1845. The indenture of his apprenticeship was formally signed on 9 February 1846, binding Carlaw to serve seven years as an apprentice mathematical instrument maker.[1] After the expiry of his apprenticeship 'he was for some years in the employment of Messrs. Hommersley [sic], of London', the optical firm.[2] Returning to Glasgow, he became manager to James White (q.v.) and worked closely with Professor William Thomson, both 'in making and fitting up many of the splendid equipments of instruments' in his laboratories at Glasgow University, and in making instruments for the Atlantic cable project.[3]

In 1860, aged about twenty-eight, Carlaw set up independently, as an instrument and model-maker, in South Portland Street, Glasgow, with one assistant, Robert Wallace.[4] However, he did not appear in the Glasgow street directory until 1865, by which time he was working in Sydney Court, 62 Argyle Street. He stayed there until about 1871; from 1872 until 1874 he was at Havelock Buildings, 75 East Howard Street; from 1875 until 1897 the business was at various periods at 52, 54, 56 and 58 Ropework Lane, with premises also at 81 Dunlop Street from 1892 until 1897.

45   David Carlaw (1832-1907). *Jack House*

In 1897 purpose-built premises in Finnieston Street were occupied.[5]

From the beginning of his independent business Carlaw is said to have made 'a special feature of instruments corresponding to those which he had constructed for Lord Kelvin - theodolites, surveying instruments, compasses, spirit levels &c'.[6] He also made working models of engines and locomotives; the increasing importance of this specialisation within his business suggests that Carlaw abandoned the instrument making discipline in which he was trained, in favour of model-making and mechanical design and construction, in both of which he and his firm excelled. In 1888 his model of a Denny & Co. engine was described in awed tones by a reviewer as 'the most beautiful engine model ever produced'.[7]

Carlaw's 'faculty for taking infinite pains which is akin to genius',[8] also showed itself in his pioneering designs for consecutive-numbering printing machines and his perfection of the envelope-making machine. The former project came to lucrative fruition in 1878 with his Glasgow Numerical Ticket and Check Book Printing Company, while from 1888, when Carlaw's first envelope machine was displayed in Glasgow, this side of the business also flourished. A commercial guide to Glasgow of 1888 not only contains a brief sketch of the organisation of the Carlaw business (including the workshop layout and the types of work undertaken) but also refers warmly to Carlaw's enthusiasm for invention and sympathy for their inventors, as well as his reputation as an authority on poultry and flowers.[9] It was also said that 'his "shop" gets the credit among workmen of being the best wage-paying in the city, and thus always secures a superior class of capable artisans'.[10]

The commercial and mechanical skill of Carlaw's three sons who joined the firm (known as David Carlaw & Sons from about 1894 onwards) contributed to a further success and diversification in 1900 into motor cars; the firm was large and prosperous by the time of Carlaw's death on 20 November 1907.[11] As the 1901 British Association handbook put it: 'Probably of all the mechanical workshopss in Glasgow no one is of so much interest to the inventive genius as that of Messrs. Carlaw, where work of all kinds requiring the greatest exactitude can be done with the intelligent aid of the members of the firm.'[12]

REFERENCES

1. Jack House, *A Family Affair. The Story of David Carlaw & Sons . . .* (Glasgow, 1960), 9-10.

2. Obituary of David Carlaw, 27 September 1907, in *The Engineer* 104 (1907), 323. The reference should be to Joseph Hammersby of London.

3. *Ibid.* The first successful transatlantic cable was laid in 1857-58 (Charles Bright, *Submarine Telegraphs: their History, Construction, and Working* (London, 1898), 77).

4. *Ibid.*; House, *op. cit.* (1), 15.

5. *Glasgow Post Office Directory* 1871-1897.

6. Obituary of David Carlaw, 4 October 1907, in *Engineering* 84 (1907), 470; *Engineer, op. cit.* (2). However, apart from the protractor in the Frank Collection (NMS T1980.176) few other signed pieces by Carlaw have been noted, although a microscope made by him is known from an illustration (House, *op. cit.* (1), 12).

7. 'Models at the Glasgow Exhibition', 8 June 1888, *Engineering* 45 (1888), 557, quoted by House, *op. cit.* (1), 23; *International Exhibition Glasgow 1888: Official Catalogue* (Glasgow, 1888), 111, 387. This was a model of a triple expansion marine engine for the *Buenos Aires*. At the same exhibition Fairfields showed a model of the triple expansion engine with Bryce Douglas valve gear they had built for the *Alle*: the two models were said to cost £1300 and £1100 respectively (*The Bailie*, 23 May 1888, 7). We are grateful to John Burnett for this information.

8. House, *op. cit.* (1), 33, quoting *Engineering, op. cit.* (6).

9. *Glasgow of Today* (Glasgow, 1888), 145, quoting an article in *The Bailie*.

10. *The Bailie Exhibition Supplement*, 10 October 1888, 4. We are grateful to John Burnett for this reference.

11. House, *op. cit.* (1), 17-31.

12. Angus McLean (ed.), *Local Industries of Glasgow and the West of Scotland* British Association for the Advancement of Science Handbook: Vol. 1 (Glasgow, 1901), 45.

86. Drawing protractor: David Carlaw, Glasgow, c.1865
(T1980.176)

6″ protractor in bronze, with a semi-circular arc divided for 60° on either side of its centre line; pivoted at the centre of the arc and clamped at its circumference is a 24″ rule with 2 parallel drawing edges. Engraved by the pivot 'DAVID CARLAW 62 ARGYLE S^T GLASGOW.' The underside of the arc checked along its diameter for locating against a raised horizontal straight edge on a drawing board. The inclination of the rule to this edge read against a silvered index to ½°, the arc calibrated 30-[90]/0-60, 60-[0]/[90]-150, 120-[180]/[90]-30.

Arm radius: 612mm.
Scale radius: 76mm.

The protractor has an early repair where the arm has broken at the clamp. It is likely that the arc, which has been subject to considerable wear, was originally silvered.

Provenance: Purchased at Sotheby's, 19 December 1966, Lot 38.

JAMES BROWN

James Brown was born at Clifton, near Mandeville in Jamaica on 23 December 1836, where his father Hugh Brown was a missionary of the London Missionary Society: Hugh Brown died in 1837 and his mother brought him back to Scotland, where he was educated in Rhynie, Aberdeenshire.[1] On first coming to Glasgow in 1857, he was employed in the silk department of J.& W. Campbell & Co.'s warehouse in Ingram Street. 'From there, he went into the service of Messrs. Gardner & Son, (q.v.), opticians, Buchanan Street, in whose employment he remained for nine years.'[2] At the age of thirty-four in 1871, he set up as an optical,

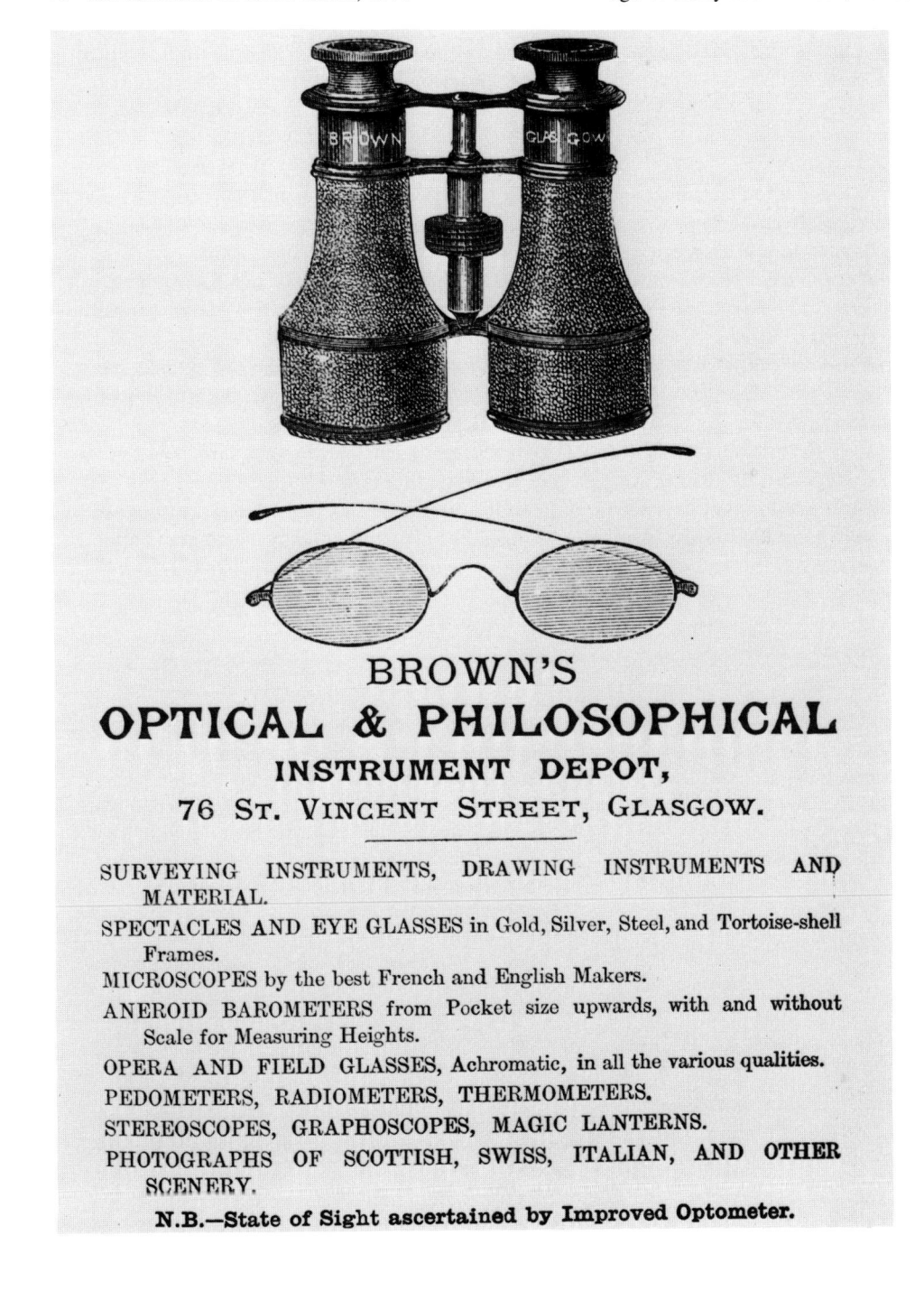

46 Advertisement for James Brown, 1876

BROWN'S
OPTICAL & PHILOSOPHICAL
INSTRUMENT DEPOT,
76 St. Vincent Street, Glasgow.

SURVEYING INSTRUMENTS, DRAWING INSTRUMENTS AND MATERIAL.
SPECTACLES AND EYE GLASSES in Gold, Silver, Steel, and Tortoise-shell Frames.
MICROSCOPES by the best French and English Makers.
ANEROID BAROMETERS from Pocket size upwards, with and without Scale for Measuring Heights.
OPERA AND FIELD GLASSES, Achromatic, in all the various qualities.
PEDOMETERS, RADIOMETERS, THERMOMETERS.
STEREOSCOPES, GRAPHOSCOPES, MAGIC LANTERNS.
PHOTOGRAPHS OF SCOTTISH, SWISS, ITALIAN, AND OTHER SCENERY.

N.B.—State of Sight ascertained by Improved Optometer.

mathematical and philosophical instrument maker at 76 St. Vincent Street, Glasgow, 'and there he remained for 41 years until, the property being pulled down, he removed to Buchanan Street'.[3] A commercial account of Glasgow in 1909, in describing the foundation of Brown's business in 1871, stated that 'For a considerable period of time prior to that he held a responsible position in another establishment',[4] and this would appear to be a reference to his position as salesman with Gardner & Co., which he held in 1861.[5]

From 1872 Brown advertised as an optician. Soon after setting up he also began to import fancy goods, and continued to do so until after 1900.[6] However, the branch he established at 33 West Nile Street in 1873 as part of his initial expansion was closed by 1877, and the optical side of the business was thenceforward probably more important. Brown in fact became one of Glasgow's leading spectacle makers and ophthalmic opticians, with a 'comprehensive' knowledge of eye troubles.[7]

In 1909 his stock, reflecting a continuing popular interest in scientific and quasi-scientific instruments, included barographs, rain-gauges, hygrometers, thermographs, anemometers, aneroid and mercurial barometers, field glasses and telescopes.[8] He was also dealing, like most other makers and retailers were by that date, in photographic materials. Brown was also said to be supplying microscopes for physicians, surgeons, chemists and mineralogists, 'at prices not one tenth of the cost of such instruments when he began business'.[9] The remark indicates the effect which mass production had on prices of apparatus. By 1910 Brown was also selling electrical instruments, particularly for medical purposes, and ten years later in 1920 the firm was advertising itself as a sundial maker. An example of this work is recorded as having been installed in Bellahouston Park, Glasgow, a pillar sundial on the site of Bellahouston House.[10]

Brown moved to 131 Buchanan Street in 1912, and the firm continued there until 1928 when it ceased to appear in the directory: the premises were taken over by the London firm of ophthalmic opticians, Dollond & Aitchison. Brown himself died on 15 December 1913, at the advanced age of seventy-seven; he 'had been able to attend to business until about three weeks' before his death.[11] Apart from his nephew Alexander N. M. Stewart, recorded in 1881 as his assistant,[12] only one other of Brown's employees is known: William Dalrymple, who by the early 1920s had set up independently as an instrument maker at 250 Buchanan Street.[13]

Brown was a member of the Glasgow Philosophical Society between 1876 and 1913.[14] That he was a member at all is unusual, for only a handful of Glasgow makers were; the fact that he was recommended for membership by three members actively interested in instrumentation[15] suggests that he was more involved in the Glasgow scientific community than the general appearance of his business might indicate. On the other hand he is not known to have actually constructed any instruments. He entered the Incorporation of Hammermen in 1879,[16] and was also involved in the Glasgow Chamber of Commerce, the Merchants' House, the Royal Scottish Geographical Society and the Scottish Meteorological Society.[17] He was married with six daughters and two sons, and lived in a large suburban house in Crosshill.[18] He and his business are not to be confused with James Brown & Son, nautical stationers and publishers, who were active around 1900.

REFERENCES

1. 'The Late Mr James Brown', *Glasgow Herald* 17 December 1913. Brown's birthplace is revealed in the 1881 census (General Register Office (Scotland), Census of 1881 (Renfrew, Cathcart), 560/5A, p.63). He may have been related to Hugh Brown, an ironmonger and maker of tools, guns and mathematical instruments in Glasgow (*Glasgow Post Office Directory* 1873).

2. *Glasgow Herald*, op. cit. (1).

3. *Ibid.*

4. *Glasgow Today* (Glasgow, 1909), 82.

5. G.R.O.(S.), Census of 1861 (Lanark, Govan), 646¹/18, p. 17.

6. *Glasgow Post Office Directory* 1872; advertisement in op. cit. 1880.

7. *Glasgow Today*, op. cit. (4). His frequent advertisements, for instance, in the street directory, mostly concerned his ophthalmic work.

8. *Ibid.* Apart from the levels (NMS T1980.128 and T1980.129), the parallel rule (NMS T1980.170), the telescope (NMS T1980.276), the microscope (NMS T1980.232) and pocket aneroid barometers (NMS T1980.214 and T1980.215) in the Frank collection, the following have been recorded: a refracting telescope with an iron tripod signed 'J. Brown Glasgow' was offered by Sotheby's, 16 October 1972, Lot 78; a 2¾ inch refracting telescope signed 'J. Brown, 76 St. Vincent St., Glasgow' with tripod and contained in a case, was offered by Phillips, 26 January 1983, Lot 151; a 4¾ inch reflecting telescope, signed 'J. Brown, 76 St. Vincent Street, Glasgow', with secondary telescope, rack and pinion focus, teak and brass mounted adjustable tripod stand, was offered by Christie's Scotland, 4 November 1987, Lot 51; a monocular microscope, signed 'James Brown, Glasgow', was offered by Christie's South Kensington, 19 November 1987, Lot 390; a Skyes hydrometer, signed 'Jas. Brown, Glasgow', was offered by Phillips Edinburgh, 31 July 1987, Lot 90; and a Lister limb microscope, signed 'J. Brown, Glasgow', was offered by Sotheby's Sussex, 15 November 1988, Lot 3371.

9. *Ibid.* As early as 1876, Brown had advertised 'microscopes by the best French and English Makers' (St.J.V.Day, *et al., Notices of Some of the Principal Manufactures of the West of Scotland* British Association for the Advancement of Science Guide Book (Glasgow, 1876), 1). His later claims must include the microscope represented in the Frank collection (NMS T1980.232) and also that in the Billings Microscope Collection (inventory AFIP M-004047-74-7353), described in H.R. Purtle (ed.), *The Billings Collection of the Medical Museum, Armed Forces Institute of Pathology* 2nd edition (Washington, D.C., 1974), 210.

10. W.B. Stevenson, 'Sundials of six Scottish Counties near Glasgow', *Transactions of the Glasgow Archaeological Society* new series, 9 (1937-40), 236.

11. *Glasgow Herald, op. cit.* (1).

12. G.R.O.(S.), Census of 1881 (Renfrew, Cathcart), 560/5A, p. 63.

13. *Glasgow Post Office Directory* 1923.

14. *Proceedings of the Glasgow Philosophical Society* 10 (1875-77), 199, 415, and *passim*.

15. *Ibid.* They were: Alexander Scott, Dr James Finlayson and Dr Andrew Fergus. Scott published 'On a Large-index Barometer', *ibid.* 13 (1882), 154-156; Finlayson's papers were mainly physiological, but had instrumental aspects to them; Fergus was an engineer.

16. H. Lumsden and P.H. Aitken, *The History of the Hammermen of Glasgow* (Paisley, 1915), 339.

17. *Glasgow Today, op. cit.* (4). Brown was elected an Annual Member of the Scottish Meteorological Society on 19 March 1900 (*Journal of the Scottish Meteorological Society* 3rd series, 11 (1900), 250).

18. G.R.O.(S.), Census of 1891 (Renfrew, Cathcart), 560/8, pp. 13-14.

---

87. Level: J. Brown, Glasgow, c.1910 (T1980.128)

10″ level in oxidised brass (stripped), with bubble tube over the telescope, rack and pinion focus and doublet objective; the limb threaded to a parallel plate levelling head with free azimuth motion. Engraved on the telescope tube 'J. BROWN / 76 ST VINCENT STREET / GLASGOW'. Components of the level and levelling head stamped with the assembly mark '1'.

Telescope length: 248mm.
Tube diameter: 28mm.
Aperture: 23mm.

Most of the surface has been mechanically cleaned and a modern lacquer applied. Lacking diaphragm wires and objective cover.

This level bears close similarities to item 131, signed J. White.

88. Level: J. Brown, Glasgow, c.1910 (T1980.129)

10″ level in oxidised brass (stripped), with bubble tube over the telescope, rack and pinion focus and doublet objective; the limb threaded to a parallel plate levelling head with free azimuth motion. Engraved on the telescope tube 'J. BROWN 76 ST VINCENT STREET / GLASGOW'. Components of the level are stamped with the assembly mark '4', but those of the levelling head are stamped '2'.

Telescope length: 247mm.
Tube diameter: 28mm.
Aperture: 23mm.

The levelling head is from a different instrument, another portion of which appears to be with item 158. Although the 4 levelling screws are individually stamped with the assembly mark '2', one is of a coarser thread (26 t.p.i. against 31 t.p.i.). The glass bubble tube is a replacement. The lens mounts have suffered damage in a fall and the limb has required re-alignment. Lacking objective cover.

This level bears close similarities to item 131, signed J. White.

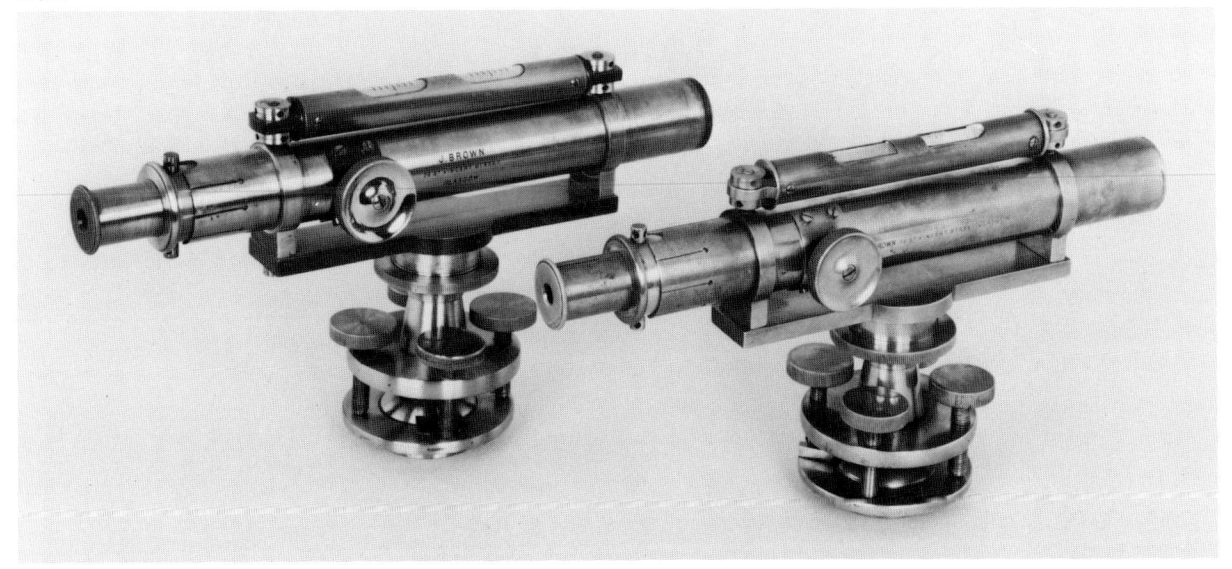

87, 88

89. Parallel rule: James Brown, Glasgow, c.1890 (T1980.170)

24" rolling parallel rule in bronze, with chamfered edges and 2 lifting knobs. Engraved 'JAMES BROWN./ GLASGOW.'

Length: 613mm.

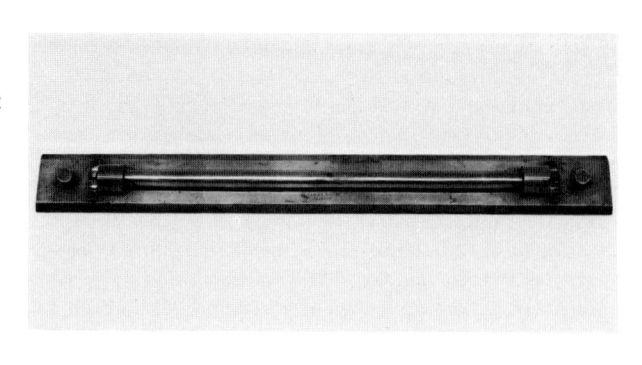

90. Pocket aneroid barometer: J. Brown, Glasgow, c.1880 (T1980.214)

Double-sided pocket aneroid barometer, temperature compensated, in a glazed brass case. Engraved on the silvered dial 'Compensated' and 'J. BROWN / Optician / GLASGOW'. The pressure scale divided in inches of mercury, 23-31 by 0.05", the reading indicated by a blued-steel hand. An altitude scale divided in feet, 8000-0 by 100ft., can be rotated to normalise the altitude when used as a pressure-indicating altimeter. On the reverse, a curved Centigrade/Fahrenheit thermometer scale (the thermometer itself lacking) and a central glazed compass (defective); also the adjustment point for the aneroid. With satin and velvet lined leather-covered fitted case, opening on both sides.

Diameter: 47mm.

Lacking the thermometer, compass needle and compass rose.

Although not identical, this instrument corresponds broadly to one of the many pocket aneroid barometers offered by the London wholesaler J.J. Hicks in his *Illustrated & Descriptive Wholesale Catalogue of Standard, Self-Recording, and other Meteorological Instruments ...* (London, n.d. [c.1880]), 52, no. 194, 'Pocket Aneroid barometer, with silvered dial, in morocco case, ordinary range scale, raised dial, curved Thermometer, compensated "Fair," "Change" and "Rain" £2 8s 0d.'

91. Pocket aneroid barometer: J. Brown, Glasgow, 1893 (T1980.215)

Mountain pocket aneroid barometer, temperature compensated, in a silver case. Engraved on the silvered dial 'Compensated' and 'J. BROWN / OPTICIAN / GLASGOW'; the silver case hallmarked London, 1893-94, with the initials of the casemaker Thomas Wheeler. The pressure scale divided in inches of mercury, 21-31 by 0.05", the reading indicated by a blued-steel hand. An altitude scale divided in feet, 10,000-0 by 100ft., can be adjusted to normalise the altitude by turning the knob at the suspension ring when used as a pressure-indicating altimeter. Adjusting point on the reverse.

Diameter: 49mm.

Although not identical, this instrument corresponds broadly to one of the many pocket aneroid barometers offered by the London wholesaler J.J. Hicks in his *Illustrated & Descriptive Wholesale Catalogue of Standard, Self-Recording, and other Meteorological Instruments ...* (London, n.d. [c.1880]), 52, no. 200, 'Mountain Aneroid Barometer, Compensated for Temperature, with silvered dial, in morocco case. Altitude scale, from 6,000 to 10,000 feet. Standard silver £1 2s 0d.'

90, 91

92. Microscope: James Brown, Glasgow, c.1885 (T1980.232)

Compound microscope, 'English Medical Microscope' model of J. Parkes & Son, in brass and oxidised brass, on an inclining pillar and 2-toed lead-weighted tripod base. Engraved on the body tube 'James Brown, / S! Vincent S!. / Glasgow.' and stamped under the stage and base with the assembly mark '28'. Single-draw body tube (the draw tube with 3 settings marked), in a leather-lined draw sleeve for coarse focus, attached to the pillar through a spring-loaded slide, operated by a fine focus screw on the top of the pillar. 2-component eyepiece assembly (engraved 'A') and objective on sliding adaptor.

Body length: 163mm.
Body diameter: 31mm.
Height: 270mm.
Base size: 84 x 120mm.

Lacking the second eyepiece, the stage diaphragm which fits flush to the stage surface, and one spring clip.

The instrument can be identified as the 'English Medical Microscope' offered for £6 to £6 10s in 1882 by J. Parkes & Son, *Wholesale Catalogue of Optical ... Instruments* (Birmingham, n.d.[1882]), 8, no. 5009, where it was noted that 'several hundred of the above Instruments have already been supplied to the various Medical Schools and Hospitals in London, Edinburgh, Glasgow, Dublin, and other places, and a number of voluntary testimonials have been received, as to their satisfactory performance'. The model was also illustrated in the *Journal of the Royal Microscopical Society* 3 (1880), 1048 (the block being modified from that of the closely related 'Portable Educational Microscope', no. 5010). An example of a microscope inscribed 'James Brown 76 St. Vincent St Glasgow' but also with the open eye trade mark symbol used by Parkes is in the Collection of Historical Scientific Instruments at Harvard University (inventory 1262), confirming that Parkes supplied Brown with microscopes. Another similar but unsigned microscope is described and illustrated in H.R. Purtle, (ed.), *The Billings Microscope Collection of the Medical Museum, Armed Forces Institute of Pathology* 2nd edition (Washington, D.C., 1974), 208. The 'patent sliding adapter' supplied by Parkes with nos. 5009 and 5010, is present on this instrument: it was described and illustrated in *Journal of the Royal Microscopical Society* 3 (1880), 1048-1049. Six months provisional protection was obtained by S.H. Parkes in March 1878 under U.K. patent 904 of 1878, but he did not proceed further (*The Engineer* 46 (1878), 305).

93. Refracting telescope: J. Brown, Glasgow, c.1890 (T1980.276)

2¾" aperture 2-draw mounted refracting telescope in brass; with doublet objective recessed from the end of the barrel, but lacking the inner draw tube and optics, and with rack and pinion focus (defective) to the second draw tube. Engraved on the barrel end plate 'J. BROWN,/ GLASGOW.' Pillar-mounted by a knuckle-joint, with a clamped steady rod to the base of the pillar and with free azimuth motion, on a folding brass table tripod.

Length closed (less eyepiece): 965mm.
Tube diameter: 77mm.
Aperture: 72mm.

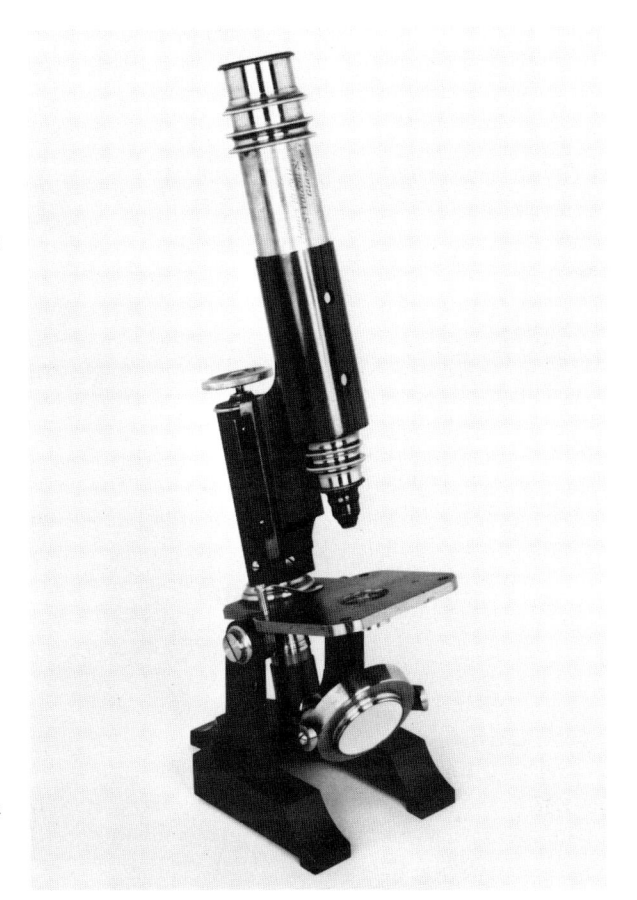

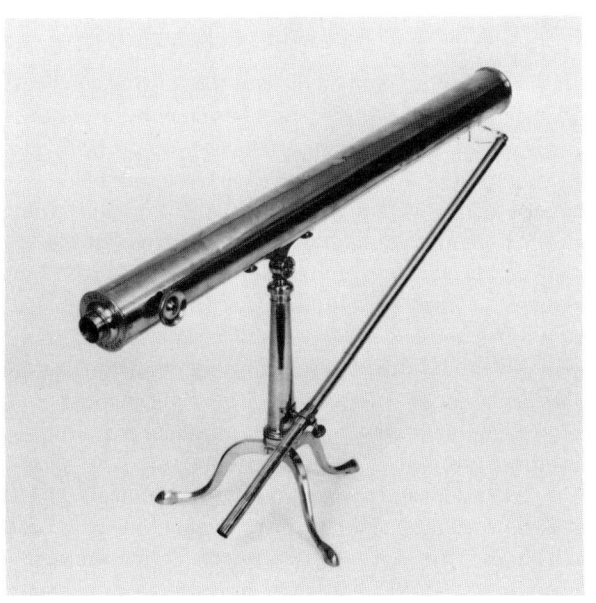

Lacking the eyepiece and erecting optics, inner draw tube and draw slide. The barrel has been marked by mechanical abrasive cleaning and the removal of dents, and a modern lacquer has been applied. The barrel end plate has been refitted. The tripod attachment holes have been enlarged to make the otherwise incompatible tripod fit, and the steady arm attachment to the barrel is missing.

Detailed similarities in design (excluding the stand) and in the mounting of optical components suggest that this item is by the same manufacturer as items 57 (E. Lennie), 66 (Lowdon), 85 (Gardner & Co.) and 161 (Whyte, Thomson & Co.).

15
THOMAS MORTON AND ANDREW BARCLAY, ENGINEERS

Thomas Morton and Andrew Barclay, the two men whose instrument making activities are discussed below, shared in common the place where they lived and worked - Kilmarnock, in Ayrshire - and the fact that professionally they were engineers rather than instrument makers. Despite these similarities, they were of different generations, and it does not appear that they knew each other personally nor that the younger man, Barclay, was directly influenced by Morton.

Both men had engineering works at their disposal, which allowed them to indulge in their taste for astronomy and telescope making. They were able to produce, from the resources available to them, professional-looking instruments and to divert money, manpower and technical expertise into a sideline which in neither case developed into a business. Both remained, strictly speaking, amateur instrument makers.

In both cases, a hobby interest of the head of the firm was being exploited. It is possible that either or both of them hoped that telescope production would become a money-spinner for his business. In a situation of economic flux, entrepreneurs were able to change the lines of commodity sold to suit the demands of the market, and this sort of evolution can be observed in a number of firms. For instance, Archibald Baird & Son (q.v.), were manufacturers of colliery apparatus, but were able to produce control instruments. Similarly, David Carlaw (q.v.) began as an instrument maker but moved gradually into light engineering, and after his death his firm abandoned instrument making altogether and concentrated on the economically rewarding manufacture of consecutive-numbering ticket machines. Two engineers who built security printing machinery and then gradually moved into instrument making were the Irish telescope maker Thomas Grubb (1800-1878),[1] who initially made banknote machinery for the Bank of Ireland, and R.W. Munro (1839-1913) of London who constructed machines for printing banknotes for the Bank of England and postal orders for the General Post Office before moving into the manufacture of seismological and meteorological apparatus, as just one part of a business with expertise in making precision machinery for industry.[2]

If this was what Morton and Barclay hoped to do, then location may have played a part in their apparant failure. Geographically speaking, Kilmarnock was poorly placed at the end of the eighteenth century for serving an instrument market. Indeed, until the coming of the mainline railways in 1850,[3] the town seemed to be dependent on its agricultural hinterland, while what industry there was - the shoe trade, tanning, gloves, bonnets, and increasingly, carpet manufacture - made the town prosperous by the standards of the day.[4] However, with ready supplies of coal and a growing population, Kilmarnock became more industrialised but until mid-century found 'the chief disadvantage [to be] the distance from the sea, (six or seven miles,) and the consequent expense of land carriage'.[5] Only with the railways could towns like Kilmarnock reach substantially beyond their local markets.

For neither Morton nor Barclay did instrument making become a successful professional concern. In Morton's case, this was possibly because he started in business too soon to catch the railway boom and he was in any case already managing a thriving and demanding carpet-machinery operation. By the time Barclay's interest had developed, the structure of the instrument production trade had moved towards a concentration in the hands of a few large successful wholesalers distributing nationally from the Midlands and South of England and from the Continent: Kilmarnock did not figure in this sort of scheme. Indeed, it was to take some very unusual firms to break this mould: concerns such as Barr & Stroud, Kelvin & James White (q.v.), the Cambridge Scientific Instrument Company and Sir Howard Grubb of Dublin, none of which needed to operate from the metropolis nor its environs, because their product was highly specialised and sought after by an international clientele. This, the other end of the instrument market, was a rarified arena to which neither Morton nor Barclay aspired.

REFERENCES

1. For an assessment of the Grubb business, see A.D. Morrison-Low and J. Burnett, *'Vulgar and Mechanick': The Scientific Instrument Trade in Ireland 1650-1921* (Edinburgh and Dublin, forthcoming).

2. Obituary of R.W. Munro, *Quarterly Journal of the Royal Meteorological Society* 39 (1913), 72. For a discussion of the Munro business see Mairi Williams, 'Technical Innovation: Examples from the Scientific Instrument Industry', in Jonathan Liebenau (ed.), *The Challenge of New Technology: Innovation in British Industry since 1850* (Aldershot, Hants., 1988), 8-29.

3. Campbell Highet, *Glasgow and Southwestern Railway* (Lingfield, 1965). We are grateful to John Burnett for this reference.

4. 'Kilmarnock', in Sir John Sinclair Bt., (ed.), *The Statistical Account of Scotland* II (Edinburgh, 1792), 87-92.

5. Robert and William Chambers, 'Kilmarnock' in *The Gazetteer of Scotland* (Edinburgh, 1832), II, 632.

THOMAS MORTON

Thomas Morton (1783-1862) was born at Mauchline, Ayrshire, the son of a brickmaker also named Thomas Morton. When aged three, the family moved to Kilmarnock; when he was about fifteen he was apprenticed to Bryce Blair, a turner and wheelwright, starting in business himself in 1806 three or four years after completing his apprenticeship.[1]

47 Thomas Morton (1783-1862) aged 50, by an unknown artist, 1833. *Dick Institute, Kilmarnock*

'Before this period, he had evinced a strong propensity for mechanical pursuits. Every scientific instrument that came within his reach he eagerly and minutely examined; in short, so great, in early life, was the bias of his mind in this way that when wandering Italians came to Kilmarnock with telescopes, barometers, &c., for sale, he might have been seen following them in the streets, admiring their various articles, or endeavouring to obtain a knowledge of the principles on which they were made. At length

Mr Morton attempted the construction of a telescope, and was successful in finishing it. But he did not rest satisfied with this achievement, for, observing that the telescopes then in use were somewhat defective, he soon applied his genius to remedy the deficiency, and ultimately succeeded in adding several conveniences to this useful instrument.'[2]

However, telescope making did not become Morton's exclusive profession, although he was able to make his living through his abilities as a mechanic. He invented a number of improvements to carpet-making machinery, the first of which initially suggested itself to him because he had been asked to repair a barrel organ: 'The examination of the internal structure of the instrument led Mr Morton to conceive the idea of applying a barrel, like that of the organ, to elevate the warp-yarns in the loom, and the invention of the barrel-carpet-loom was the result.'[3] The machine was perfected in 1811, orders came flooding into Morton Place, and his fame spread: 'in 1814, the report of his ingenuity in connection with this machine reached the ears of the Emperor of Russia, who was then on a visit to England'.[4] Despite imperial entreaties to emigrate, Morton remained in Kilmarnock where, although he did not patent his inventions, 'the carpet manufacturers of the district voluntarily came under obligation to use exclusively machines of Mr Morton's construction, and this in some measure secured him the benefit of a local patent right'.[5] Morton's manufacturing ingenuity had already won him a Board of Trade prize of £20 in 1808. In 1821 he produced his first three-ply carpet, and in 1827 he developed a modification of his carpet machine combined with the Jacquard loom. In 1816 he had considered the effect of driving a ship by propellor, an idea inspired by the windmill; but he did not pursue this, and it was left to others to claim priority and take further.

In 1817 a local banker, Mr William Anderson, asked Morton if he would repair a reflecting telescope whose speculum needed polishing. Morton's success was such that 'he immediately set to work to construct one for himself, and in the sequel to manufacture telescopes for sale'.[6] The three telescopes in this collection are representative of this work,[7] and several others are known.[8]

Although the construction of telescopes must be regarded as a sideline to Morton's main occupation as a carpet machinery manufacturer,[9] the fascination it held for him led him back to the study of practical astronomy. In 1818 he built for his own use the Astronomical Observatory at Morton Place, at a cost of about £1000. It was

> 'furnished with two excellent telescopes of different sizes. The large one, which is constructed on the Newtonian principle, is 9¾ inches in diameter. The other is of the Gregorian construction, and is seven inches in diameter. They are both the workmanship of Mr Morton, and are fine specimens of mechanism. The large one in particular, besides possessing great power, has many admirable conveniences. In the Observatory there is also a *camera obscura*, by which, in suitable weather, the visitor can behold, as in one enchanting picture, the whole of the varied and beautiful scenery of the surrounding country.'[10]

The observatory was open to the public,[11] and subsequently passed into the hands of Morton's son-in-law Thomas Lee, F.R.A.S. (1818-1892), who used it to help with his teaching of astronomy at the Kilmarnock Academy, where he was mathematics teacher between 1843 and 1875.[12] Thereafter, it fell into disrepair, and was finally demolished in slum clearance in 1957.[13] A fairly full description of the observatory and its instruments was given in the *Kilmarnock and Riccarton Directory* for 1840, in which Morton himself appeared twice, once as an 'engineer and machine maker', and again as an 'optician'.[14]

Morton was active in the Kilmarnock Philosophical Institution, of which he was a founder member in 1823, and by 1847 the society had about one hundred and seventy members, a library, and ran occasional lectures. In 1832, Morton visited, and presumably encouraged, the local shoemaker John Fulton, who was then building his third orrery. Fulton's second orrery was purchased by the Philosophical Institution, but is not known to have survived.[15]

Morton's fame as an optician was not restricted to Kilmarnock, however. The Dumfries and Maxwelltown Astronomical Society was established at a meeting on 23 January 1835, and began by resolving to convert an old windmill into an observatory.[16] At a subsequent meeting, it was reported that a Mr Robert Thomson, acting on behalf of the management committee, had been in touch with an Edinburgh lawyer, and had asked him to seek advice through James Jardine, civil engineer, about the proposed fittings for the observatory: 'Mr Adie Optician in Edinb[r]' was the name that had come forward.

> 'Mr Adie it appeared recommended the Society to purchase a Refracting Telescope as less liable to get out of order than a Reflecting one. He also recommended to the Society to purchase a Camera Obscura & agreed to furnish one of the same description as that furnished by him to the Edinb[r] Observatory for £60. A Telescope Mr Adie agreed to furnish at the price of £73 10/-.

Mr Thomson further stated to the Meeting that a letter had been received from a Mr Morton of Kilmarnock recommending a Reflecting

48 'Back of Morton Place', engraving by Thomas Smellie, c.1890

Telescope and agreeing to furnish one at various prices from £60 to £200.'[17]

It is difficult to make out from the papers whether it was Alexander Adie (q.v.) or his son John (q.v.) who was being consulted; the term 'Mr Adie Optician in Edinb.' may well have embraced the firm rather than an individual. The Meeting went on to decide to obtain a detailed estimate for the camera obscura from Adie, and to consult Sir John Ross (the Polar explorer), whose estate was at Stranraer, for an opinion on whether a reflecting or a refracting telescope would be the more suitable instrument, given their budget.[18]

Adie produced estimates and plans at the Meeting of 19 March, but when the Directors of the Society met in April, it was to decide that

> 'The Meeting was of the opinion that Mr Morton of Kilmarnock be employed to furnish a Gregorian Telescope, the speculum of which is to be Eight Inches in diameter, and a Camera Obscura of the description approved of by Mr Morton when here. These Instruments to be approved of by Sir John Ross and the prices of each to be the sums asked by Mr Morton when here namely £73 for the former and £27 10/- for the latter. The Meeting instructed the Committee to write Mr Morton to the above purpose and to procure the Instruments with as little delay as possible.'[19]

By the following July, Morton was ready to install the instruments, and towards the end of the month the Dumfries and Maxwelltown Astronomical Society were able to open their Observatory, and present Morton with a share in it 'as a Token of the Shareholders respect for him as a worthy and obliging man & as a most ingenious Mechanic'.[20]

Apart from this single commission, it is not known who else Morton supplied with telescopes, or how much of his business was invested in instrument making. A contemporary chronicler of Kilmarnock life stated that Morton had supplied Sir John Ross with his telescopes,[21] but this is not confirmed elsewhere. In 1835 his letter 'regarding an important improvement lately discovered by him in the Carpet and other Machines for Figured work, especially in Three-ply Machine' to the Edinburgh-based Society of Arts was read before one of the Society's meetings.[22] That same year he was elected an Honorary Member of the Society of Arts for Scotland, and was proud enough of this distinction to engrave it on telescopes constructed by him after this date.[23]

By 1861 he described himself as a 'Carpet Machine and Telescope Maker' and employed one journeyman and two apprentices,[24] and although the identity of one earlier apprentice is known, it is difficult to gauge how much telescope construction was done within the business.[25] Morton died, aged seventy-eight on 10 March 1862.[26] In his will of 1832, he had nominated his wife Agnes Wilson as one of his executors, but this was superseded by a later codicil of 1861 nominating one of his sons, Charles (who declined to accept the executry). Although his estate was worth a reasonable sum - £281 3s 1d - essentially the business was to be carried on, on his widow's behalf, by Morton's two sons Alexander and Charles. Most of the money due to Morton which was outstanding at his death, was to do with carpet machinery side of the business, although Messrs. Gardner & Co. (q.v.) owed £1 0s 1d.[27]

The later history of the firm is obscure. Morton's son Alexander published a small book entitled *The Art of Making Reflecting Telescopes made Simple and Easy* in the year of his father's death, and claimed 'more than fifty years' experience of the art of making Reflecting Telescopes, by his late father ... and himself'.[28] He also held a number of patents taken out from the late 1840s,[29] including a device for ascertaining longitude.[30] One telescope with Alexander Morton's signature has been recorded.[31]

REFERENCES

1. Archibald McKay (revised by W. Findlay), *History of Kilmarnock* 5th edition (Kilmarnock, 1909), 281-282.

2. *Ibid.*

3. The main account for Morton's life is the anonymous *Biographical Sketch of the Late Thomas Morton Esq., Honorary Member of the Royal Scottish Society of Arts* (Ardrossan, 1920), extracted from the *Kilmarnock Weekly Post*, 22 March 1862. Various extracts from it appear elsewhere: for example, in the *Kilmarnock Standard*, 13 August 1892; *ibid*, 12 and 19 May 1900; *ibid*, 4 June 1904. Another contemporary biography, which may have been used as a basis for this, was the 'Memoir of Mr Morton, Engineer', in James Thomson, *The Ayrshire Miscellany: or, Kilmarnock Literary Expositor* 2 (1818), 37-39.

4. *Biographical Sketch*, op.cit. (3), 10.

5. *Ibid.*, 11.

6. *Ibid.*, 14.

7. NMS T1981.38, T1981.39 and T1981.40.

8. Four other telescopes by Morton are recorded in public collections, located as follows: one at Dumfries Museum, one at the Royal Museum of Scotland (NMS T1974.97) and two in the Wellcome Museum of the History of Medicine at the Science Museum, London (inventory 1979-164 or A140421, and A164910; this latter now unlocated). Another instrument, possibly made by Morton for another Kilmarnock carpet manufacturer, James Blackwood (1823-1893) was offered for sale by Sotheby's, 18 June 1986, Lot 205. Although Blackwood was himself capable of making instruments - he read the *Edinburgh Encyclopaedia* 'from beginning to end. The fruit of such a course of study was a knowledge of many branches of science. It also appeared in various instruments made by himself - a camera obscura, a microscope, a telescope, &c.' ('James Blackwood, F.G.S., &c.', *Kilmarnock Glenfield Ramblers* 1 (1894), 29) - this ability to construct instruments does not mean that he did more than build a single example of each. Stylistically, the instrument offered by Sotheby's greatly resembled those constructed by Thomas Morton; Blackwood's scientific interests were centred on geological studies. Four further examples of Morton's telescopes have come to light: one offered for sale by Phillips, 16 March 1988, Lot 123; another unsigned example is in a private collection; two others have been noted in the hands of dealers.

9. Two contemporary accounts describe Morton's main business and its success: 'A great improvement has lately been made in the manufacture of carpets by an inhabitant of this place ... the inventor, whose name is Thomas Morton, has already made upward of 180 of these engines ...' ('Kilmarnock', *Edinburgh Encyclopaedia* (Edinburgh, 1830), XII, 455); 'the carpet manufacturers in Kilmarnock are eminently indebted to the inventive genius of Mr Thomas Morton of this town, a self-instructed merchant ...' (*The New Statistical Account of Scotland* V (Edinburgh and London, 1845), Ayr, 552). One authority states that 'Mr Thomas Morton, an ingenious weaver in Kilmarnock, invented an improved draw loom, and wove the carpet as a triple cloth, the centre warp being a kind of stuffer. So strongly did Morton revive the trade that the carpet became known on the market alternatively as the Kilmarnock or Scotch carpet' (William S. Murphy, *The Textile Industries: A Practical Guide to Fibres, Yarns and Fabrics* (London, 1910-11), VII, 17). Yet Morton does not receive a mention in the most recent overview of the history of the industry: J. Neville Bartlett, *Carpeting the Millions: The Growth of Britain's Carpet Industry* (Edinburgh, 1978). It also appears that Morton was not closely related to the other famous Mortons involved in the Ayrshire textile industry: Alexander Morton (1844-1923), (see Jocelyn W.F. Morton, 'Alexander Morton' in Anthony Slaven and Sydney Checkland (eds.), *Dictionary of Scottish Business Biography 1869-1960* Volume I The Staple Industries (Aberdeen, 1986), 369-372); Gavin Morton (1866-1954) (discussed by James Mair, 'Gavin Morton', *ibid.*, 372-373); and James Morton (1867-1943) (see Jocelyn F.W. Morton, 'Morton, Sir James' in David J. Jeremy and Christine Shaw (eds.), *Dictionary of Business Biography* IV (London, 1985), 349-352). A family tree of Thomas Morton's genealogy supplied by D.A. Rodman of Sheffield to the Dick Institute, Kilmarnock, bears this out; we are grateful to James Hunter of the Dick Institute for his assistance.

10. McKay, *op. cit.* (1), 552.

11. 'Mr M.'s Observatory is open night and day for people of all ranks, and has deservedly become a resort for the tasteful and fashionable of both sexes in Kilmarnock.' (Thomson, *op. cit.* (3), 39).

12. Obituary of Thomas Morton, *Monthly Notices of the Royal Astronomical Society* 53 (1892-93), 220-221: '... he was greatly assisted in his astronomical teaching by the erection of the observatory in Morton Place, which was provided with two

good reflecting telescopes and other astronomical instruments. This observatory was built by the late Baillie Thomas Morton, whose daughter Mr Lee married in 1841...'.

13. A summary of the history of the Morton Place Observatory, together with an illustration of the tower, is given in J.H. Hammond, *The Camera Obscura: A Chronicle* (Bristol, 1981), 106-107: this illustration also appears in Thomas Smellie, *Sketches of Old Kilmarnock* (Kilmarnock, 1898).

14. J. Angus, *Kilmarnock and Riccarton Directory* (Kilmarnock, 1840), 32, 43: the description of the 'Astronomical Observatory, Morton-place', is at pp. 68-72.

15. McKay, *op. cit.* (1), 340-341; John Kelso Hunter, *The Retrospect of an Artist's Life: Memorials of West Countrymen and Manners of the Past Half Century* (Kilmarnock, 1912), 159; for a recent assessment of John Fulton, see Hugh V. McIntyre, 'Mr Universe', *Scots Magazine* new series, 122 (1985), 587-594. Fulton's third orrery is in Glasgow Transport Museum, Kelvin Hall.

16. Dumfries Museum, Minute Book of the Dumfries and Maxwelltown Astronomical Society, volume I, Minute Book 23 January 1835 to 5 July 1842; Minutes of Meeting, 23 January 1835; also, Anon., *The Camera Obscura* (n.p., n.d. [Dumfries Museum, c.1980]); quoted by Hammond, *op. cit.* (13), 107-108.

17. Dumfries and Maxwelltown Astronomical Society, Minutes of Meeting, 24 February 1835.

18. *Ibid.* Both James Jardine and Sir John Ross had shares in the Society.

19. *Ibid.*, Minutes of Meeting, 28 April 1835.

20. *Ibid.*, 28 July 1836: 'The Meeting having inspected the Instruments this day furnished by Mr Morton, are satisfied with them so far as the state of the weather will allow them to form an opinion, but as the Meeting is far from numerous the members present cannot think of coming under the responsibility of accepting of the Instruments as sufficient and in terms of the bargain with Mr Morton, Mr Thomson is therefore instructed to pay the price of the Instruments to Mr Morton on an understanding that they are to be subject to the inspection and approval of Sir John Ross or any other competent Judge to be fixed on by Mr Thomson & Mr Morton[,] such inspection to be made any time within three months from this date.'

21. This somewhat homely piece appeared in Hunter, *op. cit.* (15), 157-158: 'Thomas Morton, of Morton Place, was one who had struggled through many phases of labour, and he had labour as a dower to the day of his death. When he might have been resting his lively labour-disciplined mind was rummaging up his past experience, wishing to send it out to the world to be taken up by others. All the discoveries he had made in any subtle or mysterious metal, such as the speculum of the telescope, he wished to lay before the world in a language so free from doubt or mystification that any follower might take the fact of his past life and put it to the front of his own. He had in his youth herded kye, held conversation with the craws, made whistles, baskets and bricks, was a turner and a wheelwright, made bagpipes of an original construction; and, along with his brother-in-law, the late John Peden, sought the aid of mechanics to weave carpets without the aid of a draw boy ... Thomas Morton added the making of telescopes to his other accomplishments. Sir John Ross had telescopes of his making with him in search of a north-west passage.'

22. National Library of Scotland, Royal Scottish Society of Arts Archives, Dep. 230/3, Minute Book, volume 2, p.325, 16 December 1835.

23. McKay, *op. cit.* (1), 243. The Society of Arts received its Royal Charter as the Royal Scottish Society of Arts in 1841, and examples of instruments inscribed by Morton as 'H.M.R.S.S.A' are in the Royal Museum of Scotland (NMS T1981.39 and T1981.40).

24. General Register Office (Scotland), Census of 1861 (Kilmarnock), 597/13, pp. 9-10.

25. 'Mr Muir was ... bound apprentice to Mr Thomas Morton, of Kilmarnock, a mechanic of great and very varied skill. The work to which he was put included the ordinary jobs which find their way into a country shop, and in addition comprised the manufacture of carpet looms, bagpipes, and telescopes, his master having a great reputation for the two latter.': 'The Late Mr William Muir', *Engineering* 46 (1888), 194. We are grateful to James Wood for this reference. Muir eventually became manager of Sir Joseph Whitworth's Manchester engineering works, and was involved in the creation of the Whitworth system of screw threads.

26. G.R.O.(S.), Register of Deaths 1862 (Kilmarnock), 597/176. The cause of death was given as 'cancerous tubers of liver of unknown duration, though only confined for 2 months'.

27. Scottish Record Office, SC 6/44/29, pp. 598-609, Inventory of personal estate of Thomas Morton, engineer, entered 1 May 1862.

28. A. Morton, *The Art of Making Reflecting Telescopes Made Simple and Easy* (Glasgow, 1862). The final page advertises that 'the Author can supply any person with Telescopes of any size, on very elegant stands, or any part necessary for the above, in any state of progress, if they have not convenience for doing it themselves. Communications addressed Alexander Morton, Observatory Works, Kilmarnock, will receive prompt attention.'

29. UK patents (to 1870): 11,643, 29 March 1847, printing warps; 266, 5 February 1855, weaving carpets; 1287, 5 June 1855, motive power machines; 3014, 4 December 1857, obtaining motive power; 702, 19 March 1859, motive power; 807, 31 March 1859, improvements in sextants; 1163, 9 May 1859, weaving figured fabrics; 3244, 3 December 1862, lawn mowing machines; 381, 12 February 1863, lawn mowing machines; 744, 20 March 1863, ejecting fluids; 2775, 7 November 1863, ejecting fluids; 2106, 18 July 1867, improvements in hydraulics; 456, 15 February 1869, hydraulic apparatus.

30. 'Rowan and Morton's Apparatus for Ascertaining Longitude', *Engineering* 2 (1866), 449. UK patent 946, 3 April 1866.

31. In a private collection. We are grateful to Dr David Gavine for this reference.

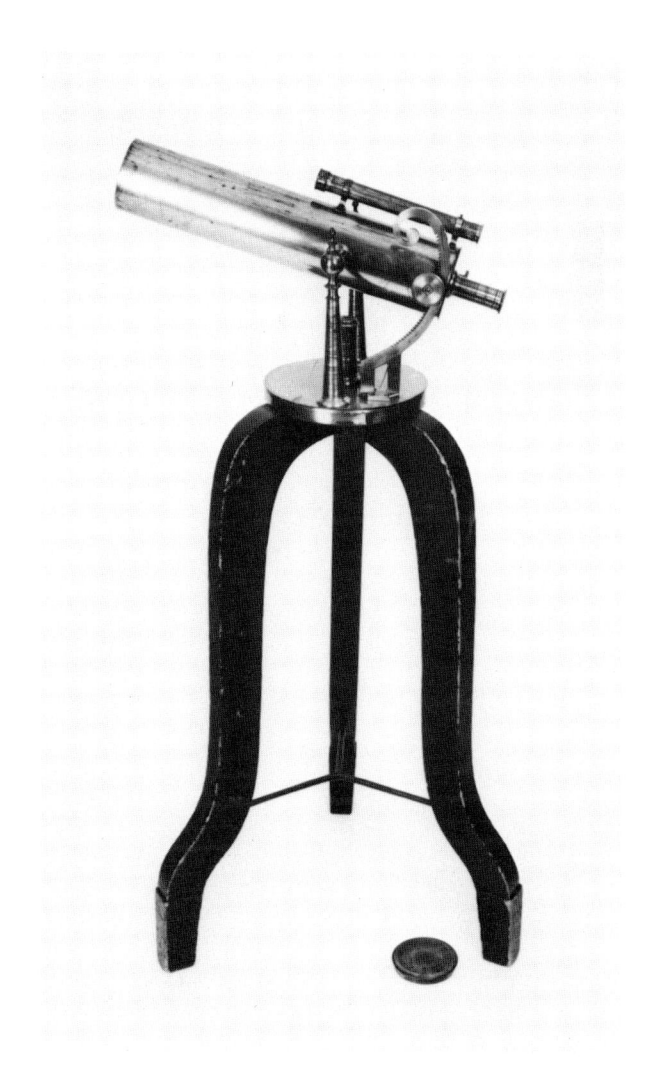

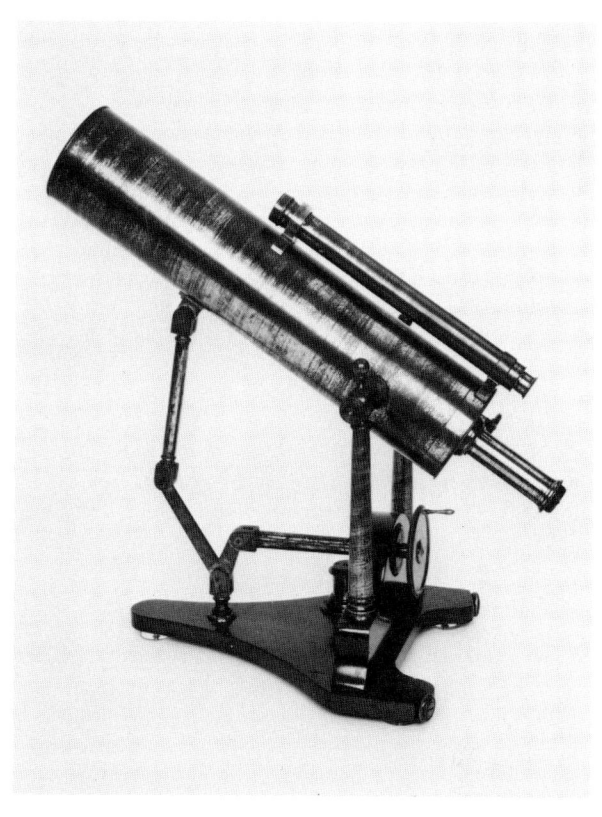

94. Reflecting telescope: Thomas Morton, Kilmarnock, c.1850 (T1981.38)

14" focus Gregorian reflecting telescope in brass with speculum metal optics. Engraved on the mirror retention plate 'Thoˢ. Morton / KILMARNOCK'. Long external focusing rod, the threaded end working on a boss connected by a slotted key (missing) to a removeable internal sliding plate, carrying the secondary speculum and tubular shade on a 3-point adjustment. The primary speculum retained by 3 leaf springs against a projecting ring in the mirror cell. 1" aperture finder telescope mounted over the barrel. The instrument is trunion mounted between 2 pillars, on a wooden turntable base revolving on a brass ring fixed on a tripod with 3 black-painted wooden ogee legs and iron stretcher. Geared altitude movement against a decorative curved brass rack mounted on the turntable. Additional eyepiece on a mount in the centre of the turntable, and one solar shade. Push-fit end cover and finder objective cover.

Length of barrel: 527mm.
Barrel diameter: 111mm.
Speculum diameter: 115mm.

The wooden turntable base is split. Lacking the locating key for the secondary speculum.

95. Reflecting telescope: Thomas Morton, Kilmarnock, 1860 (T1981.39)

14" focus Gregorian reflecting telescope in brass, with speculum metal optics. Stamped on the mirror retention plate 'THOS. MORTON. H.M.R.S.S.A. [Honorary Member of the Royal Scottish Society of Arts] / 1860'. Long external focusing rod, the threaded end working on a boss connected by a slotted key to a removeable internal sliding plate, carrying the secondary speculum on a 3-point adjustment. The primary speculum retained by 3 leaf springs against a projecting ring in the mirror cell. 1" finder telescope mounted over the barrel. The instrument is trunion mounted between 2 pillars, on a sector-shaped wooden turntable base with 2 brass wheels at the rear for running on a circular circular track 50" diameter on a stand (now missing). Wheel-operated cranked altitude movement. Two eyepieces, with provision for locating the eyepiece which is not in use in a threaded mount which is recessed in the base. Push-fit end cover, and finder objective cover.

Length of barrel: 505mm.
Barrel diameter: 103mm.
Speculum diameter: 110 mm.

Lacking the secondary speculum shade, one of the attachment rings for the finder, and the eyepiece solar shades. One of the locating screws for the altitude axis is bent.

Another instrument of the same pattern, and similarly lacking its base, was offered at Phillips, 16 March 1988, Lot 123.

96. Reflecting telescope: Thomas Morton, Kilmarnock, c.1860 (T1981.40)

26" focus Gregorian reflecting telescope in brass, with speculum metal optics. Engraved on the mirror retention plate 'THOS. MORTON H.M.R.S.S.A. / KILMARNOCK.' Long external focusing rod, the threaded end working on a boss connected by a slotted key (missing) to a removeable internal sliding plate, carrying the secondary speculum and tubular shade on a 3-point adjustment. The primary speculum retained by 3 leaf springs against a projecting ring in the mirror cell. 1" aperture finder telescope mounted over the barrel. The instrument is trunion mounted between 2 pillars, on a wooden turntable base with brass skirt, revolving on an iron ring fixed on a mahogany tripod with 3 ogee legs with carved foliate decoration and turned wooden stretcher. Geared altitude movement against decorative curved brass rack mounted on the turntable. Objective cover for finder telescope.

Length of barrel: 757mm.
Barrel diameter: 125mm.
Speculum diameter: 124mm.

The curved brass rack has been broken and restored. Lacking the second eyepiece, the housing for it in the centre of the turntable, the solar shades, the locating key for the secondary speculum, and the push fit end cover. The tube has sustained damage at the junction of the mirror cell. The focusing rod and one of the locating screws for the altitude axis are bent.

Illustrated in *Frank's Book of the Telescope* 6th edition (Glasgow, 1966), 88.

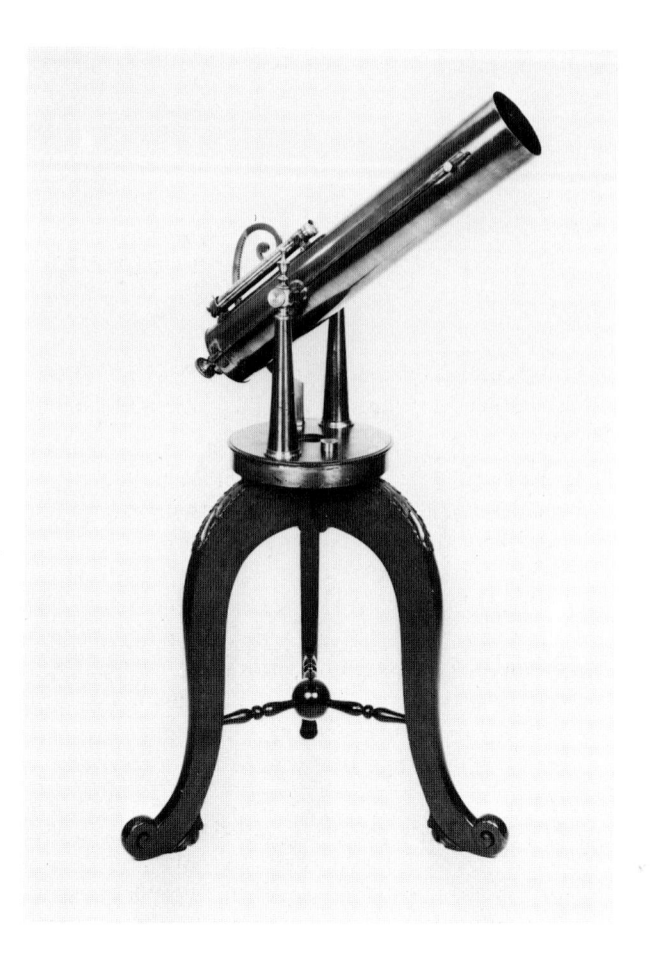

ANDREW BARCLAY

The most recent assessment of Andrew Barclay's life is worth quoting, as it sums up succinctly the problems of that colourful personality:

> 'Barclay, though an inventor with notable achievements, was with his fiercely forked whiskers, a kind of business buccaneer; he created a saga of sequestrations (four in all), falling repeatedly into the hands of indignant creditors and restraining accountants put in by them. The attempts of the latter to get Barclay to think in organisational and marketing terms and to cut out his sideshows, were to no avail. Instead of concentrating on his industrial locomotives, and moving to the production of the more lucrative main line engines, he was seized by the ambition to become a leading Cumberland ironmaster, exposing himself to the vacillations of the iron market and absorbing his slender capital. His ultimate perversity was his obsession with his giant telescope, the costs of which he levied on his firm. Barclay's career is a study in failure, much of it rooted in personality, and ending in the humiliation of destitution.'[1]

Andrew Barclay (1814-1900) was born in Dalry, Ayrshire, son of John Barclay, a millwright, who moved to Kilmarnock in 1817. In 1828, he was apprenticed as a plumber and coppersmith, but spent much of his time 'in the works amongst the machinery of Gregory, Thomson, and Company, in which firm his father was millwright and engineer'.[2] He also developed an interest in astronomy. Even more than in the case of James Veitch of Inchbonny (q.v.), Andrew Barclay's livelihood was not instrument making. His engineering skills and mechanical ingenuity meant that he was able to produce some professional-looking telescopes, but in the strict sense of the word he remained, like his fellow townsman Thomas Morton (q.v.), an amateur.

His professional career has been recounted recently by business, industrial and locomotive historians,[3] so that it is only necessary to provide a brief outline here. In 1840 Barclay went into partnership with Thomas McCulloch, a smith for whom he had previously supplied brasswork; they were in business for two years together making mill shafting and calico printing machinery.[4] Barclay then set up on his own, and began building machinery and engines for various local industries. These included rotary steam engines, high pressure boilers and winding engines for collieries. It was a natural extension of this business when he started to build colliery and shunting locomotives, the first being finished in 1859.[5] 'Between 1863 and 1865 the capacity of the works was more than doubled, allowing for increased production. By 1870 Andrew Barclay was reckoned to be the largest employer in Kilmarnock, with a workforce of 420 and an annual turnover of £70,000.'[6]

After this, however, Barclay's business career took a downward turn; he invested in ironworks in Cumberland 'in connection with which he won and lost an almost princely fortune'.[7] Having overextended his financial resources in a time of economic uncertainty, Barclay compounded the problem by speculation and parts of his concerns were sequestrated. He managed to escape bankruptcy, but despite bravely attempting to remain in business by producing more diverse products such as traction engines and tramway rolling stock, the firm was sequestrated in 1882 with debts of over £24,000. The chief creditor put in new management, and the firm returned profits between 1882 and 1884. However, the depression which affected the engineering industry generally in 1885 led to friction between Barclay and the new management who could not agree on a way forward. Work declined, and again the firm was sequestrated. By 1892, after further business troubles, the firm was converted into a limited company, but in 1893 Barclay was dismissed; by then destitute, he sued the company for part of his salary, and the dispute dragged on until 1899 when it was settled out of court.[8]

It has been claimed that 'despite his catastrophic business career, Andrew Barclay was an able engineer'.[9] He was certainly an innovative engineer, as is borne out by the numbers of his patents which were mostly connected with the main thrust of his business.[10] One patent, however, was for 'Improvements in refracting and reflecting telescopes',[11] applied for in 1854. It must have been at about this time that the Arctic explorer Sir John Ross (1777-1856), and Dr John Lee (1783-1866), then president of the Royal Astronomical Society 'when on a visit to Mr Morton, Kilmarnock's famous philosophical instrument maker, had an interview with Mr Barclay. So well pleased were these gentlemen with the view that they had of the planet Jupiter through a 7 inch Gregorian made by Mr Barclay, that immediately afterwards Sir John proposed him as a Fellow of the Society, and on the recommendation of Dr Lee the motion was carried.'[12] Barclay remained proud of this distinction all his life, although he was removed from the Society's list of Fellows for arrears in 1896, when he was in financial difficulties.

It is possible that the inspiration for Andrew Barclay's telescope-building came from the proximity of Thomas Morton's observatory. However, no connection between the Mortons and the Barclays has been established, and as they were business rivals - Barclay was successfully prosecuted for patent infringement by Alexander Morton and others in 1870[13] - they may not have been on friendly terms.

Barclay did not publish on astronomical matters until almost forty years after his election to the Royal Astronomical Society, so it is difficult to tell whether he was actively constructing instruments and making observations continually throughout that period; it appears that the instruments he did make were for his own use, so that it is probable he made them as and when they were required. His first publication in the field of astronomy was something of a baptism of fire. In a weekly journal, the *English Mechanic*, under the title 'The Unrevealed Wonders of the Heavens' Barclay announced to a disbelieving public his discovery of a number of concentric egg-shaped mountains encircling Jupiter, with illustrations of the phenomena as seen through his 9 inch and 14½ inch telescopes.[14] He also explained that he had had some difficulty in presenting his findings before the world:

> 'I brought this before the Royal Astronomical Society on the 10th December 1880, and they refused to read my paper, as they said no such

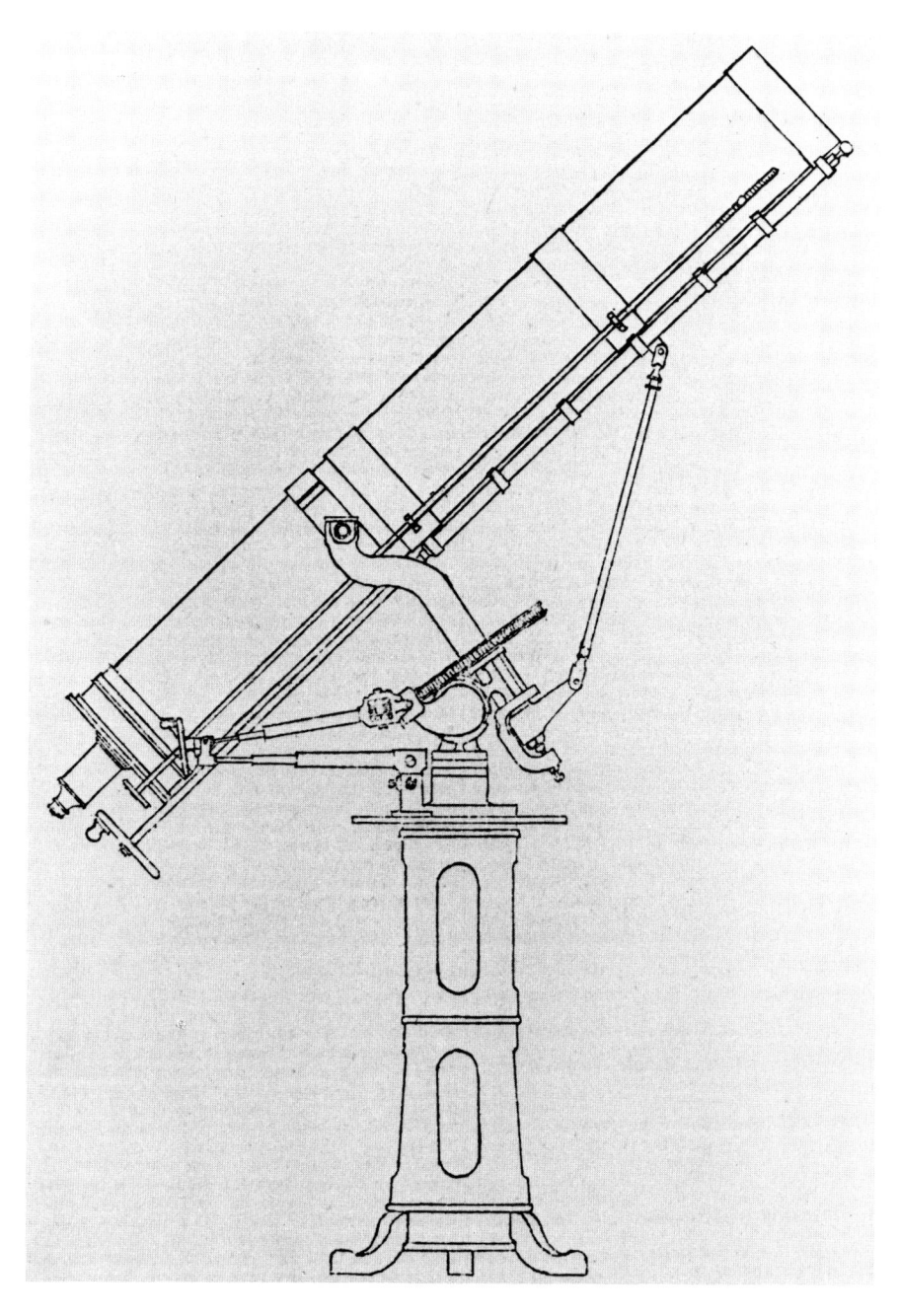

49 Gregorian telescope by Andrew Barclay, 1893. *Trustees of the National Library of Scotland*

50 Saturn, as seen by Andrew Barclay in 1897. Trustees of the National Library of Scotland

thing existed on Jupiter. I pressed the secretary to state this in writing, but he said he was not authorised to say so; but I might say that he, for himself, was of the opinion that there was something wrong with the telescope, because such a thing did not exist on Jupiter, and he had seen Jupiter through a better telescope than mine ... I also went to London on 28th November 1892, to read a paper on the above subject before the British Astronomical Association. I called upon the assistant secretary for the purpose of getting him to put it before the council. He said the telescopes must be distorting the image, as no such things existed as shown in my drawings. He put it before the council, who refused to read the paper. I also showed Lord Kelvin a model of Jupiter which I had made. He advised me to go home and think no more about it, as no such thing existed on Jupiter.'[15]

Barclay's article provoked the scorn and mockery of the *English Mechanic's* corresponding public, in particular from the fortnightly editorial column of 'F.R.A.S.', who wrote:

'I would invite any unprejudiced person simply to read [Barclay's] letter...through, and then ask himself how the Royal Astronomical Society, the British Astronomical Association, Lord Kelvin, Mr Wesley or any other human being having the slightest knowledge of the subject, could have acted otherwise than they are described by Mr Barclay as having acted in connection with his preposterous alleged discoveries! in connection with Jupiter and his system. I can only say that if I had a Gregorian telescope of 9in. or 14 ½in. diameter either that exhibited the great planet as depicted in Mr B.'s Fig.1, I would dispose of the optical part for what it would fetch, and convert the tube into a chimney cowl straightway.'[16]

'F.R.A.S.' can be identified as Captain William Noble (1828-1904), author of numerous astronomical articles.[17] Other correspondents varied from the puzzled to the critical, from the mocking to the supportive; 'F.R.A.S.' verged on the personally abusive. Perhaps the most helpful comment came from O. L. Petitdidier, Mt. Carmel, Illinois who suggested that:

'It seems probable that the metalic speculum has been finished too dry and that the squares or facets of the pitch tool have left an impression of their number and form on the surface. The portions of the surface over which the pitch-squares have been allowed to work over too dry having been overheated, and consequently cut down to a focal length differing from the general focus of the surface. When, therefore, Mr Barclay looks at Jupiter, using the average focus of the speculum, he finds his eyepiece either within or without the focus of the rays sent by the facets, and is therefore viewing Jupiter through the expanded discs of luminous portions of the image furnished by the portions of the surface of varying focal lengths.'[18]

A month after his first announcement, Barclay wrote again, with further illustrations and revelations: Jupiter had more mountain ranges, and Mars a surrounding white ring 'similar to that of Saturn ... [and] a blue spherical-looking mountain to the lower side ... which looks very like a globe attached to the main planet'. In addition, his observations of sunspots had led him to conclude that the Sun's outer surface 'appears to be composed of two thicknesses or shells', the lower one forming a platform and the sunspots acting somewhat like volcanos. He concluded with the brave statement that 'I am quite prepared to prove to the most accomplished astronomers in the world, by ocular demonstration, what I have stated in these columns.'[19] Two weeks later Barclay reported brown smoke issuing from Jupiter's mountains, and Neptune's 'golden round globular appearance, distinctly defined as spherical as a billiard ball'[20] and shortly afterwards used a good many column inches answering specific questions raised by correspondents.[21]

For the next three years Barclay maintained silence, but when he next wrote to the *English Mechanic*, it was clear that he had been busy. Before embarking on further seemingly miraculous observations, he placed before the readership the circumstances of the testing of his telescopes by an amateur scientist of some standing. Through Provost Sturrock of Kilmarnock, Barclay obtained an introduction to Sir Archibald Campbell Campbell, F.R.S., first Baron Blythswood (1835-1908), a West of Scotland politician who made valuable contributions to astronomy and physical science,[22] and was put in touch with Otto Hilger (1850-1902) who ran Blythswood's private laboratory between 1888 and 1897.[23]

In April 1894 Otto Hilger visited Kilmarnock and 'spent a whole day examining three telescopes, a 9in and a 14in Gregorian and a 4½in achromatic, for spherical aberration and definition, and he stated in the presence of my assistants that he would have to give it up, as he could find no fault with them - they were beautiful'.[24] Barclay reproduced part of the correspondence from Blythswood and Hilger, including a letter from Hilger 'which surely speaks for itself' dated 20 May 1894:

> 'I wanted to ask you if you are willing to sell some of the instruments you have on stock, to let me have the diameters, number of eyepieces of each telescope, and price. I may be able to dispose of same. I [have] got a friend who wanted me to see about a 5in achromatic. Well, I think he could not do better than to get one of your telescopes.'[25]

However, neither Blythswood nor Hilger appeared to have contacted the *English Mechanic* independently after this article appeared to corroborate or counter Barclay's claims for his telescope: 'about the mirrors no complaint was ever made: they are as bright and true to-day as on the day they were made - some of them more than forty years ago'.[26] The original correspondence is not known to have survived. In the same article Barclay described more mountains on Jupiter, a large double valley around the equator of Saturn, and further 'Unrevealed Wonders of the Heavens'. These did not produce the same vociferous reaction as his earlier piece, and it appears to have been his final contribution to amateur astronomy.

Barclay seems to have built a number of telescopes, but only two are known at present. The nine inch reflector illustrated in the *English Mechanic* in 1894 would appear to be the unsigned instrument donated by his son to the Dick Institute, Kilmarnock, and is now at the Royal Observatory, Edinburgh.[27] The 14½ inch Gregorian reflector which appears as a line drawing in the same article is similar to one offered by a Paris dealer.[28] In addition, Barclay mentions that:

> 'I am finishing a 9 inch achromatic telescope just now, the glass discs by Messrs Chance Brothers, of Birmingham. I have finished two instruments of 5in diameter, one of which I would not be afraid to put against the best telescopes yet known, as it defines objects on the planets, which no one has yet done...I have six pairs of 9in speculums, and two pairs of 14½in finished, that is large and small speculums to suit each other, all of which show what I have stated both in Jupiter and Saturn, and I have one 22in diameter in progress.'[29]

Dr Lee and Sir John Ross saw Jupiter 'through one of my 7in telescopes over thirty years ago',[30] which implies that Barclay had made a number of instruments of that size; the six pairs of completed nine-inch speculums had increased to 'nine pairs of speculums of this size' by the following year.[31] In 1897 Otto Hilger examined a nine inch and a fourteen inch reflector, and a 4½ inch refractor.[32] Barclay also made a micrometer 'to suit my 14½in Gregorian telescope, and which I consider second to none in the world. It has a clear comb divided into hundredths of an inch, with usual screws divided into hundredths; all complete with tangent screw, &c., and has a clear field of over 17½ minutes with a power of 200, all in perfect focus, and having two vertical and two horizontal spider-threads or wires.'[33]

Yet Barclay was out of touch with the contemporary world of amateur astronomy: he clung to the old-fashioned metal speculum Gregorian form of instrument when silvered glass reflectors led the field (although, to be fair, this was because he had the facilities for metal working, and not for glass); he plainly had little idea of what he was describing, and his ignorance was exacerbated by natural stubborness and the failings of old age: 'I have spent over £10,000 to find out how to finish metallic speculums and mix and cast the metals. I have made over 2000 experiments in connection therewith.'[34] Such expense and effort at a time when creditors were baying at his heels leads to the conclusion that Barclay possessed supreme self-confidence.

Andrew Barclay was twice married, firstly to Janet Campbell, and then to Effie McVean; he died at his home Argyle House on 20 April 1900, at the age of eighty five.[35]

REFERENCES

1. Sydney Checkland, 'Vehicles' in Anthony Slaven and Sydney Checkland (eds.), *Dictionary of Scottish Business Biography 1860-1960* Volume 1 The Staple Industries (Aberdeen, 1986), 249-250.

2. Obituary of Andrew Barclay, *The Engineer* 89 (1900), 437.

3. The most recent assessment of Barclay's business career is given in Michael S. Moss, 'Andrew Barclay' in Slaven and Checkland, *op.cit.* (1), 253-254; an industrial history is Michael S. Moss and J.R. Hume, *Workshop of the British Empire - Engineering and Shipbuilding in the West of Scotland* (London, 1977), Chapter 4 'Andrew Barclay Sons & Co. Ltd., Locomotive Manufacturers and Engineers', 68-84; and a locomotive history is Russell Wear, 'The Locomotive Builders of Kilmarnock', *Industrial Railway Record* No. 69 (January 1977), 325-347.

4. *Kilmarnock Standard*, 21 April 1900.

5. Wear, *op. cit.* (3).

6. Moss, *op. cit.* (3).

7. *Kilmarnock Standard, op. cit.* (4).

8. This brief sketch of Barclay's business career is derived from Moss, *op. cit.* (3).

9. *Ibid.*

10. Between March 1843 and November 1895, Barclay took out a staggering 75 UK patents.

11. UK patent 2008, 15 September 1854.

12. *Engineer, op. cit.* (2).

13. *Engineering* 9 (1870), 225, 242, 278 and 295.

14. Andrew Barclay, 'The Unrevealed Wonders of the Heavens Jupiter, Mars, Saturn and the Sun', *English Mechanic and World of Science* 58 (1893), 198-200.

15. *Ibid.*, 199.

16. *Ibid.*, 222.

17. *Kilmarnock Standard*, 3 March 1894. Noble's 'Obituary' relates how after a military career he became a dedicated amateur astronomer; he was a founder member of the British Astronomical Association and its first president; he was elected F.R.A.S. in 1855, and 'for nearly forty years he contributed each fortnight to the *English Mechanic* an article, over the signature 'A Fellow of the Royal Astronomical Society' in which he criticised with some freedom current astronomical events. It must be admitted that he was not entirely free from prejudice, but the transparent honesty of his purpose gave value to these outspoken utterances. At the same time he took pains to answer the questions of interminable correspondents and inform them on astronomical and other subjects.' *Monthly Notices of the Royal Astronomical Society* 65 (1905), 342-343.

18. *English Mechanic and World of Science* 58 (1893), 371.

19. *Ibid.*, 308-309.

20. *Ibid.*, 372.

21. *Ibid.*, 440.

22. Andrew Barclay, 'Unrevealed Wonders of the Heavens ...', *English Mechanic and World of Science* 65 (1897), 218. For Campbell, see T.E. J[ames], 'Archibald Campbell, Lord Blythswood', in S. Lee (ed.), *Dictionary of National Biography*, 2nd supplement, I (London, 1912), 298-299.

23. Obituary of Otto Hilger, *Monthly Notices of the Royal Astronomical Society* 63 (1903), 199. Otto Hilger returned to London in 1897, after the death of his brother Adam, in order to take charge of his instrument making business.

24. Barclay, *op. cit.* (22), 218.

25. *Ibid.*, 218-219.

26. *Ibid.*, 219.

27. *Ibid.*, 58 (1893), 198.

28. *Ibid.*, 199. The illustration in Alain Brieux, Paris, [Catalogùe des] Instruments Scientifiques (November, 1983), 52, item 37, signed 'Andrew Barclay, Engineer, Kilmarnock', differs in mounting and stand from the line drawing of 1893.

29. *English Mechanic and World of Science* 58 (1893), 200.

30. *Ibid.*, 309.

31. *Ibid.*, 440.

32. *Ibid.*, 58 (1897), 218.

33. *Ibid.*, 58 (1893), 308. This may be the micrometer in the Frank collection, Royal Museum of Scotland (NMS T1980.281).

34. *Ibid.*, 198.

35. General Register Office (Scotland), Register of Deaths 1900 (Kilmarnock), 597/192. His cause of death was given as dropsy and heart disease.

97. Telescope micrometer eyepiece: Andrew Barclay, Kilmarnock, c.1880 (T1980.281)

Telescope eyepiece assembly in brass, comprising: doublet erecting lens in a cylindrical housing for mounting in the focusing mechanism; focal plane micrometer; and eyepiece with eye relief stop and screw cover. The micrometer housing is engraved 'Andrew Barclay F.R.A.S./ Engineer./ KILMARNOCK./MAKER.' An external tangent screw allows for complete rotation of the micrometer against a silvered scale divided to degrees. A rack and pinion motion translates the micrometer across the field of view to position a notched scale of 2 inches subdivided to 0.01" (together with a fixed vertical and 2 horizontal wires), and also a single moveable vertical wire stretched over an inner sliding frame which is controlled by micrometer screws at each end of the carriage and further divided on their silvered perimeters into 100 parts.

Overall width: 310mm.
Body diameter: 107mm.

Lacking all micrometer wires. Only traces of the axial scale silvering remain. Eyepiece end cover appears to be from a different instrument.

This micrometer tallies with Barclay's 1893 description of a telescope micrometer made for his 14½ inch Gregorian reflector, mentioned above.

16

'CARVERS AND GILDERS'

The three firms of carvers and gilders grouped together here represent aspects of the retail outlets of the wholesale instrument trade. This trade was aimed principally at customers whose interest in buying pieces such as barometers, thermometers, telescopes and microscopes was not scientific, but, as a modern historian of the barometer has argued, stemmed instead from a widely popular taste for such items as pieces of furniture. Increased wealth among the middle classes and their desire to furnish their homes comfortably, have been ascribed as the causes of the boom in the production of the wheel barometer (the most popular form of the domestic barometer) from the early nineteenth century as a concomitant of the expansion of the furniture industry.[1]

Also of some significance may have been the popularisation of science during the same period, a movement which gripped the public imagination both with individual discoveries (including the Royal Navy's several voyages of exploration) and with co-operative endeavours such as the formation of the British Association for the Advancement of Science in 1831. Interest in the physical sciences was reflected in the many provincial philosophical societies active in the first half of the nineteenth century, and even a cursory reading of contemporary scientific periodicals reveals that serious, almost obsessive attention was being given to climatological and meteorological study throughout Europe. In both these respects Scotland was no exception. It was therefore not a coincidence that the two most easily used of meteorological instruments, the barometer and the thermometer, sold in huge numbers.

High demand led in turn to mass production and poorer standards, and to the concentration of manufacture in the hands of wholesalers, who supplied retailers in smaller centres with their goods. This was the case with London and the provinces,[2] and it has been argued that the same was true of Scotland where the local trade had to compete with London makers.[3] From about the second decade of the nineteenth century the distribution of barometers passed from the hands of scientific instrument makers into those of opticians, mirror and hardware retailers and carvers and gilders. In part this was because the domestic and the scientific barometer evolved separately from the late eighteenth century.[4] Although some Scottish makers such as the Adies (q.v.) and the Gardners (q.v.), who produced specialised instruments for professional customers, also continued to make pieces for the non-professional customer, the barometer trade in Scotland seems generally to have followed its English counterpart.

It was similar in another respect. As Goodison has pointed out, migrant Italians were already well established in England by the early nineteenth century as barometer makers, and by 1840 had come to dominate the trade. The particular skills they brought with them were glass-blowing and carving and gilding, both of which extended naturally into looking-glass and barometer manufacture.[5] Not only did they produce capillary tubing for barometers, but also blown-glass items such as specific-gravity beads. The three Glasgow carvers and gilders who are described here are not known to have been related commercially or by marriage; each had a similar business and each was probably buying much of its stock from English wholesale firms. Although it has been recently shown that most barometer makers of Italian extraction came from the mountainous area around Como in Lombardy, the Glasgow community 'had a strong presence of people from Tuscany, especially from the commune of Barga near Lucca; but ... also ... from the Liri Valley, ... from Liguria and the southern region of Abruzzi'.[6] The three businesses discussed here are probably representative of the several other Italian carvers and gilders and opticians in business during the early and middle decades of the century, and presumably formed a significant proportion of the 353 'foreigners' (out of a total population of 202,426) recorded as living in Glasgow in 1831.[7]

REFERENCES

1. Nicholas Goodison, *English Barometers 1680-1860* 2nd edition (Woodbridge, Suffolk, 1977), 83-85.

2. *Ibid.*

3. D.J. Bryden, *Scottish Scientific Instrument-Makers 1600-1900* (Edinburgh, 1972), 15.

4. Goodison, *op. cit.* (1), 83-85.

5. *Ibid.*, 86-88; and Nicholas Goodison, 'The Foreign Origins of Domestic Barometers 1800-1860', *Connoisseur* 170 (1969), 76-83.

6. Lucio Sponza, *Italian Immigrants in Nineteenth-Century Britain: Realities and Images* (Leicester, 1988), 33.

7. Quoted in Bruce Lenman, *An Economic History of Modern Scotland* (London, 1977), 106.

ANTONI AND JOHN GALLETTI

Antoni Galletti first appeared in 1805 at 10 Nelson Street, Glasgow as a carver and gilder,[1] but his trade was certainly wide enough to include mathematical, optical and philosophical instruments. Specific claims were indeed made by him, and his son John who succeeded him in 1850 or 1851, that they manufactured philosophical glassware, and that it was available wholesale as well as retail.[2]

In later years John advertised the firm as having been established in 1789. Our first record of Antoni Galletti, however, is in 1798 when he was working as a barometer maker in Edinburgh, and was required to register as an alien under the terms of a Royal Proclamation made earlier that year in response to the troubled situation in France.[3] His trade was given as 'barometers or prints' and his address as in lodgings at the foot of the Canongate. He had arrived in Britain at Dover at an unspecified date, having previously worked in Milan. He gave his age as twenty-one, which would have made him only twelve in 1789, and so 1789 is more likely to have been the date he entered training either in Italy or Scotland. Given that he was later involved in manufacture rather than merely in retail, it is possible that he was working in 1798 for one of the long established Italian glassblowers in Edinburgh: the two most likely candidates being the elderly Charles Molinari (or Moliner), originally from Como, who had come to Britain in 1752 and worked in the Netherbow as a 'barometer and philosophical glass maker', and the barometer and specific gravity bead maker Angelo Lovi, 'glass manufacturer' from Milan, who had arrived in 1772 and worked in Niddry Street.[4] The Molinari business appears to have closed on his death or retiral in 1801,[5] and this may have been the stimulus that led Antoni to set up on his own in Glasgow. He may no longer have been in Edinburgh in 1803 when Charles Galletti, print seller aged twenty-six from Como, was recorded in the Register of Aliens: Charles lodged instead with another carver and gilder, Dominick Stampa, in Leith Walk.[6] In 1807 Charles Galletti appeared briefly in the Glasgow directory as a carver and gilder, whose stock included specific gravity beads,

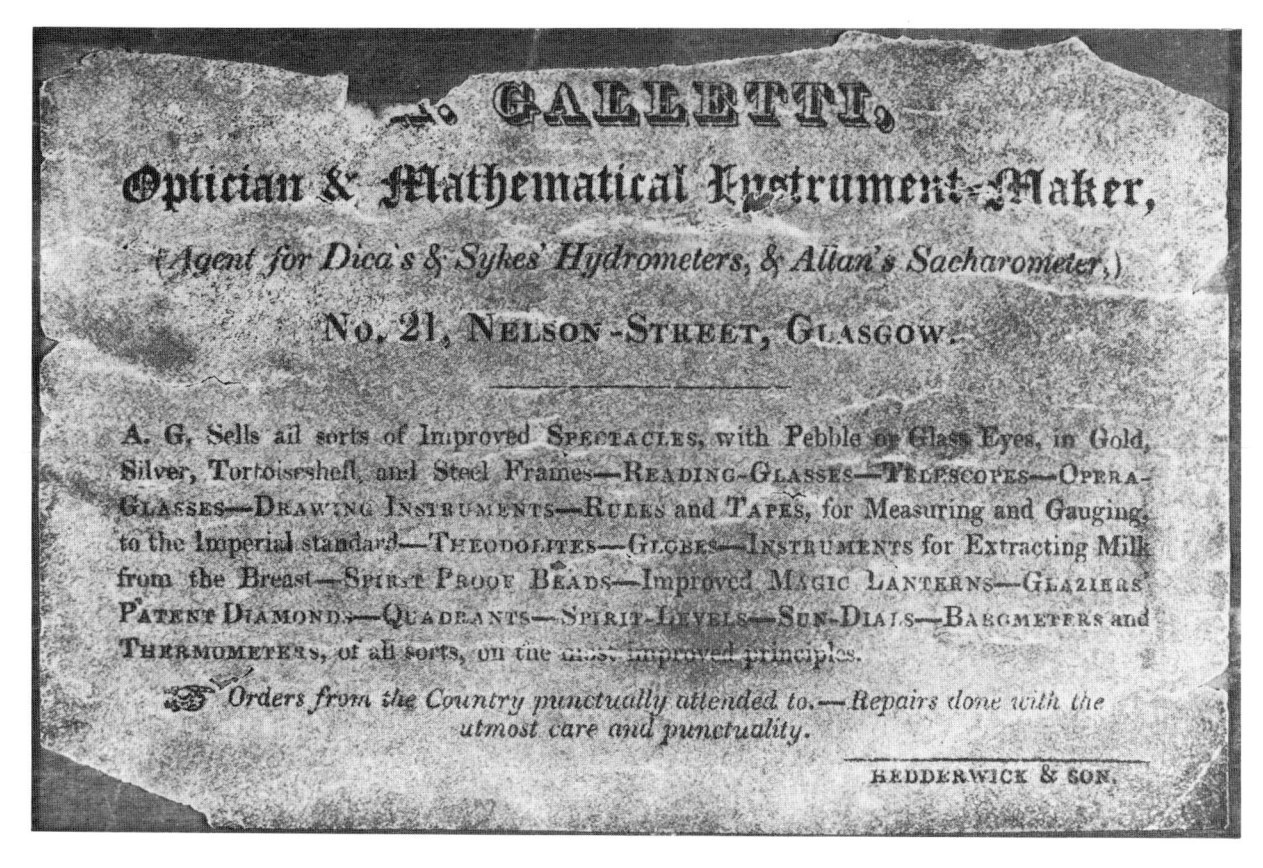

51 Antoni Galletti's trade card for his address between 1826 and 1828. Private collection

and it seems likely that he was closely related to Antoni.[7]

The first Glasgow address recorded for Galletti is 10 Nelson Street in 1805, with 21 Nelson Street in addition from 1826 to 1828. In 1828 he moved to the new Argyll Arcade development off Buchanan Street, initially occupying Nos. 24 and 25, but No. 24 alone from 1843 until his final appearance in the directory in 1850.[8] He was initially described only as a carver and gilder: 'optician' was added only from 1829, and 'mathematical instrument maker' only from 1846.[9] However, a surviving trade card for the 21 Nelson Street address of the late 1820s shows him simply as 'Optician & Mathematical Instrument-Maker'.[10] The range of stock listed in this advertisement includes optical and surveying devices and shows that he was involved with blown-glass apparatus as well as measuring equipment calibrated to the newly-introduced Imperial Standards. The importance of hydrometry instruments for his business is underlined by his description as 'Agent for Dica's & Sykes' Hydrometer, & Allan's Saccharometer'. The instruments of Bartholomew Sykes were legally sanctioned (those of John Dicas being the approved American pattern), whereas Alexander Allan's instrument, devised in Edinburgh in 1805 by the chemist Thomas Thomson, saw widespread but unofficial use by the Scottish Excise.[11] Two sets of specific gravity beads for his 10 Nelson Street address have been recorded, and both have instructions in the lid which note that they are 'Made and sold, wholesale and retail, By A. Galletti ...'.[12]

Antoni was succeeded in the business by his son John (c.1821-1894), who advertised himself in 1851 at the same premises as 'Optician, mathematical instrument-maker, barometer, thermometer, and looking-glass manufacturer, wholesale and retail, 24 Argyle Arcade'.[13] At the time of the census in 1861 John was employing one girl and one boy.[14] Amongst other instruments that have been recorded, sets of Galletti specific gravity beads from this period have survived, but it is no longer possible to say with any certainty that they were made by the firm.[15] Indeed, the emphasis of Galletti's trade appears to shift progressively away from scientific instruments in the 1870s. In 1883 for instance, he described the business as 'Optician, model dock-yard, model engine depot, and emporium of novelties'.[16]

John Galletti continued trading until his death on 25 August 1894, aged seventy-three. He married twice: first, Helen Prendergast, from Ireland and secondly Virginia Foucart, who survived him.[17]

REFERENCES

1. *Glasgow Post Office Directory* 1805.

2. See below, at references 12 and 13.

3. Edinburgh City Archives, 'Register of Aliens 1798 [-1803]', f.22. We are grateful to James Gilhooley for drawing this source to our attention.

4. *Ref. cit.* (3), ff. 46, 73.

5. D.J. Bryden, *Scottish Scientific Instrument-Makers 1600-1900* (Edinburgh, 1972), 54.

6. Edinburgh City Archives, Register of Aliens 1803-1825, f.9.

7. *Glasgow Post Office Directory* 1807. There is a set of specific gravity beads in the Royal Museum of Scotland, with trade card for 'CHAs. GALLETTI & Co/No 82. GLASSFORD-STREET GLASGOW', NMS T1967.32.

8. *Ibid.* 1805-1850.

9. *Ibid.*

10. Trade card attached to a zograscope in private hands, recorded by the Museum, 1974.

11. Bryden, *op. cit.* (5), 15; F.G.H. Tate, *Alcoholometry* (London, 1930), 7, 12.

12. NMS T1988.20, purchased at Phillips Edinburgh, 29 April 1988, Lot 98. A slightly larger set was offered for sale by Tesseract, Catalog 10 (Spring 1985), item 63.

13. *Glasgow Post Office Directory* 1852.

14. General Register Office (Scotland), Census of 1861 (Glasgow, Central), 644^1/15, p.18.

15. Examples of these are in the Science Museum, London, inventory 1909-203, and the Museum of the History of Science, Oxford, inventory 50-28. A Cary-type microscope retailed by John Galletti is in the Whipple Museum of the History of Science, Cambridge, inventory 779; it is described in O. Brown, *Catalogue 7: Microscopes* (Cambridge, 1986), item 178. A box sextant, signed 'Galletti' was offered for sale at Sotheby's Belgravia, 30 November 1979, Lot 215. Apart from the telescope in the Frank Collection, NMS T1980.252, which was almost certainly a bought-in piece, a number of barometers have been noted: wheel barometers with his signature were offered at Phillips, 16 September 1986, Lot 4, and Sotheby's Sussex, 19 June 1986, Lot 1949; a stick barometer was offered at Sotheby's Chester, 21 May 1981, Lot 257. A 3 inch retailed pocket globe in a private collection has been recorded.

16. *Glasgow Post Office Directory* 1883 and *passim*.

17. G.R.O.(S.), Register of Deaths 1894 (Glasgow, Partick), 646^3/573. The cause of death was given as apoplexy.

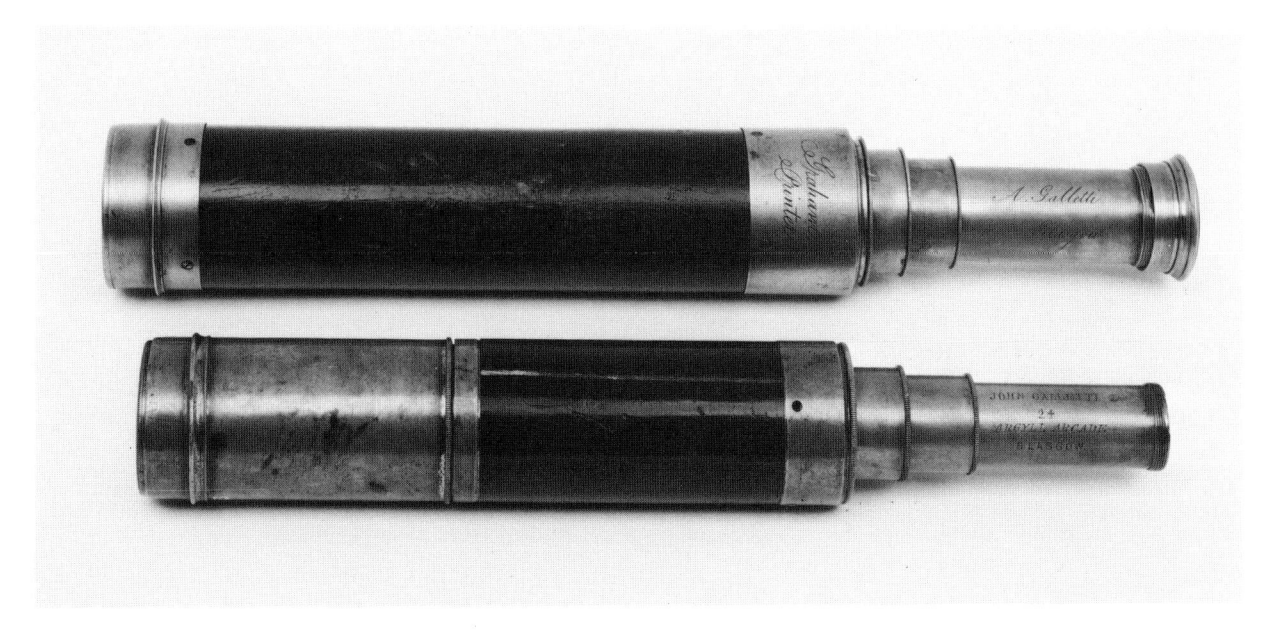

98 top, 99

98. Refracting telescope: A. Galletti, Glasgow, c.1840 (T1980.252)

1½" aperture 3-draw hand-held refracting telescope with doublet objective and 2-component eyepiece and erecting assemblies (defective); in brass, with wooden barrel. Engraved on the first draw tube 'A. Galletti / Glasgow' and on the barrel 'Graham./ Printer.' Scratched with the assembly mark 'XVII' on the draw tubes, 2 of the draw slides and the erecting assembly.

Length closed: 240mm.
Tube diameter: 46mm.
Aperture: 39mm.

Lacking one component of the image erecting assembly, and also the objective cover.

Throughout the period 1828-50 when Antony Galletti's address was in Argyll Arcade, John Graham, printer, was nearby in the Trongate, at 124 until 1839 and thereafter at 181.

99. Refracting telescope: John Galletti, Glasgow, c.1860 (T1980.253)

1¼" aperture 3-draw hand-held refracting telescope with doublet objective and 2-component eyepiece (defective) and erecting assemblies; in brass, with objective ray shade and wood-veneer covered barrel. Engraved on the first draw tube 'JOHN GALLETTI /24 / ARGYLL ARCADE / GLASGOW.' Scratched assembly mark 'V' on the first draw tube and draw slide, and 'VII' on the other two.

Length closed: 202mm.
Tube diameter: 42mm.
Aperture: 34mm.

Lacking both lens components of the eyepiece assembly, and also the external eye stop and objective cover. The veneer covering of the barrel has contracted and split.

CHARLES GERLETTI

Charles Gerletti first appeared in the directory as a looking-glass manufacturer at 156 Saltmarket, Glasgow in 1828. He worked at that address until 1832; in 1833 he was listed at 153 Saltmarket, and from 1834 until 1839 at 145 Saltmarket. He worked at 10 Candleriggs Street from 1840 until 1848, after which date he disappeared from the directory to be succeeded by Dominick Gerletti in the same premises and business, who advertised himself as, among other things, an optician and a firework artist. Others of the family included John Gerletti, a barometer and thermometer maker at 95 Candleriggs, Glasgow in 1853, who was perhaps Charles's son.[1] A number of barometers with Gerletti signatures have been noted.[2]

REFERENCES

1. *Glasgow Post Office Directory* 1828-1835.

2. Besides the example in the Frank collection, NMS T1980.212, a 10-inch wheel barometer signed 'Gerletti Glasgow' was offered for sale at Sotheby's, 10 December 1981, Lot 237; a stick barometer with a similar signature was offered by Sotheby's, 5 December 1985, Lot 152; a wheel barometer, signed 'C. Gerletti, Glasgow' was offered by Sotheby's, 4 February 1977, Lot 36; another, with a similar signature but with a mother of pearl inlaid rosewood 8-inch dial was offered by Bonham's, 22 March 1984, Lot 89; a wheel barometer signed 'G. Gerletti, Glasgow' was offered by Christie's Glasgow, 16 November 1983, Lot 32; another signed 'D. Gerletti, 10 Candleriggs Street Glasgow' was offered by Christie's Glasgow, 3 June 1987, Lot 53; and another signed 'C. Gerletti, Glasgow' was offered by Sotheby's Chester, 10 November 1988, Lot 2502.

100. Wheel barometer: C. Gerletti, Glasgow, c.1840 (T1980.212)

Mercury wheel barometer, with 8″ silvered dial with setting indicator, reading from 28″ ('Stormy') to 31″ ('Very Dry') divided to 0.01″, convex mirror, spirit thermometer calibrated 20°-130°F, hygrometer calibrated [25]-0-[25], and spirit level. Stamped on the spirit level mount 'C. GERLETTI/ GLASGOW'. Veneered pine body with a scroll pediment, and long door at rear giving access to siphon tube. Scratched 'XII' inside rear door, and 'XVI' behind thermometer and in thermometer recess.

Overall height: 970mm.

The hinges on the rear door are replacements. Lacks hygrometer oat-beard and indicator, and also top finial. The main glass and bezel replaced from another instrument.

This instrument can be identified as one of the many domestic wheel barometers offered by the London wholesaler J.J. Hicks in his *Illustrated & Descriptive Wholesale Catalogue of Standard, Self-Recording, and other Meteorological Instruments ...* (London, n.d. [c.1880]), 38, no. 92, 'Wheel Barometer, 8-in. silvered metal dial, convex mirror, spirit level, oat-beard hygrometer, and attached spirit Thermometer 15s 6d'.

J. & M. RIVA

The Riva family first appeared in the directory as A. Riva & Co., carvers and gilders at 70 High Street, Glasgow in 1823. In 1825 the business appeared as J.& M. Riva, carvers and gilders, at 143 High Street, where they remained until 1849, when they moved to 147 High Street. A final move took them to 63 John Street, in 1857, where they worked until they disappeared from the directory after 1861.[1] Like other carvers and gilders they also made looking glasses (advertising from 1835), and barometers and thermometers (advertising from 1850 and 1857 respectively).[2]

REFERENCES

1. *Glasgow Post Office Directory* 1823-1861. M. Riva was Michael Riva, who advertised under his own name in 1826. The business had a second shop at 249 Argyle Street between 1844 and 1846.

2. Three barometers with J. & M. Riva's signature have been noted: a wheel barometer offered for sale by Sotheby's, 29 November 1974, Lot 33; another, with a plaque instead of a thermometer, inscribed 'The Property of Robert Hannan of Anderstown Brewerie, Born 1777, Died 1860', offered by Phillips, 16 September 1981, Lot 6; and a third, offered by Sotheby's Chester, 10 November 1988, Lot 2501, signed 'I. & M. Riva, Glasgow'. It is possible that, like the telescope in this collection (NMS T1980.254) and another offered for sale by Christie's South Kensington, 11 August 1988, Lot 24, the barometers were retailed.

101. Refracting telescope: J. & M. Riva, Glasgow, c.1830 (T1980.254)

1½" aperture 3-draw hand-held refracting telescope with doublet objective (chipped) and 2-component eyepiece and erecting assemblies; in brass, with painted wooden barrel. Engraved on the first draw tube 'J & M Riva / Glasgow'. Objective cover with 1¼" aperture stop, and with sliding closures on the objective cover and external eye stop.

Length closed: 277mm.
Tube diameter: 62mm.
Aperture: 39mm.

The forward ferrule on the barrel has been distorted by a blow; the objective mount has also been damaged and the lens retaining ring has broken. The rear ferrule has become detached and the barrel is split.

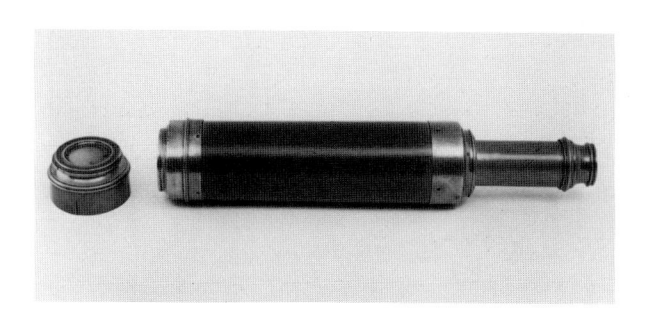

17
RICHARD MELVILLE, MAKER OF SUNDIALS

The art of dialling, that is the design and construction of sundials, has a long tradition firmly based in celestial geometry. It was as a mathematical science that the subject was promoted in early texts as well as by the eighteenth century popularisers of science such as James Ferguson and Benjamin Martin.[1] Although complex dials, and particularly portable dials, are readily understood as scientific instruments, the simple horizontal garden dials of the eighteenth and nineteenth centuries are perhaps more properly classed as luxury items or simply garden ornaments. Arguably, they are scientific pieces only in the restricted sense that they were almost invariably produced by scientific instrument makers, and characteristically they were engraved on the instrument makers' raw materials, brass and bronze. In Scotland, a dial-making tradition separate from that of the London workshops evolved: but one would hesitate to describe its impressive monumental polyhedral dials, cut by masons in stone, as scientific instruments, in spite of their undoubted mathematical foundation.[2]

52 Sundial, attributed to Richard Melville, at Polmadie, Lanarkshire

Richard Melville, a specialist sundial maker working in slate and active in central Scotland in the 1840s, might perhaps best be described as a slate draughtsman. His horizontal dials were not technically sophisticated but they clearly had novelty value at the time. They were of a very distinctive design, made in an unusual material and often of an impressive size, and incorporated a great deal of topical geographical information. Presumably, as they were cut in a comparatively soft material, they were also cheaper than equivalent dials in metal. Melville may have been fortunate in that this was a period in which there was much activity in central Scotland in garden landscaping and in house building and alteration.[3] His dials appear to have found a receptive market and a number are recorded at substantial properties.

Although sundials are notoriously prone to being removed, and in addition slate dials are particularly vulnerable to damage, enough Melville dials have survived to make it clear that he made a considerable number. The prices that he charged for his dials are, however, not known. Nor is it known whether, or for how long, dial production represented his principle source of income; his sole appearance in the Glasgow street directory, in 1846, described him as 'Sundialmaker'.[4]

It is presumed, however, that dial-making remained financially rewarding and that sufficient commissions were forthcoming, because after leaving Scotland in about 1850, Melville spent several years travelling around England before moving to Dublin in the early 1860s, and during this time sundials are the only product associated with his name. The latest of the dials we have recorded, made in Dublin in 1871, rather more than thirty years after the earliest, is numbered 2662.[5] If this is indeed a simple serial number, it implies a production rate of on average about one dial per week over his known period of work, which is at least plausible considering the amount of engraving work involved, if not the problem of securing such a large number of commissions.

Melville's dials are of a fairly standardised form. The great majority are engraved on oblong slabs of slate (although three circular or octagonal dials are known). The six examples in the Royal Museum of Scotland all have the purplish colour characteristic of imported Welsh slate.[6] In most examples (and certainly all early examples) the edge of the principal gnomon intersects the dial plate at its centre, so that the hour ring divisions are symmetrical about the 6 a.m. and 6 p.m. markings. Outside the hour ring are engraved the names of forty to eighty places around the globe so as to indicate the time of their local noon. The latitude for which the dial is constructed (and hence the angle of the gnomon) is almost invariably inscribed on the dial.[7] To a limited extent, the longitude of use may also be determined; for example, a dial for use in Glasgow would not include Glasgow in the array of place names, but probably would include Edinburgh.

A characteristic feature of Melville's dials is the use of smaller additional dials placed around the central dial displaying the local time of particular places. On the smaller sundials the four locations shown are New York, Alexandria, Malabar (or the Isle of Borneo) and New Zealand; but on larger sundials there may be eight subsidiary dials with the four corner dials each displaying the time at two locations, making a total of twelve.

Remaining dial space is taken up with a tabulated 'equation of time' for correction of the dial readings throughout the year, with decorative motifs, and with latinised mottoes. On earlier dials Melville used the mottos 'Tempus Fugit / Memento AEternitatis' (time flies, remember eternity), whereas on later dials he favoured 'Horas non numero nisi serenas' (I count the bright hours only), together with a quotation from St. Paul's Epistle to the Ephesians, chapter 4: 'Sol non occidat super iracundiam vestram' (Let not the sun go down upon your wrath).

Study of Melville's dials has been hampered not only by his movement around the country, but also because he used the two surnames 'Melville' and 'Melvin'. There is nothing sinister in this apparent use of an alias: the two are in fact equivalent and interchangeable forms of the same name.[8] He seems to have used the form that was more acceptable in the locality in which he was working, and in Glasgow this was 'Melville', whereas after his move to England it was always 'Melvin'. The first association of the Glasgow 'Melville' dials with the later 'Melvin' dials was made by Eden and Lloyd in their 1900 extended edition of Mrs Elizabeth Gatty's *Book of Sundials*, although they made no comment about a possible relationship of the two groups.[9] More recently W.A. Seaby noted 'Melvin' and 'Melville' variant names on Ulster dials, and this has led to an entry for Richard Melvin as a sundial maker in Belfast in Brian Loomes' 1976 supplement to G.H. Baillie's *Watchmakers and Clockmakers of the World*.[10]

It is now clear on stylistic and other grounds that the Ulster dial maker is the same as the Richard Melville who subsequently worked in Scotland. Indeed, an Ulster origin for Melville goes some way to explaining a specialisation in dial making. Although there was at least one notable Scots peripatetic dial maker working in slate, John Bonar of Ayr, active in Scotland and Ireland in the early seventeenth century, there is no marked tradition of slate engraving for dials.[11] By way of contrast, slate dials were popular in Ulster, and a considerable number of horizontal and vertical slate dials are preserved in several collections in Northern Ireland.[12] Some of these have a naive charm which suggests they were cut by amateur hands, but others show highly competent engraving skills. They are frequently signed and dated (the dates ranging from the 1790s) and are often engraved with the latitude for which they have been constructed; some even exhibit the subsidiary dials that were the particular hallmark of Melville's dials.[13]

Not infrequently vertical dials are found mounted on the walls of early nineteenth century country churches,[14] which perhaps suggests that the slate workers who engraved them were also employed in the engraving of the slate grave stones which found favour at this time. Although slate was produced locally in Ulster, notably in County Down, the early nineteenth century saw rapidly increasing imports from the Welsh mines of domestic roof slate (and no doubt larger pieces suitable for gravestones), with a fierce price-cutting war reaching a peak during the 1830s.[15] It was against this background that Richard Melville developed sundial making into a speciality, his clients principally being the newly prosperous merchant and banking class, whose mansions, built on the proceeds of the linen industry, were appearing in increasing numbers in Antrim and Down.[16]

The registers surveyed by the International Genealogical Index show a small concentration of Melvins in Seaforde parish, near Newcastle, County Down, with entries from the 1770s, and it seems likely that this was Richard Melville's birthplace. Although his own baptism and marriage are not in registers that have survived, the Index has noted the baptism of a daughter to a Richard and Agnes Melvin in 1835 in Seaforde.[17]

Three dials by Richard Melville of the late 1830s and early 1840s have been recorded in Ulster, but although the three are known to be stylistically similar, adequate details are available on only the last of these, dated 1842.[18] This shows many of the standard features of the Scottish dials, including the distinctive bifurcated motif which is, for example, seen on the dial in the Frank collection, but the dial design appears to be at an intermediate stage and other aspects that were to become characteristics of the fully-fledged designs are absent. In particular, the equation of time correction is described but not yet tabulated, and the band of geographical locations has yet to be introduced. Another feature which separates this group of Melville dials from later examples is that they each carry the name of his client. An 1838 dial was for the merchant Robert Calwell of Annadale, County Down.[19] William Mann, the client of an 1842 Melville dial, has not been identified.[20] William Dobbin (1800-1886) of Ternascobe, County Armagh, owned the 1842 dial described above.[21] Although surviving Scottish dials normally have a vacant area where the purchaser's name could be recorded, Melville's later clients may have been from more established families who did not feel the same need to record their ownership in this fashion.[22]

Only one earlier example of Melville's work has been located. A dial of 1832 made for John Sandwich, of Downpatrick, just a few miles from Seaforde, and signed with the name Richard Melvin, is in the Down Museum, Downpatrick.[23] It is a very simple horizontal dial, with no subsidiary dials, made for a stated latitude of 54° 20'. The dial shows little sophistication and Melville has even made the simplification of spacing the hour lines equally round the chapter ring, as would be required if it had been an inclined dial set in the equatorial plane. However, the presence of an extended fixing for a conventional gnomon makes it clear that this is indeed a horizontal dial. It is apparent therefore that in a span of about ten years, Melville's dials have progressed from fairly rudimentary devices to quite sophisticated and marketable products.

The date of Melville's move to Scotland is not known with precision, but an 1843 dial for the latitude of Ayr survives.[24] Perhaps his departure was linked to the very depressed economic and social conditions in Ireland after the years of the bad harvests, and the need to seek new markets for his dials.[25] No dials of 1844 are known, but several of 1845 have the conventional Glasgow latitude of 55° 50', including the earliest examples of the more impressive design with eight subsidiary dials. One of these larger dials was recorded and illustrated by W.B. Stevenson in 1934 at Dargavel House, Bishopton, Renfrewshire, and this commission may hold some clue to Melville's success.[26] Dargavel was the home of the advocate John Hall Maxwell younger of Dargavel (1816-1866), a prominent member of the Highland and Agricultural Society of Scotland of which he became Secretary in 1845.[27] Two other dials recorded by Stevenson were at Maxwell houses: Pollok House, seat of Sir John Maxwell of Pollok, 9th Baronet, had a smaller 1847 dial, and Aikenhead House, which was by this time in the hands of John Gordon, son of a wealthy tobacco merchant, had a large 1849 dial.[28] None of these three dials is still in place, nor are those from the three other Glasgow locations that have been identified.[29]

Dials made between 1846, when he appeared in the Glasgow street directory, and 1849 have the location 'Glasgow' added after his name, the only exception being a dial of 1848 in the Royal Museum of Scotland which is marked 'Stirling'.[30] The only other dial of 1848 that has been noted was in Bridge of Allan, immediately north of Stirling, and this tends to support the idea that Melville spent at least some time in the area of Stirling.[31] A possible association with earlier clients is suggested by the presence of Keir House near Bridge of Allan. William Stirling (1818-1878), nephew of Sir John Maxwell of Pollok and later his successor as 10th Baronet, succeeded to the Keir estates in 1847. He was a prominent agriculturalist who took an active part in the affairs of the Highland and Agricultural Society and the Glasgow Agricultural Society, and from 1849 extensively remodelled Keir.[32]

A more unusual dial, in manuscript and apparently in Melville's hand, also dates from this period.[33] The dial, in the Royal Museum of Scotland, is drawn on paper laid on an oblong wooden board, and its varnished surface is now somewhat distressed. The design of the dial plate, with its four subsidiary dials for New York, Madagascar, Malabar and New Zealand, is recognisably due to Melville. It now lacks all its gnomons, but claims to have been constructed for latitude 55° 58′, and is therefore probably intended for use in Edinburgh. It is signed 'RICARDUS MELVILLE Fecit' and dated in the 1840s (the last numeral now being illegible). Although it gives the impression of having been constructed by Melville for use as an educational tool in mathematics teaching, there are significant discrepancies in the layout of the two sides of the chapter ring of the main dial. It is therefore more likely that this is not his work but is an inaccurate copy of an actual dial, made by someone who has not sufficiently understood the projection of the dial face but has nonetheless faithfully recorded details such as the signature. The dial that was copied can however be seen to have had the edge of the gnomon offset from the centre of the chapter ring, and represents the first known instance of this construction being used by Melville.

For reasons which are not clear, Melville's prospects, and indeed his name, appear to change about this time, and his next known commission is the only one where he can be associated with any degree of confidence with an established architect. William Butterfield (1814-1900) was a pioneer of the original high Victorian phase of the gothic revival movement, whose work was predominently ecclesiastical. He handled only three Scottish commissions, of which the most substantial was the Episcopalian Cathedral for Perth, begun in 1847. His (Tractarian) College of the Holy Spirit, near Millport on the island of Cumbrae in the lower Clyde estuary, was consecrated as the Cathedral of the Isles, and was build in the period 1849-1851.[34] Both projects were under the patronage of George Frederick Boyle (1825-1890), subsequently 6th Earl of Glasgow. Melville, now however signing himself 'R. MELVIN' of Glasgow, contributed a dial in 1851 for the building complex, and the dial plinth carries Boyle's initials: it is the only Melville dial that A.R. Somerville has recorded that is still at its original location.[35]

The Cumbrae dial represents a design departure for Melville, in that it is octagonal with the eight subsidiary dials grouped around a comparatively small central dial. There are, however, strong similarities with a surviving undated circular dial from an unknown site, signed 'RICH.D MELVIN / Fecit Liverpool' constructed for latitude 53° 25′.[36] There are some specific differences, notably to the projection of the central dial, and the fact that each of the eight subsidiary dials is for two locations; but in general it is hard to escape the feeling that these two dials are closely related in time. If Melville did operate for a time from Liverpool, this has gone unrecorded in the Liverpool street directories.[37]

Another significant commission which was geographically close was probably undertaken at about this time. At Ruthin Castle in Denbighshire, North Wales, seat of the Hon. Fredrick Richard West, M.P. for Denbigh, extensive rebuilding in a late perpendicular style was undertaken in the period 1848-1853 by the architect Henry Clutton. His then partner William Burges (1827-1881), the prominent gothic revival architect and designer, worked with him in fitting up the interior at Ruthin in 1851-1856.[38] The Melville dial at Ruthin appears to have been undated, but was probably erected near the conclusion of the rebuilding work, and like the Cumbrae dial it was octagonal with eight subsidiary dials.[39] Although it has not been located or examined, it is clear from such details as have survived that it was very similar indeed to the Cumbrae dial, and (with the exception of the Liverpool dial) was unlike any other that has been recorded. Whether either commission was obtained by recommendation from the client or architect of the other property cannot be said, but it is at least possible that the Cumbrae dial, which is the only Scottish dial to be signed 'Melvin', was made after Melville's move to the Liverpool area.

Too few dials have been recorded to enable Melville's progress around the country to be charted. Eden and Lloyd noted three in unspecified locations in Warwickshire, and a fourth in Dover,[40] but apart from these, all the surviving English dials which state a latitude are for the conventional 51° 30′ or 31′ latitude of London. Typical of half a dozen that have been noted in recent years is an example in the Royal Museum of Scotland with four subsidiary dials, the later form of projection for the central dial, the second pair of latinised mottoes, and signed 'R.D MELVIN Fecit from LONDON'.[41] Only one location was recorded by Eden and Lloyd in 1900: at Ember Court near Esher in Surrey, a house demolished between the Wars.[42] Although the majority of these standardised London dials are signed but undated, at least one unsigned dial has been recorded[43] and a dated example of 1858 is known with no signature or latitude marked.[44]

In about 1860 Melville returned to Ireland. A dial dated 1864 was recorded by Eden and Lloyd at an unspecified location in Killiney, outside Dublin,[45] but this has not been located. A fine surviving Belfast dial, which in every respect (except its latitude of 54° 30′) is comparable with the best of the London dials and is from this late period rather than his earlier days in Ulster, was originally made for Mount Vernon, the demolished Belfast mansion of the merchant Hill Hamilton.[46]

No other record of his work in Ireland has been found, although an isolated 1871 dial with no provenance is in the Royal Museum of Scotland.[47] This is for a Dublin latitude of 53° 15′, but it is unusual in carrying what appears to be a serial number, mentioned earlier, and also gives a full address, namely 9 Lower Wellington Street, Dublin. Unfortunately, he is not entered in the street directories, and so his length of stay in Dublin is unknown. Melville's death date has not been discovered and the loss of Irish wills for this period makes this information difficult to pursue. It is to be hoped that further examples of his work will emerge so that a better understanding of this enigmatic specialist may develop.

REFERENCES

1. On Ferguson and Martin see J.R. Millburn, *Wheelwright of the Heavens* (London, 1988); idem, *Benjamin Martin* (Leyden, 1976); idem, *Benjamin Martin: Supplement* (London, 1986).

2. Thomas Ross, 'Ancient Sundials of Scotland', *Proceedings of the Society of Antiquaries of Scotland* 24 (1889-90), 161-273. A recent analysis is given by A.R. Somerville, 'The Ancient Sundials of Scotland', *ibid.* 117 (1987), 233-264, and fiche 3; issued May 1989.

3. See for example V. Fiddes and A. Rowan, *David Bryce 1803-1876* (Edinburgh, 1976) and A.A. Tait, *The Landscape Garden in Scotland 1735-1835* (Edinburgh, 1980).

4. *Glasgow Post Office Directory* 1846, recorded at 160 Saltmarket Street.

5. NMS T1987.134. Since no earlier numbered dials have been recorded, the series may have an arbitrary starting point, but may also embody assumptions about previous production.

6. We are grateful to Dr. A. Livingstone, Curator of Mineralogy, National Museums of Scotland, for examining these dials in September 1988. He has noted that a similar slate was also quarried at Aberfoyle, near Stirling, whereas pyrite crystals are recognisable in slate from other Scottish mines.

7. In at least one later instance no latitude is given: see below, reference 44.

8. P.H. Reaney (ed.), *A Dictionary of British Surnames* 2nd edition (London, 1977), 238.

9. H.K.F. Eden and E. Lloyd, *The Book of Sundials originally compiled by the late Mrs. Alfred Gatty now enlarged and re-edited* ... (London, 1900), 426.

10. B. Loomes, *[G.H. Baillie's] Watchmakers and Clockmakers of the World, Volume 2* (London, 1976), 158. Melvin's dates, given as 1790-1838, have been derived only from the inclusion of Washington (founded in the 1790s) among the places listed on a Melville dial of 1838 (Ulster Museum report form B/00159, 14 April 1967).

11. Five surviving Bonar dials are discussed by R.R.J. Rohr and A.R. Somerville in articles in *Antiquarian Horology* 16 (1986), 227-242.

12. The holdings of the Down Museum, the Ulster Folk Museum and the Ulster Museum have been examined.

13. Examples of fine early dials are one by William Adair, 1794, at the Ulster Folk Museum (inventory UFTM 756-1972) and another inscribed James Philips, Derriaghy, 1800, at the Ulster Museum (inventory 55-1925). An undated dial with four subsidiary dials by J. McNally is at the Down Museum (inventory 1985-85).

14. Dials of 1819 and 1826 by the same maker are attached to the Church of Ireland and the Catholic chapel respectively at Poyntzpass, Co. Armagh: these were kindly drawn to our attention by David Wright, Comber, Co. Down.

15. A. Gailey, *Rural Houses of the North of Ireland* (Edinburgh, 1984), 110.

16. L. Clarkson, 'The City and the Country' in J.C. Beckett, et al., *Belfast The Making of the City 1800-1914* (Belfast, 1983), 161.

17. International Genealogical Index: British Isles, Ireland, fiche E0026, p. H17 (August 1978).

18. Armagh County Museum, Armagh, inventory 75-1947. Central dial and 4 subsidiary dials, made for latitude 54° 21′.

19. Ulster Museum, Belfast, acquired before 1967, but unlocated in 1988. On Calwell see the entry on his son in R.M. Young, *Belfast and the Provinces of Ulster in the 20th Century* (Brighton, 1909), 540.

20. Information provided c.1975 to B. Loomes by W.A. Seaby, formerly Director of the Ulster Museum.

21. Information supplied by Roger Wetherup, Armagh County Museum, May 1986. We are grateful to Robert Heslip, Ulster Museum, for drawing this dial to our attention.

22. A possible exception may be Samuel Higginbotham of Springfield House, Lanarkshire, just to the east of Glasgow, who was a prominent Glasgow printer and partner in the firm of Todd & Higginbotham. A sundial was recorded by Stevenson with the name Samuel Higginbotham, Springfield (W.B. Stevenson, 'Sundials of Six Scottish Counties, near Glasgow', *Transactions of the Glasgow Archaeological Society* new series, 11 (1937-40), 230). The description matches that of Melville's smaller dials, but the surface appears to have been rubbed and the engraving difficult to read: the latitude, for example, has been read as 53° 50′ rather than 55° 50′. Springfield House was demolished in the 19th century, and the dial was recorded in 1934 as at Castlemilk, Renfrewshire, and about to be moved to Coulter House, near Biggar, Lanarkshire, where it is no longer to be found.

23. A full-size drawing of this dial made in 1946 was examined at the Ulster Folk Museum, Cultra (UFTM Archives D3-12-11e), and we are grateful to J.J.R. Adams for locating this. As a result of enquiries made by the Down Museum, the dial itself has been traced in Eire and was gifted to the Down Museum in November 1988 (inventory DB. 324). John Sandwich has not been identified, but an Edward Sandwich is recorded as a local land holder in a valuation of 1863 (information from Lesley Simpson, Down Museum, September 1988).

24. Now in private ownership in Kirkcudbright. Central dial and 4 subsidiary dials, constructed for latitude 55° 30'. We are grateful to Dr. A.R. Somerville for drawing this example to our attention. Melville may have made the short crossing from Ireland to Irvine.

25. Possibly also influenced by the improved level of food relief: on the comparative levels of relief in 1846 see T.C. Smout, *A Century of the Scottish People 1850-1950* (Fontana edition, London, 1987), 12. The impact on agriculture in the western Lowlands of Scotland of seasonal migrant workers from Ulster in the early 1840s is discussed in T.M. Devine, *The Great Highland Famine* (Edinburgh, 1988), 152-153.

26. Stevenson, *op. cit.* (22), 237 and plate II. This dial, which is no longer at Darvagel, can be seen to differ from the closely similar 1845 dial at the Royal Museum of Scotland (NMS T1979.103). Two 1845 examples with 4 subsidiary dials are NMS T1967.90 (which has been illustrated by D.J. Bryden in his article 'Scientific Instruments', in P. Phillips (ed.), *The Collectors' Encyclopedia of Antiques* (London, 1973), 605) and another sold at Phillips Edinburgh, 29 August 1980, Lot 47.

27. *Transactions of the Highland and Agricultural Society of Scotland* 3rd series, 2 (1847). Maxwell, whose family was a cadet branch of the Maxwells of Pollock, inherited Dargavel from his father in 1847. Other clients of Melville's were active in the Society (such as William Malcolm Fleming of Barrochan), but Melville himself was not a member.

28. Stevenson, *op. cit.* (22), 237.

29. Stevenson recorded an unsigned example for 55° 46' at Barrochan House, Renfrewshire, in 1934 (*ibid.*, 237). One of the larger dials at Polmadie House, Lanarkshire, is described (without the maker's name being given) and illustrated in D. McGibbon and T. Ross, *The Castellated and Domestic Architecture of Scotland* V (Edinburgh, 1892), 490. A third example appears in a photograph by Thomas Annan of Clober House, near Milngavie, Dunbartonshire, in J.G. Smith and J.O. Mitchell, *The Old Country Houses of the Old Glasgow Gentry* 2nd edition (Glasgow, 1878), 489-490. The dial recorded by Stevenson at Mount Charles, Alloway, near Ayr, had presumably already been moved from its original site since it also was made for a Glasgow latitude (Stevenson, *op. cit.* (22), 237).

30. A dial of 1849, understood to have come from Uddingston, is in private owenership in Kirkbean, Kirkcudbright: we are grateful to Dr A.R. Somerville for providing information about this example. The Stirling dial, for latitude 56° 8', has 8 subsidiary dials, and is NMS T1982.100: the original location has not been traced.

31. Eden and Lloyd, *op. cit.* (9), 426. The dial had Melville's name followed by 'Glasgow', and no latitude has been noted. In 1900 it was in the garden of the Royal Hotel, Bridge of Allan. This may have been its original site since a contemporary guide describing the hotel noted that 'tastefully laid out pleasure grounds are attached, decorated by some neat specimens of statuary and *jets d'eau*' (Charles Roger, *A Week at Bridge of Allan* (Edinburgh, 1851), 30: this reference was kindly provided by A. Maxwell-Irving). No other dials from outside the greater Glasgow area have been found (with the possible exception of one of 1845 for latitude 55° 55', offered by Sotheby's, 23 October 1985, Lot 323, which has not been examined).

32. Thomas Seccombe, 'Sir William Stirling-Maxwell' in S. Lee (ed.), *Dictionary of National Biography* LIV (London, 1898), 386; A. Rowan, 'Keir, Perthshire - III', *Country Life* 158 (1975), 506-510; Fiddes and Rowan, *op. cit.* (3), 123.

33. NMS T1984.234. The dial, which has no provenance, was found in central Edinburgh, and presented through the interest of Alastair McWatt Green and Thomas Walls.

34. P.R. Thompson, *William Butterfield* (London, 1971), 429.

35. Somerville, *op. cit.* (2), microfiche inventory, section beginning E13.

36. Described and illustrated in Historical Technology, Catalog 110 (Spring 1975), item 199 (where Saul Moscowitz draws parallels with the 1847 dial in the Frank Collection): this is now in the Royal Museum of Scotland, NMS T1988.79.

37. Information from Martin Suggett, National Museums and Galleries on Merseyside, July 1988.

38. E. Hubbard, *The Buildings of Wales: Clwyd* (Harmondsworth, 1986), 71, 272; J. Mordaunt Crook, *William Burges and the High Victorian Dream* (London, 1981), 43.

39. The dial plate was sold in a sale of the contents of Ruthin Castle in 1963 (Jackson-Stops & Staff, Chester, sale of 11-18 July 1963, Lot 5), although the pedestal is still in place (information from Douglas Hogg, Cadw, Cardiff, October 1988). Some descriptive information is given in Eden and Lloyd, *op. cit.* (9), 426. It is not visible on the plate published in *The Builder* 11 (1853), 578-9, and no photograph has been found.

40. Eden and Lloyd, *op. cit.* (9), 409.

41. NMS T1985.85: previously offered by Lawrence, Crewkerne, Somerset, 11 October 1984, Lot 21, and Trevor Philip & Sons, Catalogue 3 (June 1985), item 34. Other examples include Sotheby's, 2 February 1976, Lot 111; Phillips, 12 September 1978, Lot 53; Historical Technology, Catalog 128 (Summer 1985), item 129.

42. Eden and Lloyd, *op. cit.* (9), 281, 409.

43. Offered for sale by Christie's South Kensington, 30 March 1989, Lot 223.

44. In a Massachusetts collection, July 1988.

45. Eden and Lloyd, *op. cit.* (9), 282, 409.

46. In private ownership in Co. Down, August 1988. The presence of a large pyrite crystal in the slate may indicate that it is from an Argyll quarry. Information on Mount Vernon supplied by Roger Dixon, Belfast Public Libraries. A photograph of Mount Vernon by A.R. Hogg showing the dial in place is in the Local History Collection, Ulster Museum, Belfast. The house was demolished in the early 1950s.

47. NMS T1987.134; previously offered at Sotheby's, 23 June 1987, Lot 151.

102. Sundial: Richard Melville, Glasgow, 1847 (T1980.168)

Oblong horizontal pedestal sundial in slate, engraved for the latitude 55° 50', inscribed 'RICHARD MELVILLE / Fecit Glasgow AD 1847.' and with the mottoes 'Tempus Fugit' and 'Memento AEternitatis' (Time flies, remember eternity). Central 16-point compass rose, bisected by the 6a.m. and 6p.m. hour lines, within a chapter ring IV-XII-VIII divided on an outer ring to 5 mins., and with displaced chapter rings providing local time at Riga, Rome and Paris. Flanked by a 2-part tabulated 'equation of time', corrections given to the nearest minute at 10-day intervals throughout the year. Four subsidiary dials in the corners of the plate indicate local time at New York, Alexandria, Malabar and New Zealand.

Size: 266 x 321mm.

All 5 of the copper gnomons, which are pinned and leaded under the plate, have been broken off at the upper surface of the dial. A recent break in the slate exhibited at one corner.

This dial is compatible with what is known of the 1847 Melville dial once located at Pollok House, Renfrewshire, mentioned in the text above.

18

NAUTICAL INSTRUMENT MAKERS

The enormous success of the shipping and shipbuilding industries of the River Clyde, which so transformed Glasgow, created a vigorous demand for a broad range of nautical instruments. In the third quarter of the eighteenth century the most notable manufacturer and supplier was the instrument maker James Watt, better known for his improvement to the steam engine: by the early nineteenth century numerous firms, several of which were descended from Watt's business, were supported by the production and supply of nautical apparatus.[1] A number of these in time became large manufacturing concerns with considerable reputations. James White in particular occupied a commanding position towards the close of the century, and his firm, together with those of his contemporaries Alexander Dobbie, Duncan McGregor and Whyte & Thomson, are discussed in detail in succeeding sections.

The majority of businesses however were comparatively small, and the five described below represent different aspects of the nineteenth century trade. On one hand, there are the ship chandlers, and instrument retailers such as Macalister & Fyfe and S.H. Fyfe, who may have commissioned some of the lines they sold, but were largely retailers. On the other hand are the workshops of actual makers, Henry Bell, Robert Park and Cameron & Blakeney, who nonetheless relied heavily on supply from wholesalers. It is no coincidence that the latter three flourished in mid-century, for it is to that period that the main expansion of shipbuilding on the Clyde can be traced, and it was with the growth of the industry that the activities of the Clyde instrument makers were intimately connected.

The shipbuilding industry first began on a significant scale in the harbours of Greenock and Port Glasgow, where it was stimulated by heavy shipping traffic serving the import trade in tobacco, cotton and sugar from the eighteenth century, and the increasing export of industrial products to the foreign markets in the nineteenth. 'Sugaropolis', as Greenock became known because of its staple industry of sugar refining, had nine or ten shipyards in the early nineteenth century[2] (in addition to boat and yacht yards) whose output until about 1842 was steamers.[3] However, on the Clyde as a whole, wooden shipbuilding, although very labour intensive, represented a mere five per cent of the British annual tonnage in 1835. The introduction of the screw propeller and more efficient steam engines, together with relatively cheap iron production in local centres, dramatically altered this position from the 1840s, with the result that between 1851 and 1870 the Clyde produced seventy per cent of the British tonnage of iron ships.[4] During the 1880s the annual tonnage launched on the Clyde was 400,000 tons.[5] In the Greenock district in particular the annual tonnage rose from 20,000 in 1876, to 52,744 in 1882 and to 107,494 in 1909.[6]

Nautical instrument makers and ship chandlers were clearly two of the most important shipping services to be stimulated by the increase in shipbuilding. The demand for the instrument maker's skills grew rapidly, especially with the need for increasingly sophisticated navigational equipment in the form of magnetic compasses and binnacles, and sounding machines suited to fast iron vessels. Other nautical items such as sextants, telescopes, chronometers, and latterly ships telegraphs and electrical equipment, were also in huge demand. The compass was perhaps the most important single instrument produced, and it ranged from the basic boat compass sold by Fyfe to the more sophisticated instruments represented here by a Cameron & Blakeney compass. The former, like other ship chandlers, was concerned primarily to provide external and internal fittings for ships, and stocked nautical instruments bought wholesale only as part of the business. The latter by contrast were involved in solving the problem of compass variation on iron ships, and were likely also to have adjusted the compasses they supplied on board the ships in which they were fitted. This appears to have been a standard service provided by the larger workshops, and the specialist staffs of compass adjusters are often to be found in the street directory under their own names.

While ship building was the most important stimulus in sustaining the expanding instrument making business on the Clyde, education played an important role in providing the requisite mathematical knowledge for the demanding work of construction. Greenock, bathing in the reflected glory of James Watt, whose will had left £100 to establish a scientific library, boasted the Mechanics' Institution founded in 1830 and a Navigation School established by the magistrates.[7] Of more significance, perhaps, in stimulating the minds of pupils aspiring to a nautical career, was the enthusiastic teaching of Colin Lamont, a mathematics master at the grammer school. He is reputed to have been the first to introduce into a public school the application of modern astronomy to navigation. In 1785 he was teaching his pupils the use of nautical instruments for lunar observation, and of chronometers for navigation, in a 'place' fitted at his own expense with a 3½ inch telescope, an astronomical circle, a clock and other instruments.[8] It is tempting to speculate that he was responsible for teaching some of the Greenock makers who were to become prominent by the mid-nineteenth century.

REFERENCES

1. D.J. Bryden, *Scottish Scientific Instrument-Makers 1600-1900* (Edinburgh, 1972), 30, 34.

2. James Brown, *Guide to Greenock* (Greenock, 1910), reprinted in the *Greenock Telegraph*, 17 November 1979 to 9 February 1980, *passim*.

3. John Shields, *Clydebuilt* (Glasgow, 1949), 116-121.

4. Bruce Lenman, *An Economic History of Modern Scotland* (London, 1977), 177-180.

5. Quoted in W.H. Marwick, *Economic Developments in Victorian Scotland* (London, 1936), 225.

6. Brown, *op. cit.* (2), *passim*. Although there was a huge increase in tonnage launched, it is unclear exactly how far this represented a corresponding increase in the number of vessels, and therefore in the number of instruments required; because of the massive tonnage of an iron ship, this number may have been smaller than at first appears.

7. Robert Murray Smith, *The History of Greenock* (Greenock, 1931), 288, 332-333. The Mechanics' Institution, split by internal wrangles, does not appear to have had the intellectual drive of its Glasgow and Edinburgh counterparts (see *idem., A Page of Local History: Being a Record of the Origins and Progress of Greenock Mechanics' Library and Institution* (Greenock, 1904)).

8. *The New Statistical Account of Scotland* VII (Edinburgh and London, 1845), Renfrew, 468-469.

ROBERT PARK

Robert Park, a nautical instrument maker of Greenock, was born about 1819 the son of Robert Park, a wheelwright, and Jean Blair. His early career is not known, but he first appeared in the Greenock directory in 1853 as an optician and nautical instrument maker at 17 William Street, Greenock.[1] In 1861 he was working at 14 Cathcart Street, advertising also as a chronometer maker and compass adjuster. The census return of the same year lists him as an optician and nautical instrument maker employing '3 men and 2 boys'.[2] This suggests that while Park's business was more modest than those of other Clyde makers, it was similar in size to those of many watch and clockmakers, and serves as a good example of the medium-sized instrument workshop.

Nothing specific is known of the scope of Park's work and insufficient instruments survive to assess its quality. He was in business for a relatively short period, for he died on 19 April 1863 aged forty-six. He was unmarried.[3] He left the whole of his estate, worth £2,653 15s 7½d, divided into three equal portions between his two unmarried sisters, and thirdly amongst various nephews and nieces. He also left John Scott, his shopman, one of his best gold watches, gold chain and key 'on account of the great interest which he has manifested in the success of my business'.[4]

Among the individuals with possible trade connections who were listed as owing him money was a Mr Dunn, with a debt of £7 15s; and W.H. Fyfe (q.v.) with one of £13 13s 6d.[5] His business may have been continued by John McNeilage, who was listed as an optician and nautical instrument maker at 14 Cathcart Street from 1868 to about 1872, and James Fleming, listed as an optician and chronometer maker there in 1873 and 1875.[6]

REFERENCES

1. *Greenock Post Office Directory* 1853.

2. General Register Office (Scotland), Census of 1861 (Greenock, West), 564³/4, p.8.

3. G.R.O.(S.), Register of Deaths 1865 (Greenock, West), 564³/361. The cause of death was given as tuberculosis.

4. Scottish Record Office, SC 58/42/32, pp.340-345, Inventory of the Estate of Robert Park, Nautical Instrument Maker.

5. *Ibid.*, pp. 348, 350.

6. *Greenock Post Office Directory* 1868-1875.

---

103. Refracting telescope: R. Park, Greenock, c.1860 (T1980.255)

2″ aperture single-draw hand-held refracting telescope with doublet objective, and 2-component eyepiece and erecting assemblies, the latter mounted at the division of the draw tube; in brass, with objective ray shade, the barrel originally leather-covered. Engraved on the draw tube 'R. Park,/ Greenock.' Parts scratched with the assembly mark 'II' throughout, and the mount of one of the lenses in the erecting assembly scratched 'WF 1906'. Objective cover with 1½″ aperture stop, and with sliding closures on the objective cover and external eye stop.

Length closed: 521mm.
Tube diameter: 60mm.
Aperture: 47mm.

Lacking the leather covering to the barrel.

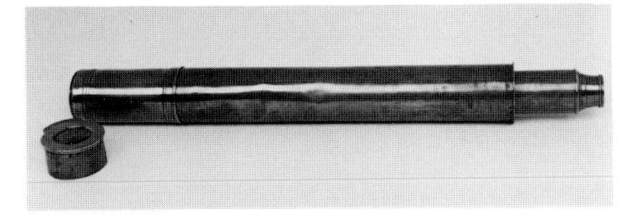

MACALISTER & FYFE

The firm of Macalister and Fyfe, ship chandlers, of Greenock was founded between 1858 and 1861 by Daniel Macalister and William H. Fyfe. Macalister was a house painter and paper hanger who had appeared in the Greenock directory by 1845, and died between 1861 and 1868.[1] He was a principal shareholder of the Greenock Union Bank founded in 1840, and was probably related to some of the numerous Macalisters listed in the Glasgow street directory as shipowners and sailmakers and in related trades.[2]

William Holborn Fyfe was born at Port Glasgow about 1812, the son of Elizabeth Holborn and John Fyfe a ship chandler of Glasgow and Greenock.[3] He was the elder brother of Samuel Holborn Fyfe whose activities as a chandler are noticed separately. William first appeared in the Glasgow street directory in 1840 as of R. & W. Fyfe, the then current name of the Glasgow business, which went bankrupt in 1844.[4] He moved to Greenock between 1846 and 1849 and appeared there in the 1853 directory. Following John Fyfe's death in about 1848 Samuel and William presumably divided the business, with Samuel continuing the Glasgow shop and William taking over in Greenock. William traded as William H. Fyfe & Co., ship chandlers and paint manufacturers at 5 William Street and 4 East Breast until 1854. At that point, Fyfe's partnership with George Burns was dissolved, and for two years he traded with D.D. Yeo as Fyfe & Yeo, and then reverted to business as W.H. Fyfe.[5] Unfortunately, Fyfe must have experienced some difficulties as sole partner, for his business was sequestrated in February 1858; however, when he came to offer his creditors composition, he did so with the support of Daniel Macalister.[6]

From about 1861, with the formation of their joint enterprise, Macalister and Fyfe traded at 4 East Breast as ship chandlers, painters and paint manufacturers.[7] They acted as agents for yacht sales, and like Duncan McGregor, their more eminent contemporary and fellow 'Sugaropolitan', they were shipowners in a small way. In 1868 they were listed as owners of the *Cabot* (599 tons) and the *City of Edinburgh* (599 tons), both engaged in foreign trade.[8] The firm continued in business long after the founders' deaths and traded at the same address until 1933.

William Fyfe was married with four daughters and a step-daughter from his wife Margery. He died insane aged fifty-six in the Gartnavel Royal Lunatic Asylum, Partick on 17 February 1868.[9]

The scope of Macalister and Fyfe's work is not known, but it is reasonable to suppose that as chandlers and paint suppliers their activities as makers or repairers of nautical instruments were negligible if they existed at all. Macalister was by training a decorator, and Fyfe seems to have acquired some of his partner's skills, for he is described in the Census of 1861 as 'Ship Chandler and Paint manufacturer'.[10] It is interesting to note that his brother S.H. Fyfe (q.v.) also moved away from the trade in which he had been trained. By 1888, when the firm exhibited at the Glasgow International Exhibition, they displayed no instruments at all.[11]

REFERENCES

1. *Greenock Post Office Directory* 1852-1868, *passim*.

2. Dugald Campbell, *Historical Sketches of the Town and Harbour of Greenock* (Greenock, 1879 & 1881), I, 259; e.g. *Glasgow Post Office Directory* 1882.

3. General Register Office (Scotland), Census of 1861 (Greenock, West), 564³/3, p.18; Register of Deaths 1868 (Partick), 646²/64.

4. Scottish Record Office, CS 280/30/42, Sequestration Papers, Robert and William H. Fyfe. The estate was sequestrated on 4 April 1844; they offered composition of 7s per pound on 2 May 1844; they were discharged on 18 July 1844.

5. 'Notice of Dissolution, Greenock, March 1, 1854; The Ship Chandlery, Ironmongery, and Painting Business carried on in Greenock by the Subscribers, as sole Partners thereof, under the Firm of W.H. FYFE & CO., was this day DISSOLVED by mutual consent. [signed] W.H.FYFE GEORGE BURNS ... In reference to the above Notice of Dissolution, Mr W.H. Fyfe takes leave to intimate that he has assumed the Subscriber, Mr D.D. Yeo, as a Partner in the Business, which will be carried on as formerly in all its Branches under the Firm of FYFE & YEO. From Mr Yeo's long practical experience as a Painter and Decorator both in London and Torquay, they hope to meet a share of that patronage so long bestowed on W.H. Fyfe & Co.' (*Edinburgh Gazette*, 3 March 1854, 185). 'The

Business carried on by the Subscribers, the only Partners, as Ironmongers and Ship Chandlers, Painters, and Paint Manufacturers in Greenock, under the Firm of FYFE & YEO, was this day DISSOLVED by mutual consent ... W.H. FYFE D.D. YEO Greenock, April 26, 1856.' (*Edinburgh Gazette*, 29 April 1856, 392).

6. S.R.O., CS 318/2/60, Sequestration Papers, William Holborn Fyfe. The Sederunt Book contains some details about the business. Fyfe declared at the hearing before the sheriff that he had begun his business in March 1849 without partners. In April 1852 he had borrowed additional capital from Mr George Burns of Dumbarton, who subsequently became Fyfe's partner in August 1852. Burns ceased to be a partner on 1 March 1854, when Dr Daniel Yeo replaced him. This in turn was ended in 1856. Fyfe offered composition of 9s in the pound to his creditors, amongst whom was Samuel H. Fyfe (q.v.), David Heron (q.v.), and D. McAllister (*sic*) (q.v.). Fyfe was discharged 12 May 1858. Notices also appeared in the *Edinburgh Gazette*: the initial notice of sequestration (5 February 1858, 239); the announcement of the Trustee and Commissioners (23 February 1858, 415); and the offer of composition, with 'Daniel Macalister, Painter and Feuar in Greenock' as Cautioner (19 March 1858, 612).

7. *Greenock Post Office Directory* 1852-1863.

8. *Ibid.* 1868, Appendix, 88.

9. G.R.O.(S.), Register of Deaths 1868 (Partick), 646²/64. The cause of death was given as 'Inflammation of the Brain and Lung'.

10. G.R.O.(S.), Census of 1861 (Greenock, West), 564³/3, p. 18.

11. Their stand comprised 'Yacht stores, ropes, blocks, buoys, fenders etc.' *International Exhibition Glasgow 1888: Official Catalogue* (Glasgow, 1888), 97.

---

104. Marine compass: McAlister & Fyfe, Greenock, c.1920 (T1980.200)

5″ liquid-filled card compass. Printed 128-point rose marked 'McALISTER & FYFE. / GREENOCK.' surrounding a central float, and suspended in a glazed and painted sealed brass bowl with lubber line, gimbal-mounted on a brass stand or bracket. The bowl flange and upper glazing ring stamped with the assembly mark '6'.

Card diameter: 125mm.
Bowl diameter: 162mm.

The gimbal ring and base flange are distorted. The join of the two parts of the stand or bracket has been re-soldered.

SAMUEL HOLBORN FYFE

Samuel Holborn Fyfe (c.1823-1905), the son of Elizabeth Holborn and John Fyfe, a ship chandler[1] with a business at 152 Broomielaw, Glasgow, and 10 William Street, Greenock, joined his father's Glasgow business, and in 1847 appeared in the street directory for the first time as a member of the firm. Shortly afterwards his father died or retired and S.H. Fyfe continued the business from new premises at 32 Clyde Place, first advertising in 1849 as an ironmonger, ship chandler and flag maker. From 1850 he advertised as a compass maker, and from 1852 as an oil and colour merchant.[2] The compass in this collection demonstrates that he supplied at least a limited range of material to other retailers because it has until recently had another name disc obscuring Fyfe's name at the centre of the card.[3]

From 1855 to 1856 he was at 42 Clyde Place, in 1857 at No. 45, and from 1858 until 1860 at 62 Clyde Place.

Significantly, in 1859 he advertised principally as a house and ship painter, and subsequent directory entries confirm that Fyfe's interests turned from ship-chandling and instrument retailing (or production) towards painting. In 1860 he went bankrupt, but appears to have recovered from this setback remarkably quickly.[4] In 1861 he appeared at 37 Morrison Street as a 'marine painter' and was not listed as a flag and compass maker; from 1862 until 1871 he was resident at 127 Paisley Road where he was a marine artist (from 1867 a photographer also). He lived and worked at 12 Pollock Street from 1872 to 1876, and at 38 Pollock Street from 1877 until 1882, where he also established himself as a publisher of marine photographs. From 1883 until his death on the 15 May 1905, he lived at 72 Houston Street. Fyfe was married, and left a son, John who was perhaps a surveyor and measurer.[5] His elder brother William H. Fyfe, a partner in Macalister and Fyfe (q.v.) is noted above.

REFERENCES

1. General Register Office (Scotland), Register of Deaths 1905 (Glasgow, Kinning Park), 644[14]/188. John Fyfe's business was advertised as R. & W. Fyfe 1840-1843, and as Fyfe & Black in 1846. R. & W. Fyfe were presumably the Robert and William Fyfe, mathematical instrument makers, admitted on 27 August 1841 into the Incorporation of Hammermen of Glasgow by virtue of being sons of a member; this was probably John Fyfe himself, listed as resident in Greenock and admitted on 28 August 1840 (H. Lumsden and P. H. Aitken, *History of the Hammermen of Glasgow* (Paisley, 1915)). John Fyfe last appeared in the Directory in 1848.

2. *Glasgow Post Office Directory* 1847-1905, *passim*. From 1849 Fyfe was flag maker to the Royal Northern Yacht Club.

3. NMS T1980.196.

4. Scottish Record Office, CS 318/5/106, Sequestration Papers, Samuel Holborn Fyfe. There is no Sederunt Book for this case, and the warrant for discharge is dated 24 August 1861. Notices also appeared in the *Edinburgh Gazette*: the initial notice of sequestration (8 May 1860, 625); the announcement of the appointment of the Trustee and Commissioners, one of whom was 'David [*sic*] Macalister, Ship Chandler, Greenock' (25 May 1860, 696); and an announcement that the Bankrupt had offered Creditors composition of 3 shillings per pound, this offer to be confirmed at a meeting called for 4 July 1860 (15 June 1860).

5. G.R.O.(S.), Register of Deaths 1905 (Glasgow, Kinning Park), 644[14]/188. S.R.O., *Calendar of Confirmations* 1905, p.206, shows that Fyfe's estate at his death was worth £484.

105. Marine compass: S.H. Fyfe, Glasgow, c.1850 (T1980.196)

6" dry card compass, with jewelled bearing and replacement broad needle rivetted under the covered mica card, with lead adjustment weight on the needle and red wax on the lower paper cover. Printed 32-point compass rose with 'S. H. FYFE./ GLASGOW.' around the centre. Suspended on a brass point in a 2-part turned and painted wooden bowl, lead weighted under the base and originally glazed, with trunions for gimbal mounting.

Card diameter: 150mm.
Bowl diameter: 175mm.

The glass and retaining putty have been absent for some time and the surface of the card is stained. The clean centre of the card has been protected by an additional name disc, removed more recently, of which a vestige survives: there is no record of the retailer's name from this disc. The 2 rivet holes for the original needle are visible. Replacement nut on the counterweight attachment and needle support. Remnants survive of white distemper and lubber line inside the bowl and a dark green outside. Lacking gimbal ring and case housing.

HENRY BELL

Henry Bell was a nautical and optical instrument maker active in Glasgow between 1856, when he first appeared in the directory, and 1881. He traded at 46 Maxwell Street from 1856 to 1857, and at 50 Maxwell Street from 1858 to 1859. In 1860 he set up with Isaac Bell at 54 St. Enoch Square, and moved with him to 70½ Great Clyde Street in 1862. Isaac Bell had begun as a brassfounder in 1838, and advertised himself as a compass maker. From 1863 Isaac and Henry Bell ceased to advertise together, but continued to trade from the same address. Isaac did not appear in the directory after 1871, and Henry's last entry was ten years later in 1881.[1]

Bell appears to have been typical of the smaller maker in that he retailed instruments bought in from wholesalers, but as nothing is known either of his own or Isaac's work no firm conclusions can be drawn.[2]

53 Trade label for Henry Bell, c.1880

REFERENCES

1. *Glasgow Post Office Directory* 1838-1881, *passim*. For more on Isaac Bell's addresses see D.J. Bryden, *Scottish Scientific Instrument-Makers 1600-1900* (Edinburgh, 1972), 44. Routine checks of sources have not revealed any further information about the pair.

2. For instance, an ebony and brass octant marked 'Henry Bell, Glasgow' was offered for sale by Christie's South Kensington, 29 April 1982, Lot 69.

106. Refracting telescope: unsigned, c.1880 (T1980.271)

1½" aperture 4-draw hand-held refracting telescope with doublet objective and 2-component eyepiece and erecting assemblies; in brass, with objective ray shade and leather-covered barrel. Trade card of Henry Bell, Glasgow, pasted inside the objective cover. Parts scratched with the assembly mark 'IIII' or 'IV' throughout. Sliding closure on external eye stop.

Length closed: 232mm.
Tube diameter: 53mm.
Aperture: 34mm.

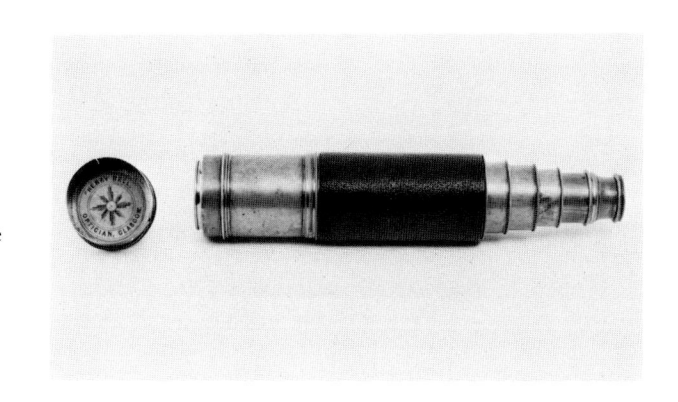

CAMERON & BLAKENEY

Cameron & Blakeney, who traded as mathematical, nautical and optical instrument makers between 1853 and 1860,[1] was the joint enterprise of Paul Cameron and John Blakeney. Cameron was born about 1814 in Dunbartonshire and first appears as a mathematical, optical and philosophical instrument maker at 87 London Street, Glasgow in 1851, previously the premises of William Green from 1846 to 1848, and of Robert Finlay from 1849 to 1850.[2] Both these men were in the same trade as Cameron, and it is therefore possible that he had worked under or with them, and continued the business under his own name.

The range of his work is reflected in the four items he sent for display in the Great Exhibition in 1851; they were an azimuth compass, an indicating level, a slide rule and a thermometer, steam and vacuum gauge.[3] While these instruments had both a civil engineering and nautical application, it was to the nautical side which Cameron appears to have leant in the early 1850s. In April 1852, for instance, he delivered his second paper to the Glasgow Philosophical Society on vapour from salt water in marine engines,[4] and in 1853 lectured to the Society on compass variation in iron vessels,[5] a phenomenon which was to provide instrument makers on the Clyde with extensive employment as compass adjusters.

This third paper was stimulated by Dr. William Scoresby's paper to the British Association in 1854; Cameron corresponded with Scoresby about his technique for permanently magnetizing the compass needle, but was criticised both for his alleged failure to

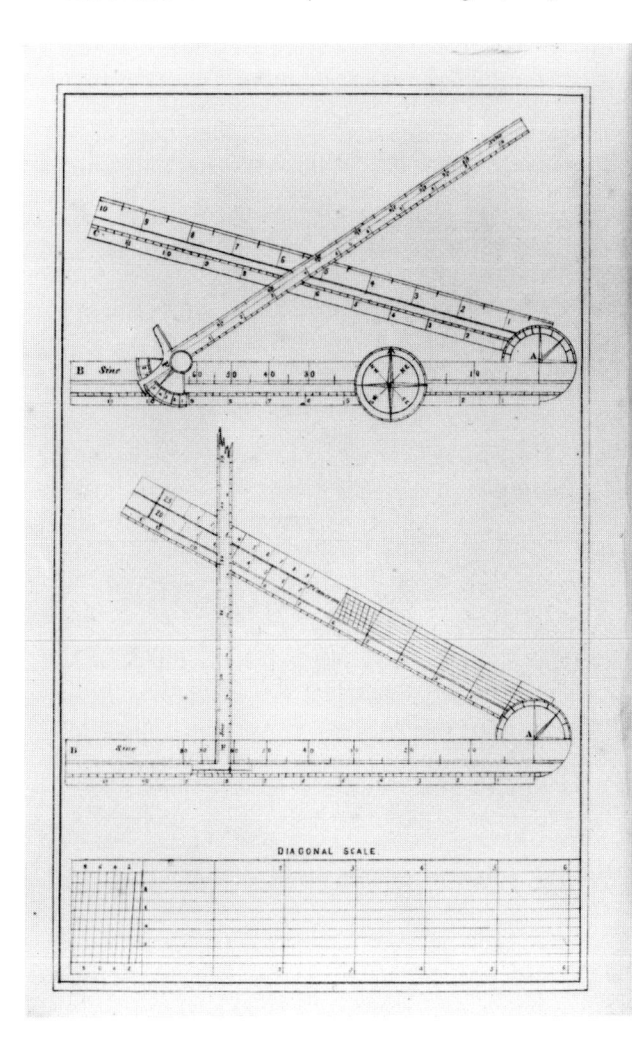

54 Frontispiece and title-page to Paul Cameron's *The Engineer's Mathematical Slide Rule*... (London and Glasgow, 1848)

apply the method properly, and for his notions of compass construction, by the Glasgow engineer James R. Napier.[6] Cameron, however, continued to produce designs and in 1868 took out a patent for a magnetic compass.[7] As a member of the Glasgow Philosophical Society he sat on its committee along with T.R. Gardner (q.v.), Malcolm McNeil Walker of D. McGregor & Co. (q.v.) and leading Glasgow scientists, to organise a good representation of Glasgow and other Scottish instrument makers at the 1855 Paris International Exhibition.[8] In the event none of the Glasgow makers exhibited at Paris.[9]

Cameron's preoccupation with compass design may have been connected with his partnership in 1853 with John Blakeney, a nautical instrument maker whose work is not known before this date and to whom he was related by marriage.[10] They traded together at 76 Great Clyde Street 1853-1855, 102 Great Clyde Street 1856-1859, and at 94-96 Jamaica Street until the partnership ended in 1860.[11] Each then set up independently. John W. Blakeney occupied the Jamaica Street premises, taking in 25 Turner's Court between 1861 and 1863 as a workshop, before transferring his business entirely to Turner's Court in 1864. He last appeared in the directory in 1873.

Paul Cameron & Co. was established at 11 and 19 Howard Street in 1861, and there appears to have been a brief hiatus in business in 1864, when one Hugh Buchan, a brassfounder and gasfitter, applied for Cameron's sequestration, as the only known partner in the firm.[12] However, this petition was withdrawn one month later, in September 1864. The business was briefly transformed into Houston & Cameron in 1865, and subsequently traded at 25 Howard Street in 1866-1867, 2 York Place in 1868 and finally at 178 Broomielaw in 1869.[13] Cameron was listed in the 1861 Census, aged forty-seven as a 'Mathematical & Nautical Instrument Maker employing 12 Men and 5 Boys'.[14] This striking piece of evidence suggests, in the absence of comparable sets of figures, that Cameron's workshop was among the largest, if not the largest, of the instrument shops in Glasgow at this date.

REFERENCES

1. *Glasgow Post Office Directory* 1853-1860.

2. D.J. Bryden, *Scottish Scientific Instrument-Makers 1600-1900* (Edinburgh, 1972), 45, 48, 49.

3. *Catalogue of the Great Exhibition* (London, 1851), I, 449, entry 356. Cameron was one of four Scottish makers who exhibited there (Bryden, *op. cit.* (2), n.195). Cameron also published on the slide rule: Paul Cameron, *The Engineer's Mathematical Slide Rule: An Early Assistant to the Study and Practice of Mathematics...* (London and Glasgow, 1848). This had two pages of recommendations by eminent Glaswegians of Cameron's Nautical and Sliding Rules.

4. Paul Cameron, 'The Force of Vapour from Saline Water, as Applied to Marine Engines', *Proceedings of the Glasgow Philosophical Society* 3 (1848-55), 246-248. The date of his first paper is unrecorded.

5. Paul Cameron, 'On the Variation of the Compass, and the Best Means of Rectifying it; and on the Best Means of Constructing Iron Ships, with a View to a more Correct Indication of the Compass', *ibid.*, 298 (read 30 March 1853); Paul Cameron, 'On the Deviations of the Compass in Wooden and Iron Ships; illustrated by Models and Experiments bearing on Dr Scoresby's Paper at the British Association, and Professor Airy's Reply', *ibid.*, 364 (read 13 December 1854).

6. James R. Napier, 'Remarks on Ships' Compasses', *ibid.*, 365-375. William Scoresby's paper was 'An Inquiry into the Principles and Measures on which Safety in the Navigation of Iron Ships may be reasonably looked for', *Report of the Twenty-Fourth Meeting of the British Association for the Advancement of Science in 1854* (London, 1855), Part II, 53-54, 161-162.

7. U.K. patent 1709, 25 May 1868. An earlier patent was taken out for a barometer in 1861 (1366, 31 May 1861). He published on both these subjects: Paul Cameron, *The Variation and Deviation of the Compass Rectified* 3rd edition, (London, 1868), and Paul Cameron, *Practical Directions for the Method of Reading the Barometer and Hygrometer; and for Observing the Bearing and Direction of the Winds and Formation of Clouds; also, the Method of Predicting the Probability of Storms* (London, Liverpool and Glasgow, 1868). In this last, Cameron described himself as 'Inventor of the azimuth compass, dial card, and mathematical slide rule, author of *The Theory and Practice of Navigation*; *Essays on Energy, Heat and Vapour*'. No copies of these two latter works have been located.

8. Minutes of meeting on 10 January 1855 in *Proceedings of the Glasgow Philosophical Society* 3 (1848-55), 364.

9. *Exposition des Produits de l'Industrie de toutes les Nations, 1855. Catalogue officiel* (Paris, 1855). The only Scots whose work was shown were Adie (q.v.), Peter Stevenson (both of Edinburgh) and Alexander Gerrard of Aberdeen (*ibid.*, 326-327).

10. General Register Office (Scotland), Census of 1861 (Glasgow, Clyde), 644^5/51, p.18.

11. Examples of instruments known to have been sold with Cameron & Blakeney's name on them include an improved sympiesometer numbered 2534, offered for sale at Sotheby's, 15 December 1972, Lot 68; a wheel barometer, offered for sale at Christie's South Kensington, 12 January 1983, Lot 279. An 8-inch octant marked 'Cameron & Blakeney, Glasgow & Sunderland' is in the National Maritime Museum, inventory NA 9088. This same museum has an example of a dry compass card signed by Cameron, from the Admiralty Compass Observatory, inventory ACO.C/88.

12. Scottish Record Office, CS 318/8/80, Sequestration Papers, Paul Cameron & Co. There is no Sederunt Book, and these

papers consist of the Application for Sequestration, with notes added. Notices appeared in the *Edinburgh Gazette*: an initial notice announcing Buchan's Petition, summoning Cameron to Court to 'shew cause why sequestration of their estates should not be awarded' (23 August 1864, 1078); a second announcement stating that the Petition of 19 August had been dismissed (20 September 1864, 1181); a further notice of sequestration (23 September 1864, 1193); and the announcement of the appointment of the Trustee and Commissioners (14 October 1864, 1297).

13. Cf. Bryden, *op. cit.* (2), 44, 45. Cameron probably died c. 1870; he was not listed among the members of the Glasgow Philosophical Society in 1874 (*Proceedings of the Glasgow Philosophical Society* 9 (1873-75), 147-152).

14. G.R.O.(S.), Census of 1861 (Glasgow, Clyde), 644⁵/51, p. 18.

---

107. Azimuth dial: Cameron & Blakeney, Glasgow, c.1855 (T1980.205)

Circular patent azimuth dial, printed on paper and on a wooden base; with a calibrated index, apparently originally attached to the needle (now lacking), on a rudimentary pivot with a solar stile; the small recessed card compass (with 32-point compass rose and azimuth scale 0-90-0-90-0) probably a later modification. The dial is enclosed by a rotating glazed brass housing with provision for a sight. The printed dial plate is marked at the centre 'CAMERON & BLAKENEY / GLASGOW' and at the east point 'CAMERON'S PATENT'. A quadrant of the dial area contains a grid 'TO FIND THE DIFFERENCE OF LATITUDE AND DEPARTMENT ALSO THE SUNS ALTITUDE AT ANY TIME'; the remaining three-quarters with a compass rose (64-point) and perimeter azimuth scale, marked 'TO FIND THE SUNS AZIMUTH VARIATION AND DEVIATION AT ANY TIME'. The needle index divided in 'DISTANCE' [5]-60, and 'SINE' [5]-[90].

Diameter of compass plate: 210mm.
Overall diameter: 232mm.

Lacking the external sight, the original correction and suspension for the index, and the attachment support for the base (the outline of which is still visible).

The instrument follows Paul Cameron's U.K. patent 665 of 1853 covering improvements in marine and surveying compasses, including the use of a shadow-casting stile attached to the needle (rather than its support) and external telescopic sights. The layout of the improved compass in his specification differs in that the grid-divided area occupies the lower half of the dial plate. The instrument in this collection is probably a form provided for surveying use since it is not adapted for gimbal suspension. The modification of inserting a small card compass in the dial plate shows that the instrument has been used with the needle removed from the rotating index.

108. Refracting telescope: Cameron & Blakeney, Glasgow, c.1855 (T1980.258)

1½" aperture 3-draw hand-held refracting telescope with doublet objective, and a further long focus lens in the external aperture stop, with 2-component eyepiece and erecting assemblies; in brass, with objective ray shade and lacquered leather-covered barrel. Engraved on the first draw tube 'Cameron & Blakeney./Glasgow.' One draw slide scratched with the assembly mark 'IIII'. Objective cover with 1¼" aperture and sliding closure, marked in MS '28/6'; sliding closure on external eye stop.

Length closed: 295mm.
Tube diameter: 61mm.
Aperture: 39mm.

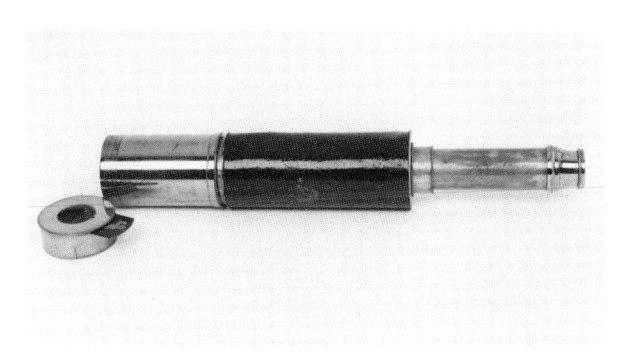

19

ALEXANDER DOBBIE

Alexander Dobbie (1815-1887) the founder of Alexander Dobbie & Son, was probably the nephew of John Dobbie, a Glasgow watchmaker active from about 1801 to 1845, and one of several members of this family involved in the watch trade.[1] Following a likely apprenticeship to John Dobbie, Alexander, then aged twenty-six, set up independently as a watch and clock maker at 20 Clyde Place in 1841, although he did not appear in the Glasgow street directory until 1842.[2] Before long, he appears to have specialised in chronometers, which he is listed as making from 1847 onwards, and from 1851 he advertised as a nautical instrument maker and chart seller.[3]

At least a proportion of his instruments were manufactured for him by London specialist makers, including octants by James Crichton.[4] He also had an early agency with Barraud & Lund of Cornhill, London for the supply of chronometers: according to an account written shortly after his death,

> 'Showing remarkable ability in his profession, and being a man of irreproachable character, he secured for his firm in 1850 the agency for Scotland of Barraud of London, the world-renowned chronometer maker, an agency that has been held ever since.'[5]

From 1884 Dobbie also described himself in the directory as a chronometer maker to the Admiralty. Dobbie's first submission of a chronometer to the Admiralty trials at Greenwich Observatory was in 1862 when his No. 2192 of 'ordinary construction' was rated, and four further instruments were submitted in the period to 1870.[6] Apart from the first instrument (which has a serial number which does not fit the Barraud series) the other four chronometers all have numbers greater than the highest number in this band recorded by Cedric Jagger, the point at which he proposed that Barraud's actual construction of chronometers ceased.[7] From at least the 1880s the firm also purchased direct from Thomas Mercer of London,[8] and the chronometer of about 1895 in the Frank collection is identified as being the work of the London chronometer springer and balance maker Robert Gardner, although this judgement may have to be qualified by the knowledge that chronometers in this same series were being offered by at least four other Glasgow chronometer makers, namely R. & W. Sorley of Argyle Street, James Weir of Buchanan Street, Whyte, Thomson & Co. (q.v.) of Broomielaw, and Laing of Brunswick Street.[9]

In 1857 Alexander Dobbie moved to 24 Clyde Place, and in 1873 took in No. 25 also. For a brief period around 1886 there was also a branch at 31 Centre Street. Dobbie employed a relation, William Dobbie, almost certainly his elderly father, who is reported to have defended the shop against looters during the Glasgow Bread Riots of 6 March 1848, when he managed to save part of the stock by locking it in the safe.[10] Another set-back for the business occurred in July 1872, when the workshops were completely destroyed by an explosion in the neighbouring Tradeston Flour Mills: eighteen people were killed, and a number of Dobbie's workmen were injured including

> 'John Dick, a boy ... employed in the shop of Mr Dobbie, compass maker, Clyde Place. Immediately upon the explosion being heard, the roof of the workshop fell in, and all in the place were buried in the ruins. They got out with some difficulty; the boy Dick escaped with a bruised left arm and a cut on the head. Thomas Gibson ... was also working in Mr Dobbie's when the accident happened, and received some severe cuts about the head and shoulders ... Samuel Forrest ... was working in Mr Dobbie's shop when the roof fell in. He has been cut and bruised about the arms, legs and shoulders.'[11]

Although the size of Dobbie's workshop in mid-century is not known, it seems from later figures to have been smaller than that of D. McGregor & Co. (q.v.), with which business Dobbie's had much in common. Moreover, it appears to have stayed at this size: in 1871 Dobbie was recorded in the Census as a master chronometer maker employing six men and two boys (i.e. apprentices), and ten years later his workforce consisted of six men and three boys.[12] Alexander was joined in the business by his youngest son, John Clark Dobbie, who was working as a ship-broker's clerk in 1881,[13] but was designated a nautical instrument maker at the time of his father's death on 18 February 1887. The firm had become Alexander Dobbie & Son in 1886, when John became Alexander's partner. After an agreement with his family and the other executors, John bought his father's share of the business for £1800.[14] Alexander Dobbie was survived by his wife Mary Wilkie Brown, and their three daughters and four sons.[15]

The subsequent expansion of the firm in the 1890s

55 Sir James Johnston Dobbie, Kt., F.R.S., L.L.D., M.A., D.Sc., (1852-1924)

seems to have been due to the talent and energy of John C. Dobbie. In 1889 he not only acquired extra premises in Commerce Street, but also established his workshop at 120 Broomielaw. In 1892 the firm moved its shop to 44 & 45 Clyde Place, formerly the premises of M. Walker & Son.[16] In the following year, 1893, it vacated the Broomielaw works, while retaining the Commerce Street premises. In 1896 Dobbie turned the firm into a limited company. An associated firm known as Dobbie, Son & Hutton was established in London in 1894; another offshoot, Dobbie, Hutton & Gebbie was formed in Cardiff in 1899. Meanwhile, branches of Dobbie & Son were established at 28 Cathcart Street, Greenock, and in South Shields in 1897. John C. Dobbie had a share in these associated firms[17], and in 1893 bought the business T.S. McInnes of 41 & 42 Clyde Place, changing its name to T.S. McInnes & Co. Ltd.[18] He managed the two businesses separately until 1903, when he amalgamated them to form Dobbie, McInnes Ltd., with a head office at 45 Bothwell Street. A Liverpool branch was opened in 1907 and in 1908 the head office moved to 57 Bothwell Street, Glasgow.

By the time of the amalgamation, Alexander Dobbie & Son was a substantial specialist firm dealing, like D. McGregor & Co. (q.v.), in most aspects of nautical instrument production and retailing, as well as producing non-maritime apparatus, acting as chemical, mathematical, optical and philosophical instrument makers; they were Admiralty Chart Agents as well as Chronometer Makers to the Admiralty, and advertised as wholesale clock manufacturers.[19] And like most other nautical makers, Dobbie's had a specialist group of compass adjusters who adjusted instruments at sea.[20] Of the greatest significance (although not perhaps in a purely commercial sense) was the design of nautical instruments, for which John C. Dobbie was largely

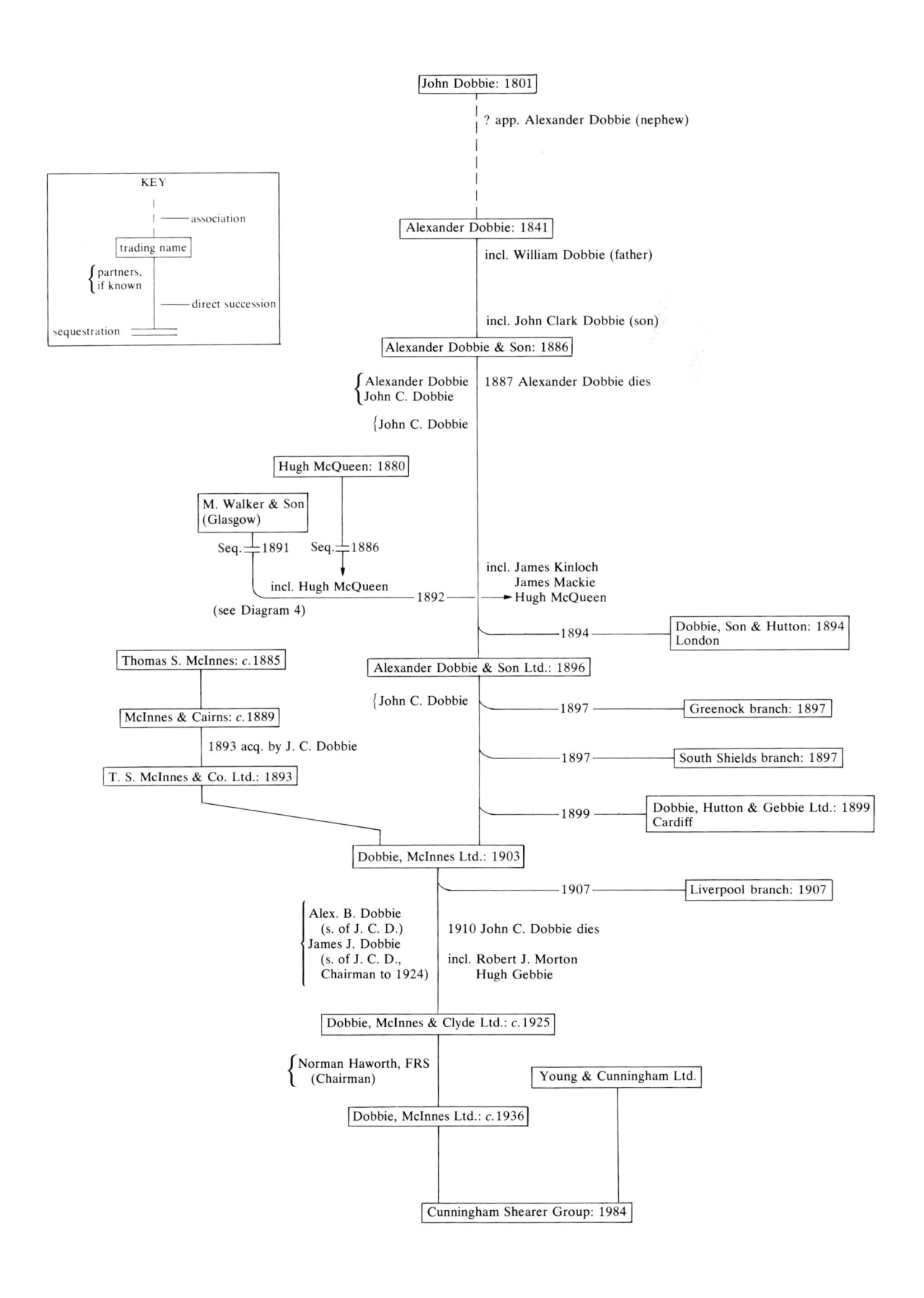

Diagram 3. Schematic development of the Dobbie business.

responsible. Between 1887 and 1910 he took out seventeen patents, five of which were for magnetic compasses and three for mariners' compasses; seven were for sounding machines, one for ships' logs and one for a bearing instrument.[21] Statistically, John C. Dobbie's technical inventiveness may be compared to that of W.D. Whyte of Whyte, Thomson & Co. (q.v.) who took out some twenty patents between 1891 and 1914. However slight some of these may have been, Dobbie has been credited with a particular innovation which subsequently altered compass design throughout the world. The wording of the patent of 23 April 1901 is modest enough: 'In fluid compasses, the card may be made smaller than usual with the same object of steadying the compass bowl.'[22] The principle of a reduced diameter card in liquid magnetic compasses was later adopted as standard by other makers.

The other half of the new company, T.S. McInnes, was founded in about 1885 and manufactured engineering and mathematical instruments, specialising in a patented steam engine indicator extensively used by the Admiralty. Following the founder's death, and Dobbie's 1893 purchase of the firm, four successful variants of the indicator were in production by 1901; other instruments being produced in quantity were steam pressure, vacuum and hydraulic gauges, revolution counters and speed indicators, calorimeters, clocks for ship and engine room use, and general engine and boiler fittings.[23] The amalgamation of 1903 does not appear to have affected the individual products of the two constituent workshops, which continued to be advertised as before. The only noticeable difference is that the engine indicators became known first as 'McInnes-Dobbie' and later from 1920 as 'Dobbie-McInnes'. An anonymous account of the firm went so far as to say 'This fruitful merger enabled them to co-ordinate and develop both sides of their business ...' in producing instruments both for the bridge and the engine-room. John Clark Dobbie was replaced as head of the firm by his eldest brother, Professor Sir James Johnston Dobbie,[24] who in 1903 had vacated his chair at Bangor to take up his appointment as Director of the Royal Scottish Museum in Edinburgh. He left the Museum in 1909 to become Principal of the Government Laboratories in London, and was knighted in 1915.[25] He was chairman of Dobbie McInnes until his death in 1924.

The First World War gave further impetus to the firm, and a logical extension of its business in the post-war years took it into the aircraft and automobile industries. It operated as Dobbie McInnes & Clyde Ltd. between about 1925 and about 1935, before reverting to the old name Dobbie McInnes Ltd.[26] After the Second World War, under the chairmanship of the Nobel Prize winner Sir Norman Haworth (who had married a daughter of Sir James J. Dobbie), a scientific research and development department was formed. The firm branched out into electronic recording and computing equipment, and its output climbed in the post-war years.[27] In 1984 it was announced that

> 'The old established Glasgow firm of Dobbie McInnes Ltd., has joined Young & Cunningham Ltd., in the Cunningham Shearer Group of companies. Dobbie McInnes will continue to handle and develop their own business in conjunction with the remote control and gauging business of Young & Cunningham.'[28]

REFERENCES

1. *Glasgow Post Office Directory, passim*. John Smith has also recorded an Andrew Dobbie, active from 1820 after serving an apprenticeship with John Dobbie, until 1848 (John Smith, *Old Scottish Clockmakers* 2nd edition (Edinburgh, 1921), 111-112). Alexander Dobbie's father William (b.1781) may also have been trained in the watchmaking trade, and in 1848 was working with Alexander; his elder brother Thomas (b. 1804) may have been the watch and clockmaker operating in Adelphi Street from 1828 until after 1845. The relationship of William, John (b.1784) and Alexander has been established from ages in census returns and entries in the International Genealogical Index, British Isles, Scotland, Fiche F0307 (May 1988), Lanarkshire.

2. *Glasgow of Today* (Glasgow, 1888), 173. The agency is also noted in C. Jagger, *Supplement to Paul Philip Barraud* (Ticehurst, Sussex, 1979), 193.

3. An example of the elaborate binnacles fashionable in the mid-nineteenth century was preserved in the firm's offices in 1953. It was made by Dobbie for the wealthy grocer Sir Thomas Lipton's yacht *Shamrock*, and was carved with the thistle, rose and shamrock, and a legend in Gaelic. Other types were often of three dolphins in cast brass (Anon., 'Dobbie McInnes', *Glasgow Chamber of Commerce Monthly Journal* 36 (1953), 243-244).

4. An ebony octant, signed 'Crihton [*sic*] of London Made for A. Dobbie, Glasgow' was offered for sale by Sotheby's, 29 April 1977, Lot 101; another, signed 'Crichton London made for A. Dobbie Glasgow' in a case with a trade card for J.R. Cameron of Liverpool, was offered by the same saleroom, 18 June 1986, Lot 132; and again, 28 October 1986, Lot 19. Other instruments signed by Dobbie alone include an ebony and brass octant, signed 'A. Dobbie, Glasgow', offered by Phillips Leeds, 22 June 1983, Lot 332; a further similar item signed 'Alexander Dobbie, Glasgow' in a case with a trade label for J. Sewill, 61 South Castle Street, Liverpool, was offered for sale by Christie's South Kensington, 18 July 1985, Lot 179.

5. *Glasgow of Today, op. cit.* (2).

6. *Rates of Chronometers on Trial for Purchase by the Board of Admiralty at the Royal Observatory, Greenwich, 1862* (London,

1862). The other instruments in the Greenwich Trials were 3715 in 1864 and 1865, 3987 in 1866 and 1867, 3989 in 1869, and 4323 in 1870.

7. C. Jagger, *Paul Philip Barraud* (London, 1968), 136; idem, *op. cit* (2), 203-4, 246.

8. T. Mercer, *Mercer Chronometers* (Ashford, Kent, 1978), 170-172. The first purchases recorded in surviving Mercer records, in 1885, are of instruments M/3376 - M/3381, using the 'M' numerator also found on McGregor (q.v.) chronometers. Instruments 3382-86 initiated purchases in 1886, which continued with chronometers numbered in the main Mercer series, from 4328.

9. The Frank collection chronometer (NMS T1980.208) has serial number 5/3987: it can be dated by name and address information to 1892-96, and carries a repair date of 1896. The Greenwich Trials included 5/3933 submitted by Sorley in 1889, and 5/3931, with a Loseby balance (R.T. Gould, *The Marine Chronometer* (London, 1923), 198-200) and improved mount by Robert Gardner of Lloyd's Square in 1892. Subsequently, 5/3947 and 5/3963 (Weir) in 1893, 5/3947 (Gardner), 5/4050 and 5/4051 (Whyte, Thomson & Co.) in 1895, and 5/4005 (Weir) in 1896, were described in almost identical terms as having balance compensation acting in extremes and reverse detents. Gardner's 5/4162 was the best performing instrument in 1897 (*op. cit.* (6) 1893-1897).

10. 'Mr William Dobbie, however, with admirable foresight and courage, had locked some of the most valuable articles in the premises into the safe and hid the key ...' (*Glasgow of Today, op. cit.* (2)); Smith, *op. cit.* (1), 111 quotes the *Edinburgh Evening Courant*: 'Serious Riot on Monday 5th March 1848. Mob attacked the shop of Mr Dobbie, Clyde Place, watchmaker, and presented a pistol at the head of the shop boy, threatening to shoot him if he dared to offer any resistance. They then proceeded to rifle the shop of the gold and silver watches it contained; some of the marauders were observed running out of the shop with their pockets full of watches. Mr Dobbie himself was wounded in defending his property.' The report in the *Glasgow Herald*, 10 March 1848, was more sober: '... The shop of Mr Dobbie, watchmaker, in this neighbourhood [Clyde Place] was also attacked and rifled.'

11. *Glasgow Herald*, 11 July 1872.

12. General Register Office (Scotland), Census of 1871 (Glasgow, Tradeston), 644⁹/64, p.10; Census of 1881 (Glasgow, Kelvin), 644⁹/44, pp. 10-11.

13. G.R.O.(S.), Census of 1881 (Glasgow, Kelvin), 644⁹/44, pp. 10-11.

14. Scottish Record Office, SC 6/44/48, pp. 943-949. Inventory of Alexander Dobbie's estate in Record of Inventories, Sheriff Court of Ayr. This was also announced: 'The Partnership interest of the late Alexander Dobbie, Chronometer and Nautical Instrument Maker in Glasgow, in the Firm of ALEXANDER DOBBIE & SON, Chronometer and Nautical Instrument Makers there, has been acquired by John Clark Dobbie, sole remaining Partner of said Firm, who will continue to carry on the said Business under the same Firm Name for his own behoof. [Signed] MARY W. DOBBIE. JAMES J. DOBBIE. JOHN C. DOBBIE. ALEXR. B. DOBBIE. Sole Trustees of the late ALEXANDER DOBBIE. JOHN C. DOBBIE. As an Individual.' *Edinburgh Gazette*, 29 November 1887, 1209. The original of this document was witnessed by two compass makers, who were presumably employed by the firm, Alexander S. Wood, and George Allwood; however, nothing further is known of them.

15. Mary Wilkie Brown was born c.1820 at Eastwood, Renfrewshire. Their eldest son William Wilson Dobbie died in Australia shortly after his father's death. James Johnston Dobbie graduated at Glasgow University in 1875, became Professor of Chemistry at the University of Bangor, North Wales, in 1884, and was later head of the firm (see below). The third son, Alexander Brown Dobbie, began as a merchant's clerk, was later an engineering draughtsman and assistant to the Professor of Engineering at Glasgow University in the 1880s (S.R.O., SC 6/45/19, pp.92-97; Alexander Dobbie's Trust Disposition and Settlement in Records of Testamentary Deeds, Sheriff Court of Ayr, 27 September; G.R.O.(S.), Census of 1871 (Glasgow, Tradeston), 644⁹/64, p.10, and 1881 (Glasgow, Kelvin), 644⁹/44, pp.10-11; *Glasgow Post Office Directory, passim*).

16. See Chapter 20, 'Duncan McGregor' (q.v.).

17. *Glasgow Post Office Directory, passim*.

18. Angus McLean (ed.), *Local Industries of Glasgow and the West of Scotland* British Association for the Advancement of Science Handbook (Glasgow, 1901), 89.

19. *Glasgow Post Office Directory* 1890-1901. At the Glasgow International Exhibition in 1888, they exhibited 'binnacles, compasses, chronometers, aneroid and mercurial barometers, clocks, patent logs, sounding machines, fog-horns, sextants, binocular glasses, night glasses, opera glasses, telescopes, thermometers, palinuruses, charts and nautical publications': *International Exhibition Glasgow 1888: Official Catalogue* (Glasgow, 1888), 95. Apart from the sextant in the Frank Collection (NMS T1980.194), a combined sympiesometer and stick barometer was offered for sale by Phillips Edinburgh, 27 September 1985, Lot 89.

20. They included, with their known dates of activity bracketed: James Kinloch (1887-1900+), James Mackie (1890-1920+), Robert James Morton (1909-10), Hugh Gebbie (1909) and Hugh McQueen (1892-1909). McQueen had previously been with M. Walker & Son, an offshoot of D. McGregor & Co. (q.v.), as had W.W. Gebbie, a compass adjuster flourishing in Greenock c.1900, who may have been related to Hugh Gebbie (*Glasgow Post Office Directory, passim*). McQueen appears previously to have been in business at 63 Clyde Place as a chronometer maker from at least 1880 until his bankrupcy in 1886 (Mercer, *op. cit.* (8), 205).

21. U.K. patents 14,783, 31 October 1887; 1915, 27 January 1899; 12,658, 17 June 1899; 25,712, 30 December 1899; 8305, 23 April 1901; 9661, 29 April 1903; 24,881, 16 November 1903; 10,314, 5 May 1904; 2720, 10 February 1905; 4702, 7 March 1905; 13,527, 1 July 1905; 19,224, 27 August 1907; 28,197, 21 December 1907; 24,248, 11 November 1908; 7926, 2 April 1909 (addition to 19,224/07); 20,563, 8 September 1909; 2277, 29 January 1910.

22. U.K. patent 8305, 23 April 1901. The Dobbie patented compass and sounding machine were being sold in 1903 (*Glasgow Post Office Directory* 1903).

23. McLean, *op. cit.* (18), 89.

24. *Chamber of Commerce, op. cit.* (3), 243.

25. *Who was Who. Vol.II (1916-1928)* 4th edition (London, 1967), 291; also obituaries in the *Glasgow Herald*, 21 June 1924

and the *Scotsman*, 21 June 1924: 'It was mainly through his efforts that the Egyptian collection there [i.e. the Royal Scottish Museum] was increased in size'.

26. *The Engineer* 147 (1927), 494; *ibid.*, 160 (1935), 315; and *ibid.*, 164 (1937), supplement no. 1, x.

27. *Chamber of Commerce, op. cit.* (3), 243. C.A. Oakley (ed.), *Scottish Industry: An Account of What Scotland Makes and Where she makes it* (n.p.[Glasgow], 1953), 102-103.

28. *Glasgow Chamber of Commerce Journal* 69 (1984), 260.

---

109. Sextant: Alexander Dobbie & Son, Glasgow, c.1890 (T1980.194)

6½" sextant (incomplete) with cast 3-circle frame, in oxidised bronze and brass (distressed). Scale divided on silver, [-5]-[160], the reinforced index arm with clamp, tangent screw and vernier reading to 10 secs. Engraved on the arc 'Alexander Dobbie & Son, Glasgow.' and stamped '3915'. Screw adjustment to the ring support for the sight tubes. The frame with rear handle, and supported on feet.

Scale radius: 156mm.

Lacking sight tubes and telescopes, index mirror, both sets of shades and swinging magnifier.

110. Marine chronometer: Alexander Dobbie & Son, Glasgow, c.1895 (T1980.208)

Two-day marine chronometer, with Earnshaw's spring detent escapement, free-sprung steel balance spring and 2-arm bimetallic balance with cylindrical weights. Spotted movement with fusee and chain and jewelling to the balance and escape arbors. Stamped '3987' on the dial support ring, on the front plate and inside the bowl. Silvered bezel and dial; the dial with Roman chapter, subsidiary seconds, 56-hour winding indication, and engraved 'Alex[r] Dobbie & Son, / MAKERS TO THE ADMIRALTY, / 45, CLYDE PLACE, GLASGOW' and '5/3987'. The reverse of the dial stamped '3987' and scratched 'C Perks[?]/ HONG KONG / 17/6/96 / 10 9 1901'. Glazed chronometer bowl in gimbals, engraved on the base 'H.S. [broad arrow] 1'. In a mahogany deck case with a glazed top, originally with an upper lid, and with an inset name plate 'ALEX[R] DOBBIE & SON, / 5/3987 / GLASGOW.' The case containing a repair label of Motion Smith & Son Ltd., Singapore, February 1940, and an H.M. Chronometer Depot, Bradford-on-Avon, issue label of 18 July 1945, No. 217. Winding key.

Movement plate diameter: 82mm.
Bezel diameter: 126mm.
Case size: 185 x 185 x 170mm.

The upper lid of the case is lacking, and the present upper surface restored.

Provenance: Recorded in the Admiralty Chronometer Records at the National Maritime Museum, Greenwich, as 'Hydrographic Property (ex Min[istry] of Sea Transport)', first noted on H.M.S. *Bulan* in July 1943 and taken into stock in August 1943; subsequently purchased as surplus from the Admiralty in 1964 by C. Frank Ltd. (We are grateful to Beresford Hutchinson for providing these details). The engraved 'H.S.1' mark indicates that it has been an ocean-going chronometer (as opposed to a deck watch or hack watch). This instrument is quite unrelated to chronometer No. 3987 submitted to Greenwich Trials by Alexander Dobbie in 1866 and 1867 (*Rates of Chronometers on Trial for Purchase by the Board of Admiralty, at the Royal Observatory, Greenwich, 1866, 1867*).

111. Refracting telescope: Dobbie McInnes, Glasgow, Greenock, South Shields & London, c.1905 (T1980.268)

2" aperture single-draw hand-held refracting telescope with doublet objective and 2-component eyepiece and erecting assemblies; in brass, with objective ray shade, the barrel originally leather-covered. Engraved on the draw tube 'Dobbie, McInnes L^{td}/ Glasgow, Greenock, / South Shields & London.' Sliding closure on external eye stop.

Length closed: 555mm.
Tube diameter: 57mm.
Aperture: 49mm.

Lacking the leather covering to the barrel, and also the objective cover. A reinforcing sleeve has been added where the rear ferrule has become detacted from the barrel, and an inexpert soldered repair has been made to the eye stop.

112. Refracting telescope: Dobbie McInnes, South Shields & Liverpool, c.1910 (T1980.269)

1⅜" aperture single-draw hand-held refracting telescope with doublet objective, and 2-component eyepiece and erecting assemblies (defective), the latter mounted at the division of the draw tube; in brass, with objective ray shade and attached external aperture (the same as that of the objective), and with leather-covered barrel. Engraved on the draw tube 'Dobbie McInnes Ltd, / GLASGOW, SOUTH SHIELDS & LIVERPOOL.' The objective mount scratched with assembly mark 'IX' or 'XI'. Sliding closure on external eye stop.

Length closed: 517mm.
Tube diameter: 59mm.
Aperture: 35mm.

Lacking the forward lens of the image erecting assembly, and the closure slide of the objective aperture.

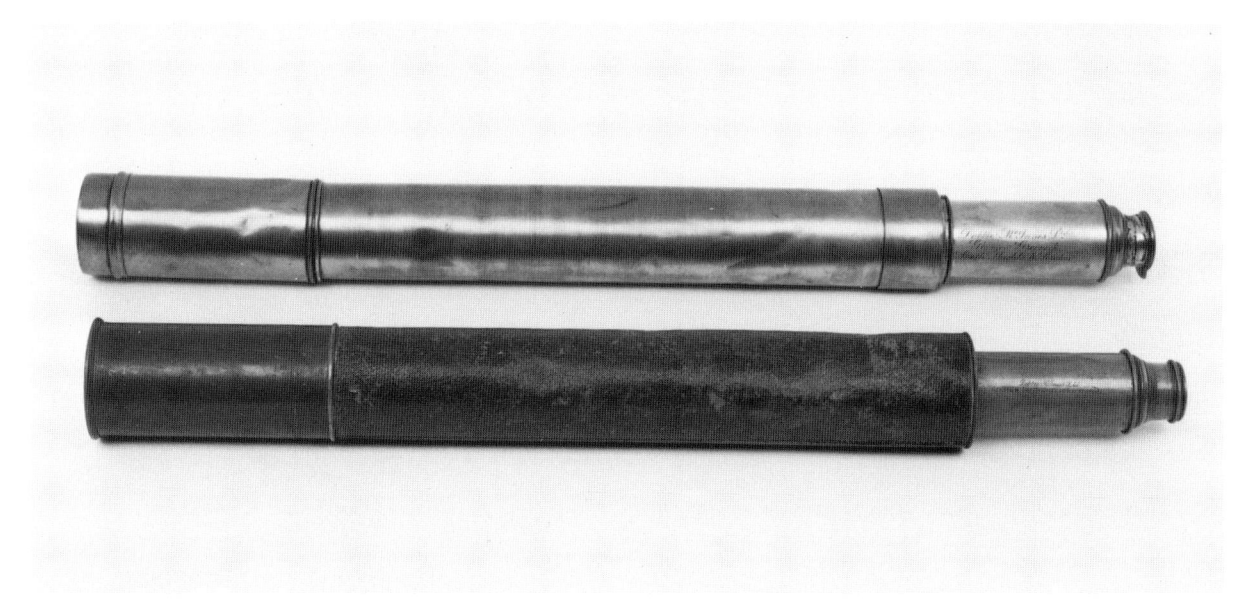

111 top, 112

20

DUNCAN McGREGOR

The firm originated with Duncan McGregor senior (c.1803-1867), who first appeared in the 1844 Glasgow street directory as a nautical instrument and chronometer maker and chart seller at 24 Clyde Place.[1] He was presumably the Mr McGregor, ship chandler in Clyde Place, from whose shop chronometers worth £400, telescopes and nautical publications were looted during the Glasgow Bread Riots on 6 March 1848.[2] McGregor lived initially in Greenock and may have been apprenticed to the Heron business (see Whyte & Thomson (q.v.)).[3] He was subsequently in partnership with David Heron in Greenock, possibly from 1827 when Heron opened his Glasgow outlet: the partnership was dissolved in 1836, but McGregor continued at the same address, advertising himself as 'successor to Heron, nautical, optical and stationery warehouse and exporter of nautical instruments'.[4] The premises at 1 William Street are also listed for the watchmakers Heron & Son, founded by John Heron, who is believed to have been David's father, together with a further son James.

McGregor's original Glasgow premises were moved in 1855 to 38 Clyde Place, and in 1858 took in Nos. 39 and 40. The expansion of business which this suggests is confirmed by the change of name in 1856 to D. McGregor & Co. From 1869 until 1890 the street directories record the firm operating from 44 and 45 Clyde Place; it was listed as at 37 and 38 in 1885, but it was not apparently until 1891 that it returned permanently to that address. From 1902 to 1908 the premises consisted only of 37 Clyde Place; in 1909 the firm moved to 57 Bothwell Street. The Greenock shop had meanwhile been maintained; it was at 1 William Street in 1836, 8 William Street in 1853, 32 Cathcart Street from 1861 or earlier until 1885, 36 Brymner Street from 1886, and 33 Cathcart Street in 1923. English branches were opened at Liverpool in 1879 and in London in 1886.[5]

If in 1848 McGregor could be described merely as a ship chandler, by 1860 (when the firm had begun advertising as compass adjusters) the business was already substantial, and census return evidence suggests that the Glasgow workshop was larger than that of a near neighbour Alexander Dobbie (q.v.), and of comparable size with that of James White (q.v.). In 1861 Malcolm McNeil Walker, who was McGregor's manager by this date, was listed as a nautical instrument maker employing '9 men & 6 boys';[6]

Dobbie's workshop is not given at the 1861 census, but was '6 men & 2 Boys' a decade later, while the size of White's labour force was similarly not mentioned in 1861, but he was 'employing 14 men 4 boys' in 1871, and '60 men' in 1881.[7]

As with other firms, the range of their classified entries in the street directory widened, which can be taken at least as an indication of a growing confidence in their own work, if not an absolute increase in workshop production to meet demand.[8] By 1870 the firm was advertising as makers of optical and mathematical instruments, as well as of chronometers and nautical instruments, and ten years later as makers also of chemical and philosophical instruments. In 1868 they won a silver medal at the International Maritime Exhibition at Le Havre, 'for the excellent make of their marine binnacles and liquid compasses' and twenty years later they exhibited a range of nautical instruments at the International Exhibition in Glasgow.[9] The firm advertised themselves as chronometer makers to the Admiralty, and from 1876 acted as agents for Sir William Thomson's patent compass. Much of their attention was devoted to compasses and binnacles, and to chronometers supplied not only to the Royal Navy, but also to Greenwich Observatory and 'foreign governments'. By the 1890s they were also wholesale clock manufacturers and makers of ships' telegraphs, in addition to acting as 'ship furnishers', an echo of the early days of the firm.[10]

After a period in retirement,[11] Duncan McGregor senior died in 1867[12] a very wealthy man. His movable estate alone was worth £7,858 14s 7d, and in addition he left properties in Greenock and Glasgow.[13] Duncan junior, then aged about twenty-five, had been working as an optician and nautical instrument maker since before 1861.[14] He appears to have inherited the Glasgow stock and business, and to have continued to run it; his elder brother Malcom (sic) inherited the Greenock shop and stock,[15] but seems to have sold out to his brother so that both shops were run together as before. By 1873, there were three partners: Duncan McGregor junior, Malcolm NcNeil Walker, and John G. Walker.[16] Duncan McGregor junior was active outside his workshop both as a Fellow of the Royal Geographical Society and a member of the Glasgow Philosophical Society from about 1874 to 1898.[17] He was also unusual among his fellow Glasgow makers for

235

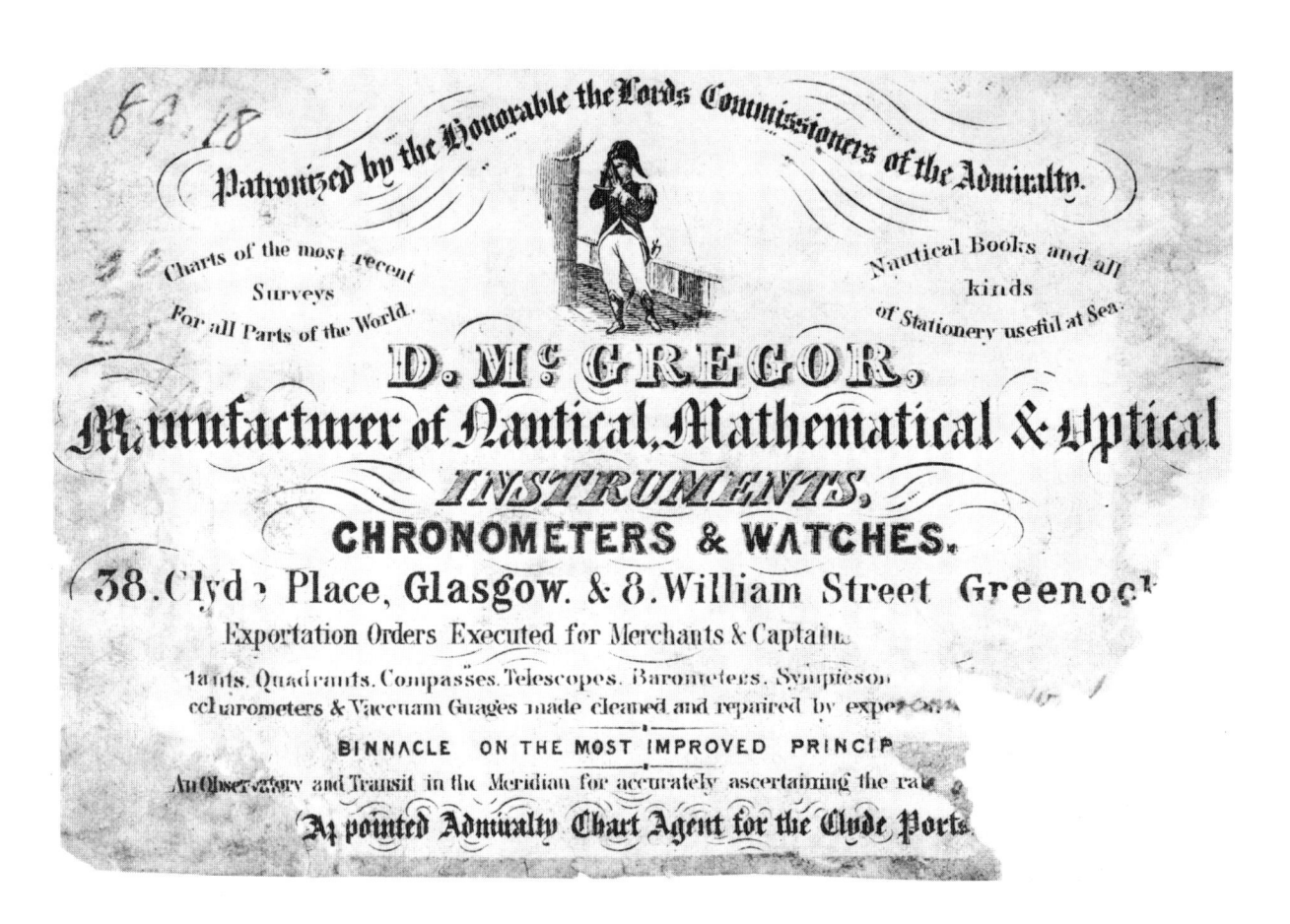

56 Duncan McGregor's trade card for his addresses in 1855-6.
Trustees of the Science Museum, London

having contributed a paper to the Edinburgh-based Royal Scottish Society of Arts in 1861, on the mechanism for a system of tide signals, first proposed by the Secretary to the Northern Lighthouse Board. He was advised by another Fellow of the Royal Geographical Society, Commander E.J. Bedford of the Royal Navy's West Coast Survey.[18]

Duncan McGregor's early promise was fulfilled, for he was responsible for a series of patented improvements to the magnetic compass which earned his firm an enviable reputation. He took out three patents for compasses in 1875, 1880 and 1893[19] and if a not disinterested account of the firm in 1882 can be taken at face value,

> 'The compass of D. McGregor ... is reported to fulfil all the conditions of a good instrument, and is as geometrically and mathematically correct as a compass can well be. This instrument is the outcome of many years' studious labour and widely extended experience.'[20]

The same source credits McGregor with an 'inventive faculty' and goes on to describe the business. The various types of liquid compass made by the firm were apparently in great demand, and had been supplied to eleven shipping lines and the Argentine Navy. They included the 'Anti-Vibration Polar Compass', in which the card was steadied by reducing frictional error to a minimum, and the Patent Standard Compass. McGregor also designed a patented Azimuth Reflecting Circle, which fitted to the top of a compass glass and was detachable, and which took azimuths by double reflection. The firm also produced a Compass Verifier for checking a ship's course by its latitude and the declination of the sun.[21]

Surviving records of the firm's activities as chronometer makers begin towards the end of Duncan McGregor senior's life. In 1859 McGregor & Co. had a two-day chronometer, No. 2760, rated in the Greenwich Observatory trials for Admiralty purchase, and in 1861 their No. 2776 had the best performance of the instruments tested.[22] Even better rates were achieved in 1866 when their No. 3795 again out-performed other instruments.[23] All these instruments, and others submitted by McGregor & Co. up to their last instrument rated at Greenwich in 1876, had some form of auxiliary compensation to the balance, and frequently the form introduced by the London chronometer manufacturer, John Poole.

A good trial result leading to purchase by the Admiralty would entitle the maker to describe himself as 'Chronometer Maker to the Admiralty' even though the chronometer need not have been made or even finished by him.[24] The extent to which McGregor actually manufactured these instruments is unknown, but it is likely that the majority would have been bought from specialist makers. The serial numbers of seventeen chronometers submitted to the Greenwich trials have a range from approximately 2500 to 4500 in the period from 1860 to 1873, but the series appears to be discontinuous with some large gaps: these may be from the number series (perhaps itself discontinuous) of another manufacturer.[25] There are indeed some similarities with the number series of Poole chronometers apparent from the Greenwich trials.

One wholesale supplier who was certainly used by McGregor was Thomas Mercer of London, for which there are surviving records.[26] Between the mid-1860s and 1872, approximately one hundred two-day chronometers were sold to McGregor & Co. at £20 to £22 each. These were given serial numbers from about 1900 to 2000, in a special series which is separate from Mercer's main series. In 1873 the numbers continue from 2500 to about 2650 in 1885 (and include chronometers ordered for the Greenock branch), and thereafter they continue the separate series from 7500 which Mercer had begun for McGregor's Liverpool branch in 1879 (the initial batch being charged to Glasgow), reaching about 8220 at the last recorded order in 1917. It is possible that these instruments were then sold under a totally separate McGregor numbering series, or else that (with the exceptions noted below) none of the Mercer instruments were entered for Greenwich trials.

As well as the major discontinuity of Mercer numbers in 1873, McGregor chronometers start appearing from this date with serial numbers in the form of fractions, with the usual four figure number as a denominator and a lettered numerator. Since only 'M' numbers are recorded in the Mercer records it may be that the 'M' denotes supply by Mercer: two such chronometers which can be identified as Mercer instruments were submitted by McGregor to the Greenwich Trials of 1874 and 1876.[27] By analogy it is conceivable that the other series of numbers that has been recorded, prefixed by 'F', were of Frodsham-supplied chronometers, although it may equally be a code for another manufacturer.[28]

Business under Duncan junior prospered. By 1882 production of instruments was divided between Liverpool and Glasgow. Attached to the Liverpool showroom at 72 South Castle Street were workshops where 'a considerable amount of manufacturing is done'; there, the firm was privileged to have a time signal communication relayed from the Royal Observatory, Greenwich.[29] By contrast only 'a certain proportion' of manufacture was done at Glasgow; this included production of the Anti-Vibration Compass, the Azimuth Reflecting Circle and the Compass Verifier. The works were 'well equipped and efficiently laid out'. A large brass foundry, which could produce castings of considerable size, adjoined a brass-finishing and erecting shop with equipment for producing 'that fine finish which is always found on the Messrs.

McGregor's productions. Numerous binnacle and compass makers find here steady employment.' A large room contained completed instruments, ready to be sent to the vessels for which they were made. Upstairs were workshops producing chronometers, ships' clocks and engine counters.

The 'saleshop' displayed chronometers, sextants, marine glasses, telescopes, barometers, charts, watches, and a large stock of nautical books. McGregor's were also Admiralty chart agents, and agents for the Meteorological Department of the Board of Trade.[30] Like James White (q.v.) they published tide tables, and also *The Engineers Annual*, and in 1888 were running what they called 'McGregor's Naval Academy' at 37 and 38 Clyde Place, which provided lessons for candidates for Marine Board Examinations and other certificates.[31]

On 29 October 1878 McGregor and Malcolm McNeil Walker together entered the Glasgow Incorporation of Hammermen,[32] an event which underlined their partnership in the firm, and which marked their status in the commercial and industrial community. Because the Incorporation was in no sense a trade organisation by this date, their membership of it as instrument makers is insignificant. Walker was also a member of the Glasgow Philosophical Society from 1853 to 1895, and was listed there as a Fellow of the Royal Astronomical Society.[33] He was born the son of a cooper from Luss, Dunbartonshire, in about 1825,[34] and was with McGregor's from 1852 or earlier.[35] By the 1880s he was the only partner other than Duncan McGregor with the Liverpool side of the firm, and with his son, John Gray Walker, one of the three partners of the Glasgow and Greenock end. In 1885, aged about sixty, he set up independently as M. Walker & Son, a business which appears almost identical to McGregor's.[36]

M. Walker & Son were based at 44 and 45 Clyde Place, Glasgow and 32 Cathcart Street, Greenock (which necessitated McGregor's removal to 36 Brymner Street), and opened a branch at Bath Street, Liverpool in 1887.[37] Walker's independence was short-lived and troubled; firstly, his two partners - his son and son-in-law - left him,[38] and although he found others to replace them, he subsequently went bankrupt in April 1891, and

> 'the estates of the said M. Walker & Son, Malcolm McNeill Walker, William Williamson Gebbie and Robert Weir [individual partners], were sequestrated by the Sheriff of the County of Lanark at Glasgow ... The bankrupts' liability amounted to £8758 12s 11d.'[39]

A major creditor was the instrument firm James White (q.v.), which was owed £1202 13s 5½d.[40] The stock, which was assessed by James Thomson of Whyte, Thomson & Co. (q.v.), was valued at £1468 18s 10d for the Glasgow shop and £578 2s 8d for the Greenock branch.[41] Walker was examined on 8 May 1891, and some detail about the operation of both D. McGregor's business and his own is revealed in his statement:

> 'I am the senior partner of the firm of M. Walker & Son. That firm was constituted about March 1888. I first began business in 1854 when I was assumed a partner of the firm of D. McGregor & Company; my senior and only partner at that time was laid aside through illness. I did not put any capital into the firm of D. McGregor & Company ... Between the date of the dissolution of D. McGregor & Company [which Walker could not remember; he had fulfilled a ten years' contract from 1854, and a subsequent one of about 'five or seven years ... to about 1871'] I did not carry on business in my own name: I formed a partnership with my son and son-in-law, and carried on business under the name of M. Walker & Sons. That continued till the constitution of the present partnership in 1888. When the firm of D. McGregor & Company dissolved my capital in that concern was ... somewhere about ten thousand pounds, and on the dissolution of that firm I acquired the business.'[42]

Walker was uncertain how much of his £10,000 was in cash, but said that no money was paid to him on that occasion; it was assets in the business which he had taken over.

> 'After 1871 I took in the son of D. McGregor and carried on business till 1885, and after 1885 I took in my son and son-in-law and carried on business under the firm of M. Walker & Son. At this time the present firm was constituted ... by my assuming two old employees, Mr Weir and Mr Gebbie. These gentlemen had been in my employment for a good number of years ... The whole assets consisted of goods, tools, furniture and sundry debts in my shops in Glasgow and Greenock. The whole amount of cash put into the concern by me is about £45.
> Q: So that this firm seems to have started with a great amount of stock and no working capital.
> A: Precisely. One of the provisions of the contract was that the firm was to take over the liabilities of my previous firms of M. Walker & Son.'[43]

Among these 'liabilities' was an overdraft with the Clydesdale Bank of £1166 3s 11d; another with the British Linen Bank for £1097; a third with the Commercial Bank for £91 13s, besides problems with Malcolm Walker's personal estate and debts impinging on the business. It seemed that Walker's brother-in-law, John C. Yuill, had taken charge of Walker's finances since May or June 1888. Walker's partners had not been consulted; neither did they know that Yuill was being paid £20 a month for his services out of the firm's profits.[44] It further emerged that Yuill had been given power of attorney and was able to act for Walker both as an individual, and on behalf of the firm. Since 1888, Walker had withdrawn £3170, when the deed of copartnery had allowed for a maximum of £200 *per annum*.

William Williamson Gebbie, one of Walker's partners, who had no estate except his household goods and life assurance, declared under examination that he had heard Walker's statement 'and I generally concur with him as to what he said regarding the business ... I have practically no knowledge with regard to the financial affairs of the firm. These were left in the hands of Mr Yuill.'[45] He stated that he and Weir shared responsibility for the trading, which carried on at a loss; Gebbie was still employed by the Greenock business. In summary, it was agreed that the immediate reasons for the trading loss were fourfold: firstly, large sums drawn by Walker for private use; secondly, large unauthorised sums drawn by Yuill; thirdly, the large amount of contracts left by the firm's predecessors; and fourthly, £600 of watches sold to Walker but not credited to sales.[46] The third partner, Robert Weir, who had no estate except his household goods, concurred with Gebbie's account.[47]

In the long list of the firm's creditors were mentioned: James White (q.v.), was owed £1202 13s 5d; James Imray & Son, 89 Minories, London, £152 18s 2d; Joseph Levi & Co., 40 Furnival Square, London, £107 5s 1d; J.D. Potter, 31 Poultry, London, £78 18s 1d and a further £10 14s 4d; A. Dobbie & Son (q.v.), £29 9s 11d; Edward Cetti, 36 Brooke Street, Holborn, £6 16s; Victor Kullberg, 105 Liverpool Road, London, £1 15s; Whyte, Thomson & Co., (q.v.), £1 4s; James Chesterman & Co., Eccelsall Road, Sheffield, £2 10s 11d; The London & Paris Optical Co., 22 Edmund Place, Aldersgate Street, London, £86 2s 3d; Heath & Co., Observatory Works, Crayford, £26; Thomas Mercer, 81 Westmorland Place, City Road, London, £598; Norie & Wilson, 156 Minories, London, £55 12s 4d; and H. Hughes & Son, Fenchurch Street, London, £47 7s 3d. The firm's total liabilities came to £6962 10s 8d; its assets to £2383, leaving a deficiency of £4579 10s 8d. This was worked out to yield a repayment of 6s 8d per pound.[48]

In fact, a first dividend of five shillings per pound was paid to creditors on 10 October 1891, a second dividend of 3d per pound on 10 May 1893 and a third and final dividend of 1½d on 10 November 1897. On 20 January 1904, there was a petition for a new trustee:

> '... That a dividend of 5s 4½d per £1 sterling was declared and paid to the creditors of the said M. Walker & Son, Malcolm McNeill Walker, William Williamson Gebbie, and Robert Weir, and their debts, under deduction of said dividend, are still due and resting owing. The trustee, on 1st July 1898, was discharged of his whole actings and intromissions as trustee of

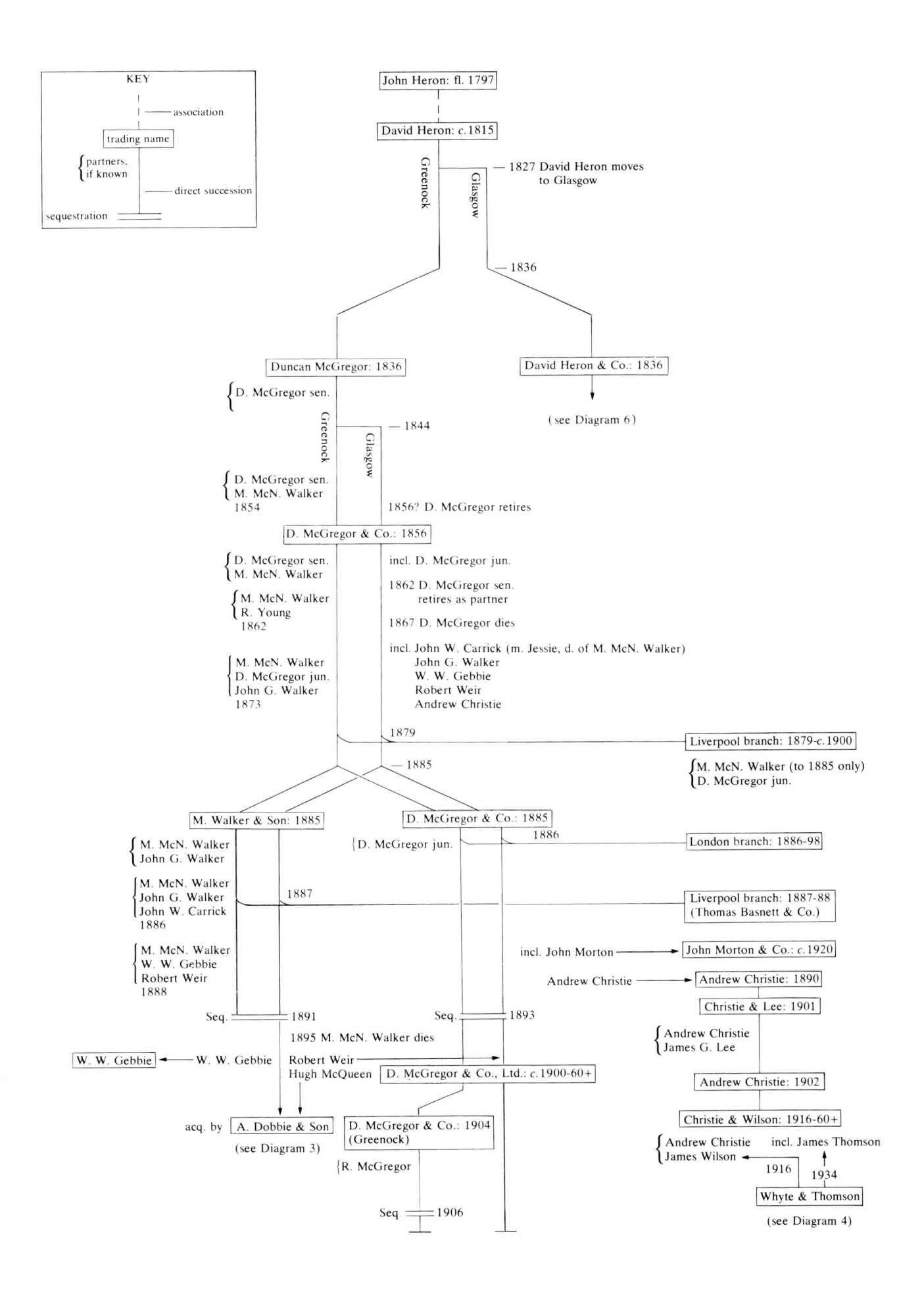

Diagram 4. Schematic development of the McGregor business.

said sequestered estates. The said Malcolm McNeill Walker died undischarged, his heir-at-law being John Gray Walker ... the said William Williamson Gebbie was discharged on 7th March 1892, without composition; the said William Williamson Gebbie at present carries on business as an optician and nautical instrument maker in Greenock ... The said Robert Weir is now in the employment of Messrs. D. McGregor & Company Ltd., nautical instrument makers, 37 Clyde Place, Glasgow'.

The petition went on to state that some of Walker's property had been sold in 1887 for £420 15s 1d.[49] The new trustee was discharged 21 November 1905, after various creditors had been awarded proportional amounts.[50] By this time, Malcolm McNeill Walker had died in 1895 of 'lung disease and senility' aged seventy-three.[51] The Glasgow premises (and possibly the business) were acquired by Alexander Dobbie & Son (q.v.).

Walker's son John Gray Walker, who was recorded as an instrument maker at the age of twenty-two,[52] and who partnered the early stages of his father's venture, may previously have been employed by McGregor's. Other names associated with McGregor's are known, such as Robert Weir, mentioned above, a compass adjuster who moved to McGregor's after Walker's sequestration.[53] Another adjuster was John Morton, secretary of the firm in 1910, who appears to have gone on to found John Morton & Co., nautical instrument makers and nautical publishers.[54] Most important perhaps, was Andrew Christie, who probably began as a clerk under Duncan McGregor junior, and remained in the firm until 1890, when he left to set up independently. Later he co-founded Christie & Wilson (q.v.) with James Wilson of Whyte, Thomson & Co.(q.v.).

Meanwhile, business was not going well for the parent firm. Duncan McGregor junior, sole partner of D. McGregor & Co., went bankrupt himself in 1893, and was discharged in 1897.[55] The Trustee appointed to wind up his affairs reported that McGregor's bankruptcy was 'caused by losses in business and depreciation in the value of his heritable property'[56] beyond his control. The Glasgow and London shops together owed over £5000, the Liverpool shop about £560 and the Greenock shop £100.[57] McGregor had obtained loans on the security of his extensive properties in Glasgow to their full valuation of £60,000, and unsecured debts of some £34,000 ensured his bankruptcy.[58] It is unclear just what happened after this, although it would seem that the London business was sold off,[59] and that the Glasgow, Greenock and Liverpool concerns were similarly split up. More details can be gleaned from the demise of the Greenock branch in 1906, when Robert McGregor, apparently a son of Duncan McGregor junior (by this date deceased),[60] also went bankrupt.[61] It is worth quoting Robert McGregor's testimony from the Sederunt Book to show how easily bad business practices could run a concern into deep trouble:

> '... I am the sole Partner of the firm of D. McGregor & Co., chronometer makers, nautical opticians and photographic dealers, 33 Cathcart Street, Greenock. I started the business on 1st October 1904. I had practically no capital. Prior to said starting I was in that same business as managing shopman to D. McGregor & Co., Limited. I purchased said business for £1000 - of which, £600 represented goodwill, and £400 stock and part of the book debts: the book debts were of very small amount. I paid no portion of said purchase price of £1000. I first found myself in difficulties in March, 1905. The state of affairs which I have lodged in process shows a deficiency of £3149 7s 10d. Said deficiency has arisen from 1st October 1904 - a period of nineteen months. I have made up a statement showing how I account for said deficiency ... My cash book shows an excess of £1023 19s 3d of payments over receipts. Said excess arose through the paid-away side being complete, and through the received-side being incomplete in respect that it did not contain entries of the sums that I received by said pledging transactions ...'[62]

It appeared that McGregor obtained goods on approval, and pawned them; in addition, he took items from stock and put them into pawn also:

> '... I have not paid for any parts of the goods which I received on approbation and pledged; nor for any part of the goods which I received on sale or return and which I pledged. I intend to redeem all the goods that I pledged. Immediately I got certain goods I pledged them. All the goods that I took out of stock and pledged had been paid for: in particular, the Purchase Ledger is not complete. Of my deficiency spoken to, £337 arose through my keeping dogs, which are now in Paisley. My cash book contains no record of my payments for the keeping of my dogs...'[63]

Later in his testimony, McGregor admitted to owning, at one point, some twenty dogs; unfortunately, 'they got diseased'.[64] Presumably these were greyhounds kept for racing, but for McGregor they proved a poor investment.[65] He explained:

> 'At the time of my puchase of the business it was being carried on as a Limited Company, and the directors were Mr R.S. Allan and Mr McLeod. No statement of how the £1000 of purchase price was arrived at was made up, beyond stock checks and book debts amounting together to £400 and the balance between that and the £1000 which was asked and I agreed to pay was taken to represent goodwill ... At the time of the deal the

principal shareholder was Mr J.M. McLeod, and he as representing the Company made the bargain with me: I was at that time manager in Greenock for the Company. My mother is not the principal shareholder of D. McGregor & Co., Limited. I have no interest in my late father's estate: he had nothing to leave: I do not mean that he died insolvent, but he left no money and my mother got everything that there was. It was previous to my father's death that the business of D. McGregor & Co. was turned into a Limited liability Company. I got married in October 1904 - the same month as I bought the business. It was not my intention to live on my creditors.
Q. But you did so, as a matter of fact?
A. Practically.'[66]

The case was reported, in outraged journalese, under a headline 'Raising the Wind' in the *Greenock Telegraph*, and in more sober tones in the *Glasgow Herald*.[67]

Amongst his creditors were listed Mr James Hamilton, a Director of D. McGregor & Co. Ltd., owed £1348 7s; and Macalister & Fyfe (q.v.), owed £25 16s 8d. Robert McGregor was discharged in early 1907 after a single dividend payment.[68]

Although in its day McGregors was clearly a productive and successful concern, and went on to spawn a trio of independent firms, so that it must be considered to be of no small significance in the nineteenth century history of the Scottish instrument making trade, yet both its founder and his son remain shadowy figures. Perhaps in the light of the subsequent financial disasters which seemed to haunt the business after the retiral of the founder this is not altogether surprising: all too frequently history is made up of success stories rather than dismal failures. The Glasgow business flourished under the name of McGregor & Co. until the 1960s, although apparently unconnected with the family.[69]

REFERENCES

1. *Glasgow Post Office Directory* 1844.

2. *Glasgow Constitutional*, 8 March 1848. Another report gives a different picture: '... A few doors further west they attacked the shop of Mr D. McGregor, chronometer maker, which they completely demolished.' (*Glasgow Herald*, 10 March 1848).

3. McGregor's were said in 1882 to have experience going back over 80 years and this presumably included an allowance for the Heron business (Anon., 'The New Mariner's Compass by D. McGregor F.R.G.S.', *The Mercantile Age* 7 (1882), 378-380).

4. *Fowler's Greenock Directory* 1836-37. The dissolution of the Copartnery was announced: 'The Business carried on by the Subscribers in Greenock, as Dealers in Nautical, Mathematical, and Optical Instruments, &c. was DISSOLVED by mutual consent upon the 1st of January last ... [signed] DAVID HERON. DUNCAN McGREGOR.' (*Edinburgh Gazette*, 3 May 1836, 122).

5. *Glasgow Post Office Directory* 1844-1909, *passim*; *Fowler's Greenock Directory* 1836-37; *Hutcheson's Greenock Register, Directory and General Advertiser* 1845-46; *Greenock Post Office Directory* 1853, 1861; D.J. Bryden, *Scottish Scientific Instrument-Makers 1600-1900* (Edinburgh, 1972), 53.

6. General Register Office (Scotland), Census of 1861 (Glasgow, Govan), 646¹/14, p.14.

7. G.R.O.(S.), Census of 1861 (Glasgow, Govan), 644⁹/72, pp. 11-12; Census of 1871 (Glasgow, Tradeston), 644⁹/64, p. 10 for Dobbie; Census of 1861 (Glasgow, Govan), 644⁹/75, p. 8; Census of 1871 (Glasgow, Barony), 644⁶/47, p. 8; Census of 1881 (Glasgow, Partick), 646²/33, p. 56 for White.

8. This can be argued even taking into account the appearance of new trade classifications in the Directories.

9. *Exposition Maritime Internationale du Havre, 1868. Rapports du Jury International, et Catalogue Officiel des Exposants Récompensés* (London, 1868), 50. We are grateful to Dr Anita McConnell for this reference. In 1888, they displayed 'marine chronometers, compasses, sextants, meteorological instruments, position finders, charts etc.' (*International Exhibition Glasgow 1888: Official Catalogue* (Glasgow, 1888), 97).

10. *Glasgow Post Office Directory* 1870-1900, *passim*. Apart from the items in the Frank collection, the following instruments have been recorded: an ebony octant, signed 'Spencer Browning & Co., London' and 'Made for D. McGregor, Greenock & Glasgow', offered by Sotheby's, 9 December 1977, Lot 87; a brass sextant, signed 'Spencer, Browning & Co., London, Made for D. McGregor, Greenock & Glasgow', offered for sale by Sotheby's West Sussex, 21 February 1985, Lot 1415; an ebony octant, signed 'D. McGregor, Greenock & Glasgow', offered for sale by Sotheby's, 28 February 1984, Lot 206; a brass octant, signed 'D. McGregor, Greenock & Glasgow', offered by Phillips, 8 May 1985, Lot 215; a sextant, signed 'McGregor & Co., Greenock & Glasgow' offered by Christie's, 19 July 1978, Lot 50; a brass sextant, signed ' D. McGregor & Co., Glasgow, Greenock & Liverpool' offered by Christie's, 19 December 1973, Lot 40; a sextant, signed 'D. McGregor & Co., Glasgow, Greenock, Liverpool & London', offered by Sotheby's, 7 July 1978, Lot 5; a nickel-plated nautical parallel rule, signed 'Capt. Field's Improved, Registered Jan. 1854, No 3549, McGregor & Co., Glasgow, Liverpool, Greenock & London', offered by Phillips, 3 June 1987, Lot 114; and a brass cased aneroid barometer by 'McGregor & Co Greenock', offered by Christie's South Kensington, 8 December 1988, Lot 9.

11. 'Mr DUNCAN McGREGOR, Nautical Instrument Maker in Greenock, presently residing in Oban, ceased on the 30th April 1862 to be a Partner of the Company carrying on Business in Greenock and Glasgow, as Nautical Instrument Makers, under the Firm of DUNCAN MACGREGOR [sic] & COMPANY.' (*Edinburgh Gazette*, 17 March 1863, 369).

12. G.R.O.(S.), Register of Deaths 1867 (Greenock, West), 564³/372. He was a widower, and died of 'cardiac disease 12 years Paralysis & dropsy'. His father, also Duncan McGregor, had been a master cooper.

13. Scottish Record Office: McGregor's Testament is recorded

241

in *Register of Deeds*, 4 July 1867, 1294, pp. 5-11. The Inventory of his personal estate is to be found in *Record of Settlements and Inventories. Sheriff Court of Paisley*, 9 August 1867 (S.R.O., SC 58/42/34, pp. 693-700). The bulk of his estate consisted of large investments in railway companies. When making out his testament, dated 2 June 1854, Duncan McGregor called himself 'Nautical Instrument Maker and Ship Owner'. Duncan junior was served heir to the Greenock properties in 1887 (*Services of Heirs in Scotland* 2nd series, 3 (1880-89), 16). His father was listed as one of the first principal shareholders of the Greenock Union Bank, founded in 1840 (Dougald Campbell, *Historical Sketches of the Town and Harbours of Greenock* (Greenock, 1879 & 1881), I, 259).

14. G.R.O.(S.), Census of 1861 (Glasgow, Govan), 644⁹/72, p.3.

15. S.R.O., *Register of Deeds, op. cit.* (13), by the terms of their father's testament. However, Malcom's name does not appear again in connection with the business.

16. 'The former Copartnery of D. McGREGOR & CO., Nautical Instrument Makers in Glasgow and Greenock, was DISSOLVED as upon the 30th April last, by the retiral therefrom of Robert Young, Nautical Instrument Maker in Greenock; and since that date the Subscribers hereto, as the sole Partners, continue to carry on the Business under the same firm of D.McGREGOR & CO. [signed] M. McN. WALKER, D. McGREGOR, JOHN G. WALKER [witnessed by] L. MACKINTOSH, shopman to D. McGregor & Co., [and] ROBT. GLANFORD, salesman to D. McGregor & Co.' (*Edinburgh Gazette*, 5 September 1873, 546).

17. *Proceedings of the Glasgow Philosophical Society* 9-30 (1873-1899), *passim*. McGregor was admitted a member of the Society in 1867, *ibid.* 9 (1873-75), 369.

18. 'On a Plan of the Practical Arrangement for Working a Uniform Code of Tidal Signals for Day and Night', *Transactions of the Royal Scottish Society of Arts* 6 (1864), 131-136.

19. U.K. patents 3991, 17 November 1875; 2392, 12 June 1880; 16798, 7 September 1893; also 1397, 14 January 1884 for instrument dials and 6137, 10 April 1889 for ships' logs. Examples of McGregor's compasses have been recorded: one described as 'McGregor's Patent Polar Compass' was offered for sale by Sotheby's, 28 October 1986, Lot 18; another, described as a 'Tell-Tale' compass, signed 'D. McGregor & Co., Glasgow & Greenock' was offered for sale by Sotheby's, 1 June 1988, Lot 313.

20. *Mercantile Age, op. cit.* (3), 379.

21. *Ibid.*, 379-380: there is an example of this in the Frank collection (NMS T1980.204). McGregor's liquid compasses were compared favourably to those made by Ritchie of Boston.

22. *Rates of Chronometers on Trial for Purchase by the Board of Admiralty at the Royal Observatory, Greenwich, 1859* (London, 1859); *ibid.* 1861.

23. *Ibid.* 1866.

24. R.T. Gould, *The Marine Chronometer* (London, 1923), 258.

25. *Op. cit.* (22) 1860-1873, *passim*.

26. The figures given are extracted from Tony Mercer, *Mercer Chronometers* (Ashford, Kent, 1978), 204-205.

27. *Op. cit.* (22) 1874, 1876. The first 'M' numbers recorded are M/2509-M/2514 in 1873, but no more are recorded in the Mercer records for 3 years. Surprisingly, M/2519, which was rated at Greenwich in 1879, is not listed and is the only number missing from the series. M/5003, rated in 1876 and 1877, appears to have a McGregor rather than a Mercer denominator, but is presumably the same as W/5003 listed in the Mercer records in 1877. An isolated 'N' chronometer appears under McGregor's Greenock address in 1882: this and the 'W' perhaps represent transcription errors (Mercer, *op. cit.* (26), 204-205).

28. Chronometer F/3038 in the Frank Collection (NMS T1980.207) can be dated approximately by the submission by McGregor of F/3073 to the Geenwich Trials of 1874 (*op. cit.* (22) 1874)). The movement is stamped 'J.P', probably indicating original manufacture by Joseph Preston and Sons of Prescot, Lancashire (C. Jagger, *Supplement to Paul Philip Barraud* (Ticehurst, Sussex, 1979), 195), and the number on the movement is compatible with the series of the London chronometer maker Victor Kullberg, who supplied chronometer movements to Charles Frodsham amongst others (V. Mercer, *The Frodshams* (n.p. [Ticehurst, Sussex], 1981), 263, Appendix XI: no records are available for the period 1872-79). Kullberg movements were also supplied to Bassnett of Liverpool which was controlled (see below) by Malcolm Walker (Christie's, 7 October 1987, Lot 26).

29. *Mercantile Age, op. cit.* (3), 378-379.

30. *Ibid.*

31. *McGregor's Almanac and Tide Tables for the year 1888*, published by D. McGregor & Co. (Glasgow, 1887), 3 and *passim*. An advertisement for the firm (p. 251) lists several types of engine-room fitting, including Richards', Elliott's and Darkes' indicators. The firm displayed various chronometers, a compass and binnacle, and a range of other nautical and meteorological instruments in Glasgow in 1880, described in *Catalogue of the Naval and Marine Engineering Exhibition* (Glasgow, 1880), 80-81.

32. H. Lumsden and P.H. Aitken, *History of the Hammermen of Glasgow* (Paisley, 1915), 338.

33. *Proceedings of the Glasgow Philosophical Society* 3-17 (1848-1886), *passim*. However, Walker was no longer a Fellow of the Royal Astronomical Society at his death, and thus did not receive an obituary in the *Monthly Notices*.

34. G.R.O.(S.), Census of 1861 (Glasgow, Govan), 646¹/14, p. 14; Register of Deaths 1895 (Gourock), 567²/162.

35. *Glasgow Post Office Directory* 1852.

36. 'The Copartnery of D. McGREGOR & CO., Nautical Instrument Makers and Dealers in Nautical Instruments in Liverpool, of which the Subscribers were the sole Partners, was DISSOLVED upon the 15th day of April last, by efflux of time. The subscriber Duncan McGregor continues to carry on Business in the same premises ... [signed] D. McGREGOR. M.McN. WALKER. [signed] JOHN W. CARRICK, Book keeper to the late Firm of D. McGregor & Co., 45 Clyde Place, Glasgow; JOHN D. SILLARS, Clerk to M. Walker & Son, 45 Clyde Place, Glasgow: witnesses to Walker's signature. [Signed] ALEXR. GLENDINNING, Manager to the present Firm of D. McGregor & Co., 36 Brymner Street, Greenock; JOHN STEWART, Clerk to the present Firm of D. McGregor & Co., 37 and 38 Clyde Place, Glasgow: witnesses to McGregor's signature.'
'The Copartnery of D.McGREGOR & CO., Nautical Instrument Makers, and Dealers in Nautical Instruments in

Glasgow and Greenock, of which the Subscribers were the sole Partners, was DISSOLVED upon the 30th day of April last, [by] efflux of time. The Subscribers Malcolm McNeil Walker and John Gray Walker, ... will carry on Business in the same premises in Glasgow and Greenock on their own behalf under the Firm of M. WALKER & SON ... The Subscriber Duncan McGregor will carry on Business at 37 and 38 Clyde Place, Glasgow, and 36 Brymner Street, Greenock, under the Firm of D. McGREGOR & COMPANY. [signed] M. McN. WALKER. D. McGREGOR. JOHN G. WALKER. [witnesses as above] Glasgow 15 September 1885.' (*Edinburgh Gazette*, 18 September 1885, 757).

37. *Glasgow Post Office Directory* 1886, and *passim*. The Greenock branch was latterly at 28 Cathcart Street (*ibid.* 1890). 44 Clyde Place may have been vacated after 1888. A sextant signed 'M. Walker & Son, Glasgow & Greenock' was offered for sale by Christie's Scotland, 1 April 1987, Lot 3.

38. '7th March 1888. The Subscriber John Wilson Carrick has this day retired from the Firm of M. WALKER & SON, Nautical Instrument Makers, Clyde Place, Glasgow, and Cathcart Street, Greenock. The Business of M. Walker & Son will be continued as formerly under the same name ...' (*Edinburgh Gazette*, 15 June 1888, 618).
'9th March 1888. The Subscriber John Gray Walker has this day retired from the Firm of M. WALKER & SON ...' (*Ibid.*, 619).
'The Firm of THOMAS BASSNETT & COMPANY, Nautical Instrument Makers, 10 Bath Street, Liverpool, of which Firm the Subscribers were sole Partners, has this day been DISSOLVED of mutual consent ... [signed] JOHN W. CARRICK. M.McN. WALKER.' (*Ibid.*).

39. S.R.O., CS 318/49/322, Sequestration Papers, M. Walker & Son. These are fairly detailed, and give some idea of the structure of the firm: two Sederunt Books survive with these papers. Notices also appeared in the *Edinburgh Gazette*: the initial notice of sequestration (10 April 1891, 395); the announcement of the appointment of the Trustee and Commissioners (1 May 1891, 483); announcement of the payment of the first dividend on 10 October 1891 (25 August 1891, 925); a Petition presented by William Williamson Gebbie 'to be finally discharged of all debts contracted by him before the date of the Sequestration of his Estates' (2 February 1892, 105); a second and equalising dividend to be paid on 10 May 1893 (24 March 1893, 310); a third and final dividend to be paid on 10 November 1897 (3 September 1897, 857); and a Creditors' meeting called for 12 May 1898 to consider discharging the Trustee (19 April 1898, 378).

40. S.R.O., *op.cit.* (39), Sederunt Book 1, p.6.

41. *Ibid.*, p. 14.

42. *Ibid.*, pp. 19-21.

43. *Ibid.*, pp. 26-28.

44. *Ibid.*, pp. 28-33.

45. *ibid.*, p. 77.

46. *Ibid.*, p. 80.

47. *Ibid.*, p. 89.

48. *Ibid.*, pp. 99-109.

49. *Ibid.*, and S.R.O., CS 318/42/293, Sequestration Papers, M. Walker & Son.

50. *Ibid.* Notices also appeared in the *Edinburgh Gazette*: the Petition for a new Trustee, and a Creditors' meeting called for 17 February 1904 (9 February 1904, 146); the announcement of the appointment of the new Trustee and Commissioners (23 February 1904, 204); and the announcement of an equalising and supplementary dividend to be paid on 4 August 1904 (21 June 1904, 672).

51. G.R.O.(S.), Register of Deaths 1895 (Gourock), 567[2]/162. Walker married firstly, Elizabeth Andrews and secondly, Eliza Nevatt; John Gray Walker and Jessie Walker were offspring of the first marriage.

52. G.R.O.(S.), Census of 1871 (Glasgow, Govan), 646[1]/15, p.6.

53. *Glasgow Post Office Directory* 1891 and *passim*. Walker employed at least two other compass adjusters, William W. Gebbie (who was also a partner) and Hugh McQueen, who later worked for A. Dobbie & Son and Dobbie McInnes (q.v.).

54. *Ibid.* 1910 and 1920.

55. S.R.O., CS 318/41/269, Sequestration Papers of D. McGregor and D. McGregor & Co. There is a Sederunt Book with these papers. Notices also appeared in the *Edinburgh Gazette*: the initial notice of sequestration: 'The Estates of DUNCAN McGREGOR, residing at 48 Maxwell Drive, Pollokshields, Glasgow, sole Partner of the Firm of D. McGREGOR & CO., 37 Clyde Place Glasgow, as such Partner, and as an Individual, and the Estates of the said D. McGREGOR & CO., carrying on Business at 37 Clyde Place, Glasgow, and in London, Liverpool and Greenock, as a Company, were sequestrated on 14th June 1893 ...'(16 June 1893, 620); the announcement of the appointment of the Trustee and Commissioners (4 July 1893, 707); the postponement of a dividend payment (31 October 1893, 1142); an announcement of accelerated payment: a dividend to be paid on 14 February 1894 (12 January 1894, 67); the first and final dividend to be paid on 14 February 1894 (26 January 1894, 128); Creditors' meeting called for on 4 March 1897 to consider discharging the Trustee (9 February 1897, 137); and Duncan McGregor and D. McGregor & Co., have presented a Petition to be finally discharged (16 April 1897, 369).

56. S.R.O., CS 318/41/269, Report by J.M. Macleod, Trustee, 1897.

57. *Ibid.*, Sederunt Book, p. 32. The debts owed to the Glasgow shop amounted to some £1740, Liverpool £1220, Greenock £300 and London £260. The stock in trade at Glasgow was £1160, Liverpool £810, London £365 and Greenock £205 (*ibid.*, pp. 29-30). The firm's principal creditors included the wholesale suppliers Heath & Co., S.J. Levi & Co., Edward Clarke, Seagrove and Woods, and Joseph White & Sons of Coventry (*ibid.*, pp. 12, 123-132).

58. *Ibid.*, pp. 27-28, 32-33; Final return by J.M. Macleod, Trustee, 31 October 1897.

59. McGregor & Co., chronometer makers were at 57 Fenchurch Street, E.C.: *Kelly's London Post Office Directory* 1887-1890; McGregor & Co., nautical instrument makers, 14A London Street, E.C.: *ibid.* 1891-1898, and do not appear in the directory after that. Our thanks to Dr Anita McConnell for this reference.

60. G.R.O.(S.), Census of 1881 (Glasgow, Tradeston), 644[13]/53, pp. 5-6. This lists Duncan McGregor [junior], Nautical Instrument Manufacturer, aged 39; his wife Marion, aged 36; their three sons Duncan, Robert and John, aged respectively 12, 7 and 5; and their three daughters Bessie, Marion and Annie, aged 14, 3 and 1.

61. S.R.O., CS 318/51/228, Sequestration Papers, D. McGregor & Co. There is a Sederunt Book with these papers, from which most of the details about the firm have been taken. Notices also appeared in the *Edinburgh Gazette*: the petition for sequestration presented by John Baird, Wholesale and Manufacturing Optician (27 March 1906, 363); the notice of sequestration (6 April 1906, 409); the announcement of the appointment of the Trustee and Commissioners (24 April 1906, 458); the first dividend to be paid on 3 October 1906 (17 August 1906, 871); and a Creditors' meeting called for 26 February, to consider discharging the Trustee (1 February 1907, 121).

62. S.R.O., CS 318/51/228.

63. *Ibid.*

64. *Ibid.*

65. The Gaming Act, extended to Scotland in 1874, made betting difficult (except for the very wealthy) outside racecourses; and the popularity of greyhound racing appears to have become popular only after 1926, when the electric hare was introduced (T.C. Smout, *A Century of the Scottish People 1830-1950* (Fontana edition, London, 1987), 156-157).

66. S.R.O., CS 318/51/228, Sederunt Book, pp. 38-39.

67. '"RAISING THE WIND!" "Raising the wind" has ever since the earlies times, both in Oriental and Occidental story been a chosen and favourite employment with a certain class of men... This method of keeping the human head above society's engulfing waters was adopted by Mr Robert McGregor, optician, of Greenock, who was this week examined in bankruptcy before Sheriff Neish, and whose confessions were a regular hair-raiser. Indeed, the exposure of Robert McGregor's mode of 'raising the wind,' and keeping it as long as possible at half a gale at least, is by far and away the leading local sensation of the week. It is such that had our report been liable to question, had it not been as good as official, we should have hesitated to give it publication. But there it is, as accurate, as much a fact, as Ailsa Craig itself...' (*Greenock Telegraph*, 4 May 1906). A more sober account is 'GREENOCK OPTICIAN'S SEQUESTRATION', *Glasgow Herald*, 8 May 1906.

68. S.R.O., CS 318/51/228.

69. *Glasgow Post Office Directory* 1900-1960, *passim*.

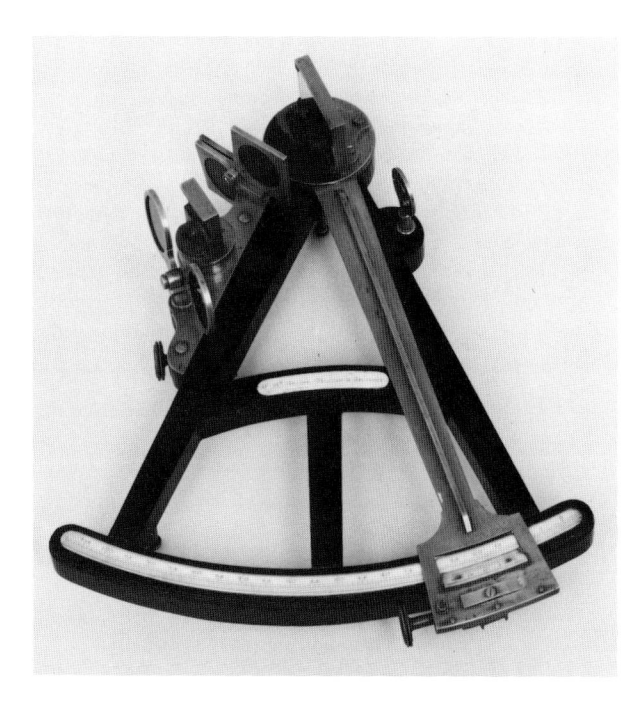

---

113. Octant: D. McGregor, Glasgow & Greenock, c.1850 (T1980.188)

10" octant with ebony frame and brass fitments. Scale divided on inset ivory arc [-3]-[109] by 20 mins., the reinforced index arm with clamp (defective), tangent screw and vernier reading to ½ min. Stamped on the inset ivory name plate 'D. McGregor, Glasgow & Greenock.' The frame with single vertical strut and bowed horizontal strut; and provided with double pin-hole sights, 5 shades, adjustable horizon glass, noteplate on reverse and provision for pencil in vertical strut, and supported on feet. With fitted keystone-shaped stepped case.

Scale radius: 247mm.
Case size: 300 x 325 x 115mm.

Lacking the clamping plate screw for the index arm, and also the pencil. The ivory scale and name plate are discoloured and are detached at the ends.

114. Sextant: D. McGregor, Glasgow & Greenock, c.1850 (T1980.190)

7½" sextant with cast lattice frame, in oxidised bronze and brass. Scale divided on silver, [-5]-[150], the reinforced index arm with clamp, tangent screw and vernier reading to 10 secs., and ground glass shade. Engraved on the arc 'D. McGregor, Glasgow & Greenock.' Screw adjustment to the ring support for the sight tubes. The frame with 7 shades and rear handle, and supported on feet. With fitted keystone-shaped case, an attached paper label marked in MS 'XANTHE / SEXTANT'.

Scale radius: 188mm.
Case size: 325 x 275 x 125mm.

Lacking sight tubes and telescopes, swinging magnifier and attachment for tangent screw.

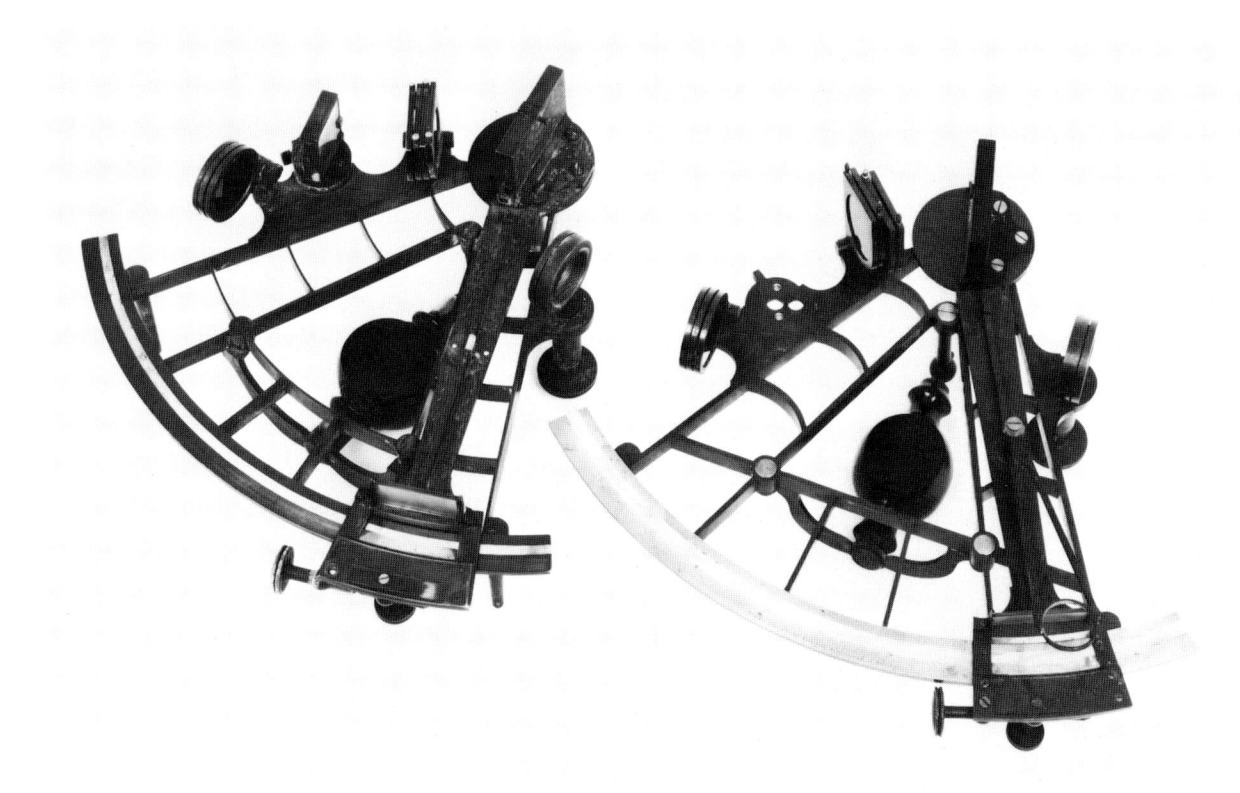

114, 115

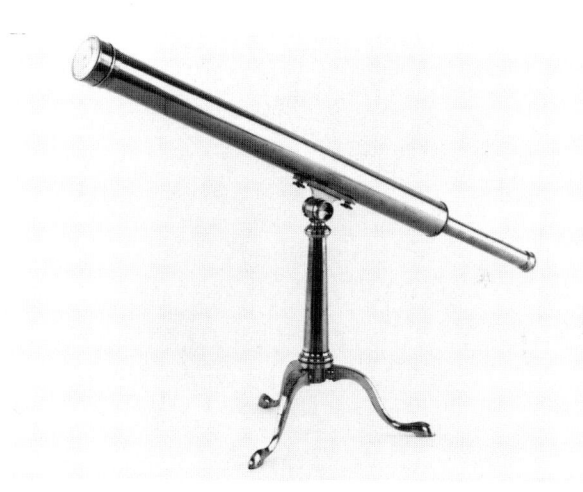

115. Sextant: D. McGregor, Glasgow & Greenock, c.1850 (T1980.193)

8" sextant (incomplete) with cast lattice frame, in oxidised bronze and brass. Scale divided on silver, [-5]-[155], the reinforced index arm with clamp, tangent screw and vernier reading to 10 secs., ground glass shade and swinging support for magnifier. Engraved on the arc 'Spencer, Browning & Co. London /Made for D..MᶜGregor, Greenock & Glasgow,' and '3854'. Screw adjustment to the ring support for the sight tubes. The frame with 7 shades and rear handle, and supported on feet.

Scale radius: 190mm.

Lacking sight tubes and telescopes, horizon glass and magnifying glass for vernier.

Dates for Spencer, Browning & Co., of London are given as 1840-1870 (M.A. Crawforth, Letter, 'Simon Speaks' in *Bulletin of the Scientific Instrument Society* No. 16 (1988), 18). Beginning as Spencer & Browning in the late 18th century, and becoming Spencer, Browning & Rust in 1784, the firm produced monogrammed octant scales from 1791 (A. Stimson, 'Some Board of Longitude Instruments in the Nineteenth Century', in P.R. de Clercq (ed.), *Nineteenth-Century Scientific Instruments and Their Makers* (Leiden, 1985), 113) and there is evidence that the firm produced scale plates for nautical instruments well into the century (W.D. Hackmann, 'The Nineteenth-Century Trade in Natural Philosophy Instruments in Britain', *ibid.*, 57). Among other Scottish makers who retailed instruments with such scales were Adie (q.v.) (NMS T1982.57), and John Bon of Dundee (Sotheby's, 12 June 1984, Lot 307).

116. Refracting telescope: D. McGregor, Glasgow & Greenock, c.1855 (T1980.280)

2" aperture single-draw mounted refracting telescope in brass with doublet objective, and with rack and pinion focus. Supplied with low-power 2-component astronomical eyepiece, and draw tube extension section with 2-component image erecting assembly for use on terrestrial objects. Engraved on the draw tube extenstion 'D..MᶜGregor,/ Greenock & Glasgow.' Objective cover. Pillar-mounted by a knuckle-joint with free azimuth motion, and on a folding brass table tripod. With fitted case (distressed).

Length closed (with erecting assembly): 956mm.
Tube diameter: 58mm.
Aperture: 53mm.
Case size: 765 x 285 x 100mm.

Detailed similarities in design and in the mounting of optical components suggest that this item is by the same manufacturer as items 31 (Adie & Son) and 81 (Gardner & Co.).

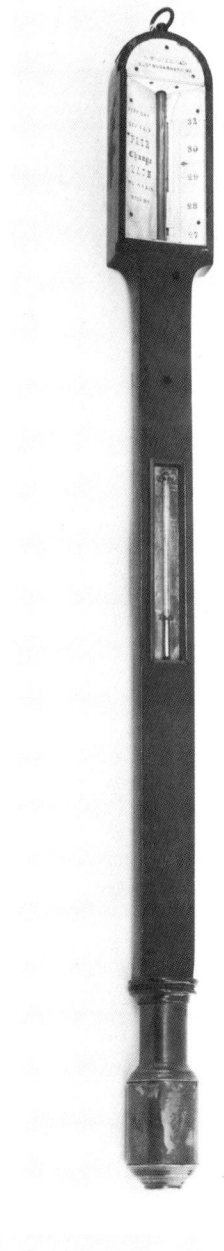

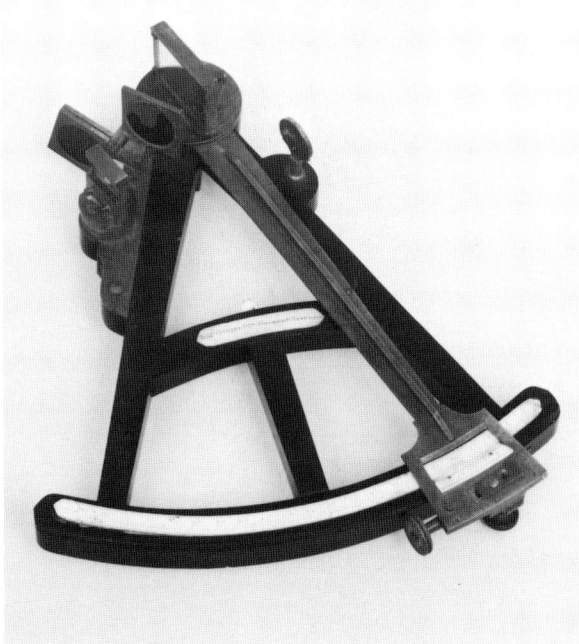

117. Stick barometer: D. McGregor & Co., Glasgow & Greenock, c.1875 (T1980.211)

Marine stick barometer in a wooden case with brass cistern and ivory register plates. Signed above the register 'D. M^C GREGOR & Co / GLASGOW & GREENOCK'. Scale marked in inches of mercury 27-31 by 0.1", with rack and pinion operated vernier reading to 0.01", the reading distinguished as 'STORMY' (at 28"), 'MUCH RAIN', 'RAIN', 'Change', 'FAIR', 'SET FAIR' or 'VERY DRY' (at 31"). Ivory-backed mercury thermometer divided 20-100°F, and 'Freez/ing', 'Tempe/rate', 'Sum^r / Heat' and 'Blood Heat'. The glass barometer tube sealed in a screw-top turned wooden cistern with leather base. Support ring at top of case.

Overall height of case and cistern: 936mm.

The barometer is designed to be supported in a gimbal on a projecting hinged bracket: neither is present. Most of the wooden trim around the register plates and the thermometer is missing or damaged, and neither is now glazed. The barometer tube has broken at an old repair, and a small hole had to be drilled at the Museum in July 1987 to release the remaining mercury. The thread on the brass cistern cover has become stretched with repeated cross-threading and is no longer adequate to support the weight of the filled tube. Lacking the vernier adjustment knob.

This instrument can be identified as one of the many marine barometers offered by the London wholesaler J.J. Hicks in his *Illustrated & Descriptive Wholesale Catalogue of Standard, Self-Recording, and other Meteorological Instruments ...* (London, n.d. [c.1880]), 24, no. 28, 'Marine Barometer of simple construction, quite trustworthy, ivory scales and gimbals, sliding vernier and attached Thermometer, in solid mahogany £1 10 0'.

118. Octant: D. McGregor & Co., Glasgow & Greenock, c.1860 (T1980.186)

10" octant with ebony frame and brass fitments. Scale divided on inset ivory arc [-2° 20']-[106] by 20 mins., the reinforced index arm with clamp, tangent screw and vernier reading to 1 min. Stamped on the inset ivory name plate 'D Mc Gregor & Co Glasgow & Greenock'. The frame with single vertical strut and bowed horizontal strut; and provided with pin-hole sights, 3 shades, adjustable horizon glass, noteplate on reverse, and pencil in vertical strut, and supported on feet.

Scale radius: 247mm.

The retaining slide for the tangent screw block is a replacement, now lacking one attachment screw. The wooden limb is split at the end of the arc. The ivory scale and name plate are discoloured and are detached at the ends.

119. Octant: D. McGregor & Co., Glasgow & Greenock, c.1860 (T1980.187)

9" octant with ebony frame and brass fitments. Scale divided on inset ivory arc [-3]-[108] by 20 mins., the reinforced index arm with clamp, tangent screw and vernier reading to ½ min. Stamped on the inset ivory name plate ' D McGregor & Co Glasgow & Greenock'. The frame with single vertical strut and bowed horizontal strut; and provided with double pin-hole sights, 6 shades, adjustable horizon glass, and supported on feet. With fitted keystone-shaped case: pencil notes inside the lid include 'Mr D Black / Clarabella' and '4352 / Wales / 'Napier'.'

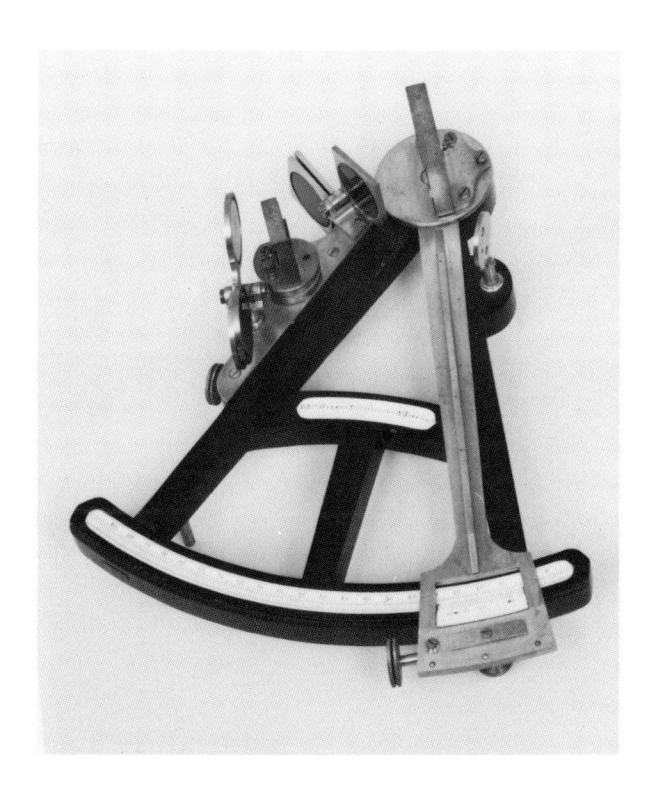

Scale radius: 220mm.
Case size: 290 x 305 x 105mm.

Lacking the cover slip for the sights.

120. Sextant: D. McGregor & Co., Glasgow & Greenock, c.1860 (T1980.191)

7½" sextant (incomplete) with cast lattice frame, in oxidised bronze and brass. Scale divided on silver, [-5]-150, the reinforced index arm with clamp, tangent screw and vernier reading to 10 secs., and swinging magnifier with ground glass shade. Engraved on the arc 'D. M^c.Gregor & Co. Glasgow & Greenock.' and twice stamped '790'. Screw adjustment to the ring support for the sight tube, and with sighting telescope with 2 eyepieces and shade. The frame with 7 shades and rear handle, and supported on feet. With fitted case.

Scale radius: 185mm.
Case size: 265 x 255 x 125mm.

Lacking horizon glass, and one of the 3 feet.

121. Marine chronometer: D. McGregor & Co., Glasgow & Greenock, c.1875 (T1980.207)

Two-day marine chronometer, with Earnshaw's spring detent escapement, free-sprung blued steel balance spring and 2-arm bimetallic balance with cylindrical weights. Spotted movement with fusee and chain and jewelling to the balance and escape arbors. Stamped '3305' on the dial support ring, on both plates and inside the bowl; the front plate additionally stamped 'J.P'. Silvered bezel and dial; the dial with Roman chapter, subsidiary seconds, 56-hour winding indication, and engraved 'D. M^CGREGOR & C^O / MAKERS TO THE ADMIRALTY,/ GLASGOW & GREENOCK' and 'F/3038'. The reverse of the dial stamped 'E Langton / Bush[?]/ 10/88' and '3386', and in a different hand '3038'. The glazed chronometer bowl in gimbals in a rosewood deck case with glazed top and lid, and with an inset name plate 'D. M^CGREGOR & C^O/ MAKERS TO THE ADMIRALTY,/ GLASGOW & GREENOCK./ F/3038'. Winding key.

Movement plate diameter: 82mm.
Bezel Diameter: 123mm.
Case size: 180 x 180 x 185mm.

The punch for the third numeral of the serial number is damaged: it is very probably '0', but possibly '9'.

The single stamp 'J.P' on the front plate is of the Prescot, Lancashire, manufacturer Joseph Preston & Sons (C. Jagger, *Supplement to Paul Philip Barraud* (Ticehurst, Sussex, 1979), 1950).

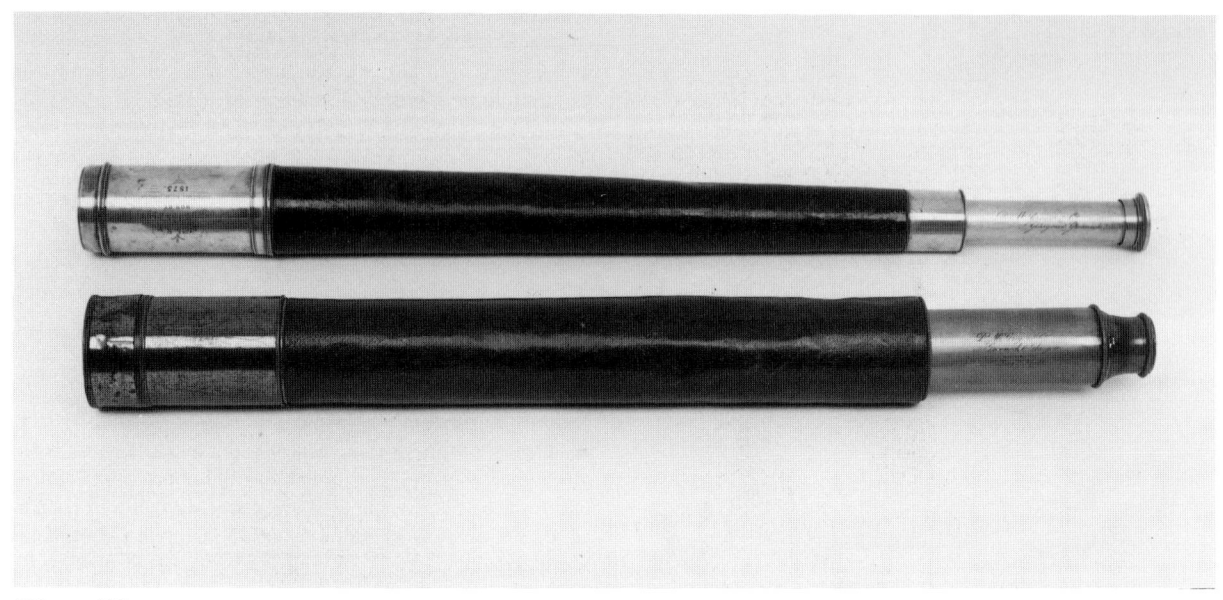

122 top, 123

122. Refracting telescope: D. McGregor & Co., Glasgow & Greenock, 1873 (T1980.266)

1¾" aperture single-draw hand-held refracting telescope with doublet objective and 2-component eyepiece and erecting assemblies: in nickel-plated brass, with objective ray shade and tapered leather-covered barrel. Engraved on the draw tube 'D. McGregor & Co /Glasgow & Greenock.' and on the ray shade '25TH L.R.V./ SHOOTING CLUB,/ WON BY / SERGEANT J.W. M^cFARLANE./1873.' Scratched assembly marks 'XI' or 'IX' on the draw tube, 'X' on the sliding closure of the eye stop, and 'III' on the ray shade and its attachment ferrule.

Length closed: 596mm.
Maximum tube diameter: 56mm.
Aperture: 44mm.

The draw slide, originally leather-lined, has been reduced in length and crudely modified in an attempt to provide grip to the draw tube. Lacking the objective cover.

Probably by James Parkes & Son of Birmingham, cf. *Wholesale Catalogue of Optical, Mathematical, Meteorological & Surveying Instruments manufactured by James Parkes and Son* (Birmingham, n.d.[1882]), 55, 'First-Class one-draw ship telescope, taper body, leather covered, with sliding sun shade, nickel-plated tube and mounts', no. 5548. The 25th Lanarkshire Rifle Volunteers, with headquarters in Kelvinhaugh Road, Glasgow, was an 1861 consolidation of the 8 original companies of the 6th Battalion of the Lanarkshire Rifle Volunteers, which first appeared in the Army list in 1860. The battalion was known locally as the 'Clyde Artisans' because the companies were all raised from the Clyde shipbuilding and engineering yards. It was renumbered the 6th Lanark in 1880, and with 2 additional companies, as the 2nd Volunteer Battalion of the Highland Light Infantry in 1887 (J.M. Grierson, *Records of the Scottish Volunteer Force 1859-1908* (Edinburgh and London, 1909), 262-263). Another comparable item is the 4-draw telescope signed 'Gardner & Co. Opticians, 53 Buchanan Street, Glasgow' and '1st Lanarkshire Artillery Volunteers, 8th Battery, Carbine Competition 1864, John Whitehead, First Prize' now in Glasgow Museums and Art Galleries (inventory A8408), acquired at Phillips Glasgow, 12 April 1984, Lot 724. We are grateful to Robert Woosnam-Savage for this information.

123. Refracting telescope: D. McGregor & Co., Glasgow & Greenock, c.1870 (T1980.267)

2¹/₈" aperture single-draw hand-held refracting telescope with doublet objective and 2-component eyepiece and erecting assemblies: in brass, with objective ray shade and leather-covered barrel. Engraved on the draw tube 'D. McGregor & Co/Greenock and Glasgow./ Day and Night' Scratched with the assembly mark 'III' on the draw tube, draw slide, barrel and ray shade ferrule, 'VII' on the objective mount. Objective cover with 1½" aperture stop and sliding closure.

Length closed: 524mm.
Tube diameter: 60mm.
Aperture: 54mm.

Lacking closure on external eye stop.

Probably by James Parkes & Son of Birmingham, cf. *Wholesale Catalogue of Optical, Mathematical, Meteorological & Surveying Instruments, manufactured by James Parkes and Son* (Birmingham, n.d.[1882]), 55, 'First-class one-draw ship telescope', no. 5561.

124. Octant: D. McGregor & Co., Glasgow, Greenock & Liverpool, c.1880 (T1980.192)

7½" octant with cast frame, in oxidised bronze and brass. Scale divided on silver, [-5]-[125], the reinforced index arm with clamp, tangent screw and vernier reading to 15secs., and swinging magnifier. Engraved on the arc 'D. McGregor & Co Glasgow Greenock & Liverpool.' The frame with 7 shades and rear handle, and supported on feet.

Scale radius: 184mm.

Lacking sight tubes and telescopes.

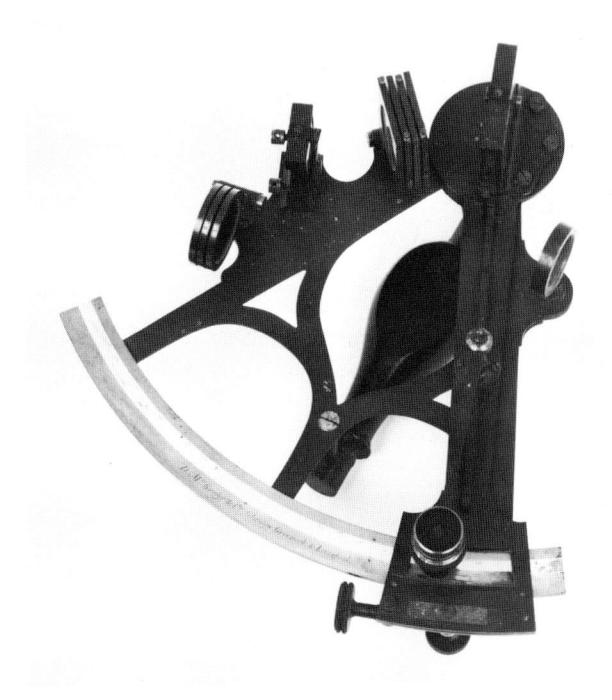

reflector for solar observations, rotating over the dial scales and with pointers to indicate the bearing. The dial is decorated at the north point with the Royal Arms and marked round the centre '"COMPASS VERIFIER" / D. M^CGREGOR & C^O PATENTEES & MANUFACTURERS./ DIAL PROTECTED BY PATENT 1397 / GLASGOW GREENOCK / LIVERPOOL': the name of the instrument obscured by an extension of the central disc which secures the enamelled plate against its back plate.

Diameter of dial and azimuth rotation ring: 215mm.
Case size: 275 x 275 x 210mm.

D. McGregor's U.K. patent 1397 of 1884 covered the use of vitrified enamel, porcelain, china and earthenware for the dials and scales of instruments for determining compass error, courses and bearings. It is unclear why the instrument's name (Compass Verifier) should have been obscured, unless this description was felt to imply a restriction in its use to compass adjusting alone. Presumably the instrument was available on a wholesale basis in a version in which McGregor's name was obscured also.

125. Marine bearing indicator: D. McGregor & Co., Glasgow, Greenock & Liverpool, c.1885 (T1980.204)

Pelorus dial for use in conjunction with a ship's compass to take bearings or determine compass error. Black on white enamelled copper plate reinforced by a brass plate underneath, with a vitrified printed 64-point compass rose and an azimuth scale in degrees 0-180-0, and around the north point a scale of east and west magnetic variation 45-0-45 by 1°. The dial plate can be rotated within a brass ring stamped with the direction of 'SHIPS HEAD', and is weighted and mounted in a brass gimbal, set in a wooden case with lid. Azimuth slit and window sight in brass, incorporating a black-glass

126. Marine compass: D. McGregor & Co., Glasgow, Greenock, Liverpool & London, c.1890 (T1980.197)

5" dry card compass, with jewelled bearing and broad needle rivetted under a covered mica card, stamped 'N' at one end and with red adjusting wax. Printed 64-point compass rose divided at the circumference in degrees 0-180-0, with 'D. M^CGREGOR & C^O L^{TD}/ GLASGOW, GREENOCK, LIVERPOOL, LONDON.' printed around the centre, and at the north point 'M^CGREGOR & C^O L^{TD}/ GLASGOW / GREENOCK / LIVERPOOL'. Suspended on a brass point in a glazed and painted brass bowl with lubber line, lead-weighted in the base of the bowl and gimbal mounted in a wooden case.

Card diameter: 128mm.
Bowl diameter: 171mm.
Case size: 205 x 205 x 145mm.

Lacking the sliding lid to the case. The gimbal ring is attached in the case by modern screws; the case is of comparatively low quality and may not belong to this compass.

CHRISTIE & WILSON

Christie & Wilson, nautical opticians and compass adjusters, was founded by Andrew Christie and James Wilson in 1916. Christie already had long experience as an instrument maker. He was born about 1856, the son of Andrew Christie, a law clerk who had moved to Glasgow from his native Dysart in Fife, and his wife Mary, from Crieff.[1] Andrew Christie junior is recorded at the age of fifteen as a 'Nautical Instrument Maker's Clerk' in the Census of 1871,[2] and since he was later working for Duncan McGregor & Co. (q.v.), it is reasonable to assume that he was with the firm at this date. By 1886 he was sufficiently senior in McGregor's to have a separate entry in the Glasgow street directory as a nautical instrument maker.

In 1890 Christie left McGregor's and set up independently at 27 Clyde Place as a nautical and mathematical instrument maker, a compass adjuster and maker, a chronometer maker, a clock and watch maker and wholesale clock manufacturer, and an optician.[3] In 1896 he expanded into 28 Clyde Place, and in 1901 entered a brief partnership with James Gilchrist Lee, whose principal experience had been in shipbroking.[4] From 1902 Christie was once more independent, working as a compass adjuster at 192 Hyndland Road until 1907, at 34 Robertson Street from 1908 until 1910, and at 54 Broomielaw from 1911 onwards.

James Wilson came from Whyte, Thomson & Co. (q.v.), with whom he had been a compass adjuster from at least 1903, and continued to advertise himself as such.[5] The staff of the firm included James Thomson, a compass adjuster (who had been with Whyte, Thomson & Co. from before 1929 until 1934, when he moved to Christie & Wilson), and Angus McNair, an adjuster with the firm in 1920. The type of work the firm carried out appears to have been the same as that during Christie's independence; that is, predominantly the selling of nautical instruments and compass adjustment. Apart from the instruments in the Frank collection, others with the firm's signature have been noted.[6] In 1925 the firm moved to 130 Broomielaw, in 1928 to 132 Broomielaw, in 1940 to 90 Broomielaw, and in 1960 to 130 Broomielaw.[7]

REFERENCES

1. General Register Office (Scotland), Census of 1871 (Glasgow, Tradeston), 644⁹/58, p.9.

2. *Ibid.*

3. *Glasgow Post Office Directory* 1886.

4. *Ibid.* 1890, 1896, 1900. Lee died or retired in about 1902.

5. *Ibid.* 1903 and *passim*. He advertised as an adjuster until at least 1930.

6. A sextant, signed 'Andrew Christie' was offered for sale by Sotheby's Belgravia, 21 December 1976, Lot 229; another, signed 'A. Christie, Glasgow, L010' was offered by Christie's, 19 December 1973, Lot 42; a third, described as 'made for Andrew Christie, Clyde Place, Glasgow, ... serial No. 4099' was offered by Christie's Glasgow, 14 December 1983, Lot 40; and a fourth, signed 'Christie & Wilson, Glasgow, No. 2676' was offered by Phillips, 20 April 1983, Lot 109. A brass dry card ship's compass, signed 'Christie & Wilson, Glasgow' pattern no. C29, no. 941, was offered for sale at Christie's South Kensington, 22-23 September 1988, Lot 449. A ship's clock, signed 'Christie & Wilson, Glasgow' was offered by Christie's Scotland, 8 June 1988, Lot 57.

7. *Glasgow Post Office Directory* 1925-1970, *passim*.

127. Marine compass: Christie & Wilson, Glasgow, c.1930 (T1980.199)

4″ liquid-filled card compass. Printed 128-point compass rose marked 'CHRISTIE & WILSON / GLASGOW.' surrounding a central float, and suspended in a glazed and painted sealed brass bowl with lubber line, for gimbal mounting.

Card diameter: 101mm.
Bowl diameter: 129mm.

Lacking gimbal ring and case housing.

128. Marine azimuth mirror: unsigned, c.1935 (T1980.203)

Azimuth sight, in aluminium and plated brass (stripped), for use in conjunction with an 8″ compass. Sighting prism, with solar shades and a transverse bubble tube mounted over a shielded viewing lens for reading the compass scale; set on a base, leadweighted at the toe, to span the compass bowl. Stamped on the base with the assembly mark '2'. With fitted case containing a printed instruction sheet in the lid for 'THE THOMSON SYSTEM AZIMUTH MIRROR' issued by Christie & Wilson, 132 Broomielaw, Glasgow.

Overall length: 184mm.
Case size: 215 x 125 x 135mm.

21

JAMES WHITE AND LORD KELVIN

James White, the founder of the business which carried his name, was born in 1824 in Port Ellen, Islay, the son of William White a yarn merchant there, and Margaret Adam or Adams.[1] While still a boy he settled in Glasgow and in about 1839 was apprenticed to Gardner & Co. (q.v.),[2] for whom he probably also worked as a journeyman before setting up independently in 1850 at the age of about twenty-five.[3] His first shop was at 24 Renfield Street from 1850 to 1852, then at 14 Renfield Street from 1853 to 1856.[4] He partnered the otherwise unknown John Haddin Barr, as White & Barr, at 1 Renfield Street and 60 Gordon Street in 1857, and 1 Renfield Street from 1858 to 1859.[5] In 1860 he resumed independence working at both the latter addresses until 1863 (despite going bankrupt in 1861). He was subsequently at 95 Buchanan Street from 1864 to 1868, 78 Union Street from 1869 to 1875, 241 Sauchiehall Street from 1876 to 1883, 209 Sauchiehall Street and 16, 18 and 20 Cambridge Street from 1884 to 1890. From 1891 the Sauchiehall Street premises was occupied by Mathew Edwards (q.v.); and after the business became Kelvin & James White Ltd. in 1900, and Kelvin, Bottomley & Baird Ltd. in 1914, it continued in the Cambridge Street shop.[6] In 1947 the firm amalgamated with the London instrument makers Henry Hughes & Son, and was renamed Kelvin & Hughes Ltd.,[7] and in 1965 became the Kelvin & Hughes Division of Smith's Industries, Ltd.[8]

According to an account of 1901, James White set up independently as an optical and philosophical instrument-maker 'on the strong recommendation' of the then Professor William Thomson,[9] who was created Baron Kelvin of Largs in 1892. This suggests that Thomson, who had been appointed to the chair of natural philosophy at the University of Glasgow in 1846, recognised White's talent and encouraged him with promises of work. Thus although the exact point at which their association began is not known, it is likely that a period of experimental construction in the early 1850s preceded the start in about 1854 of the production of Thomson's electrical instruments, although even these were only experimental pieces for class demonstration. Thomson's laboratory was set up at the Old College in 1850, and although White supplied items in 1854, he does not appear to have become regularly involved in supplying and mending apparatus until 1858. He was largely responsible for equipping Thomson's second laboratory at the University's new premises at Gilmorehill, the Gilmorehill laboratory of 1870 being the first of its kind in the world.[10] He was subsequently made Optician by appointment to the University.[11] For the most part, however, White's output in the 1850s appears to have followed that of his former master, T.R. Gardner (q.v.). Often advertising simply as 'Optician', White produced standard optical instruments, and by about 1860 was also venturing into making photographic apparatus and other specialised lines.[12]

Meanwhile, his professional relationship with Thomson became increasingly one of mutual dependence. However, closer inspection reveals that it was surprisingly informal. There appears to have been no legal deed of copartnery, and none of Thomson's patents were held jointly with White, even though in practice White was the sole manufacturer. This is not to say there was no monetary investment by either party, but the more obvious and expected professional ties were not in evidence.[13] A brief outline of the firm's nineteenth century business history demonstrates this. The year 1859 saw the dissolution of White's brief partnership with John Barr,[14] but after slightly less than two years on his own he petitioned for the sequestration of his estate.[15] White's liabilities amounted to virtually £3000 (of which open accounts and bills accounted for £1691 8s 0d and £1306 in borrowed money) while his assets were estimated at £1384 3s 2d.[16]

The breakdown of trade suppliers to whom he owed money at this stage demonstrates that in 1861 his relationship with Thomson was by no means exclusive and that he conducted a broad-based retailing business. Thus, among White's minor creditors who had open accounts and were due small sums of money (none were owed more than £40, but the total came to £712 19s 1d) were the following identified instrument firms: Alexander Adie & Son (q.v.), £11 2s 6d; J.M. Bryson (q.v.), £2 2s 4d; Edward Cetti of London (barometer maker), £11 9s 9d; L.P. Casella of London (maker of meteorological instruments), £9 17s 6d; Elliott Brothers of London (instrument manufacturers, successors to Watkins & Hill), £3 19s 6d; R. Griffin & Co., Glasgow (manufacturer of chemical apparatus), 10s 11½d; Ladd & Oertling of London (philosophical instrument makers), 18s; John Moffat of Edinburgh (photographer), £6 15s 3d; Negretti & Zambra of London (instrument wholesalers), £2 16s 9d; J.D. Potter of London (nautical instrument maker; Admiralty chart

57   James White (1824-1884)

agent), £2 12s 6d; John Spencer (q.v.), £1 2s 5½d; Smith & Beck of London (microscope makers), £2 10s 5d; J. Pastorelli of London (glassblowers), £18 2s 7d; J.P. Cutts, Sutton & Son of Sheffield (instrument wholesalers and manufacturers), £11 10s 7d; James Parkes & Son of Birmingham (instrument wholesalers and retailers), £14 12s 3d; Peter Frith & Co., of Sheffield (instrument wholesalers and retailers), £12 14 11d; Joseph Levi of London (instrument wholesalers and retailers), £34; Powell & Lealand of London (optical instrument manufacturers and microscope specialists), £2 2s.[17]

The summary of his stock reveals that at this point in his career that there was nothing to indicate that White's business would become either nationally or internationally important; indeed, the similarities between it and the stock of other bankrupt instrument firms discussed in this work is remarkable. Here again, on 24 August 1861, there are enumerated the telescopes, binoculars, barometers, thermometers and spectacles; the mathematical instruments, photographic albums, microscopes, T-squares, curves, lenses, stereoscope stands, photographic cases and technical apparatus, prismatic compasses, lantern slides, magnetic machines, theodolites, chemicals, glass apparatus, prisms: in total, a detailed inventory amounting to £1317 4s 3½d.[18]

However, bills were received from a number of people, including David Carlaw (q.v.), for £6 6s 11d; William Gardner (q.v.), for £2 0s 9d; John Lizars (q.v.), for 10s 1d; John H. Barr of New Zealand (presumably his erstwhile partner), for £14 6s 2d; Alexander Morton of Morton Place, Kilmarnock (q.v.), for £7 8s 6d; and William Hart of Edinburgh (instrument maker), for 9s 4½d.[19] White had to undergo public examination on 18 September 1861, in which he revealed that:

> 'I began my present business in Glasgow in March 1850. I had then a capital of £160, a portion of which I borrowed from friends. My state of affairs shews my liabilities to be £2997-8-6, and my assets to be £1384-3-2. I account for the deficiency between these to sums by losses in business through various unprofitable contracts, and by a patent connected with an improved Sextant.'[20]

Further claims were made on White's estate by Frederick Newcombe, philosophical instrument makers, London, £24 6s 4d (he was already owed £11 7s 4d on account); Joseph Hammersby, telescope and opera glass manufacturer, London, £35 15s 1d (and a further £35 2s 7d on account); William Henry Brown, rule maker, Birmingham, £17 4s 11d (and a further £4 2s 6d on account); Peter Stevenson, instrument maker, Edinburgh, £3 11s; and Alfred Sloper, mathematical instrument maker, London, £81 18s 9d (and a further £40 2s 7d on account).[21]

Fortunately, White was able to offer composition of 6s 8d per pound, in equal instalments at four and eight months after his discharge, and his creditors accepted this.[22] The accounts in the Sederunt Book demonstrate that White's workmen continued to be paid by the Trustee while the estate was sequestrated, and White

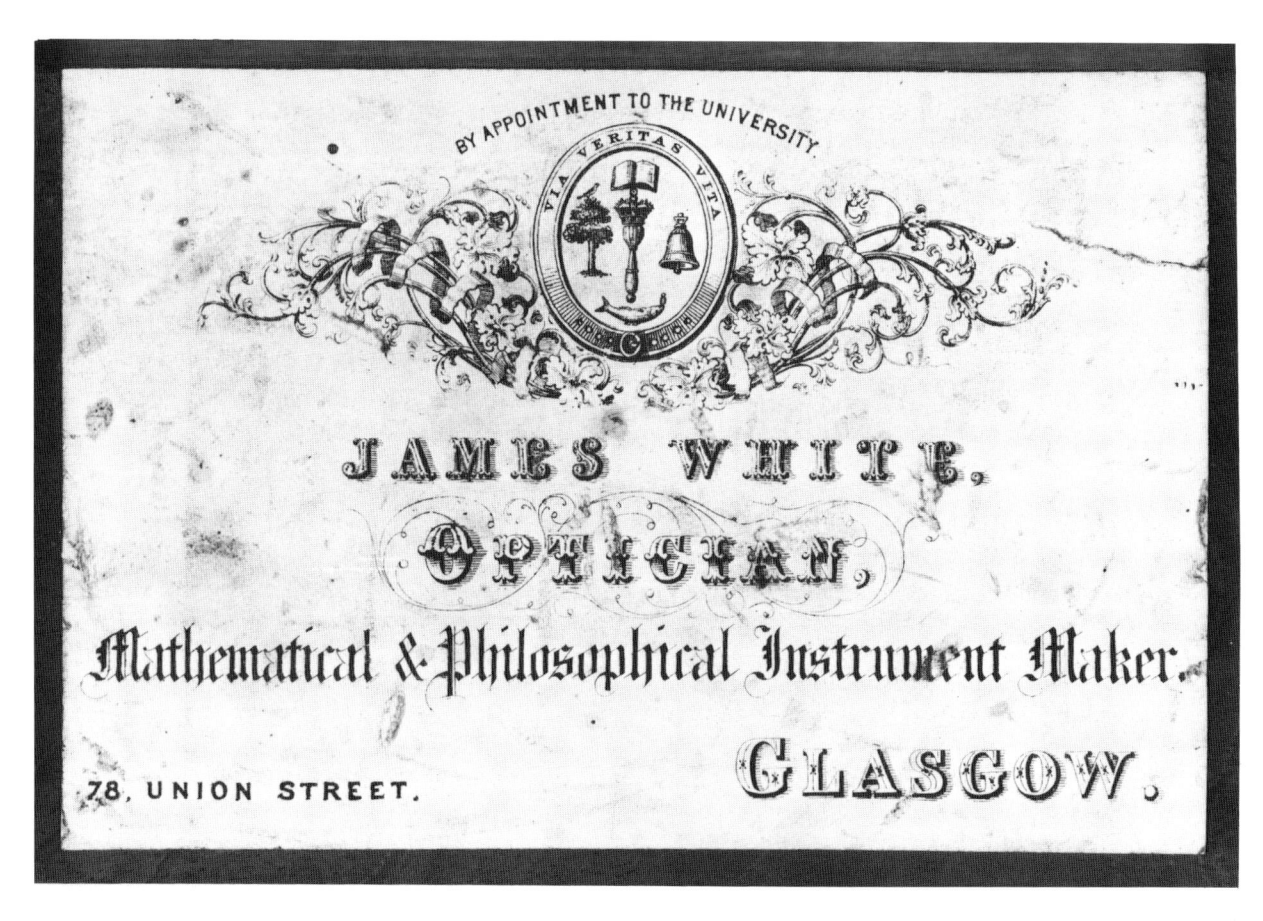

58 James White's trade card for his address between 1869 and 1875

was discharged on 19 December 1861, almost four months to the day after his bankruptcy.[23] Nowhere is the name of William Thomson mentioned.

One interesting name occurs in the sequestration papers: Robert McCracken, dentist, had sold White a life assurance policy, and was owed £194 1s 5d plus a further £200.[24] In early 1877 a notice announcing the dissolution of a partnership between McCracken and White appeared.[25] In this case, as elsewhere, the snag (for historians) with legal partnerships is that official notice is taken only when they are dissolved, so that there is no way of knowing how long they have been in operation. Presumably McCracken's input to the business had been financial rather than practical or theoretical; nothing else about him has been uncovered.

In 1884, James White died, and his obituary stated that:

> 'The business will be continued by Mr Matthew [*sic*] Edwards and Mr David Reid, both of whom were long in the employment of Mr White and latterly in partnership with him.'[26]

This - one of the earliest secondary accounts of White's career - was careful not to mention the word 'partnership' (with its legal connotations) to connect White with Thomson, and put White's independence at 'about thirty years ago' and the start of his work with Thomson five years later at 'about twenty-five years since'; that is, around 1859.[27] A legal notice was published in 1884, announcing the continuation in copartnership of Edwards and Reid.[28] A few years later this partnership was dissolved, when in 1891 Mathew Edwards (q.v.) became independent, taking part of the business with him, and leaving the capable David Reid to run the remainder under the name of James White.[29]

Although it is clear from these partnership details that Thomson did not himself become a partner in 1884, he had by this stage recently made a substantial investment in new premises for the company, as will be discussed below. Considering that White had suffered from an extended debilitating illness it may be assumed that Thomson had developed a controlling interest in the firm even though he was not formally a partner.

In 1899 Thomson, by now Lord Kelvin, retired from the chair of natural philosophy which he had held since 1846, and became a research fellow of the University. His nephew, James Thomson Bottomley, who was experimental demonstrator (and gave many of lectures for his uncle, as Thomson explained to his Cambridge colleague G.G. Stokes),[30] resigned at the same time. Both men became partners of Kelvin & James White Ltd., which was formed in 1900, with Kelvin holding some £20,000 in shares.[31]

The reason for this emergence as partner may simply be that the firm had finally been established as a

254

limited liability company, so that there was no longer any danger of Thomson's personal estate being sequestrated. Up to this point, the financial interest of Thomson and Bottomley can only be deduced from the use of the 'WTB' trademark (presumably standing for White, Thomson and Bottomley) which appears on marine instruments covered by Thomson patents and was certainly in use in the 1880s. We conjecture that this implies an early agreement between Thomson and Bottomley on one hand and the White business on the other about royalty payments. A similar arrangement in early 1884 between Thomson, Bottomley and Reid (on the firm's behalf) for a commissioned electrometer is discussed below.

So, although Thomson had no formal legal partnership rights in the business before 1900, he does appear to have had an increasingly close working relationship with the founder, James White. Indeed, in their method of work can be discerned some of the reasons for their successful co-operative association, and for the prosperity of what was to become their joint business. Thomson's genius for invention lay in the empirical solution of problems of applied physics, and in James White he found a man capable of executing what were often initially simply ideas, sometimes spontaneously generated in unlikely locations.[32] According to one memoir, Thomson frequently entered White's workshop to give him directions, and would 'scribble on a torn envelope the only 'working drawing' that in those days ever guided the birth of his inventions'.[33] Another writer stated that 'while White was no doubt a most skilful mechanic, he had further a singular aptitude in grasping Thomson's ideas and giving them practical shape'.[34] One of Thomson's biographers wrote that Thomson 'thought in steel and brass, and must continually have the incipient instrument before him when working out the details ...', and that the sequence of tests and alterations was 'both slow and costly'.[35]

This design and construction process is confirmed in an anonymous account of 1898 by a member of the workshop, which although it dates from after White's death, may be taken to refer to the whole period of Thomson's active design work. Following an inspiration Thomson would consider its practicability, then:

> 'Immediately, a rough sketch is made, instructions given to a workman of the experimental staff, who starts blindly to follow his instructions and make something he does not know the name of. The result, as might be expected, is not exactly beautiful to look upon, but there must be no slipshod work about it. Everything must be accurately and well made in all its working principles, but without any degree of finish.'[36]

Thomson would inspect and pick out half a dozen faults which would eventually be eliminated and the instrument perfected. Sometimes 'the workman may be called upon to do something he deemed impossible'.

Close contact between inventor and maker was maintained by Thomson's almost daily visits to the workshop, which was near the University. Later, on the University's removal to Gilmorehill, Thomson still laid aside a part of each day for workshop business, and this was facilitated by the installation of a telephone link between Thomson's university laboratory and the Cambridge Street works: this was possibly the first such line in Britain, and formed the nucleus for the Glasgow telephone exchange.[37] Thomson's need to supervise instrument construction at every stage until the perfection of the design must therefore have placed some strains on White and his assistants, and it is reasonable to suppose White's good nature was tested by this, as much as Thomson's patience was tried by occasional delays in the workshop.[38] However, in his lecture room Thomson 'often spoke of and praised the excellence and exactness of White's work,'[39] and in about November 1883, a student later recollected, he described one of his electrical machines to the class, 'apostrophising the machine as "charming" and "perfectly beautiful"'.[40] No doubt he was enthusing not only over the physical phenomena it measured, but also over White's craftsmanship. On the occasion when he commissioned another maker, Légé of London, to finish his tide predictor, he regretted that he had not asked White to make the whole because

> 'with his genius to help, the details would have been much more satisfactorily carried out - and I should have had *infinitely* less trouble... They did their best but their dynamical intelligence was not superior to that of ordinary instrument makers and, on that account, not from any wish on their part to obstruct me, and because they were 400 miles from my home, I had infinitely more trouble and a less satisfactory result than I should have had with White.'[41]

Electrical instruments were the first major productions of White's association with Thomson, and from 1854 the pioneering series of electrometers and electrical balances began to be produced,[42] which not only established Thomson's international reputation but also brought White's workshop into great prominence. By 1890 the firm was manufacturing six or seven types of instrument for standard measurement, and in addition produced a range of ammeters, voltmeters and ohmmeters for power and lighting circuits and for laboratory work.[43]

Thomson's preoccupation with electricity led to his involvement from 1856 in the Atlantic Telegraph project. It was in response to the need for an alternative instrument to replace the comparatively insensitive land telegraph, that he began designs which resulted first in the mirror galvanometer patented in

1858, which enabled transatlantic contact to be made for the first time, and culminated in the siphon recorder of 1867 which produced accurate automatic recording of telegraph messages. James White was deeply involved in the evolution of the galvanometer, and was credited by Thomson with the important suggestion that a light-weight silvered piece of microscope glass should be substituted for the metal mirror.[44] One later source dated Thomson's 'permanent attachment' to White from the time of the galvanometer's success,[45] and it is clear that the workshop was still producing the sophisticated and delicate siphon recorder in significant numbers as late as 1898.[46] An indication of the overall importance of electrical instruments within the workshop's production, is given by the fact that twenty-four out of Thomson's sixty patents were for generating, regulating, measuring, recording and integrating electrical circuits.[47] Moreover, the firm's surviving records show that whereas in 1885 electrical apparatus earned less than half of what compass production brought in (and half of it was earned by the siphon recorder), yet fifteen years later in 1900, electrical apparatus sales were apparently about twice as lucrative as compass sales.[48]

It was not surprising that as a fertile inventor and a keen yachtsman, Thomson should have turned his attention to marine instruments. Following the formation in 1867 of a Committee of the British Association, to undertake a harmonic analysis of tides at several points on the globe, White constucted several tide gauges under Thomson's directions; these were sent to Australia, Italy, Madeira and other places. In 1872 White built the first Model Tide Predictor (although it was completed by A. Légé), and sometime after 1878 Thomson's harmonic analyser of tides was produced.[49] Meanwhile Thomson had become aware of the need for a highly accurate compass for metal ships, when asked to contribute an article on the compass for the magazine *Good Words* in 1874. His first patented design followed in 1876, and a stream of patented improvements were made until his death in 1907. The main features of the design were in the reduction of the weight of the card, in the use of a sapphire bearing on an irridium-tipped pivot, and in the development of highly accurate deflectors. From the beginning White was the sole maker of the compass and the binnacle on which it was mounted, and the prosperity of the firm from the late 1870s can be attributed in large part to the sales of the instrument, which in 1885 accounted for more than twice the income from electrical apparatus. Nor should its relative commercial decline compared to the latter necessarily be taken as an indication of falling business; rather it suggests that the huge success of electrical instruments was even greater than that of compasses.

The first compass was ready in June 1876 and by 1 January 1877 a modest fifteen had been produced. Gradually, however, annual sales picked up, climbing to 350 in 1885, and peaking at 580 in 1892. The approximate annual average for the years 1876 to 1889 was 288 compasses, and it increased for the period from 1900 until the records ceased in April 1918.[50] By 1900 6,900 compasses had been produced, and the last recorded compass supplied on 29 April 1918 was numbered 13,474.[51] Not surprisingly the largest customer was the Admiralty, which was supplied with large numbers from 1883 onwards. Most of the other purchasers were instrument makers who specialised in marine apparatus, and who fitted the Thomson patent compass in their capacity as compass adjusters or supplied it as ships' chandlers. D. McGregor & Co. (q.v.) was the first Scottish maker supplied with a compass, in December 1876, and the firm was a frequent customer from that date.[52] Whyte & Co. (q.v.) followed in July 1877 and appear to have become the biggest Scottish purchaser.[53] Other Scottish customers included P. Feathers & Son (q.v.), Alexander Dobbie (q.v.), M. Walker & Son (q.v.), and F. Sewill, a Glasgow chronometer and nautical instrument maker.[54] However, if the sales for 1880 reflect the overall pattern, White despatched more compasses to England: in that year seventeen went to Whyte & Co.(q.v.), but thirty-six to F. Sewill (mostly to his Liverpool branch and only some to the Glasgow one), twenty-eight to the London maker J. Lilley & Son, and nine to J. Salleron in Paris.[55]

Despite the closeness of the relationship between Thomson and White, but perhaps as result of his earlier experiences that had led to his bankruptcy in 1861, White was not formally involved in the patenting of any of the instruments which were constructed by his business.[56] It is therefore incorrect to claim, as one recent author has done, that 'through his patents Thomson soon became a major shareholder in the firm, which in turn provided research problems for Thomson's laboratory students and jobs for his laboratory graduates for more than four decades'.[57] Another writer has pointed out the large sums of money reaped from the successful patenting of inventions: 'some idea of the overall gains from his patents and from Kelvin and White may be formed from the fact that he left investments worth £162,000.'[58] Certainly Thomson was keen to foster the revenues derived from his inventions, an attitude shared by Fleeming Jenkin, his co-patentee in telegraph improvements. In 1860 Jenkin wrote to White on their behalf, requesting changes in White's accounting methods, including regular accounts of the production of instruments under their joint patents.[59]

On the other hand, clever inventions and water-tight patent rights did not, on their own, make a successful economic venture, and this was to prove the case with Thomson's improvements to the mariner's compass. It needed good marketing, and in this instance, strategic and well-placed influence in scientific and naval circles.[60] He also had to be prepared to defend his

59 Sir William Thomson, Lord Kelvin (1824-1907), engraved by C. H. Jeens, from a photograph by Fergus

patents through the law courts, an expensive business, even for the victor. Thomson's patent no. 1339 of 30 November 1876 on the mariner's compass, was of such a broad nature that a year later he filed a disclaimer, following it with a series of further patents to improve the instrument.[61] His subsequent involvement in patent litigation has provoked the comment that

> 'the details of these cases ... does little for the image of either Sir William or British justice. Although the ideas patented by Sir William were sound and their application excellent, when taken individually it is doubtful whether most justified their patents, the majority ... having been tried and documented in former years.'[62]

Action was taken by Thomson in defence of his amended patent 1339 of 1876 against a Scottish defendant named McGregor (who subsequently acted as Thomson's agent); this case never reached the courts, as the patent was found to be insufficiently specified.[63] Shortly afterwards, Thomson successfully sued William Batty of Liverpool,[64] then F.M. Moore, a Dublin nautical instrument maker,[65] and the London instrument makers Henry Hughes & Son, although here he was less successful, because an injunction was not granted.[66] A final, and again unsuccessful action was made by Thomson (by now Lord Kelvin) regarding his patent 22,0031 of 1902 against Whyte, Thomson & Co. (q.v.).[67]

Dating from about the same time as the genesis of his compass designs, came Thomson's development of a sophisticated sounding machine, to which he was led by the problems associated with cable-laying at sea. His device could be operated at speed, unlike previous types which required the vessel to be stationary. The design involved a powerful but delicate winch, and the use of pianoforte wire for suspending the pressure-sensitive sounding head.[68] He was testing the prototype from his yacht in the Bay of Biscay in June 1872, and its first successful active use was on the U.S.S. *Tuscarora* in the Pacific in 1873-1874.[69] Thomson referred to these episodes when in 1874 he addressed the Glasgow Philosophical Society 'On Deep-Sea Sounding by Piano-forte Wire'.[70] The series of patents which protected this device and later improved versions began in 1874, and like that covering the compass, continued until Thomson's death in 1907, and was thereafter continued by his successors.[71] And like his compass, Thomson's sounding machine was a phenomenal success; more than 7,500 had been sold by late 1902.[72] Together these devices were of the highest importance in aiding safe navigation; one P.& O. captain described Thomson as 'the greatest friend of the sailor who ever lived'.[73]

Something of the breadth of Thomson's mechanical inventiveness has been outlined, and no evidence has been found that the many instruments he invented were devised or manufactured by any maker other than James White.[74] Thus, including the output of the most popular items, the compass and binnacle, and the various types of electrometer, White's workshop was kept busy. Nonetheless he also continued to produce instruments of a more conventional type. For instance, among his exhibits at an exhibition in Glasgow in 1880 were not only Thomson's compass and sounding machine, and his deflector, eclipsing lights and integrating machine, but also a marine mercurial barometer, and a maximum and minimum thermometer.[75] Similarly, an advertisement of 1888 lists not only the main marine and electrical instruments then available, but also a range of instruments and equipment no different from that of other Glasgow makers; these included telescopes, spectroscopes, microscopes and spectacles, and surveying equipment, theodolites and levels.[76] The group of levels in the Frank collection[77] serves as a reminder that if White was the 'highly-skilled coadjutor'[78] of one of the greatest-ever physicists, he was also a maker serving the more mundane requirements of professional and amateur customers.

Nevertheless, it is in White and Thomson's almost symbiotic relationship that the key to the history of the firm is to be found. For while White enabled Thomson to realise his notions in three dimensions, the huge market for the resultant instruments enabled steady expansion in the workshop, and presumably considerable profit to both men, although contemporary sources are too polite to say so. They are simply presumed to have reaped the reward for their technological brilliance.

Although no figures are extant for the size of White's workshop during the first twenty years of its existence, it is known that in 1871 he was employing fourteen men and four boys, and that in 1881 the figure had risen to sixty men.[79] Significant changes occurred soon afterwards. For the past eighteen months, wrote White's obituarist in September 1884, 'a really important manufacture has been established in the extensive suite of workshops purchased by Mr White's firm in Cambridge Street'.[80] This expansion can be linked to the increased output of the patent compass, particularly through contracts with the Admiralty which began in 1883.[81] Although not formally a partner, it was Thomson who raised the capital needed to construct and equip the first Cambridge Street workshops. The premises were purchased jointly by White, Edwards and Reid, and owned by the latter two after White's death; it was not until Edwards' departure in 1891 that Thomson acquired a share in the property alongside Reid.[82] As a direct result of the move to Cambridge Street the workforce was increased to one hundred men by September 1884[83] and doubled to two hundred persons by 1891.[84] Ten years later it had again doubled to over 400,[85] following the construction of a second five-storey building to the rear of the 1883-84 works, which was apparently under construction in 1891, and which 'bids fair to be the largest as well as the most complete electrical laboratory in the country'. The total depth of the combined building was to be from 150 to 200 feet.[86] The new building was occupied in 1892, but the premises had to be extended again in 1907 to accommodate the increased workforce.[87]

Seen overall, the growth of the workshop represents the transition from craft-based manufacture to factory-based production, although in the absence of adequate evidence for the period 1849 to 1870 it is difficult to say when this occurred. Certainly the size of the workforce in 1881 suggests that specialisation on separate components of particular instruments within the shop was occurring. Three descriptions of the works in the 1890s show that their organisation and layout were highly developed. In 1897-98 the factory was on five storeys, with electric lighting throughout, and with a floor space of 34,000 square feet. The front ground floor comprised offices and showrooms (the main shop being at 209 Sauchiehall Street) and behind in the newer part lay a well-equipped laboratory, with a 'large staff of electrical experts' responsible for standardising all the electrical instruments produced.[88] On the first floor were the joiners, cabinet-makers and polishers (all female), working on the woodwork of various instruments, including the binnacle and the sounding machine. The second floor 'is the seat of management'. In one wing lay the drawing office[89] and the experimental flat 'immediately under the foreman's eye'; in the other lay most of the heavy machinery. This included lathes for cutting cast-iron drums,[90] universal milling machines for making copper coils, and drills for boring marble slabs for switchboards; 'the visitor imagines himself in a churchyard or monumental sculptor's, with marble slabs on every side'. In this department also, 1200 foot lengths of piano wire were

60   Kelvin, Bottomley & Baird's Glasgow works, c.1930

wound onto sounding drums. The third floor contained the binnacle shop where 'a score of lusty fellows are busy pounding away on brass binnacle domes with large round-faced hammers ... Here the newspaper men, insurance and advertising agents, and bores of every description are interviewed'. In the front of the third floor large numbers of instruments were assembled, after the machined parts had been made elsewhere in the works. On the fourth floor was a second machine shop for making terminals and screws, and for cutting brass rods and so forth.[91] Power was supplied by an 80hp engine,[92] and since most of the machine tools were specially constructed by the firm, they could produce items such as screws better and cheaper than specialist makers. Also on the fourth floor were a small shop of girls producing capillary tube; the japanning department; an engraving shop, with two English engraving machines; and the clockmaking department. This last provided a range of timepieces from chronometers to sundials and turret clocks, as well as the time-drums of electrical recording instruments. In an adjoining building on the second floor was a store for material and the electrical coil-winding department, with 'an expert staff of girls'.[93]

Soon after the establishment of the Cambridge Street works, James White died aged fifty-nine on 15 August 1884, of a 'softening of the brain' which had been apparent for four years and four months, according to his doctor,[94] although the exact nature of the illness is not known. He left a widow, Jane Reid, but no surviving issue; and it was to her that he left the bulk of his estate, the net value of which was £7,088, of which £5,244 was the balance of his credit in the firm.[95] An obituary described him as

> 'a man whose character was exceedingly simple and altogether devoid of ostentation. He delighted greatly in his business, and especially that carried on in the workshop, while nothing pleased him better than to find that he had been successful in providing his customers with such instruments as they were in search of, or the construction of which they had entrusted to his care.'[96]

Another writer referred to him as 'an amiable and worthy man as well as a skilful mechanician'.[97]

After his death the firm continued to trade under his name, with Mathew Edwards and David Reid as managing partners, both of whom had been partners with White.[98] Some changes took place in 1891, when Mathew Edwards (q.v.) and another senior figure James More (q.v.) left to set up their own independent businesses. Reid continued as the sole managing partner, and by 1897 James Ferguson was in charge of the electrical instruments.[99] In 1900 the firm became a limited company, ostensibly so that several of the employees might obtain an interest in its ownership.[100]

At the same time its name was changed to Kelvin & James White Ltd. to prevent confusion with firms who were marketing compasses and instruments in imitation of Kelvin-White products under similar names.[101] This was still a problem in 1902, when readers of the firm's *Standard Tide Tables* were cautioned 'against the confusion of mind likely to be caused by certain advertisements drawn up and illustrated in a form resembling our own as closely as possible ... made by certain firms who were formerly content to act as our agents for the sale of Sir William Thomson's Compasses, but who are now endeavouring to push the sale of compasses made by themselves, and described as "Sir William Thomson's", "Sir William Thomson's system", "Thomson's pattern" &c'.[102] The implication is clearly that Whyte, Thomson & Co. (q.v.) were guilty of misleading advertising, which is confirmed by their entries in the street directory.[103]

Nor was business always smooth, despite the fact that the extra workshops in Cambridge Street had been built to cope with the huge demand for electrical instruments in the early 1890s, and the alleged fact that in 1898 the firm did not need to employ a traveller to obtain orders.[104] In 1894 David Reid was regretting that fewer orders than he would have liked were coming in, and Kelvin hoped that the visit to England of one of the senior staff, Andrew Meikle Kemp, would bring in fresh orders.[105]

In December 1906 there was alarm among Kelvin and his staff that the Admiralty was about to replace his compass throughout the Navy with a liquid compass designed by Captain the Hon. Louis Wentworth Chetwynd. No Admiralty orders had been placed for two or three years, and as Kelvin wrote, 'It is very important for us now to keep *in* with the Admiralty all we can.'[106] Incidentally, something of the scale of White's production is indicated by a figure Kelvin quoted, that the Admiralty then had about 3000 unused Kelvin & White binnacles lying in its dockyards.[107] Kelvin, for all his unassuming attitude and flights of abstract thought, could be hard-headed when it came to the defence of his patents and the royalties they produced. He told Reid in 1884 concerning his royalty for an electrometer he had ordered for someone: 'It should be not less than ten per cent; indeed I think it ought to be not quite 20 per cent, but if you show this letter to Mr Bottomley you & he will decide.'[108] In 1895 Kelvin wanted Bottomley to check if his patent for the siphon recorder of July 1867 was specific enough to be defensible against a rival patent by a Mr Dickenson.[109] Combined pride in the reputation of the firm and commercial caution is revealed by his remarks on the excellent condition of the Eastern Co.'s siphon records at Syra and Athens, seen on a trip in 1894, although on inspecting the Athens instrument he wrote, 'I did not see the name "James White" on the wheel work. You should put it, and the trademark, on every separable part. Yours truly Kelvin.'[110]

As has been indicated, White's workshop of necessity included many talented mechanicians, and it is worth recording the names of those who are known to have assisted White and his successors. In the 1850s David Carlaw (q.v.) is said to have been White's manager, and in 1857 William Gray (who had probably attended the Edinburgh School of Arts) was described as 'optician' at White's.[111] Mathew Edwards (q.v.), James More (q.v.) and David Reid probably joined at this time. James Thomson Bottomley (1845-1926), a talented graduate of Belfast and Dublin, was Kelvin's nephew. He became his uncle's private assistant in 1870, subsequently becoming a demonstrator and then ultimately deputising for him in the Natural Philosophy lecture rooms at the University of Glasgow.[112] As a researcher into electrical engineering he was useful to Kelvin in the workshop as well as the lecture-room,[113] and in addition to being a dexterous researcher, he was a

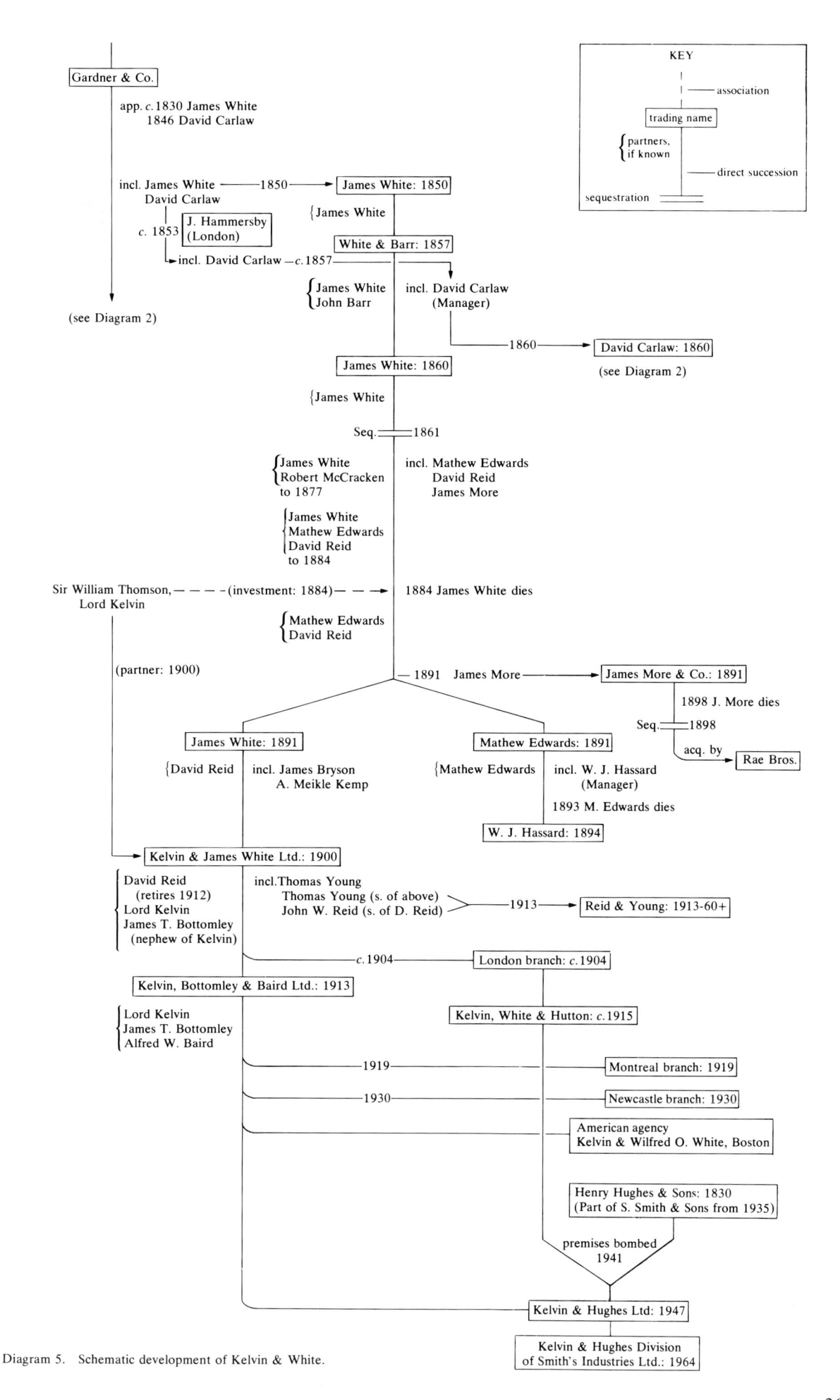

Diagram 5. Schematic development of Kelvin & White.

skilful glass-blower. From 1889 he called himself 'consulting electrician',[114] and retired from lecturing at the same time as his uncle in 1899. Although closely involved in the White business, Bottomley became a partner in the firm only when it became Kelvin & James White Ltd. in 1900. After Kelvin's death in 1907 he was elected chairman of the directors, which he remained until his own death at the age of eighty-one in 1926.[115]

In about 1912 David Reid retired; he was one of the most senior members of the firm and had managed the workshop since the 1880s.[116] Surviving correspondence provides evidence that he was held in great trust by Kelvin,[117] and a handbook of 1891 credits him not only with 'the highest capability' as an electrician but also with the qualities necessary in the managing principle of such a workshop, namely 'scientific knowledge, mechanical skill, and adaptability to the ever varying requirements of an astonishingly progressive age such as can scarcely be paralleled in any other vocation'.[118]

Following Reid's retirement, a reshuffle occurred in the firm, which became Kelvin, Bottomley & Baird Ltd. in 1914, with the assumption into directorship of Alfred W. Baird, who had been a compass adjuster with the firm since 1884. Subsequently Baird had been a personal assistant to Kelvin, with whom he worked for over fifteen years, having charge of the department which developed Kelvin's inventions. He was a partner in Kelvin & James White Ltd.[119] At the same time, David Reid's son, John W. Reid, left the firm to set up independently with a Thomas Young as Reid & Young (q.v.). Thomas Young's father, also Thomas, was one of four compass adjusters listed as employed by Kelvin & James White Ltd. in 1910; the others were A.W. Baird, Robert Ferguson and Duncan McLennan. Another adjuster, Athole Clark, was listed in 1905,[120] and among other employees were George Green and Frank W. Clark. Green was Kelvin's last personal assistant (for the period 1905-1907), and was involved with the firm's work on sounding machines for battleships and the optical projection of compass card images.[121] Clark, who was a senior member of the firm in 1906, had considerable mechanical ability, as is shown by the series of patents under his own and Lord Kelvin's (and sometimes the firm's) name, taken out between 1901 and 1911;[122] for the most part they cover the sounding machine and the magnetic compass. A group of other patentees active in the same period, whose names appear alongside the firm's,[123] were probably also members of Kelvin & James White Ltd., and latterly of Kelvin, Bottomley & Baird Ltd. At any rate they demonstrate that the inventiveness through which James White's business became one of the most important instrument making concerns of modern times, was continued, although the chief inspirer, and his 'highly-skilled coadjutor' were long since dead.

REFERENCES

1. General Register Office (Scotland), Register of Deaths 1884 (Glasgow, Partick), 646³/500; Census of 1871 (Glasgow, Barony), 644⁶/47, p.8.

2. Obituary of James White, 12 September 1884, in *Engineering* 38 (1884), 245.

3. Angus McLean (ed.), *Local Industries of Glasgow and the West of Scotland* (Glasgow, 1901), 51. D.J. Bryden, *Scottish Scientific Instrument-Makers 1600-1900* (Edinburgh, 1972), 59, gives the date as 1850.

4. *Glasgow Post Office Directory* 1850-1856.

5. *Ibid.* 1856-1859; the dissolution of the partnership was noted: 'The Copartnery hitherto carrying on Business as Opticians and Mathematical Instrument makers at No 1 Renfield Street, and 60, Gordon Street, Glasgow, under the Firm of WHITE & BARR, was this day DISSOLVED by mutual consent ...[signed] JAMES WHITE JOHN HADDIN BARR [witnesses] David Carlaw John Weir' (*Edinburgh Gazette*, 4 November 1859, 1457).

6. *Glasgow Post Office Directory* 1860-1914, *passim*. A 1931 piece of trade literature, *Navigational Instruments: Kelvin Bottomley & Baird Ltd*, 5, gives a 'genealogy' of the business, together with its current branches: Lord Kelvin (Sir Wm. Thomson) James White, Optical Instrument Maker; Kelvin & James White, Ltd.; Kelvin Bottomley & Baird, Ltd., Kelvin Works, Kelvin Avenue, Hillington, Glasgow, S.W.2. The six branches were: Kelvin, White & Hutton, 11 Billiter Street, London, E.C.3; Kelvin, Bottomley & Baird, Ltd., 303 Shellmex House, Strand, London, W.C.2; Kelvin, Bottomley & Baird, Ltd., Clayton Chambers, Westgate Road, Newcastle-upon-Tyne; Kelvin, Bottomley & Baird (Canada) Ltd., Shaughnessy Building, 401 McGill Street, Montreal, Quebec, Canada; Kelvin & Wilfred O. White Co., 38 Water St. (Corner of Coenties Slip), New York; Kelvin & Wilfred O. White Co., 90 State Street, Boston, Mass., U.S.A.

7. J. Cunnison and J.B.S. Gilfillan, *The Third Statistical Account of Scotland: Glasgow* (Glasgow, 1958), 235-236.

8. George Green and John T. Lloyd, *Kelvin's Instruments and the Kelvin Museum* (Glasgow, 1970), 12. A discussion of Smith's Industries' founder, is given by J.A. Holloway, 'Sir Allan Gordon-Smith 1881-1951' in David J. Jeremy (ed.) *Dictionary of Business Biography* (London, 1984), II, 614-616.

9. McLean, *op. cit.* (3), 51.

10. Glasgow University, Department of Natural Philosophy (Sir William Thomson Collection) Laboratory accounts, National Register of Archives (Scotland) survey 3061; for an account of the laboratory, see Romualdas Sviedrys, 'The Rise of Physics Laboratories in Britain', *Historical Studies in the Physical Sciences* 7 (1976), 409-415; see also A. Gray, 'Famous Scientific Workshops 1. Lord Kelvin's Laboratory in the University of Glasgow', *Nature* 55 (1897), 486-492.

11. White's shop sign proclaiming his appointment is preserved in the Natural Philosophy Department of the University of Glasgow (Green and Lloyd, *op. cit.* (8), 12 and Plate 3a).

12. These included a spinning top designed by Edmund Hunt in 1859 (Edmund Hunt, 'On the Cinephantic Colour Top', *Proceedings of the Glasgow Philosophical Society* 4 (1855-60), 255; Hunt explained that 'Mr White, Optician, 1 Renfield Street, Glasgow, made the top shown, and can supply copies of it'). Cf. the work of James Mackay Bryson (q.v.).

13. Sviedrys, *op. cit.* (10), 413, states that 'Thomson had obtained a share in the telegraph industry early in the 1850s by entering into partnership with James White', supporting this with the statement that 'After White's death in 1884, the business was taken over by William Thomson, his nephew and personal assistant James Thomson Bottomley, and David Reid.' (*ibid.*, 413 n15). The firm's business history is less straight forward than this.

14. *Edinburgh Gazette*, *op. cit.* (5).

15. Scottish Record Office, CS 318/6/362, Sequestration Papers, James White. This is a particularly well documented case, and the Sederunt Book survives. Notices also appeared in the *Edinburgh Gazette*: the initial notice of sequestration (23 August 1861, 1082); the announcement of the appointment of the Trustee and Commissioners (10 September 1861, 1132); the announcement that the Bankrupt has 'intimated his intention to make an offer of composition' at a Creditors' meeting to be held on 29 October (22 October 1861, 1268); this offer accepted, but to be confirmed at a further Creditors' meeting called for 20 November 1861 (1 November 1861, 1304).

16. S.R.O., CS 318/6/362, Sederunt Book, p. 11.

17. *Ibid.*, pp. 12-17.

18. *Ibid.*, pp. 17-41.

19. *Ibid.*, pp. 42-49.

20. *Ibid.*, p. 57. The patent for the improved sextant has not been traced: it was perhaps abandoned on cost grounds before the submission of the formal application.

21. *Ibid.*, pp. 12-13, 76-87, *passim*.

22. *Ibid.*, pp. 79-80.

23. *Ibid.*, p. 109.

24. *Ibid.*, p. 6.

25. 'The Partnership carried on by the Subscribers, as Sole Partners thereof, under the Firm of JAMES WHITE, as Opticians, Optical, Mathematical, Philosophical, Photographic, and Telegraph Instrument Makers in Glasgow, has been DISSOLVED as on the 30th day of December 1876, by mutual consent. All debts due to or by the Firm will be received and paid by the Subscriber James White, by whom the Business will be continued. [signed] JAMES WHITE ROBERT McCRACKEN' (*Edinburgh Gazette*, 5 January 1877, 18).

26. *Engineering*, *op. cit.* (2), 245.

27. *Ibid.*

28. 'The Subscribers David Hay Reid, Charles A. Smith, and Robert Annan, Trustees and Executors of the deceased James White, Optical and Philosophical Instrument maker in Glasgow, have no interest as Partners in the Firm of JAMES WHITE, Opticians and Philosophical Instrument Makers, Glasgow, which has been since the date of Mr White's death on 15th August last, and continues to be, carried on by his Copartners, the Subscribers Mathew Edwards and David Reid [signed] DAVID HAY REID CHAS. A. SMITH ROBERT ANNAN MATHEW EDWARDS DAVID REID.' (*Edinburgh Gazette*, 9 December 1884, 1040).

29. 'The Copartnery carried on by the Subscribers under the Firm of JAMES WHITE, as Opticians and Philosophical Instrument Makers, at 16, 18 & 20 Cambridge Street, and 209 Sauchiehall Street, Glasgow, was, by mutual consent, DISSOLVED on 31st January 1891.
The portion of the Business carried on at the Shop 209 Sauchiehall Street is retained by the Subscriber Mathew Edwards, who will continue under his own name the Business hitherto carried on there ... The portion of the Business carried on at the Works in Cambridge Street is retained by the Subscriber David Reid, who will continue, under the name of JAMES WHITE, the Business carried on there ... [signed] MATHEW EDWARDS DAVID REID.' (*Edinburgh Gazette*, 3 February 1891, 165).

30. Cambridge University Library, Stokes Collection, K 278, Thomson to Stokes, 8 November 1886, quoted in David B. Wilson, *Kelvin and Stokes: A Comparative Study in Victorian Physics* (Bristol, 1987), 53.

31. Sydney Checkland, 'William Thomson, (Lord Kelvin)', in Anthony Slaven and Sydney Checkland (eds.), *Dictionary of Scottish Business Biography 1860-1960* I (Aberdeen, 1986), 190.

32. White told the story that in 1868 Thomson had hurried him into a neighbouring barber's shop in Buchanan Street, to demonstrate the kinetic rigidity of a rubber drive-belt which he had observed while in the shop, and promptly ordered White to set one up in the University classroom at once (Silvanus P. Thompson, *Life of Lord Kelvin* (London, 1910), 741-742).

33. Sir Alfred J. Ewing, 'Lord Kelvin: A Centenary Address', *Proceedings of the Royal Philosophical Society of Glasgow* 53 (1924-5), 3.

34. David Murray, *Memories of the Old College of Glasgow* (Glasgow, 1927), 137.

35. Thompson, *op. cit.* (32), 755.

36. Anon., 'Where Lord Kelvin's Instruments are Made', *The Ludgate* 7 (1898), 149.

37. Gray, *op. cit.* (10), 490; also *Engineering*, *op. cit.* (2), 245.

38. The London mail train from Glasgow was more than once made to wait for an instrument urgently required for the Atlantic telegraph project (Thompson, *op. cit.* (32), 489).

39. Murray, *op. cit.* (34), 137.

40. David Wilson, *William Thomson Lord Kelvin. His Way of Teaching Natural Philosophy* (Glasgow, 1910), 137.

41. Thomson to G.G. Stokes, 9 July 1879, quoted in Harold Issadore Sharlin, *Lord Kelvin, the Dynamic Victorian* (Pennsylvania and London, 1979), 197.

42. Thomson's electrical devices are described in Green and Lloyd, *op. cit.* (8), *passim*.

43. [Stratten's] *Glasgow and Its Environs* (Glasgow, 1891), 109.

44. Thomson's article in the *Encyclopaedia Britannica* 8th edition (Edinburgh, 1860), quoted in Thompson, *op. cit.* (32), 348.

45. McLean, *op. cit.* (3), 51.

46. *The Ludgate*, *op. cit.* (36), 151-152.

47. Magnus Maclean, 'Lord Kelvin's Patents', *Proceedings of the Glasgow Philosophical Society* 29 (1897-98), 145-192.

48. Glasgow University Archives, Kelvin Collection, D 33/2/1: Monthly Cash 'Account Book, 1879-1900' [of James White], November 1885, pp. 131-132; 1900, pp. 353-356. In 1881 Thomson secured for White the sole agency and right to manufacture and sell Fauré accumulators in the United Kingdom (Thompson, *op. cit.* (32), 769). As the first efficient battery, it was expected to prove very lucrative.

49. Green and Lloyd, *op. cit.* (8), 45-48.

50. The number, date of supply, name of purchaser and normally the vessel into which the compass was to be fitted, is given for each compass in four MS volumes in G.U.A., Kelvin Collection, D 33/4/1-4: 'Sir William Thomson's Patent Compass Book, 1876-1883'; ditto, '1883-1893'; 'Lord Kelvin's Patent Compass Book, 1893-1904'; ditto, '1904-1918'. The annual production averages for 1900-1918 were: 1900-1904: 376; 1905-1909: 350; 1910-1914: 344; 1915-1918: 306 (Kelvin Collection, *passim*).

51. G.U.A., Kelvin Collection, D 33/4/4, pp. 200-201.

52. G.U.A., Kelvin Collection, D 33/4/1, p. 2 and *passim*.

53. *Ibid.*, p. 10 and *passim*.

54. *Ibid.*, pp. 25, 35, 68, 29 respectively, and *passim*.

55. *Ibid.*, pp. 47-69.

56. Science Museum Library, London: Kelvin & James White Ltd., Expired Patents. These 2 volumes contain the firm's copies of their patents, the first dating from 1858 and the last 1907, held by Lord Kelvin's executors. None were held, or held jointly, by James White.

57. Sviedrys, *op. cit.* (10), 413.

58. Checkland, *op. cit.* (31), 191.

59. Glasgow University Library, Kelvin Papers, J.48, Jenkin to White, 23 July 1863, quoted in Crosbie Smith and M. Norton Wise, *Energy and Empire: A Biographical Study of Lord Kelvin* (Cambridge, 1989), 701. We are grateful to Cambridge University Press for allowing us to examine this book at proof stage.

60. A very critical account of how Thomson's compass came to be the Admiralty Standard - 'for the requirements of the [Royal] Navy of the 1890s its introduction was a retrograde step' - is given in A.E. Fanning, *Steady As She Goes: A History of the Compass Department of the Admiralty* (London, 1986), 99-123.

61. Thomson's patents pertaining to the mariner's compass were: UK patent 1339, 30 November 1876; 4876, 18 December 1876; 679, 20 February 1879; 5676, 8 December 1883; 4923*, 25 January 1890; 8959, 18 April 1891; 24,841, 10 November 1894; 16,990, 3 July 1902; 22,023, 8 October 1902; and 22,695, 31 August 1905.

62. Fanning, *op. cit.* (60), 115.

63. 'Thomson v. Hughes', *Reports of Patent, Design and Trade Mark Cases* 7 (1890), 72-73; this was probably Duncan McGregor & Co. (q.v.).

64. 'Thomson v. Batty', *ibid.* 6 (1889), 84-101.

65. 'Thomson v. Moore', *ibid.* 6 (1889), 426-464; Moore unsuccessfully appealed, in 'Moore v. Thomson', *ibid.* 7 (1890), 325-336. We are grateful to Dr Anita McConnell for this reference. Moore had advertised his 'Patent Compass', with a mention of his Clyde agents, Alexander Dobbie & Son (q.v.), in the *Glasgow International Exhibition, 1888. Catalogue* 2nd edition (Glasgow, 1888), 404.

66. 'Thomson v. Hughes', *Reports, op. cit.* (63), 7 (1890), 71-101, 187-190.

67. 'Kelvin v. Whyte, Thomson & Co.', *ibid.* 25 (1908), 177-194. See Chapter 22, 'Whyte and Thomson'.

68. Thomson was working on the sounding machine by 1868, as is shown by an anecdote often told 'with great gusto' by James White as having taken place in his shop in Buchanan Street. Thomson entered with a guest, the physicist James Prescott Joule of Manchester. 'Joule's attention was called to a bundle of the pianoforte wire lying in the shop, and Thomson explained that he intended it for "sounding purposes"; "What note?" innocently inquired Joule, and was promptly answered, "The deep C."' (John Munro, *Lord Kelvin* (London, 1902), 50-51).

69. Anita McConnell, *No Sea Too Deep. The History of Oceanographic Instruments* (Bristol, 1982), 60-63. In August 1872 White promised Thomson the 'sounding drum' soon, and referred to telegraphic apparatus (Glasgow University Library, Kelvin Papers W.10, James White to Sir William Thomson, 6 August 1872). This possibly refers to the sounding machine fitted on H.M.S. *Challenger* towards the end of 1872, prior to its unsuccessful use on the *Challenger* oceanographic voyage (McConnell, *ibid.*, 61).

70. William Thomson, 'On Deep-Sea Sounding by Piano-forte Wire', *Proceedings of the Glasgow Philosophical Society* 9 (1873-5), 111-117.

71. The firm held the following patents for sounding apparatus: U.K. Patent 3452, 1 September 1876; 781, 23 February 1880; 5675, 8 December 1883; 12,240, 14 October 1885; 25,178, 30 October 1897; 22,030, 10 October 1902; 9853, 1 May 1903; 24,526, 12 November 1903; 20,813 and 20,813A, 14 October 1905; 6471, 17 March 1906; 26,132, 19 November 1906; 23,833, 29 October 1907; 22,976, 29 October 1908; 1157, 17 January 1910; 9067, 9068 and 9069, 12 April 1911; 5964, 9 March 1912; 21,409, 20 September 1912. Thomson devised three types of pressure gauge as depth sounders, of which the piston type was still in production in the 1900s (McConnell, *op. cit.* (69), 73).

72. Figure given in advertisement in Kelvin & James White Ltd., *Standard Tide Tables ... for 1903* (Glasgow, 1902).

73. Quoted in editorial article by Dr Donald Macleod, 'The Right Hon. Lord Kelvin', *Good Words* 17 (1896), 383.

74. Sydney Checkland has noted that Kelvin 'always regretted the few occasions on which he used someone else', but we have not located his source (Checkland, *op. cit.* (31), 190).

75. *Catalogue [of the] Naval & Marine Engineering Exhibition* (Glasgow, 1880), 80.

76. *Catalogue, op. cit.* (65), 403. Amongst the many staple unpatented items bearing a 'James White' signature, the following have been recorded: a miner's dial, offered by Christie's South Kensington, 14 June 1984, Lot 265; a transit theodolite, offered by Phillips, 5 April 1984, Lot 106; a Dumpy level, offered by Phillips, 8 May 1985, Lot 159; a microscope, offered by Tesseract, Catalog 19, (Winter 1987-88), item 16; a parallel rule, offered by Sotheby's Belgravia, 22 February 1980, Lot 283. Some items, other than those in the Frank collection,

with a James White signature are in the collection of the Royal Museum of Scotland: two miner's dials (NMS T1959.47 and T1974.314); a level (NMS T1966.59); a plotting protractor (NMS T1966.62); an air meter (NMS T1897.353.1) and a pit barometer (NMS T1897.353.1).

77. NMS T1980.130-136.

78. *Engineering, op. cit.* (2), 245.

79. G.R.O.(S.), Census of 1871 (Glasgow, Barony), 644⁶/47, p. 8; Census of 1881 (Glasgow, Partick), 646²/33, p. 56.

80. *Engineering, op. cit.* (2), 245.

81. Annual compass production rose quickly as a result of expansion. In 1879 77 are recorded; 1880: 122; 1881: 191; 1882: 188; 1883: 256; 1884: 282; 1885: 350.(G.U.A., Kelvin Collection, D 33/4/1-2, *passim*, Sir William Thomson's Patent Compass Book, 1876-1883, and 1883-1893.)

82. Thompson, *op. cit.* (32), 994. S.R.O., VR 102/323, 391, 403, Valuation rolls of Barony Parish, Glasgow, 1884-5, 1890-1, 1891-2.

83. *Engineering, op. cit.* (2), 245.

84. [Stratten], *op. cit.* (43), 109. In 1895 the firm was described as providing 'employment at present to over 200 operatives' ('Messrs. James White, Instrument factory, Cambridge Street', *Proceedings of the Institution of Mechanical Engineers* 48 (1895), 501). The rateable valuation of the premises doubled from 450 to 900 between 1891 and 1892 (S.R.O., VR 102/403/2, p. 55, 415/3, p. 53).

85. McLean, *op. cit.* (3), 51.

86. [Stratten], *op. cit.* (43), 109.

87. McLean, *op. cit.* (3), 51.

88. *The Ludgate, op. cit.* (36), 150. In August 1898 at about the time of this account, the laboratory staff numbered twelve; in February 1902, 23 (G.U.L., Kelvin Papers, App.66, Laboratory Staff Ledger, 1897-1902).

89. This was employing 10 men in September 1905 (G.U.L., Kelvin Papers, App.67, Drawing Office Ledger, 1905-1911).

90. The drums were to hold sounding wire.

91. *The Ludgate, op. cit.* (36), 150-152.

92. Gray, *op. cit.* (10), 490.

93. *The Ludgate, op. cit.* (36), 152-154.

94. G.R.O.(S.) Register of Deaths 1884 (Glasgow, Partick), 646³/500.

95. White's Settlement was recorded 9 December 1884: S.R.O., Sheriff Court of Glasgow, Records of Settlements, SC 36/51/89, pp. 573-576. His trustees were his two brothers-in-law, David Reid, a Glasgow muslin manufacturer, and Charles A. Smith, and Robert Annan, the Glasgow photographer. The inventory of his estate was recorded 9 December 1884: S.R.O., SC 36/48/108, pp. 142-146, Inventory Book of Sheriff Court of Glasgow.

96. *Engineering, op. cit.* (2), 245.

97. Murray, *op. cit.* (34), 50. White was a member of the Glasgow Philosophical Society between 1876 and 1884, to whom he supplied 'Barometers & c.' to the value of £7 7s in 1882 (*Proceedings of the Glasgow Philosophical Society* 10 (1875-7), 192; *ibid.* 14 (1882-3), 550).

98. *Engineering, op. cit.* (2), 245; see *Edinburgh Gazette*, 9 December 1884, 1040, quoted in reference 28 above.

99. Thompson, *op. cit.* (32), 994.

100. *Ibid.*, 155.

101. *Ibid.*

102. 'Caution' accompanying prefatial advertisements in Kelvin & James White Ltd., *Standard Tide Tables ... for 1903* (Glasgow, 1902). The firm's series of Tables ran from at least 1899.

103. See Chapter 22, Whyte, Thomson & Co.

104. *The Ludgate, op. cit.* (36), 150.

105. G.U.L., Kelvin Papers, R.6, Kelvin to David Reid, 10 September 1894.

106. G.U.L., Kelvin Papers, B.26, Kelvin to J.T. Bottomley, n.d. [c.14 December 1906]. An accommodation was reached, and F.W. Clark and Chetwynd co-patented various types of magnetic compass in 1908-9. U.K. patents 18,509, 3 September 1908; 18,510, 3 September 1908; 20,185, 25 September 1908; 21,634, 13 October 1908; 16,607, 16 July 1909; 19,057, 19 August 1909; 25,718, 8 November 1909; 29,719, 19 August 1909; 9347, 15 April 1911. An example of one of these was offered for sale by Sotheby's, 10 March 1987, Lot 123.

107. *Ibid.*

108. G.U.L., Kelvin Papers, R.4, Sir William Thomson to David Reid, 8 May 1884.

109. G.U.L., Kelvin Papers, R.7, Kelvin to David Reid, 11 August 1895.

110. G.U.L., Kelvin Papers, R.6, Kelvin to David Reid, 10 September 1894.

111. William Wood, *Report upon the Affairs of the Edinburgh School of Arts Friendly Society as at 31st December 1855* (Edinburgh, 1857), 17.

112. Obituary of J.T. Bottomley, *The Electrician* 96 (1926), 556; *Who was Who, Vol.II (1916-1928)* 4th edition (London, 1967), 109; obituary, *Glasgow Herald*, 19 May 1926.

113. Thomson and Bottomley gave a joint paper on electric incandescent lamps in 1881 (William Thomson and James T. Bottomley, 'On the Illuminating Powers of Incandescent Vacuum Lamps with measured Potentials and measured Currents', *Report ... of the British Association ... 1881* (London, 1882), 559-561).

114. *Glasgow Post Office Directory* 1889.

115. Alexander Russell, *Lord Kelvin* (London, 1938), 141-142.

116. Reid first appeared in the street directory in 1882 and was subsequently listed as a maker of the major types of instrument constructed by the firm. He last appeared in the *Glasgow Post Office Directory* for 1911. He seems to have been the son of Robert Reid, a house property inspector: G.R.O.(S.), Census of 1891 (Glasgow, Kelvin), 644⁹/1, p.8.

117. G.U.L., Kelvin Papers, *passim*.

118. [Stratten], *op. cit.* (43), 109. Reid was elected a member of the Glasgow Philosophical Society on 16 November 1887, the same day as Mathew Edwards (*Proceedings of the Glasgow Philosophical Society* 19 (1887-88), 397), and remained a member until 1926-27.

119. *Glasgow Post Office Directory* 1884-1914; obituary of A.W. Baird, *The Engineer* 155 (1928), 63.

120. *Glasgow Post Office Directory* 1884-1910. Baird first appeared in 1884-1889, reappearing in 1901.

121. Russell, *op. cit.* (115), 152-153.

122. U.K. patents 16,990, 24 August 1901; 2351, 29 January 1902; 22,030, 22,031 and 22,033, 10 October 1902; 9853, 1 May 1903; 24,526, 12 November 1903; 9514, 5 May 1905; 6471, 17 March 1906; 26,132, 19 November 1906; 1397, 19 January 1907; 3566, 13 February 1907; 10,200, 2 May 1907; 12,783, 3 June 1907; 23,579, 25 April 1907; 17,911, 26 August 1908; 18,509 and 18,510, 3 September 1908; 20,185, 25 September 1908; 21,634, 13 October 1908; 22,976, 29 October 1908; 7021 and 7022, 24 March 1909; 16,607, 16 July 1909; 19,057, 19 August 1909; 25,718, 8 November 1909; 27,181, 23 November 1909; 29,719, 19 August 1909; 1157, 17 January 1910; 29,414, 19 December 1910; 9067, 9068 and 9069, 12 April 1911; 9347, 15 April 1911.

123. They were J. Kean, active c.1904 (U.K. patent 22,695, 21 October 1904); M.B. Field, active c.1912-15 (U.K. patents 5964, 9 March 1912; 21,409, 20 September 1912; 11,503, 17 May 1913; 11,561, 19 May 1913; 14,191 and 14,192, 19 June 1913; 29,782, 24 December 1913; 11,881, 14 May 1914; 14,171, 12 June 1914; 12,973, 10 September 1915); F.A.King, active c.1913 (U.K. patent 11,158, 13 May 1913); G.H. Alexander, active c.1913 (U.K. patent 11,158, 13 May 1913); D. Renfrew, active c.1913 (U.K. patent 11,508, 17 May 1913); A.E. Conrady, active c.1915 (U.K. patent 3610, 6 March 1915; 8979, 18 June 1915; 16,776, 29 November 1915; 110,290, 15 March 1917; 124,528, 13 March 1917). Conrady's patents concerned periscopes, which during the war years preoccupied optical firms who often co-operated over the patenting of ideas; the London firm W. Watson & Sons Ltd. was also named along with Kelvin, Bottomley & Baird Ltd. in the later of his two 1915 patents. Conrady ran the Theoretical Department of Watson's in 1907, acting 'as chief Mathematician and Optical Advisor' (Watson & Sons, *Catalogue of Microscopes and Accessories* 19th edition (London, 1907-8), 12). We are grateful to Dr R.H. Nuttall for this reference. In 1929, Conrady was the author of *Applied Optics and Optical Design* (London, 1929).

129. Level: J. White, Glasgow, c.1855 (T1980.130)

11″ level in brass and oxidised brass, with bubble tube mounted alongside the telescope, slide focus operated by a knob projecting from a slot in the side of the outer tube, the eyepiece assembly incorporating an erecting system, and with doublet objective; the limb attached to a parallel plate levelling head with free azimuth motion. Engraved on the telescope tube 'J. WHITE, GLASGOW.' The limb and levelling head components stamped with the assembly mark '4', the bubble tube case scratched 'IIII'.

Telescope length: 277mm.
Tube diameter: 29mm.
Aperture: 22mm.

Lacking the fixing screws for the eyepiece assembly mounting flange and the retaining screw and plate for the azimuth bearing. The objective cover (damaged) does not seem to be original. The limb was bent on receipt and has been straightened.

130. Level: James White, Glasgow, c.1880 (T1980.134)

10½″ dumpy level in oxidised brass and bronze, with bubble tube over the telescope, rack and pinion focus, doublet objective, and objective ray shade; the limb threaded to a parallel plate levelling head with free azimuth motion. Engraved on the telescope tube 'JAMES WHITE, GLASGOW'. Components of the level stamped with the assembly mark '2' or scratched 'II', but those of the levelling head stamped '1'.

Telescope length: 270mm.
Tube diameter: 36mm.
Aperture: 35mm.

Lacking objective shade closure plate. One levelling screw is badly bent.

130, 131

131. Level: J. White, Glasgow, c.1890 (T1980.132)

10″ level in oxidised brass, with bubble tube above the telescope, rack and pinion focus and doublet objective; the limb threaded to a parallel plate levelling head with free azimuth motion. Engraved on the telescope tube 'J. WHITE / GLASGOW'. The limb and components on the levelling head stamped with the assembly mark '10'. With objective cover.

Telescope length: 248mm.
Tube diameter: 28mm.
Aperture: 24mm.

132. Level: J. White, Glasgow, c.1890 (T1980.136)

14″ dumpy level in oxidised brass and bronze (stripped), with bubble tube (glass replaced) and circular level over the telescope, rack and pinion focus, and doublet objective; the limb threaded to a parallel plate levelling head (defective) with free azimuth motion. Engraved on the telescope tube 'J. WHITE, GLASGOW'. Components of the level and levelling head stamped with the assembly mark '3' or scratched 'III'.

Telescope length: 345mm.
Tube diameter: 43mm.
Aperture: 40mm.

Lacking eyepiece (the mount a recent fabrication) and objective ray shade. The upper (threaded) plate of the levelling head has been substituted from another instrument, although the original screws are fitted to it. The glass bubble tube has been replaced improperly by two separate 3″ tubes, one of which is broken.

133. Level: J. White, Glasgow, c.1890 (T1980.131)

15″ dumpy level in oxidised brass and bronze (stripped), with bubble tube (renewed) and circular level over the telescope, rack and pinion focus, and triplet objective; the limb threaded to a parallel plate levelling head with free azimuth motion. Engraved on the telescope tube 'J. WHITE, GLASGOW'. Level components stamped with the assembly mark '9', and the objective mount scratched '15'. The limb and parallel plates have an engine-turned finish.

Telescope length: 387mm.
Tube diameter: 43mm.
Aperture: 40mm.

Lacking objective ray shade and diaphragm wires. The glass bubble tube (broken) is a replacement which is too short for the casing. The eyepiece is a recent fabrication. The surface has been marked by mechanical abrasive cleaning, and a modern lacquer has been applied.

134. Level: James White, Glasgow, c.1870 (T1980.133)

14″ dumpy level in oxidised brass and bronze, with bubble tube and circular level over the telescope, rack and pinion focus, triplet objective, and objective ray shade; the limb threaded to a parallel plate levelling head with free azimuth motion. Engraved on the telescope tube 'J WHITE 78 UNION ST GLASGOW' and 'HUTCHESONS HOSPITAL'. Components of the level and levelling head stamped with assembly mark '7' or scratched 'VII'.

Telescope length: 352mm.
Tube diameter: 43mm.
Aperture: 40mm.

Lacking rear component of eyepiece.

Hutchesons' Hospital, originally situated on the north side of Glasgow's Trongate, was founded for the relief of pensioners by two brothers, George and Thomas Hutcheson, in the mid 17th century. By the beginning of the 19th century, the original building was sold because of its commercial situation, and a new hospital built in Ingram Street 'in which, however, neither pensioners nor boys reside ... which is partly let and partly used by the patrons for meetings' (W.H. Macdonald, 'Hutchesons' Hospital', in Magnus Maclean (ed.), *British Association for the Advancement of Science Handbook on Archaeology, Education and Miscellaneous Subjects* (Glasgow, 1901), 216-223).

Top right 132, top left 133, bottom left 134, bottom right 135

135. Level: James White, Glasgow, c.1900 (T1980.135)

14″ dumpy level in oxidised brass, with bubble tube (renewed) and circular level over the telescope, rack and pinion focus, triplet objective, and objective ray shade; on a later tribrach mount. Engraved on the telescope tube 'JAMES WHITE, GLASGOW'.

Telescope length (less eyepiece): 351mm.
Tube diameter: 43mm.
Aperture: 40mm.

Lacking eyepiece, diaphragm wires, and one cover for the limb adjustment screws. The glass bubble tube has been renewed and is divided in a later style. The casing for the bubble tube has had to be modified to accommodate the attachment points and must be from a different instrument. The assembly number on the objective mount differs from that on the ray shade. The instrument has been adapted to take a later limb and tribrach stand in black lacquered brass, the base plate engraved 'STANLEY'S PATENT'.

W.F. Stanley's U.K. patent 14934 of 1892 relates to the gripping of the ball feet of the three adjusting screws.

136. Mining dial: James White, Glasgow, c.1890 (T1980.154)

Glazed circular compass in brass and oxidised brass with silvered dial, recessed circular level and two folding opposed slit and window sights. Engraved on the dial plate 'JAMES WHITE /GLASGOW'. The dial with the 4 cardinal compass points (transposed arrangement for reading bearings against the north point of the needle), the scales divided in degrees 0-90-0-90-0 and 360-[0]. Triangular-section needle on a relieved jewelled bearing. Socket mount with clamp, and separately clamped azimuth rotation of the dial and sights, read to 2 mins. against a silvered vernier within the dial.

Compass housing diameter: 170mm.

Lacking sight wires.

137. Mining dial: James White, Glasgow, 1884 (T1980.155)

Glazed circular compass in brass with silvered dial, recessed circular level and two folding opposed slit and window sights. Engraved on the dial plate 'JAMES WHITE / GLASGOW'. The dial with the 4 cardinal compass points (transposed arrangement for reading bearings against the north point of the needle), the scales divided in degrees 0-90-0-90-0 and 360-[0]. Triangular-section steel needle on a relieved jewelled bearing. Socket mount with clamp, and separately clamped azimuth rotation of the dial and sights, read to 2 mins. against a silvered vernier within the dial. The arm of the vernier is engraved 'TO MR ANDREW WINNING, along with / an Anemometer, from the workmen / of Frame Colliery, Rutherglen as a / mark of their esteem, on his leaving / 21ST Augt. 1884'. Bubble tube beneath dial for using the instrument on its side as a level. With fitted case.

Compass housing diameter: 170mm.
Case size: 235 x 210 x 100mm.

The original glass has been replaced in a gauge which is too heavy for the clearance provided in the housing. Lacking the sight wires and also the plumb line accessories for which provision is made in the case.

The Farme Colliery, Farme Cross, was one of 26 coal pits in the neighbourhood of Rutherglen, near Glasgow. It opened in 1805, and closed in 1931 (*Rutherglen Reformer*, 24 July 1931). The local newspaper reported the presentation: 'At a meeting of the employees in Farme Colliery, Mr Andrew Winning was presented with a compass and armature, and a gold watch for Mrs Winning. Mr James Anderson, manager of Farme Colliery, paid a tribute of respect to the recipient who had been in the employ for over 15 years' (*Rutherglen Reformer*, 22 August 1884). We are grateful to Alistair Gordon, Rutherglen Museum, for this information.

137,136

138. Theodolite: J. White, Glasgow, c.1880 (T1980.160)

5″ reversing transit theodolite in oxidised brass and bronze, the bevelled edge azimuth circle divided on silver [0]-360 by ½°, with clamp and tangent screw adjustment to the upper plate and verniers reading to 1 min., with rotating magnifier. Crossed levels and silvered compass, engraved 'J. White./ Glasgow.'; the compass with the 8 cardinal points (transposed arrangement for reading bearings against the north point of the needle) and divided 360-[0] by 1°; with needle on a relieved jewelled bearing. A-frames support the telescope and vertical circle, the latter divided 90-0/[90]-[0]/90-0/[90]-[0] by ½°, with clamp and tangent screw adjustment on a clipping arm and verniers reading to 1 min., with magnifiers; and on the reverse 30-0-30 'Diff. of Hypo. & Base'. The telescope with rack and pinion focus to the 1⅛″ doublet objective; with objective ray shade but lacking the bubble tube over the telescope. Integral parallel plate levelling head with axis collar clamp and tangent screw.

Telescope length: 258mm.
Aperture: 28mm.
Azimuth scale diameter: 126mm.

The vertical circle has been deformed and partly straightened, and the retaining flange to the verniers is detached; the casting for the A-frames is slightly distorted, as is the azimuth magnifier support. Lacking the main bubble tube, one transit axis cover screw, and the graticule wires; the eyepiece is a modified replacement.

139. Theodolite: James White, Glasgow, c.1870 (T1980.161)

5″ reversing theodolite in oxidised brass and bronze (stripped), the bevelled edge azimuth circle divided on silver [0]-360 by ½°, with clamp and tangent screw adjustment to the upper plate and verniers reading to 1 min., with provision for rotating magnifier. Crossed levels and silvered compass, engraved 'James White./ Glasgow.'; the compass with the 8 cardinal points (transposed arrangement for reading bearings against the north point of the needle) and divided 360-[0] by 1°, with the needle on a relieved jewelled bearing. A-frames support the altitude semi-circle and telescope mounting limb; the scale divided on silver [75]-0-90 by ½° with clamp and tangent screw, and vernier reading to 1 min.; and on the reverse 30-0-30 'Diff. of Hypo. & Base.' The telescope with a level slung beneath it, in Y-mounts over the limb; rack and pinion focus to the ⅞″ doublet objective. Integral parallel plate levelling head with axis collar clamp and tangent screw.

Telescope length: 261mm.
Aperture: 24mm.
Azimuth scale diameter: 129mm.

Lacking the magnifier for the azimuth scale; its support distorted and corresponding damage on one of the vernier scales. The telescope bubble tube broken. The support plate for the azimuth clamp screw has been replaced. Lacking graticule wires and objective cover.

139, 138

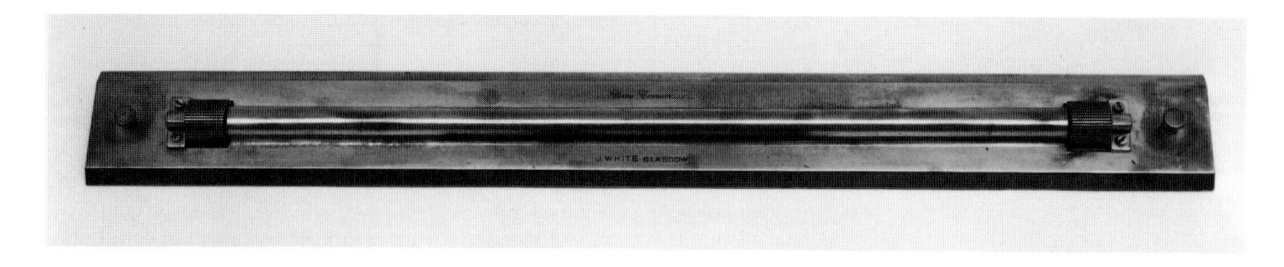

140. Parallel rule: J. White, Glasgow, c.1870 (T1980.171)

27" rolling parallel rule in oxidised bronze (stripped), with chamfered edges and 2 lifting knobs. Stamped 'J WHITE GLASGOW' and engraved 'Ben Connor'. With fitted case.

Length: 689mm.
Case size: 715 x 100 x 60mm.

Four of the 6 fixing screws securing the roller spindle cover to the rule are missing.

The owner of this rule may have been Benjamin Connor (?1813-1876) the Scots railway engineer, who for the last 20 years of his life was locomotive superintendent of the Caledonian Railway (*Engineering* 21 (1876), 117). He served an apprenticeship in Glasgow with a mechanical engineer, moving into locomotive engineering later in his career, spending some time in engine works in Liverpool and Manchester. He also worked for a time in Robert Napier's marine engineering works, gaining practical experience from voyages to Spain (*ibid.*, 135-136). He introduced 'Crewe' type framing in locomotives built for the Caledonian Railway from 1858 (John Marshall, *A Biographical Dictionary of Railway Engineers* (Newton Abbott, 1978), 55-56).

141. Drawing protractor: James White, Glasgow, c.1880 (T1980.174)

150mm. circular protractor in electrum metal, with geared motion to an arm with a radial drawing edge, and with a circular scale divided to ½° and read against a vernier on the arm to 1 min. Engraved on the arm 'JAMES WHITE / GLASGOW'.

Arm radius: 325mm.
Scale radius: 75mm.

Lacking the engraved glass centre mark.

Probably of French manufacture.

142. Drawing Protractor: J. White, Glasgow, c.1860 (T1980.175)

8" protractor in bronze, with a semi-circular arc divided on silver for 45° on either side of its centre line; pivoted at the centre of the arc, with a clamp and tangent screw adjustment to its circumference, is a 24" rule with two parallel drawing edges. Engraved at the pivot 'J. White, Glasgow.' The underside of the arc checked along its diameter for locating against a raised horizontal straight edge on a drawing board. The inclination of the rule to this edge read to 1 min. against a silvered vernier, the arc divided to ½°, and calibrated [45]-[90]/0-[45], [135]-[180]/90-[135], [225]-[270]/180-[225], [315]-[360]/270-[315]. A cross-piece (replacement) slides along a recessed groove in the upper surface of the arm. With fitted case.

Arm radius: 730mm.
Scale radius: 98mm.
Case size: 785 x 245 x 60mm.

The cross-piece has been substituted from another instrument of the same type: the slide is too narrow for the groove and will not hold the cross-piece orthogonal to the rule.

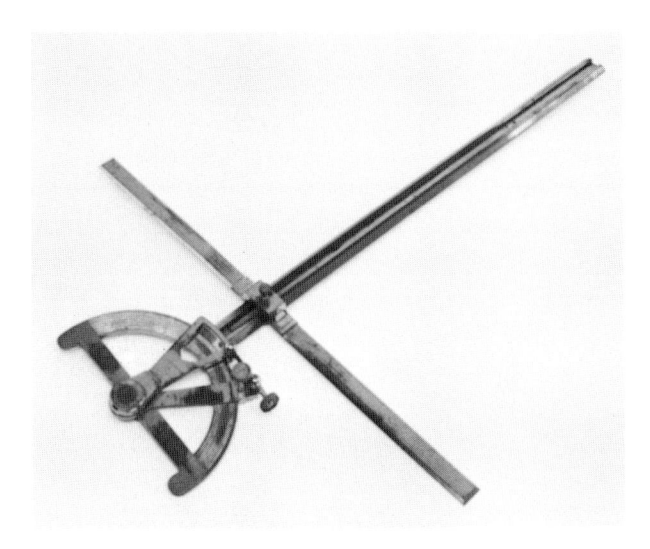

143. Marine azimuth mirror: J. White, Glasgow, c.1890 (T1980.201)

Azimuth sight in black-lacquered brass for use in conjunction with an 8″ Kelvin-type dry card compass. Sighting prism, with solar shades, mounted over a viewing tube with a lens for reading the compass scale; set on a base, lead-weighted at the toe, with a location pin for the centre of the compass bowl glazing and supported on the glass by 2 feet and a spring arm beneath the base. The base with a circular level and stamped 'REGISTERED' with the 'WTB' trade mark, and 'J. WHITE / GLASGOW / SIR W. THOMSON'S / PATENT N⁰ 3616'. Components stamped with the assembly mark '5'.

Overall length: 222mm.
Viewing tube diameter: 35mm.

The instrument follows Thomson's U.K. patent 5676 of 1883.

144. Marine azimuth mirror: J. White, Glasgow, c.1895 (T1980.202)

Azimuth sight in oxidised brass for use in conjunction with an 8″ Kelvin-type dry card compass. Sighting prism, with solar shades, mounted over a viewing tube with a lens for reading the compass scale; set on a base, lead-weighted at the toe, with a location pin for the centre of the compass bowl glazing and supported on the glass by 2 feet and a spring arm beneath the base. The base with a circular level and stamped 'REGISTERED' with the 'WTB' trade mark, and 'J. WHITE / GLASGOW / LORD KELVIN'S / SIR W^M THOMSON) / PATENTS / N⁰ 4395'. Components stamped with the assembly mark '2'.

Overall length: 222mm.
Viewing tube diameter: 35mm.

Lacking the centre sighting pin.

The patents referred to are U.K. patents nos. 5676 of 1883 and 8959 of 1890, although the instrument is of the design specified in the earlier patent. The modified design, which had a T-shaped base which spanned the compass glazing so that 'the breaking of its glass during the firing of heavy guns &c. is prevented', was added to the specification of Thomson's 1890 patent, although the model was being made at an earlier date: an example (No. 2184) from the Admiralty Compass Observatory which pre-dates the patent is in the Royal Museum of Scotland (NMS T1983.252).

145. Microscope: J. White, Glasgow, c.1880 (T1980.233)

Compound student microscope, in brass, bronze and oxidised brass, on an inclining pillar and bent claw footed tripod base. Engraved at the rear of the base 'J. WHITE./ 241 SAUCHIEHALL S^T /GLASGOW.' Single-draw body tube, in a leather-lined draw sleeve for coarse focus adjustment, attached to the pillar through a spring-loaded slide, operated by a fine focus screw at the top of the pillar. 2-component eyepiece assembly, and objective engraved '¼ in.'. The stage hinged to the base, with the pillar mounted above, and a plane/concave mirror sliding on an extension beneath; the circular stage with glass upper plate, internal rotating 4-aperture diaphragm and sub-stage sleeve, and spring clips. The base with clamping adjustment to the inclination, and a stop to prevent the instrument tilting forward beyond the vertical.

Body length: 136mm.
Body diameter: 30mm.
Height: 243mm.
Base size: 119 x 130mm.

The instrument can be identified with a microscope still on sale in 1894, 'The College Microscope' offered for £5 13s 6d by John J. Griffin & Sons, *Chemical Recreations* (London, 1894), 302-303, no. 3203: the illustration shows a draw tube with regular settings marked in the style of Parkes (see items 71 and 92 in the Frank collection). It also matches an example signed 'J. Brown, 76 Vincent Street [*sic.*], Glasgow.' described and illustrated in H.R. Purtle, *et al., The Billings Microscope Collection of the Medical Museum, Armed Forces Institute of Pathology* 2nd edition (Washington, D.C., 1974), 210. An apparently identical base is on a microscope offered in 1895 by Townson & Mercer, *Catalogue of Chemical Apparatus* (London, 1895), 159, no. 1021. The clamping adjustment on

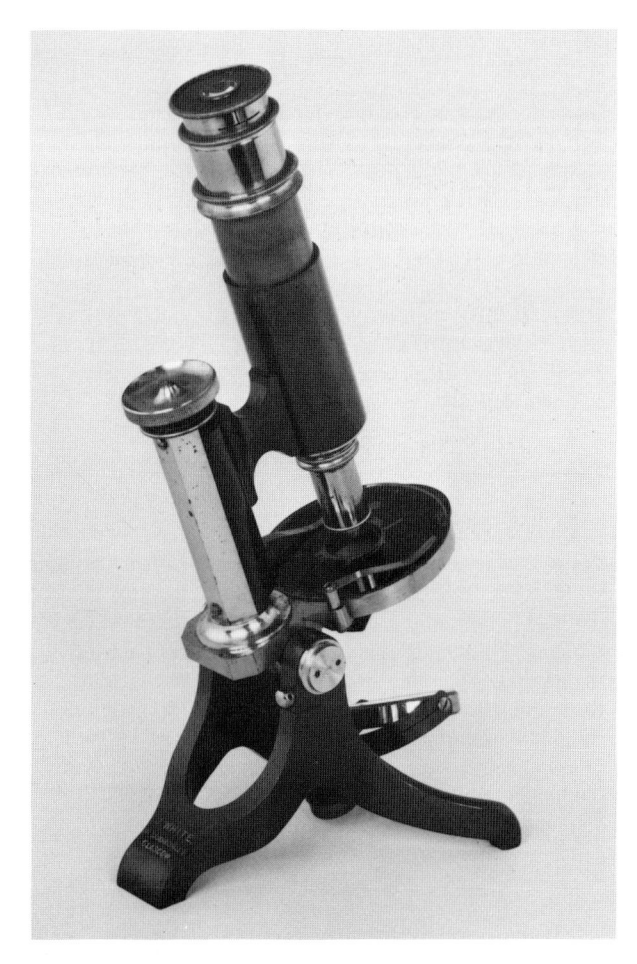

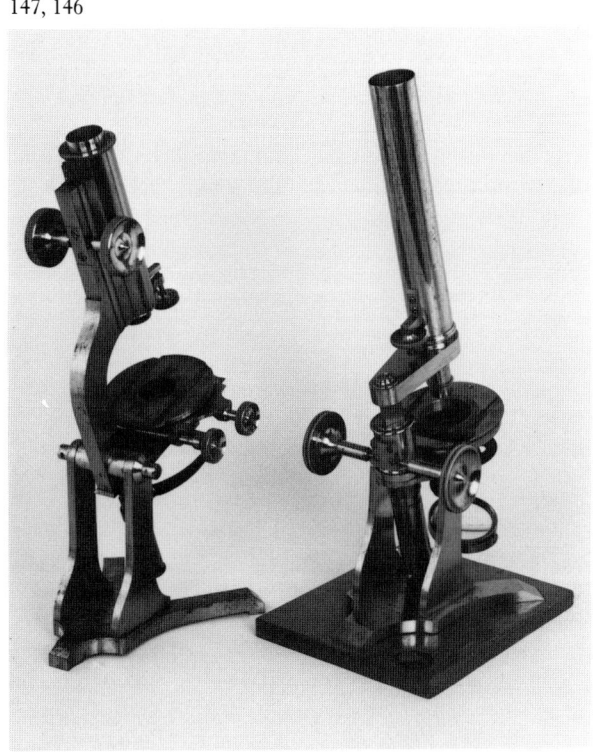

146. Microscope: J. White, Glasgow, c.1890 (T1980.234)

Compound bar-limb microscope, in brass, bronze and oxidised brass, on an inclining pillar and flat 2-toed tripod base. Engraved at the rear of the base 'J. WHITE./ GLASGOW.' The body tube screws into a bar-limb on a triangular-section slide within the pillar, a pinion engaging a rack cut into one edge for coarse focus adjustment; fine focus is by an external screw at the lower end of the body tube advancing a spring-loaded nosepiece (defective) for the objective through a lever arm. With objective. The stage hinged between uprights rising from the base, which is attached to a board. Circular rotating stage plate with slide carrier (both stamped '5') and rotating 4-aperture diaphragm, and plane/concave sub-stage mirror on an extension of the pillar.

Body length (less draw tube and eyepiece): 207mm.
Body diameter: 32mm.
Height (less draw tube and eyepiece): 374mm.
Base size (casting): 148 x 167mm.

Lacking eyepiece and body draw tube. The sliding tube operated by the fine focus lever arm is seized.

the tilt axis is characteristic of stands produced by James Swift & Son of London; see, for example, *Journal of the Royal Microscopical Society* 3 (1880), 868-869, 1053-1054, and J. Swift & Son, *A Catalogue of Microscopes* (London, 1880), 7-9, 11-12, 14.

147, 146

147. Microscope: J. White, Glasgow, c.1870 (T1980.235)

Compound Lister-limb microscope in brass and bronze, on an inclining pillar and lead-weighted flat 2-toed tripod base. Engraved on the side of the base 'J WHITE./ GLASGOW.' Single-draw body tube with coarse focus rack and pinion adjustment to the limb; fine focus is by an external screw at the lower end of the body tube advancing a spring-loaded nosepiece for the objective through a lever arm. The limb hinged between uprights rising from the base, with the stage extended forward from the lower end of the limb, and provision for a concave mirror sliding on an extension beneath. The mechanical stage with slow-motion in both directions, and manually rotated circular stage plate, with slide holder (defective).

Body length (less eyepiece): 142mm.
Body diameter: 34mm.
Height (less eyepiece): 335mm.
Base size: 155 x 161mm.

Lacking objective, eyepiece, sub-stage mirror, slide holder, and stage diaphragm.

148. Level: Kelvin & James White, Glasgow, c.1910 (T1980.138)

10½" dumpy level in oxidised brass and bronze, with bubble tube over the telescope, rack and pinion focus, doublet objective, and objective ray shade; the limb threaded to a parallel plate levelling head with free azimuth motion. Engraved on the telescope tube 'KELVIN & JAMES WHITE LTD. GLASGOW'. With fitted case containing repair label for G. Hutchison & Sons, 18 Forrest Road, Edinburgh, 1956.

Telescope length: 275mm.
Tube diameter: 36mm.
Aperture: 35mm.
Case size: 325 x 115 x 185mm.

Lacking objective shade closure plate. The forward attachment screw to the limb replaced.

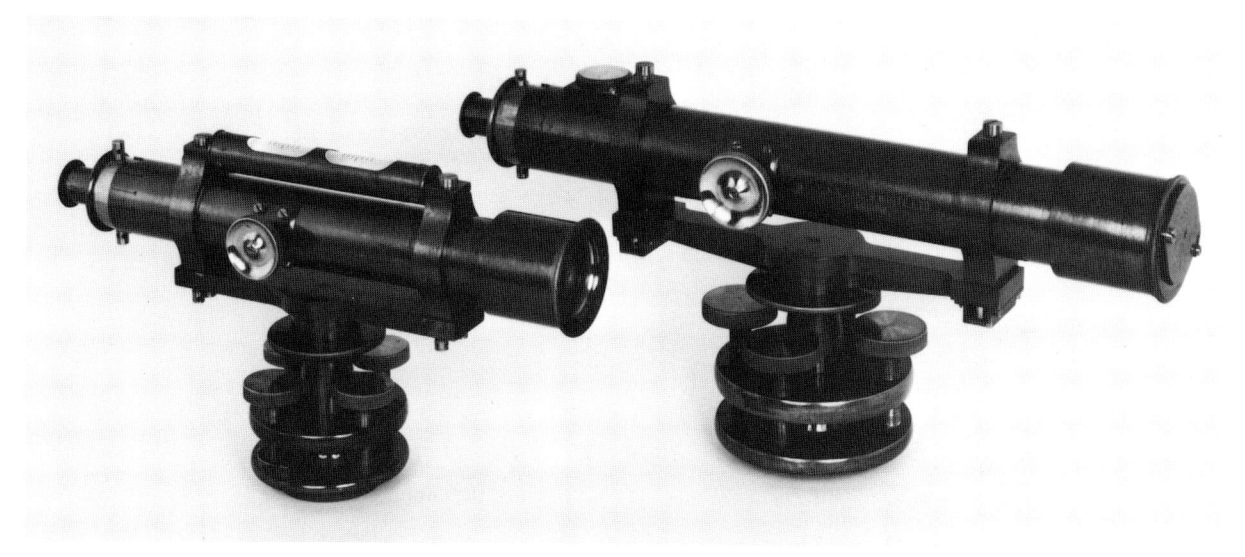

148, 149

149. Level: Kelvin & James White, Glasgow, c.1910 (T1980.141)

14″ dumpy level in oxidised brass and bronze, lacking the main level tube but with a circular level over the telescope, rack and pinion focus, triplet objective, and objective ray shade; the limb threaded to a parallel plate levelling head with free azimuth motion. Engraved on the telescope tube 'KELVIN & JAMES WHITE LTD / GLASGOW.' Components stamped with the assembly mark '1', the base of the levelling head additionally stamped '6'.

Telescope length: 347mm.
Tube diameter: 43mm.
Aperture: 40mm.

Lacking the main level tube. The objective ray shade, stamped '2', is a particularly tight fit and is probably from a different instrument.

150. Level: Kelvin & James White, Glasgow, c.1910 (T1980.137)

16″ dumpy level in oxidised brass and bronze (stripped), with bubble tube (casing only) and circular level over the telescope, rack and pinion focus, triplet objective, and objective ray shade; the limb threaded to a parallel plate levelling head with free azimuth motion. Engraved on the telescope tube 'KELVIN & JAMES WHITE LTD. GLASGOW.' Level components stamped with the assembly mark '57', or scratched 'IV' or 'VI'; levelling head components stamped '4'. The limb and parallel plates have an engine-turned finish.

Telescope length: 401mm.
Tube diameter: 43mm.
Aperture: 41mm.

Lacking the glass bubble tube, diaphragm wires and ray shade closure plate. The ray shade is stamped '58'.

151. Level: Kelvin & James White, Glasgow, c.1910 (T1980.139)

15″ dumpy level in oxidised brass and bronze (stripped), with bubble tube (casing only) and circular level over the telescope, rack and pinion focus, and triplet objective; the limb threaded to a parallel plate levelling head with free azimuth motion. Engraved on the telescope tube 'KELVIN & JAMES WHITE LTD. GLASGOW.' Level components stamped with the assembly mark '32', or scratched 'XXXII' or 'VI'; levelling head components stamped '5'. With fitted case.

150, 151, 152

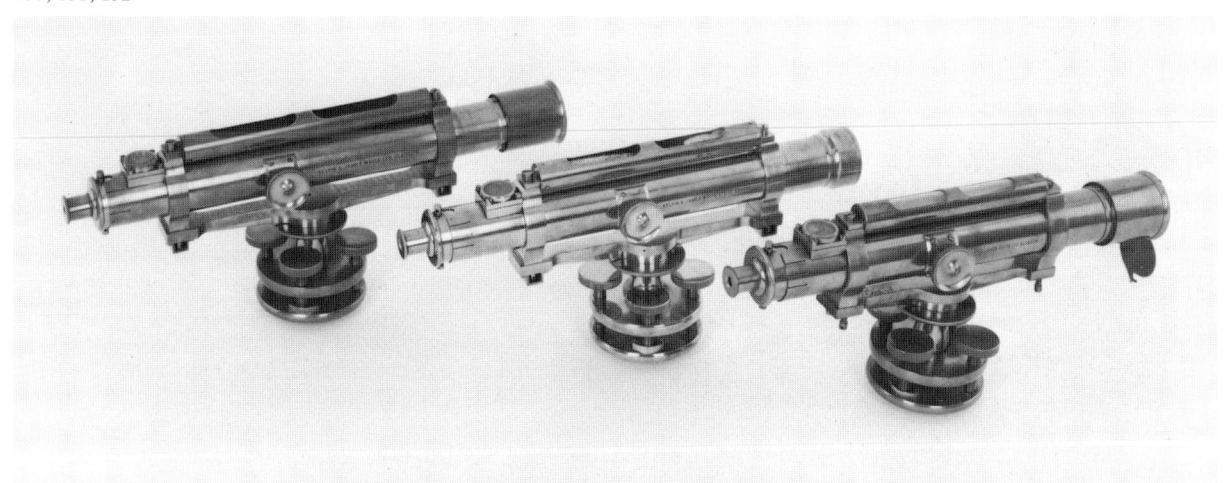

Telescope length: 383mm.
Tube diameter: 43mm.
Aperture: 40mm.
Case size: 455 x 132 x 210mm.

Lacking the glass bubble tube and objective ray shade. One of the eyelens components fractured.

152. Level: Kelvin & James White, Glasgow, c.1910 (T1980.140)

14″ dumpy level in oxidised brass and bronze (stripped), with bubble tube and circular level over the telescope, rack and pinion focus, triplet objective, and objective ray shade; the limb threaded to a parallel plate levelling head with free azimuth motion. Engraved on the telescope tube 'KELVIN & JAMES WHITE LTD. GLASGOW.' Level components stamped with the assembly mark '59'; levelling head components stamped '3'.

Telescope length: 347mm.
Tube diameter: 43mm.
Aperture: 40mm.

The limb is from a different (but compatible) instrument, and is stamped '4'. The levelling screw locating collar on the lower parallel plate has been moved to a new position and the plate counterbored for the other three screw feet. Lacking the limb adjustment screw covers and the diaphragm wires. The bubble tube glass is broken. The metal surface has been marked by mechanical abrasive cleaning, performed without dismantling the instrument.

153. Theodolite: Kelvin & James White, Glasgow, c.1920 (T1980.162)

5″ reversing transit theodolite in oxidised brass and bronze (stripped), the bevelled edge azimuth circle divided on silver [0]-360 by 20 mins., with clamp and free tangent screw adjustment to the upper plate which covers the circle except at the verniers, which are read to 20 secs., with magnifiers. Crossed levels and silvered compass (defective), engraved 'KELVIN & JAMES WHITE L^{TD} / GLASGOW'; the compass with the 8 cardinal points (transposed arrangement for reading bearings against the north point of the needle) and divided 360-[90] by 1°. A-frames support the telescope and vertical circle, the latter divided 90-0/[90]-[0]/90-0/[90]-[0] by 20 mins., with clamp and free tangent screw adjustment on a clipping arm and verniers reading to 20 secs., with magnifiers. The telescope with rack and pinion focus to the 1⅛″ doublet objective; with objective ray shade and bubble tube over the telescope. Integral tribrach stand with shifting centre, axis collar clamp and free tangent screw, stamped under the base '5'. With fitted case (distressed) containing repair label by Alex. Mabon & Sons, 128 Bothwell Street, Glasgow for two dates, the second being 28/10/1947.

Telescope length: 251mm.
Aperture: 29mm.
Azimuth scale diameter: 128mm.
Case size: 195 x 200 x 460mm.

Lacking the compass needle, needle relief arm and compass box glazing. The vertical circle vernier may be from a different instrument as it does not seat properly on the axis. The crossed levels have been remounted over spring plates.

The eyepiece is a replacement, and the graticule wires are missing. One azimuth magnifier defective.

Card pasted in case for The Clydeside Construction Co. Ltd., Public Works Contractors, Bridge of Weir.

154. Level: Kelvin & James White, Glasgow, c.1910 (T1980.172)

18″ rolling parallel rule in bronze with coloured lacquer finish, with chamfered edges and 2 lifting knobs. Engraved 'KELVIN & JAMES WHITE LTD./ GLASGOW.' and 'U. C. LTD.' With fitted case.

Length: 457mm.
Case size: 480 x 85 x 55mm.

United Collieries Ltd. was a large conglomerate formed in 1898 with an initial capital of £200,000 as the result of the amalgamation of 8 coal-mining companies in Lanarkshire and the Lothians. Four years later a further 23 companies were acquired, and the capital increased to £2 million (H. Stanley Jevons, *The British Coal Trade* (London, 1915), 319-320; Sheila Hamilton, 'James Wood', in Anthony Slaven and Sydney Checkland (eds.), *Dictionary of Scottish Business Biography* I (Aberdeen, 1986), 8081).

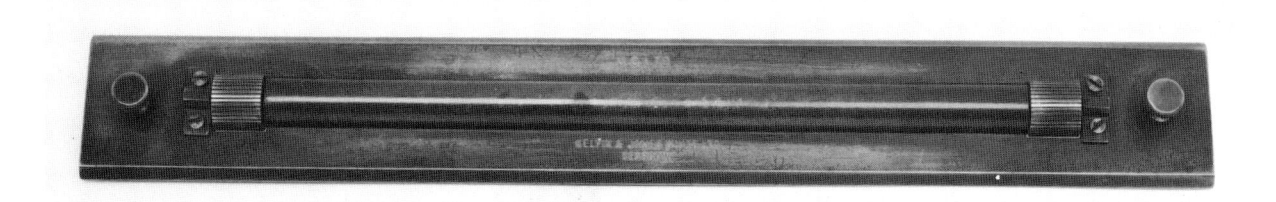

155. Sextant: Kelvin & James White, Glasgow, c.1910 (T1980.195)

7½″ sextant with cast 3-circle frame, in oxidised bronze and brass. Scale divided on silver, [-5]-[160], the reinforced index arm with clamp, tangent screw and vernier reading to 10 secs., and swinging magnifier with ground glass shade. Engraved on the arc 'Kelvin & Ja^s White L^{td} Glasgow.' and stamped '2014'. Screw adjustment to the support ring for the sight tubes. The frame with 7 shades and rear handle, and supported on feet.

Scale radius: 184mm.

Lacking sight tubes and telescopes, and one of the 3 feet.

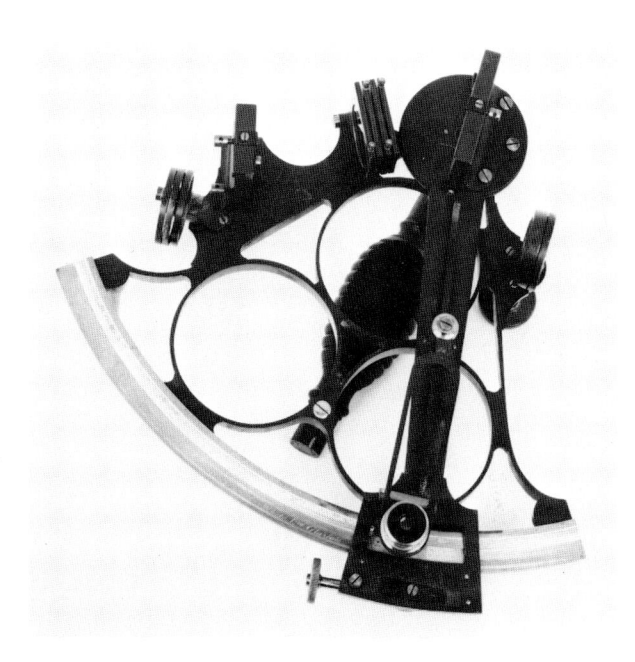

MATHEW EDWARDS

In 1891 Mathew Edwards (c.1833-1893) left the firm of James White (q.v.), of which he was one of the two surviving partners, and set up independently as a scientific instrument maker.[1] Edwards was an Irishman, born about 1833 and the son of Frederick Edwards, a tanner, and his wife Elizabeth Thomson.[2] His early career is not known, and he was first entered in the Glasgow street directory in 1881, as a maker of all types of scientific instruments at White's address, 241 Sauchiehall Street. This entry suggests that Edwards was by then of some standing in the firm, and it is known that for several years before 1890 he was a managing partner.[3]

On becoming independent in 1891, Edwards took over what had been James White's premises at 209 Sauchiehall Street, White continuing at Cambridge Street. His new premises were described as comprising a large shop (i.e. workshop) or warehouse, a showroom, a counting house and stores, together with a dark room for the use of amateur photographers.[4] The 1891 account, presumably because it was produced shortly after Edwards began in business is vague about the firm's composition and products: 'The various sections of Mr Edwards' business are severally so comprehensive and important that nothing short of a detailed catalogue would suffice to do justice to each.'[5] He may have stockpiled some instruments before he left the Cambridge Street works, or may possibly have contracted-out construction work; at any rate it is clear that he was acting principally as a retailer on an apparently ambitious scale. His microscope catalogue ran to sixty-four pages, including not only his own pieces,[6] but also those of leading English and continental makers, with a speciality in Henry Crouch of London's Photo-Micrographic Camera.[7]

Edwards' entries in the street directory show that he continued most of the lines in which White dealt, covering optical, mathematical, nautical, electrical and photographic appliances. What is unclear, is the size of his workforce, and how it related to James White's continuing, and presumably rival, 'parent' business. As the 1891 account explains,

> '... as an optician and philosophical instrument maker, Mr White attained an almost unique celebrity, and to the successes he achieved his manager Mr Edwards, of course contributed. Who, then, more fit to succeed the founder than Mr Edwards?'[8]

Amidst the hyperbole and the tenuous claim to be White's successor, lies the obvious fact that as White's former lieutenant Edwards had skill and experience. Also, according to the account, he 'has a large and high-class connection with medical men, scientists, photographers, and public institutions both at home and abroad'.[9] No doubt some of his connections were due to his membership of the Glasgow Philosophical Society, to which he was elected on 16 November 1887, at the same time as David Reid, his colleague at White's.[10]

However, Edwards did not live long to reap the benefits of his enterprise. He died on 26 December 1893, aged sixty. His son Matthew, by his wife Margaret Lauchlan, was a tobacconist and hardware merchant; he did not continue the business.[11] It was acquired and continued by his manager, W.J. Hassard.[12]

REFERENCES

1. The Notice of Dissolution of the Copartnery is worth quoting, as it reveals details about both businesses: 'The Copartnery carried on by the Subscribers under the Firm of JAMES WHITE, as Opticians and Philosophical Instrument Makers, at 16, 18 and 20 Cambridge Street, and 209 Sauchiehall Street, Glasgow, was, by mutual consent, DISSOLVED on 31st January 1891. The portion of the Business carried on at the Shop 209 Sauchiehall Street is retained by the Subscriber Mathew Edwards, who will continue under his own name the Business hitherto carried on there ... The portion of the Business carried on at the Works in Cambridge Street is retained by the Subscriber David Reid, who will continue, under the name of JAMES WHITE, the Business carried on there ...' (*Edinburgh Gazette*, 3 February 1891).

2. General Register Office (Scotland), Census of 1891 (Glasgow, Blythswood), 644[7]/46, p.22; Register of Deaths 1893 (Glasgow, Blythswood), 644[2]/550.

3. [Stratten's] *Glasgow and Its Environs* (Glasgow, 1891), 153. This is also mentioned in James White's obituary, *Engineering* 38 (1884), 245, and in the following notice: 'The Subscribers

David Hay Reid, Charles A. Smith, and Robert Annan, Trustees and Executors of the deceased James White, Optical and Philosophical Instrument Maker in Glasgow, have no interest as Partners in the Firm of JAMES WHITE, Opticians and Philosophical Instrument Makers, Glasgow, which has been since the date of Mr White's death on 15th August last, and continue to be, carried on by his Copartners, the Subscribers Mathew Edwards and David Reid ...' (*Edinburgh Gazette*, 9 December 1884, 1040).

4. [Stratten], *op. cit.* (3), 153.

5. *Ibid.*

6. None of these have been recorded.

7. *Ibid.*. They were: Beck, Crouch, Powell & Lealand, Ross, Swift, Watson (all of London), Hartnack (Potsdam), Leitz (Wetzlar), Prazmowski (Paris), Reichert (Vienna). No copies of Edwards' catalogue have been traced.

8. *Ibid.*

9. *Ibid.*

10. *Proceedings of the Glasgow Philosophical Society* 19 (18878), 397.

11. G.R.O.(S.), Register of Deaths 1893 (Glasgow, Blythswood), 644²/550. The cause of death was given as congested lungs.

12. 'The Business of Optician and Philosophical Instrument Maker carried on by the late MATHEW EDWARDS at 209 Sauchiehall Street, Glasgow, was on 21st February last acquired by Mr W.J. Hassard, his Manager, and the Trustees of the deceased ceased as at that date to have any connection with said Business. [Signed] JAMES McMICHAEL JR. ROB. PHILIPS. FRED. MACKENZIE. D.M.ALEXANDER. trustees of the late Mathew Edwards. W.J.HASSARD.' (*Edinburgh Gazette*, 6 March 1894, 202).

---

156. Pantograph: M. Edwards, Glasgow, c.1890 (T1980.179)

18" copying pantograph in brass, comprising 4 hinged arms with 2 sliding and 1 fixed sockets, conventionally identified on the arms as 'B', 'C' and 'D', and supported on ivory wheels. Engraved 'M. EDWARDS, OPTICIAN, 209 SAUCHIEHALL ST.' The 2 arms with the sliding heads calibrated with the ratios 1:2 to 1:12.

Length closed: 530mm.

Lacking anchor weight, tracing point and pencil holder, clamp screw for one of the sliding heads and clamp screws for 2 of the sockets.

JAMES MORE & CO.

James More was the son of David More, an engineer, millwright and maker of hydraulic presses and the like, whose own father had also been a millwright.[1] James was working for James White (q.v.) from 1873 or earlier, and first appeared in the Glasgow street directory in 1885 as an optician at White's. In 1891 he left the firm to establish James More & Co., opticians, in smart new premises at 77 Renfield Street. The business ceased in 1898: he died on 31 October, and his heirs petitioned for the sequestration of his estate almost immediately.[2]

It emerged from the Trustee's summary of More's state of affairs, that amongst his creditors were the following instrument suppliers: his greatest debt was owed to Walter Braham & Co. of Birmingham (instrument engineers), £158 8s 11d; minor debts were owed to Baird & Tatlock (q.v.), Glasgow, £1 13s 5d; James Brown (q.v.), Glasgow, £2 2s 11d; R.& J. Beck, London, £2 4s 5d; Jas. Chesterman & Co., Ltd., Sheffield, 13s 5d; Chadburn Bros., Sheffield, £3 16s 1d; F. Darton & Co., London, £27 1s 4d; Eastman Photo. Co., London, £4 4s 3d; John J. Griffin & Sons Ltd., London, £3 16s 4d; J. F. Shew & Co., London, £8 8s 3d; G.W. Wilson & Co., Aberdeen (photographic suppliers), £2 9s 4d. His liabilities amounted to £1413 12s 1d, while his assets were valued at £442 4s 11d.[3] The whole estate was realised by 19 September 1900, and discharged on 1 May 1901.[4]

From the beginning More advertised all types of scientific instruments, but he specialised in adjustable spectacles, designing and making his own types, notably the 'Kaliston'.[5] A second specialisation was in photographic apparatus; he constructed his own cameras, named with suitable Glaswegian emphasis the 'St. Mungo', the 'Renfield' and the 'Blythswood'.[6] More's general stock included mathematical, mining, navigating and surveying instruments, and thermometers, barometers and other standard items in this line, a proportion of which was certainly manufactured or assembled in his own works.[7] It is apparent from the list of the creditors to his sequestrated estate that he was also buying from specialist manufacturers: microscopes were clearly being bought from Beck, measuring tapes from Chesterman and cameras from Shew. However, More was proud not only of his own glass stereoscopic transparencies of the new municipal buildings in Glasgow, but also of his 'valuable assortment of artificial eyes'.[8] More is said to have a large staff for manufacture and repair work.[9] According to an account of 1891, 'Mr More is Official Lanternist to the Glasgow Amateur Photographic Society, and is universally popular in trade and general circles.'[10] A more revealing paragraph was published just before his death:

> 'Much regret will be felt among a large number of photographers, as well as those taking an interest in lantern work, at the serious indisposition of Mr James More. For many years Mr More held what may be termed a unique position in the West of Scotland in connexion with limelight projection, and his services in this respect were always eagerly sought where any important function was concerned. The able manner in which he carried through the limelight arrangements for the two International Photographic Exhibitions, held in Glasgow under the auspices of the Glasgow and West of Scotland Amateur Association, will long be remembered by those who had an opportunity of judging of his ability. During these two functions alone something like sixty limelight lectures were delivered by entirely different lecturers, but not a single hitch from first to last occurred under Mr More's expert manipulation. Some time ago, when travelling in England, Mr More caught a severe cold, from which, we regret to say, he never recovered, and his business in Renfield Street has been recently acquired by Messrs. Rae Brothers, of Glasgow.'[11]

REFERENCES

1. General Register Office (Scotland), Register of Deaths 1873 (Glasgow, High Church), 644²/964.

2. Scottish Record Office, CS 315/45/174, Sequestration Papers, James More, Sederunt Book, p.1. The Petition for Sequestration was brought by 'Margaret Scott or More, ... sole surviving trustee under deed of trust granted by Alexander More, Engineer, ... a creditor of the deceased James More.' The debt amounted to £300. Notices appeared in the *Edinburgh Gazette*: the initial notice of sequestration (11 November 1898, 1102); the announcement of the appointment of the Trustee and Commissioners (2 December 1898, 1324); the acceleration of payment of the first dividend (10 January 1899, 41); the payment of the second and final dividend to be paid on 10 November 1899 (26 September 1899, 913); and a Creditors' meeting called for 7 November 1900, to consider discharging the Trustee (16 October 1900, 1006).

3. S.R.O., CS 315/45/174, pp. 19-22. However, money was coming in from, among others: James White (q.v.), 7s 6d; A. Arrol & Sons, £2 13s 6d; and the University, 15s. (*ibid.*, p. 23).

4. *Ibid.* The creditors each received a dividend of about 6s 3d per pound that they were owed.

5. Advertised in the *Glasgow Post Office Directory* 1894 and 1897.

6. [Stratten's] *Glasgow and Its Environs* (Glasgow, 1891), 106.

7. Apart from the level in this collection, NMS T1980.142, a four-pole chain with brass handles signed 'James More, Chain Maker, Glasgow' was offered for sale by Tesseract, Catalog 9 (Winter 1985), item 56, and a theodolite signed 'James Moore [*sic*], Glasgow' was offered by Phillips Edinburgh, 25 October 1985, Lot 87.

8. [Stratten], *op. cit.* (6).

9. *Ibid.*

10. *Ibid.* 'The death is announced of Mr James More, optician and photographic materials dealer of Glasgow. The deceased was well known locally, and had a high reputation as a lantern operator. Prior to setting up in business for himself some few years ago, he had been for many years with the late James White' (*Practical Photographer* 9 (1898), 367: we are grateful to Sara Stevenson for this reference).

11. 'Notes from the West of Scotland', *British Journal of Photography* 45 (1898), 696.

---

157. Level: James More & Co., Glasgow, c.1895 (T1980.142)

14″ dumpy level in brown oxidised brass, with hinged bubble tube and crossed level over the telescope, rack and pinion focus, triplet objective, and objective ray shade; the limb threaded to a parallel plate levelling head with free azimuth motion. Engraved on the telescope tube 'JAMES MORE & CO GLASGOW.'

Telescope length: 347mm.
Tube diameter: 44mm.
Aperture: 41mm.

The eyepiece is a replacement. The glass bubble tube has been broken.

REID & YOUNG

Reid & Young, opticians and mathematical instrument makers, was founded in 1913 by two sons of employees of Kelvin and James White Ltd. (q.v.). John W. Reid was the son of David Reid, a managing partner of the firm, and first appeared in the street directory in 1903, listed as also of the firm. Thomas Young was the son of Thomas Young, a compass adjuster with Kelvin and James White Ltd. (q.v.) in 1910; and may himself have worked for the firm.[1]

Reid & Young began business at 88 St. Vincent Street, where they remained until 1963. Like other similar businesses their stock covered chemical, mathematical and optical instruments, and photographic equipment; mining and surveying instruments may have been a specialisation in the middle part of this century.[2] In 1964 the firm moved to 1787 Paisley Road West, where it was described as 'ophthalmic opticians', and disappeared from the directory after the following year. Little else is known of their activities.

REFERENCES

1. *Glasgow Post Office Directory, passim*. Young's first given home address in 1913 at 35 Rose Street, Garnethill, was in the same tenement as Mathew Edwards (q.v.) had lived in up to 1892.

2. *Ibid.*, trades entries in 1920, 1930, 1940, 1950, 1960, in all of which Reid & Young are described as 'late with Kelvin & James White Ltd.' A theodolite signed 'Reid & Young' was offered for sale by Sotheby's Belgravia, 9 May 1980, Lot 165.

158. Level: Reid & Young, Glasgow, c.1920 (T1980.143)

10″ level in oxidised brass (stripped), with bubble tube over the telescope, rack and pinion focus, and doublet objective; the limb threaded to a parallel plate levelling head with free azimuth motion. Engraved on the telescope tube 'REID & YOUNG /ST VINCENT STREET / GLASGOW'. Components of the level and levelling head stamped with the assembly mark '6' or scratched 'VT'. With objective cover.

Telescope length: 246mm.
Tube diameter: 28mm.
Aperture: 23mm.

Lacking eyepiece and diaphragm wires. The bubble tube is from another instrument with slightly more widely separated attachment points: it is stamped '2' and may therefore be a survival of the instrument that provided the levelling head for item 88. The limb was bent and required re-alignment.

22

WHYTE AND THOMSON

James Whyte, whose business became Whyte, Thomson & Co., probably began his work as an instrument maker under David Heron of Greenock. Whyte married Heron's daughter Jessie in 1860,[1] and after Heron's final entry in the Glasgow street directory in 1864, Whyte appeared at his father-in-law's address.

David Heron began making and retailing nautical instruments in Greenock,[2] and was presumably related to members of the watch and chronometer making firm of the same name: indeed it is conjectured that John Heron of The Square, Greenock, from at least the 1790s, and subsequently of 1 William Street, was David's father.[3] He may also have been related to the William Heron who was a lecturer in natural philosophy at the Greenock Institution of Arts and Sciences in 1825.[4] After David Heron removed to Glasgow in 1827[5] he continued to retain, in his own name, the 'nautical, optical, and stationery warehouse' at 1 William Street, where John Heron, the feuar and 'chronometer, watchmaker jeweller' had had his shop since 1815.[6] The premises were taken over in 1836 by Duncan McGregor (q.v.) who described himself as 'successor to Heron ...', and in fact for an unknown period before this Heron and McGregor had been in partnership together.[7] It is likely that either John or David Heron was responsible for the building of an observatory at Greenock in 1819 to provide accurate time for the town and the port.[8]

David Heron's business first appeared in the Glasgow street directory in 1834 as Heron & Co., ship chandlers and nautical warehouse at 128 Broomielaw, on the Clyde waterfront. On his removal to 212 Broomielaw in 1836 he began trading as David Heron & Co., nautical instrument makers, until in 1840 he went bankrupt.[9] He was again in business in 1841, and in about 1844 he formed the brief partnerhip of Heron and Johnston. They traded as ironmongers, ship chandlers and flag makers, but by 1845 Heron had resumed business independently as David Heron.[10] In 1848 he was again bankrupt,[11] but by 1849 he had moved to 4 Carrick Street, where he began advertising as a compass adjuster as well as a nautical instrument maker. He ceased to appear in the directory after 1863.[12]

It is likely that by this date James Whyte, then aged about twenty-seven, had been working under Heron for several years. In addition to the Carrick Street premises he occupied 144 Broomielaw, and in 1871 he moved from Carrick Street to 3 James Watt Street. When, in about 1875, the firm became Whyte & Co., he left the latter address and moved to 102 and 104 Dale Street, all the while retaining 144 Broomielaw. From 1881 to 1888 he worked from the Broomielaw premises only. In 1889, when the firm became Whyte, Thomson & Co., it took in 142 Broomielaw and moved its workshops to the Neptune Works, 123 Harmony Row, Govan. In 1912 the shop moved to 96 Hope Street, and three years later moved its workshops to 142 North Woodside Road. In 1923 the shop moved to 159 Queen Street, in 1927 to 47 Cadogan Street, in 1934 to 57 Bothwell Street, in 1948 to 191-3 Broomloan Road, and in 1951 to 57 Bothwell Road where they ceased trading by 1953. They became a limited company in 1934.[13]

Like other Clydeside makers, Whyte, Thomson & Co. had a branch at another major shipbuilding centre; theirs was opened in South Shields in 1902, and lasted until 1916.[14] And like other makers, the scope of their workshop's production appears to have increased, so that by 1890, for instance, the firm advertised as makers of chemical, mathematical, nautical, optical and philosophical instruments, as chronometer makers to the Admiralty and makers of all types of clocks, and as compass makers and adjusters.[15] Apart from the items in this collection, which include telescopes and sextants, other instruments signed by this firm have been recorded.[16] A contemporary commercial account describes the 'splendidly spacious warehouse stocked with all descriptions of nautical instruments ... The premises are elegantly appointed in mahogany and ebony-gold background, the whole presenting a most interesting and attractive appearance.'[17] More revealing is the reference to the modern and efficient Neptune Works at Govan, where light engineering work was carried out: it appears that Whyte, Thomson & Co., like Dobbie McInnes (q.v.), produced steam pressure gauges, vacuum gauges, engine counters and indicators, lamps and cabin fittings. The workforce needed to produce these items, because 'everything is made from the raw material', was necessarily large. In 1891 the firm employed seventy skilled mechanics and four compass adjusters,[18] figures which form a significant contrast with the smaller concern twenty years before of the '10 Men & 5 Boys', recorded in the Census of 1871 as the employees of James Whyte.[19]

Among the compass adjusters attached to the firm was James Thomson, a contemporary of Whyte, and like him, a native of Glasgow. He was working for James Whyte

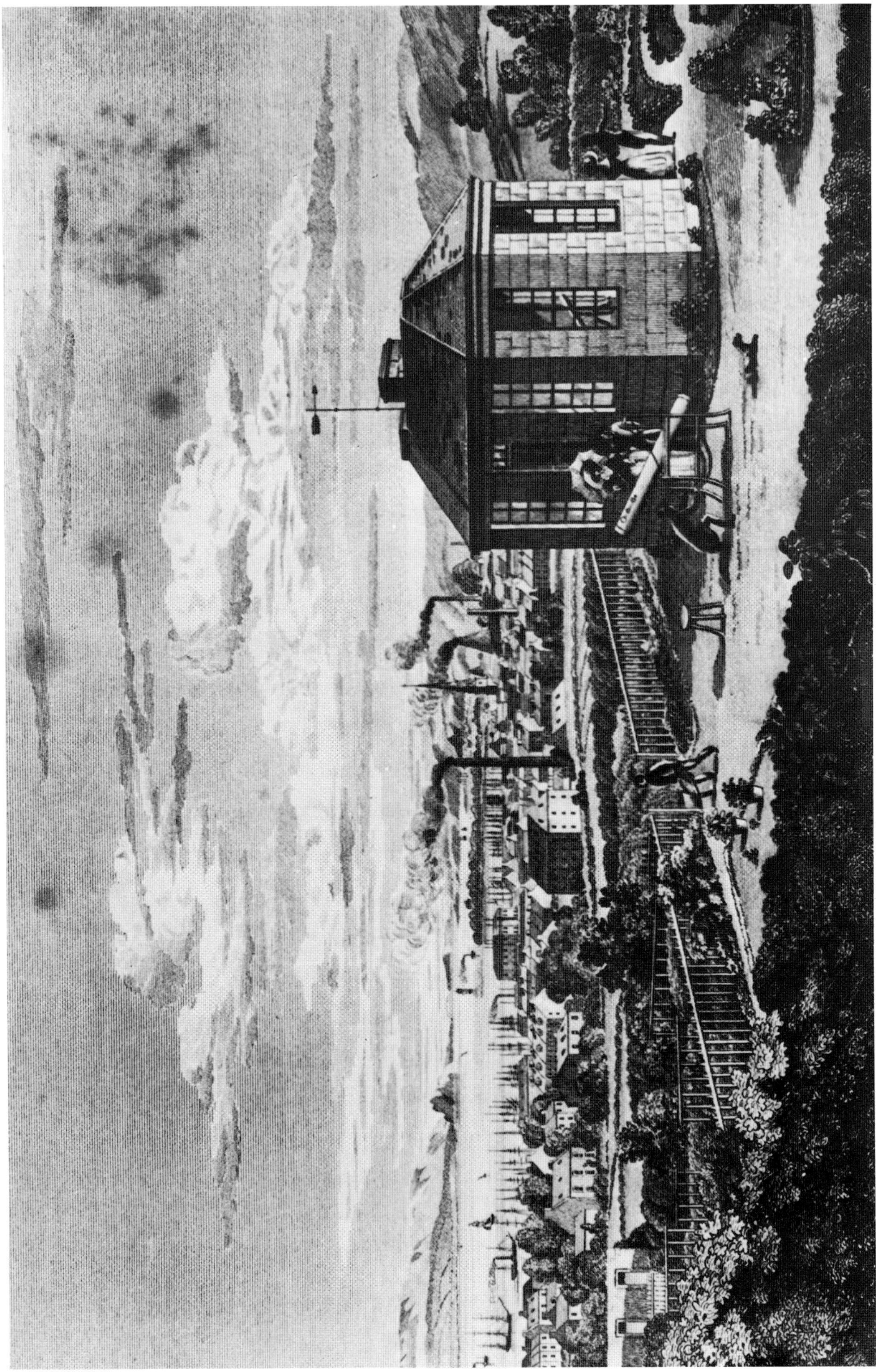

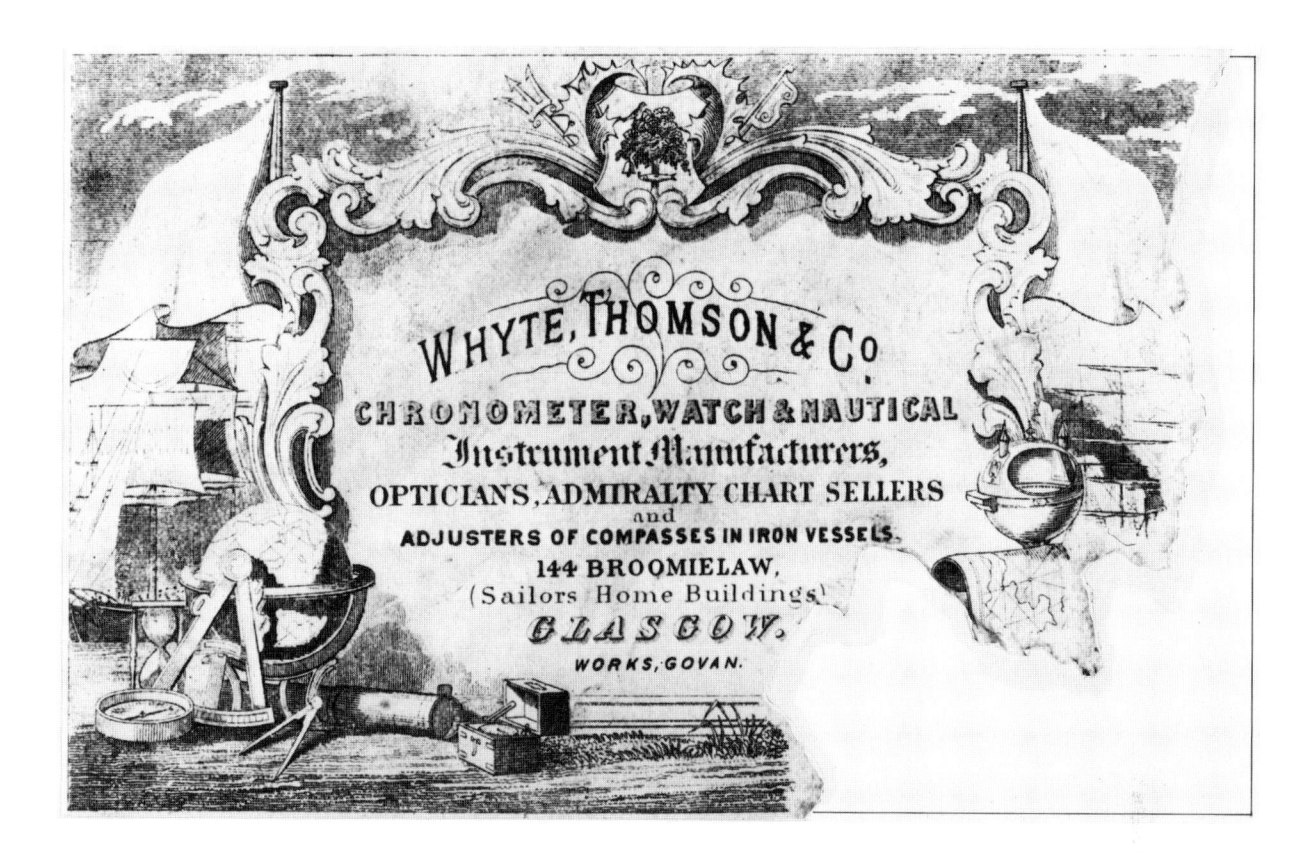

62 Whyte, Thomson & Co.'s trade card for their address from 1889

senior before 1875, but it was apparently with his son, James junior, that he formed Whyte, Thomson & Co. in 1889.[20] Other adjusters included John Graham,[21] David Young,[22] John R. Inglis, W.D. Whyte,[23] James Thomson junior, and James Wilson;[24] one specialist, James Gilchrist, who constructed compasses and binnacles, was also recorded.[25] Of these men, James Wilson is one of the more significant, for he went on to found Christie and Wilson (q.v.) with Andrew Christie, formerly of D. McGregor & Co.(q.v.).

The reputation of the firm was high. It won the silver medal at the 1886 Edinburgh Exhibition, and the top award of the diploma at Glasgow in 1888, and its trade was apparently extensive.[26] The somewhat pompous 1891 account named seven shipping lines (and hints at several others) whose vessels it fitted out, boasted of its position as chronometer makers to the Admiralty, and referred obliquely to the lucrative foreign and domestic business which its 'pre-eminent position' brought it.[27]

While this claim cannot readily be tested, the assertion that Whyte's patent standard compass (introduced about 1886) 'is now being fitted to all the leading cargo and passenger steamers now being built, and there is not another instrument in the market to compare with it'[28] is

61 'Mr Heron's Observatory' drawn by J. Stuart and engraved by James Kerr for Daniel Weir's *History of the Town of Greenock* (Greenock, 1829). *Trustees of the National Library of Scotland*

ambiguous, and could be considered misleading. The account makes no mention of the fact that the firm acted, according to their own advertisements, as 'principal agents' for Sir William Thomson's patent compass and sounding machine,[29] which were probably the most sophisticated on the market. It is clear from the records of James White (q.v.) that Whyte, Thomson & Co. were one of the two main Scottish customers for the Thomson compasses.[30] Moreover, by 1902 Kelvin and James White Ltd. (q.v.) were warning their customers against compasses made to resemble present or formerly patented designs, by makers 'who were formerly content to act as our agents', and who advertised their instruments misleadingly as genuine Kelvin and White products.[31] Examples of such advertisements have not been seen, except in Glasgow street directories, where in 1901, for instance, Whyte, Thomson & Co. were listed as 'makers of "The Sir William Thomson" compass and sounding machine'. The previous year they had designated themselves in unexceptionable fashion 'principal agents' for the instruments. The claim to be makers of these instruments, or similar ones designated ambiguously 'Thomson', continued until 1910.[32]

In 1907 Lord Kelvin and the foreman of Kelvin & James White Ltd., Francis Wood Clark, to whom patent 22,031 of 1902 had been jointly granted for 'Improvements in the Mariner's Compass', sued William David Whyte and James Thomson, as partners of Whyte, Thomson & Co. and as individuals, for patent infringement. In the judgement, however, it was

> 'held (first), that no sufficient evidence had been brought to rebut the presumption that F.W.

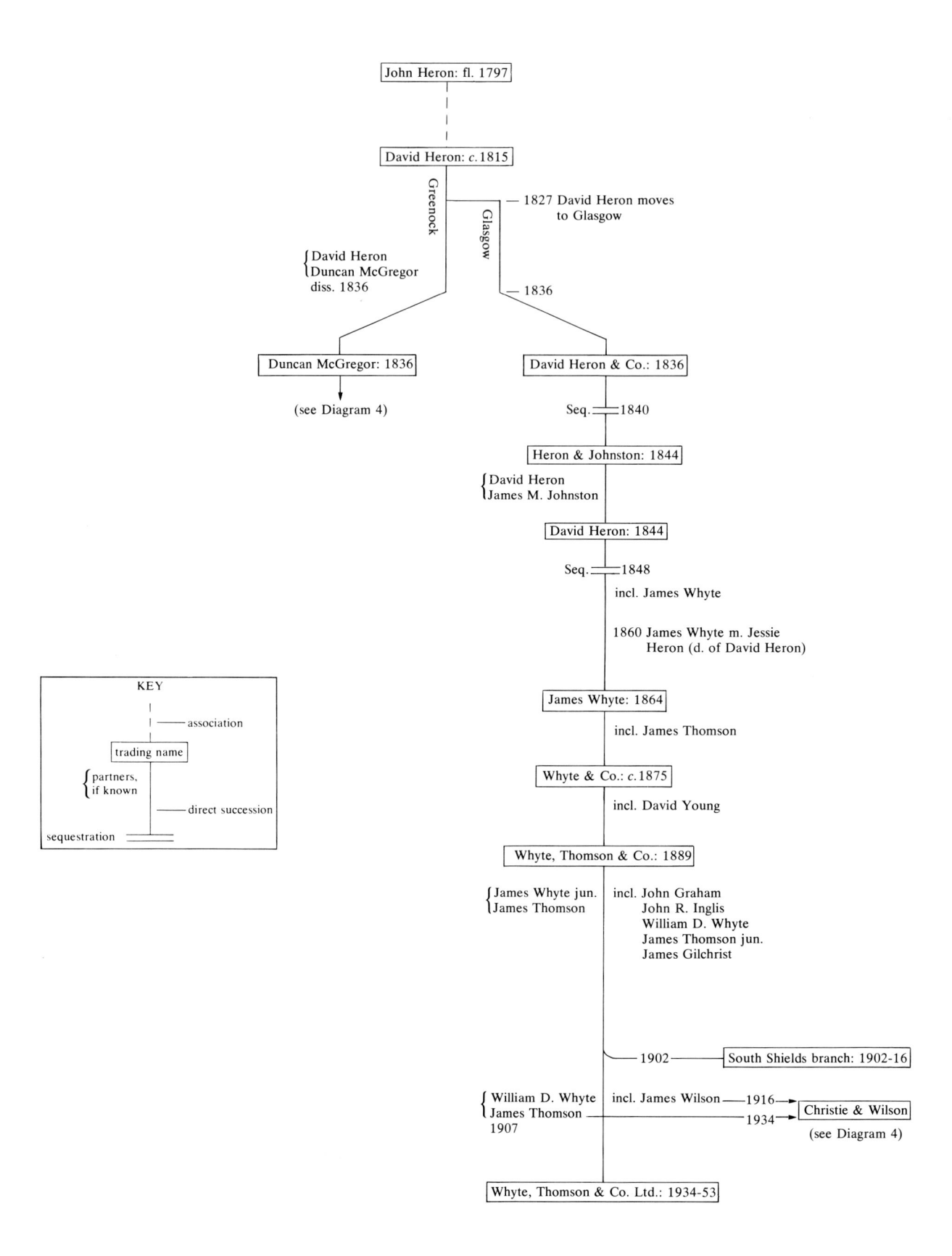

```
                    ┌──────────────────────┐
                    │ John Heron: fl. 1797 │
                    └──────────┬───────────┘
                               ┆
                    ┌──────────┴───────────┐
                    │ David Heron: c.1815  │
                    └──┬────────────────┬──┘
                   Greenock          Glasgow — 1827 David Heron moves
                      │                 │                to Glasgow
     ⎧ David Heron    │                 │
     ⎨ Duncan McGregor│                 │— 1836
     ⎩ diss. 1836     │                 │
                      │                 │
         ┌────────────┴─────────┐  ┌────┴──────────────────┐
         │ Duncan McGregor: 1836│  │ David Heron & Co.: 1836│
         └──────────┬───────────┘  └────────────┬──────────┘
                    ↓                      Seq. ══ 1840
            (see Diagram 4)                     │
                                    ┌───────────┴──────────┐
                                    │ Heron & Johnston: 1844│
                                    └───────────┬──────────┘
                                 ⎧ David Heron
                                 ⎨ James M. Johnston
                                    ┌───────────┴──────────┐
                                    │   David Heron: 1844  │
                                    └───────────┬──────────┘
                                           Seq. ══ 1848
                                                │
                                                incl. James Whyte

                                                1860 James Whyte m. Jessie
                                                     Heron (d. of David Heron)

                                    ┌───────────┴──────────┐
                                    │   James Whyte: 1864  │
                                    └───────────┬──────────┘
                                                incl. James Thomson

                                    ┌───────────┴──────────┐
                                    │   Whyte & Co.: c.1875│
                                    └───────────┬──────────┘
                                                incl. David Young

                                    ┌───────────┴──────────┐
                                    │ Whyte, Thomson & Co.: 1889│
                                    └───────────┬──────────┘
      ⎧ James Whyte jun.           incl. John Graham
      ⎨ James Thomson                    John R. Inglis
                                         William D. Whyte
                                         James Thomson jun.
                                         James Gilchrist

                                                │
                                                ├─1902──────┌──────────────────────┐
                                                │           │ South Shields branch: 1902-16│
                                                │           └──────────────────────┘
      ⎧ William D. Whyte                         │
      ⎨ James Thomson ──── incl. James Wilson ──1916─→ ┌────────────────┐
        1907                                 ──1934─→ │ Christie & Wilson│
                                                      └────────────────┘
                                                        (see Diagram 4)
                                    ┌───────────┴──────────────┐
                                    │ Whyte, Thomson & Co. Ltd.: 1934-53│
                                    └──────────────────────────┘
```

KEY

| — association
trading name
{ partners, if known
— direct succession
sequestration ══

Diagram 6. Schematic development of Whyte Thomson.

C[lark], the applicant for the Patent, was the first and true inventor: (second) that the addition of the links in the Complete Specification was a legitimate development of the invention disclosed generally in and protected by the Provisional Specification, and that there was no disconformity; (third), that having regard to the state of the art, there was subject-matter for invention [as Whyte, Thomson & Co. laid claim] in the combination of horizontal spiral springs under tension with free suspension of the gimbal ring, of which combination there had been no prior use or anticipation; (fourth), that, whether or not the Complainers' suspension [i.e. that of Kelvin & James White Ltd.] was more satisfactory than the grummit ring, it had utility as offering an useful choice; (fifth), that the Claims should be construed benevolently, so as, if possible, to support the Patent; but that, even if so construed, the Complainers' second Claim was too wide and did not with sufficient clearness differentiate the invention claimed from previous appliances, and that their Patent was invalid as it stood. The issue of infringement was not decided. The prayer of the Note was refused.'[33]

It seems that the similarity between the names of the two firms Whyte, Thomson & Co. and Kelvin & James White Ltd. was a co-incidence and not a deliberate attempt at deception. If James White had felt otherwise, they could have refused to wholesale William Thomson instruments to Whyte, Thomson & Co., but clearly the latter could and did achieve sales that James White was not confident they themselves could manage directly.

However, it is arguable that while the similarity of names apparently benefited the smaller business, it may also have been felt by Whyte, Thomson & Co. to have had the effect of preventing its name from being known in its own right. The evidence of the patent literature does indicate that the firm was actively improving the design of its own range of nautical instruments. This was mainly the work of W.D. Whyte, who was probably a younger son of James Whyte senior.[34] Between 1891 and 1914 he took out twenty patents (four of them jointly): twelve covered magnetic and mariners' compasses; four were for sounding machines; two for compass binnacles, and one each for ships' logs and a bearing instrument.[35] These figures suggest that the claims made in the 1891 account of the firm, as to the extensiveness of the Whyte compass's use, are more creditable than might at first be thought. Certainly, on the evidence that we have surveyed, the firm was not just a cheap imitation of an internationally famous name.

REFERENCES

1. Her death certificate confirms this: she died, aged 52, of pneumonia, the widow of James Whyte, Nautical Instrument Maker (master), daughter of David Heron, Nautical Instrument Maker (deceased) (General Register Office (Scotland), Register of Deaths 1895 (Glasgow, Blythswood), 644^7/147).

2. There is a 12 inch octant marked 'Heron, Greenock' at the Town Docks Museum, Hull (inventory D.B. 969), besides the refracting telescope in the Frank collection (NMS T1980.252). We are grateful to Arthur Credland of the Town Docks Museum, Hull, for this information.

3. John Smith, *Old Scottish Clockmakers* 2nd edition (Edinburgh, 1921), 189, records a John Heron of Greenock fl.1797-1822 and a James Heron at William Street, Greenock in 1836. John is listed at Forrest's Land, Square, in the *Greenock and Port Glasgow Directory* of 1805. A long-case clock by the firm is in Paisley Museum, and a gold-cased watch hall-marked 1813, signed 'Heron, Greenock' was offered for sale at Sotheby's, 26 July 1971, Lot 104.

4. *Glasgow Mechanics' Magazine* 3 (1825), 217. Heron was a Glasgow medical graduate of 1830 and for the following three years held the chair of natural philosophy in Anderson's College, Glasgow (W.I. Addison, *A Roll of the Graduates of the University of Glasgow* (Glasgow, 1898), 264). He may be the surgeon and apothecary who practiced in Glasgow for at least 1805-1820, and it is conjectured he was John Heron's brother. A William Heron was involved in the trials of Robert Jamieson's marine thermometer case (*Repertory of Arts, Manufactures and Agriculture* 2nd series, 39 (1821), 168), and may have been involved in instrument retailing since a sympiesometer made by Alexander Adie and signed 'Patent A. Adie Edinburgh No. 383 Wm. Heron Agent Greenock' was offered for sale by Sotheby's, 10 December 1981, Lot 245, and later at Phillips, 10 December 1986, Lot 5, and is now in the Royal Museum of Scotland (NMS T1987.30).

5. [Stratten's] *Glasgow and Its Environs* (Glasgow, 1891), 75.

6. *Fowler's Commercial Directory of the Lower Ward of Renfrewshire* 1834-5.

7. *Fowler's Greenock Directory* 1836-37. The notice of the Dissolution of Copartnery was published: 'The Business carried on by the Subscribers in Greenock, as Dealers in Nautical, Mathematical, and Optical Instruments, &c. was DISSOLVED by mutual consent upon the 1st January last ... [signed] DAVID HERON. DUNCAN McGREGOR.' (*Edinburgh Gazette*, 3 May 1836, 122).

8. D. Weir, *The History of the Town of Greenock* (Greenock, 1829), 107.

9. From 1839, 210 Broomielaw was also listed as one of Heron's addresses: Scottish Record Office, CS 279/979, Sequestration Papers, David Heron. Unfortunately, there is no Sederunt Book with these papers. Notices regarding his bankruptcy appeared in the *Edinburgh Gazette*: the initial notice of sequestration (11 December 1840, 394); the announcement of the appointment of the Trustee and Commissioners, and that the Bankrupt had offered Creditors composition of six shillings per pound, this offer to be confirmed at a meeting called for on 1 March 1841 (29 January 1841, 30); and an adjourned general meeting to be held on 22 March 1841 'to decide upon the offer of composition

made by the Bankrupt' (2 March 1841, 72). Notices of this process also appeared in the *Scotsman*: 12 December 1840, 30 January 1841 and 3 March 1841.

10. 'The Business carried on by Us, under the Firm of HERON and JOHNSTON, as Ironmongers and Ship Chandlers, was this day DISSOLVED by mutual consent ... [signed] DAVID HERON. JAS. MELDRUM JOHNSTON. Glasgow, September 28, 1844.' (*Edinburgh Gazette*, 18 October 1844, 344).

11. S.R.O., CS 279/1038, Sequestration Papers, David Heron: again, information is scanty, and there is no Sederunt Book. Notices appeared in *ibid.*: the initial notice of sequestration (7 March 1848, 123); the announcement of the appointment of the Trustee and Commissioners (14 April 1848, 186); the postponement of the dividend payment (22 September 1848, 472); and the announcement of the payment of the first and final dividend on 21 November 1849 (5 October 1849, 929).

12. *Glasgow Post Office Directory* 1834-1863, *passim*. Glasgow Art Gallery and Museum have an 8 inch dry card marine compass signed 'DAVID HERON BROOMIELAW GLASGOW' and there is a 12 inch octant marked 'D Heron Glasgow' with 'SBR' on the scale (i.e. supplied by Spencer, Browning & Rust of London) in the National Museum of Wales (inventory 65.132). We are grateful to Dr David Jenkins of the Welsh Industrial and Maritime Museum for this information.

13. *Glasgow Post Office Directory* 1864-1953, *passim*.

14. *Ibid.* 1902-1916.

15. *Ibid.* 1890.

16. The items in the Frank collection are: a telescope (NMS T1980.161) and a compass (NMS T1980.198). A magnetic compass, with the signature on the card as 'White [sic], Thomson & Co., Glasgow' was offered for sale by Christie's, 5 July 1971, Lot 3; a pocket combined compass and aneroid barometer signed 'Whyte Thomson & Co., GLASGOW' was offered by Christie's South Kensington, 17 April 1986, Lot 120; a sextant signed 'Whyte & Co. Glasgow' was offered by Sotheby's, 10 March 1987, Lot 109; and a ship's clock signed 'Whyte, Thomson & Co., Glasgow' was offered by Christie's Scotland, 3 June 1987, Lot 2.

17. [Stratten], *op. cit.* (5), 75.

18. *Ibid.*

19. G.R.O.(S.), Census of 1871 (Glasgow, Blythswood), 644[6]/45, p.5.

20. *Glasgow Post Office Directory* 1875-1889.

21. John Graham was born at Jordanhill about 1839 (G.R.O.(S.), Census of 1881 (Glasgow, Kinning Park), 644[14]/2, p.1); he was at Whyte & Co. from before 1875 to 1893 (*Glasgow Post Office Directory*, *passim*).

22. David Young worked at Whyte & Co. in 1880 but was also listed in the directory at his address independently at 63 Clyde Place, Glasgow, and in South Shields (*ibid.*).

23. Both John R. Inglis and W.D. Whyte were with the firm in 1890 (*ibid.* 1890). For W.D. Whyte see reference 34 below.

24. James Wilson was employed by the firm from 1902 or earlier (*ibid.* 1902).

25. *Ibid.* 1885.

26. At the Edinburgh exhibition, the firm showed 'Nautical instruments various, for ships' use. Binnacle Stands and Compasses for yachts etc. Lifeboat Binnacles and Spirit Compasses. Palinurius and Polaris for detecting deviation of compasses by aid of the sun. Clinometer, Salinometer, Sextants, Chronometer Clocks, Telescopes, Thermometers, etc.' (*International Exhibition of Industry, Science and Art, Edinburgh, 1886. The Official Catalogue* 4th edition (Edinburgh, 1886), 186). Their display at Glasgow in 1888 included compasses, binnacles, sextants, chronometers, logs, barometers, azimuth dials, night binoculars, telescopes, steam gauges and timepieces (*Glasgow International Exhibition 1888: Official Catalogue* (Glasgow, 1888), 95). They exhibited a similar range of nautical instruments at the 1911 Glasgow Exhibition (*The Scottish Exhibition ... Official Catalogue. (Industrial Section)* (Glasgow, 1911), 177).

27. 'The firm also fits out the Cunard Line, Donald Currie and Co., British India, Orient, Royal Mail Steam Packet Co., Union Steamship Co., of New Zealand, the Donaldson Line, etc., etc.' ([Stratten], *op. cit.* (5), 75).

28. *Ibid.* Despite the description, this instrument does not appear to have been given full patent protection.

29. For example, *Glasgow Post Office Directory* 1890.

30. The other was D. McGregor (q.v.): Glasgow University Archives, Kelvin Collection, D 33/4/1-4, 'Sir William Thomson's Patent Compass Book, 1876-1883'; ditto, 1883-1893; 'Lord Kelvin's Patent Compass Book, 1893-1904'; ditto, 1904-1918. See Chapter 21, 'James White and Lord Kelvin'.

31. 'Caution' accompanying preliminary advertisements in Kelvin & James White Ltd., *Standard Tide Tables ... for 1903*, published by Kelvin and James White Ltd. (Glasgow, 1902).

32. *Glasgow Post Office Directory* 1900, 1901, 1910.

33. 'Kelvin v. Whyte, Thomson & Co.', *Reports of Patent, Design, and Trade Mark Cases* 25 (1908), 178. For the entire case of Kelvin v. Whyte, Thomson & Co., see *ibid.*, 177-194. In the following year, the respondents unsuccessfully appealed against the costs of the case (*ibid.* 26 (1909), 734-736).

34. G.R.O.(S.), Census of 1871 (Glasgow, Blythswood), 644[6]/45, p.5. James Whyte's family included William aged eleven.

35. U.K. patents, 17,699, 16 October 1891; 19,220, 10 October 1894; 24,917, 30 December 1896; 19,605, 15 September 1898; 7281, 9 April 1901; 5346, 7 March 1903; 11,807, 24 May 1904; 4885, 28 February 1906; 5709, 9 March 1906; 326, 5 February 1907; 7666, 2 April 1907; 5068, 6 March 1908; 20,537, 8 September 1909; 25,936, 10 November 1909; 7718, 28 March 1911; 8955, 16 April 1912; 25,028, 1 November 1912; 27,058, 25 November 1912; 6112, 12 March 1913; 18,378, 7 August 1914. The joint patents were with James Thomson, D. Muirhead and James Ferguson.

159. Refracting telescope: D. Heron, Greenock, c.1825 (T1980.244)

¾″ aperture 6-draw hand-held refracting telescope in brass with triplet objective, and 2-component eyepiece and erecting assemblies (mounted in the first and second draw tubes respectively). Engraved on the first draw tube 'D Heron / Greenock'. Scratched assembly mark 'II' on the draw tubes (but 'III' on the first) and erecting assembly, 'I' on the outer tube and forward ferrule. Objective cover, and closure on external eye stop. Two-part leather-covered card case.

Length closed: 90mm.
Tube diameter: 25mm.
Aperture: 21mm.
Case size: 100 x 35mm. diameter.

160. Marine compass: Whyte, Thomson & Co., Glasgow, c.1910 (T1980.198)

5½″ dry card compass, with jewelled bearing and broad needle rivetted under a covered mica card, stamped 'N' at one end and with adjusting weight inserted under lower paper cover. Printed 64-point compass rose divided at the circumference in degrees 0-[360] with 'WHYTE, THOMSON & Cº / GLASGOW.' around the centre. Suspended on a brass point in a glazed and painted bowl with lubber line, lead-weighted and gimbal mounted in a wooden case with sliding lid.

Card diameter: 134mm.
Bowl diameter: 147mm.
Case size: 205 x 205 x 160mm.

The gimbal ring is attached to the case by modern screws.

161. Refracting telescope: Whyte, Thomson & Co., Glasgow, c.1900 (T1980.279)

2¾″ aperture 2-draw mounted refracting telescope in brass; with doublet objective, and rack and pinion focus to the second draw tube. Engraved on the barrel end plate 'WHYTE, THOMSON & Cº / GLASGOW.' Objective cover, and closure on external eye stop. Pillar-mounted by a knuckle-joint with free azimuth motion, and on a folding brass table tripod.

Length closed: 978mm.
Tube diameter: 77mm.
Aperture: 70mm.

The barrel has a bad dent at its mid-point.

Detailed similarities in the design and in the mounting of optical components suggest that this item is by the same manufacturer as items 57 (Lennie), 66 (Lowdon), 85 (Gardner & Co.) and 93 (J. Brown).

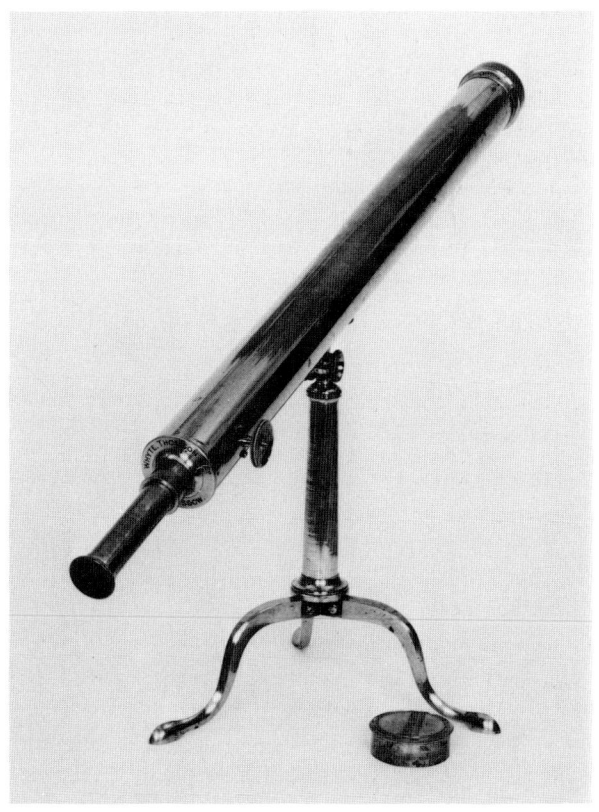

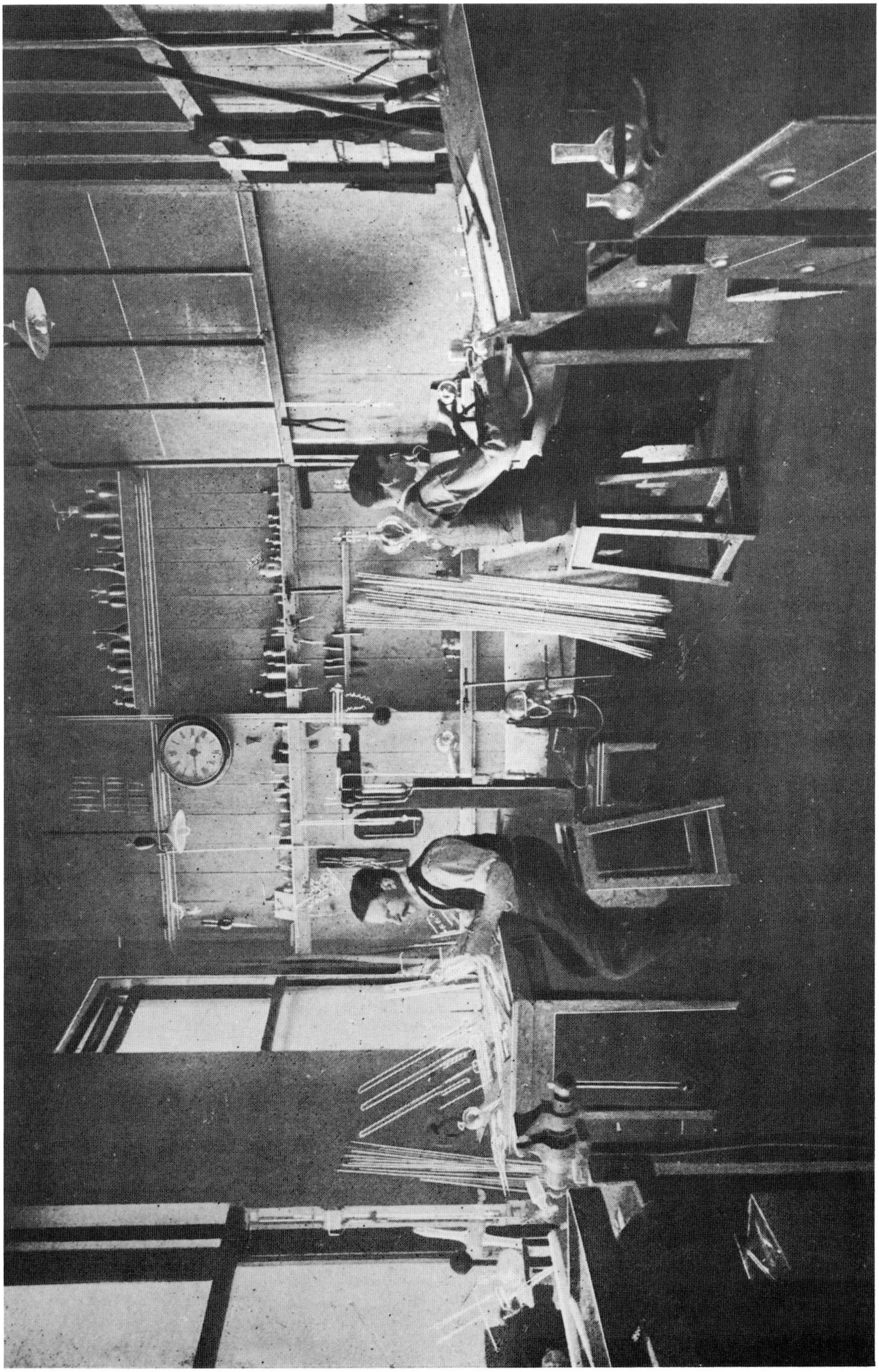

23

BAIRD & TATLOCK

Baird & Tatlock, which first appeared in the Glasgow street directory in 1881 as laboratory furnishers and chemical, mathematical and philosophical instrument makers at 100 Sauchiehall Street, was founded by Hugh Harper Baird and John Tatlock.[1] Baird (d.1911) was an agent for the life assurance National Provident Institution in Glasgow from 1877 or before, but was listed as 'of Baird & Tatlock' in the 1881 Directory. John Tatlock (c.1842-1909) was the brother of Robert Tatlock (1837-1934), an analytical and consulting chemist, and public analyst for the city of Glasgow.[2] John's business address from 1876 to 1881 was the same as Robert's, which suggests that he was working for him at that period.[3]

John Tatlock's previous experience had been with a firm of textile finishers in Glasgow from the age of eleven, and it was probably as a result of his brother's recommendation in 1862 or shortly after, that he became private assistant to Professor William Thomson, later Lord Kelvin (q.v.). Although Robert Tatlock initially acknowledged his brother's unfamiliarity with the use of philosophical instruments in natural philosophy experiments, he pointed out that John had 'a very fair education' and was willing and anxious to learn.[4] After twelve years as Thomson's assistant John Tatlock was well equipped to take advantage of the 'complete revolution' then taking place in school science teaching, and the greatly increased business of equipping and supplying school laboratories.[5] His experience shaped the specialisation of the firm in laboratory furnishings and the provision of educational scientific equipment, particularly chemical apparatus;[6] for instance, at the 1886 International Exhibition in Edinburgh the firm exhibited chemical balances.[7] At his death in 1909 Tatlock was described as an 'Analytical Chemist, Scientific Instrument Maker & Laboratory Furnisher'.[8]

By this date the business had undergone massive expansion from its small beginnings. It manufactured for both the Admiralty and the War Office, and for public laboratories; it advertised its appointments with the General Post Office, and as contractors to the India Office and Crown Agents for the Colonies; and its premises in Renfrew Street occupied five storeys.[9] The firm had moved to 40 Renfrew Street in 1889, and established a branch in Edinburgh in 1897 at 10 Drummond Street. In 1898 this branch moved to 2 Teviot Place, and in Glasgow 50 Renfrew Street was acquired in addition to No. 40. Again, in 1901, the firm moved to 45 Renfrew Street, and the Edinburgh branch into 7 Teviot Place.[10]

Meanwhile in 1889 Hugh Harper Baird had moved to London with 'some of the key men from Glasgow' where he opened offices in Cross Street under the name of Baird & Tatlock, London and Glasgow, and a few years later started up a London factory.[11] The company issued a catalogue in 1891, which was fairly slender,[12] a more substantial volume appearing in 1894 offering 1500 items.[13]

A major change in the firm's structure came in 1896, when

> 'the two partners resolved, amicably, to dissolve their partnership ... on the footing of Mr Baird acquiring as his own concern the London branch and Mr Tatlock acquiring as his own concern the original Glasgow house ... The reason for this provision would seem to have been mainly that the firm had put out a considerable amount of expense on catalogues and other business "literature" under the name of Baird & Tatlock, and that it was desired to use these for the common benefit of the new independent businesses.'[14]

A mid-twentieth century account of the firm states that:

> 'shortly after the turn of the century the partners separated, and in 1903 the London firm was turned into a private company ... Baird & Tatlock (London) Ltd. The range of instruments produced now increased even more rapidly than before, and a flourishing business grew up in designing and equipping complete laboratories, benches, fume cupboards and exhaust systems, all being manufactured in the B.T.L. factory, which by then had been transferred from London to new premises at Walthamstow.'[15]

Hugh Harper Baird died in 1911 and was succeeded as Chairman by Douglas Heriot Baird,[16] under whom the firm flourished, and it continues to do so today.[17]

From 1896 John Tatlock was the sole partner of the original Glasgow firm. He was a member of the

63   Baird & Tatlock's glass-blowing shop, c.1904-1911

Glasgow Philosophical Society from 1875[18] until his death in 1909, at the age of sixty-seven.[19] His son, J. Douglas Tatlock, who probably joined the firm before his father's death, continued it afterwards. Some indication of the elder Tatlock's prosperity is given by the value of his estate at his death at £18,627 15s 4d.[20]

The Edinburgh branch of the Glasgow business was founded in 1897, another in Liverpool was established in 1904, and one in Manchester in 1911. In 1915 the firm became a limited company, and in registering the name of 'Baird & Tatlock Ltd.' incurred the wrath of the London company, which took them to court because of the alleged confusion between the two firms caused by their similar names. The London business was unsuccessful in its attempt to oblige the Glasgow concern to differentiate its name further.[21] However, in 1925 the Glasgow company merged with John J. Griffin & Sons Ltd., a London firm with Glaswegian origins involved in the supply of chemical apparatus and chemicals, trading as Griffin & Tatlock from 1929.[22] In 1954 Griffin & Tatlock Ltd. merged with W. & J. George & Becker Ltd. and Standley Belcher & Mason Ltd. to form Griffin & George Ltd.,[23] under which name the company still trades.

REFERENCES

1. *Glasgow Post Office Directory*, passim.

2. On Robert Tatlock, see *Proceedings of the Royal Society of Edinburgh* 55 (1934-35), 172.

3. *Glasgow Post Office Directory* 1876-1881.

4. Glasgow University Library, Kelvin Papers, T.106. Robert Tatlock to William Thomson, 17 May 1862.

5. *Glasgow of Today* (Glasgow, 1909), 88.

6. Glasgow and West of Scotland Technical College, Minute Books, 1905, pp. 58-60, 20 June 1905. Amongst contractors fitting up the new laboratories was Baird & Tatlock, paid £30 13s 11d for 'chemicals and apparatus'. We are grateful to Dr R.H. Nuttall for this reference.

7. *International Exhibition of Industry, Science & Art, Edinburgh, 1886. The Official Catalogue* (Edinburgh, 1886), 199: they displayed 'Becker's Chemical Analytical Balances and Weights'.

8. Scottish Record Office, *Calendar of Confirmations* 1909, p.638.

9. *Op. cit.* (5), 88. See also the title page of their trade catalogue, Baird & Tatlock, *Physical Apparatus ... List no. 31* (n.p., n.d. [Glasgow, 1911-1916]).

10. *Glasgow Post Office Directory* 1889-1901.

11. Anon., 'The History of British Scientific Instrument Manufacturers: Sixty Years of Laboratory Furnishing', *Instrument Practice* 2 (1947-48), 489. We are grateful to John Burnett for this reference. This new branch was also noticed in the *Glasgow Post Office Directory* 1891.

12. *Instrument Practice*, op. cit. (11), 489. Two copies of the Glasgow firm's *Price List ...* (Glasgow, 1885) are located at the Science Museum, London; the earliest surviving version of the London business's *Illustrated Catalogue...* (London, 1897) is in the Baker Library, Harvard, Massachusetts. So far, no copies of either the 1891 nor the 1894 publication have come to light. We are grateful to John Burnett for these references.

13. *Instrument Practice*, op. cit. (11), 489.

14. 'Baird & Tatlock (London) Ltd. v. Baird & Tatlock Ltd.', *Reports of Patent, Design and Trade Mark Cases* 34 (1917), 86.

15. *Instrument Practice*, op. cit., (11), 489. The Royal Museum of Scotland has a Baird & Tatlock (London) laboratory bench and fume cupboard (NMS T1985.31 and 32) as illustrated in their *Standard Catalogue of Scientific Apparatus 1922 Vol. II Physiology* (London, 1922), 4, 12, which once belonged to Nobel prizewinner Sir William Ramsay (1852-1916), famous for his discovery and isolation of the family of inert gases of the atmosphere. Ramsay worked with Robert Tatlock for 18 months from October 1869, in the City Analyst's Laboratory (M.W. Travers, *A Life of Sir William Ramsay K.C.B., F.R.S.* (London, 1956), 8-10).

16. *Instrument Practice*, op. cit. (11), 489.

17. Accounts of the firm's activities are given in John Langdon-Davies, *Measuring the Future: Scientific Progress Illustrated by the Wartime experiences of Messrs Baird & Tatlock (London) Ltd.* (n.p., n.d. [London, 1948]) and in their centenary booklet, *Baird & Tatlock: Suppliers of Laboratory Products for 100 Years* (Romford, Essex, 1981).

18. Tatlock was elected a member 1 December 1875 (*Proceedings of the Glasgow Philosophical Society*, 10 (1875-77), 190 and *passim*).

19. General Register Office (Scotland), Register of Deaths 1909 (Dunoon, Argyll), 510¹/86. The cause of death was given as chronic laryngal catarrh.

20. *Op. cit.* (8), p. 638.

21. *Op. cit.* (14), 85-94.

22. *Glasgow Post Office Directory* 1904-1929. Griffin & Tatlock's *Chemical Apparatus ... Catalogue No 12A* (Glasgow, 1929) has a title page which reads: 'Griffin & Tatlock Ltd. combining the former business of John J. Griffin & Sons Ltd. and Baird & Tatlock Ltd.' and lists branches in London, Glasgow, Manchester, Edinburgh and Liverpool. A later edition of *Chemical Apparatus ... Catalogue 15B* (n.p., n.d.[?1939]) has a short introduction on the history of the firm: 'The House of Griffin was first established in Glasgow in 1826, when John Joseph Griffin, F.C.S., a writer on technical subjects, received requests from his readers for scientific apparatus ... The business has removed to London and conducted there by the sons of its founder at various addresses under the name of John J. Griffin & Sons (incorporated as a private limited company in 1889) ... In 1925 old-established businesses in Glasgow, Manchester, Edinburgh and Liverpool were brought into the organisation and since 1928 the united businesses, with branches and agencies in many countries abroad have been conducted solely under the name of Griffin & Tatlock Limited.'

23. Discussed in the introduction to *Chemical Laboratory Apparatus Catalogue 56S The Griffin & George Group* (Glasgow, 1956), v. Further details of the firm's history and later mergers are to be found in 'Griffin & George: A Short History', prefaced to Griffin & George Ltd., *A Catalogue of Modern Laboratory Equipment* (London, 1962).

---

162. Polariscope: Baird & Tatlock, Glasgow & Edinburgh, c.1900 (T1980.218)

Norremberg polariscope in brass with black-painted details. Engraved on the circular base 'BAIRD & TATLOCK / GLASGOW & EDINBURGH'. On the square-section pillar are clamped a reflection polariser of 9 thin glass sheets, a convergent condenser rotating in its mount, an inverted convergent condenser acting as an objective, an auxiliary lens in a sliding mount attached to the objective, and a mount for an analyser.

Overall height of pillar: 384mm.
Base diameter: 138mm.

Lacking the mirror which hinged to the lower edge of the polariser, the analyser containing a nicol prism, and also the means of attaching the specimen to the upper surface of the condenser. The pressure pad in the clamp for the analyser mount is lacking, as is one screw in the condenser mount.

Norremberg's polariscope was exhibited in Karlsruhe in 1858, and was described in A. Bertin, 'Note sur le microscope polarisant de Norremberg', *Annales de Chimie et de Physique* 3rd series, 69 (1863), 87-96. This example has been published by R.H. Nuttall in 'A Note on the Albert Polariscope', *Microscopy* 32 (1971), 64-66, in which he identified it as the conoscope of J.W. Albert of Frankfurt, based on the illustration and description of Albert's instrument in P. Harting, *Das Microscope* III (Braunschweig, 1866), 332-333. The instrument retailed by Baird & Tatlock is of a later date and differs principally in that it is designed to take individual specimens rather than a rotating disc of specimens, and that the Harting illustration shows the multi-layer polariser hinged below a dark-glass mirror. A later version of Norremberg's polariser was offered in Baird & Tatlock's *Scientific Apparatus List No. 31* (n.p., n.d. [Glasgow?, c.1910?]), 219, no. 2052, and the same instrument was still being offered by F.E. Becker of London in 1931.

24
RETAILING AND MANUFACTURING INSTRUMENT MAKERS

The seven instrument retailing and manufacturing firms grouped here represent the broad development of the Glasgow instrument trade from the mid-nineteenth century into the early twentieth century. As discussed above, by about 1840, barometer manufacture was largely in the hands of wholesale manufacturers in England,[1] and the same came to be true of other types of quasi-scientific instruments and of scientific apparatus proper.[2] However, the extent to which this domination of the market impinged on Scotland has not yet been determined, and in the absence of business records, only partly-informed guesses can be made.

Judging from the types of instruments advertised by the firms in this group both in actual advertisements and in entries in the local street directories, most of their stock was bought from wholesale makers. It should be noted, however, that both Abraham & Co. and Norton & Gregory exemplify the retail outlets of wholesale manufacturers based in England, and because they were essentially branches they were therefore not subject to the same conditions of supply as the others in this group, which were all native enterprises. These had to compete for a share of the market with outlets like Abraham & Co. The latter's relatively short period of activity may have been due to increasing competition from Scottish firms, but such a speculation underlines the fact that the size and the detailed structure of the instrument market, and its Anglo-Scottish aspect in particular, have not been examined. That Scottish retailers had a market for their goods is demonstrated by the firms in this group, and others besides, but the role which Scottish wholesalers played in competition with English firms is not yet understood. Some makers, such as Alexander Dobbie & Son (q.v.), D. McGregor & Co. (q.v.) and James White (q.v.) (all of whom produced timepieces for wholesale) have been discussed, but it is the retailers who appear most clearly in the surviving sources.

In themselves they are a diverse enough group, with persistent evidence of the historically close links between instrument making and light mechanical engineering in the cases of W.A.C. Smith and Archibald Baird, whose activities echoed those of Andrew Barclay (q.v.) and Thomas Morton (q.v.) earlier in the century. Most of the group catered for the current vogue in photography from the middle of the century onwards, and some appear to have specialised in spectacle-making and fitting, which by degrees supplanted their original function as opticians, thereby providing a transition to the modern usage of optician as ophthalmic optician.

Something of the extent to which their trade was sustained by the supply of wholesale goods from both England and the Continent can be seen in the case of microscopes. Instruments to fulfil both amateur curiosity and professional demands were supplied from London, Paris, Vienna and other production centres, of a good enough quality and at a low enough price to stifle any Scottish enterprise. Only one, or perhaps two, Scottish makers were actively making microscopes by the 1880s, and it is not known if Lizars made his instruments *in toto* or from bought-in parts; and no examples signed by Edwards (q.v.) have so far been recorded. Comparisons with the position of Edinburgh firms are possible in this case, for instance with that of Lennie (q.v.) and Bryson (q.v.), whose importation of optical goods is discussed above. The seven firms described here thus represent not only a sizeable cross-section of Glasgow firms, but also reflect the wider position of their type in Scotland as a whole.

REFERENCES

1. Nicholas Goodison, *English Barometers 1680-1860* 2nd edition (Woodbridge, Suffolk, 1977), 83-5; Anita McConnell, *Barometers* (Aylesbury, 1988), 26.

2. Some indication of the general trends and developments in the production of instruments in Britain, together with the difficulties in assessing the available information, is given in W.D. Hackmann, 'The Nineteenth-Century Trade in Natural Philosophy Instruments in Britain', in P.R. de Clercq (ed.), *Nineteenth-Century Scientific Instruments and their Makers* (Leiden, 1985), 53-91.

JOHN SPENCER

John Spencer first appeared in the Glasgow street directory in 1835 as a hardware merchant at 16 Saltmarket Street. He may have been related to a jeweller, J.L. Spencer who was working at 136 Trongate and 16 Saltmarket in 1828 and to William Spencer, a hardware merchant at 11 Trongate in 1828 and 56 Trongate in 1834. Around 1856, in addition to hardware, Spencer began to deal in photographic materials, advertising as a dealer in 1856, and the following year calling his business a 'wholesale photographic and stereoscopic depot'. From 1859 with new premises at 30 St. Enoch Square, he practised as an optician, mathematical instrument maker and photographic apparatus maker. Like other photographic specialists he also dealt in chemicals and chemical apparatus. From 1865 until 1869, when he ceased to appear in the street directory, Spencer worked at 39 Union Street as an optical, chemical and photographic apparatus manufacturer.[1]

Spencer was succeeded in the business by his assistant or partner George Mason, and, as George Mason & Co., the firm prospered.[2]

REFERENCES

1. *Glasgow Post Office Directory* 1835-1869, *passim*. He was also described as an 'optician, photographic and philosophical apparatus manufacturer' by J. Steinhardt (ed.), *The Illustrated Guide to the Manufacturers, Engineers, and Merchants of England, Scotland, Ireland and Wales* (London, 1869), 275. However, as yet only 2 pieces, other than the telescope in the Frank collection (NMS T1980.264) have been recorded which may have come from Spencer's workshop: a microscope, signed 'F. Spencer, Glasgow' (*sic*), was offered for sale by Phillips, 18 February 1987, Lot 113; and a 2-draw 3 inch refracting telescope with tripod, signed 'Spencer, Glasgow', was offered by Christie's Scotland, 8 June 1988, Lot 87.

2. [Stratten's] *Glasgow and Its Environs* (Glasgow, 1891), 94. 'A bulky catalogue from George Mason and Co. has just been received. This old established firm will shortly change their address from Sauchiehall Street to Buchanan Street. We hope the change may lead to profitable developments of the business. The catalogue is one especially interesting to professionals.' (*Practical Photographer* 9 (1898), 185: we are grateful to Sara Stevenson for this reference).

163. Refracting telescope: J. Spencer, Glasgow, c.1855 (T1980.256)

1⁵⁄₈" aperture hand-held refracting telescope with doublet objective and variable power eyepiece comprising 2-component eyepiece and erecting assemblies of variable separation calibrated from 25 to 35 times; in oxidised brass (with leather-lined draw slides), and with leather-covered barrel. Engraved on the second draw tube 'J. Spencer,/ 39. Union Street,/ Glasgow.' Scratched assembly mark 'X' on draw tubes, draw slides and erecting assembly. Closure on external eye stop.

Length closed: 260mm.
Tube diameter: 48mm.
Aperture: 40mm.

Marks on the barrel indicate where loops have been sewn for a carrying strap which is now missing, as are the leather end caps.

Probably by James Parkes & Son of Birmingham, cf. *Wholesale Catalogue of Optical, Mathematical and Philosophical Instruments manufactured by James Parkes & Son* (Birmingham, n.d.[1867]), 8. 'Tourist telescopes, black leather sling body, superior and extra quality', nos. 1050-54: by 1867 the particular combination of 1⁵⁄₈" objective, 3-draw plus pancratic eyepiece and without shade was not available.

W. A. C. SMITH

William A. C. Smith first appeared in the 1890 Glasgow street directory as a mechanical engineer, probably employed in a firm rather than working independently. That year he displayed a monocular microscope in the Artisan Section of the Glasgow East End Exhibition, for which he won a bronze medal.[1] From 1894 Smith was described as an optician and a photographic and scientific instrument maker at 53 Dundas Street. Subsequent entries indicate a broadening business in the 1890s, including electrical instrument manufacture, and optical work, and the expansion of the premises in 1898 to include 51 Dundas Street confirms this.[2]

Smith advertised in 1897 as an 'Optical Lantern & Camera Manufacturer ... All kinds of Apparatus skillfully repaired or made to order,'[3] and three years later was offering 'A HIGH-GRADE BRITISH MADE CAMERA ... THE INVINCIBLE MAGAZINE CAMERA ... Price 42s [and] THE R.R. INVINCIBLE MAGAZINE CAMERA ... Price £4 10s.'[4] In 1906 Smith moved to 236 Argyle Street, retaining the 53 Dundas Street premises until 1909. He had disappeared from the directory by 1926.[5]

64 Bronze medal awarded to W. A. C. Smith, for making a microscope, at the Glasgow East End Exhibition 1890-91

REFERENCES

1. *Glasgow East End Industrial Exhibition of Manufactures, Science and Art Official Catalogue* (n.p., n.d.[Glasgow, 1890]), 55. Both microscope - with Watson objectives and eyepieces - and medal are in the Royal Museum of Scotland, NMS T1984.160. The instrument is signed on the circular base in Gothic script 'W. A. C. Smith', and is also signed on the stage. The medal, inscribed 'GLASGOW 1890-91. EAST END EXHIBITION Wm. A. C. Smith' is signed 'R. SCOTT', a local diesinker and seal engraver (C. Forrer, *Biographical Dictionary of Medalists* V (London, 1912), 445).

2. *Glasgow Post Office Directory, passim*.

3. *British Journal of Photography Almanac 1897* (London, 1897), 1175. We are grateful to John Ward of the Science Museum, London, for this reference.

4. *British Journal of Photography Almanac 1900* (London, 1900), 414. There is an example of one of these cameras, marked with Smith's name, in the Science Museum, London, inventory 1979-559/29, from the Frank collection. We are grateful to John Ward of the Science Museum for this reference and information about the camera.

5. *Glasgow Post Office Directory, passim*.

---

164. Microscope: W. A. C. Smith, Glasgow, c.1885 (T1980.238)

Compound microscope in brass, bronze and oxidised brass, on an inclining pillar and flat 2-toed tripod base. Engraved on the base 'W. A. C. Smith / GLASGOW'; the components (including the base) are stamped with the assembly mark '6' or '9'. Single-draw body tube, with rack and pinion coarse focus adjustment to the pillar; fine focus is by a spring-loaded sliding inner triangular-section bar, attached through the stage to the stand, and operated by a screw at the top of the pillar, graduated [0]-20. The rectangular stage plate with stepped aperture. Compass joint, and plane/concave mirror on a swinging tail beneath the stage. Double nosepiece (with contemporary Leitz objective).

Body length (less eyepiece and nosepiece): 151mm.
Body diameter: 31mm.
Height (less eyepiece): 277mm.
Base size: 144 x 151mm.

Lacking objectives and eyepiece. The fine focus screw arbor is bent. Also lacking the sub-stage mirror and stage spring clips. The support for the swinging tail for the mirror is bent, and the lower end piece of the coarse focus rack is detached.

The instrument bears a resemblance to item 167 (signed J. Lizars) and the two may have a common source.

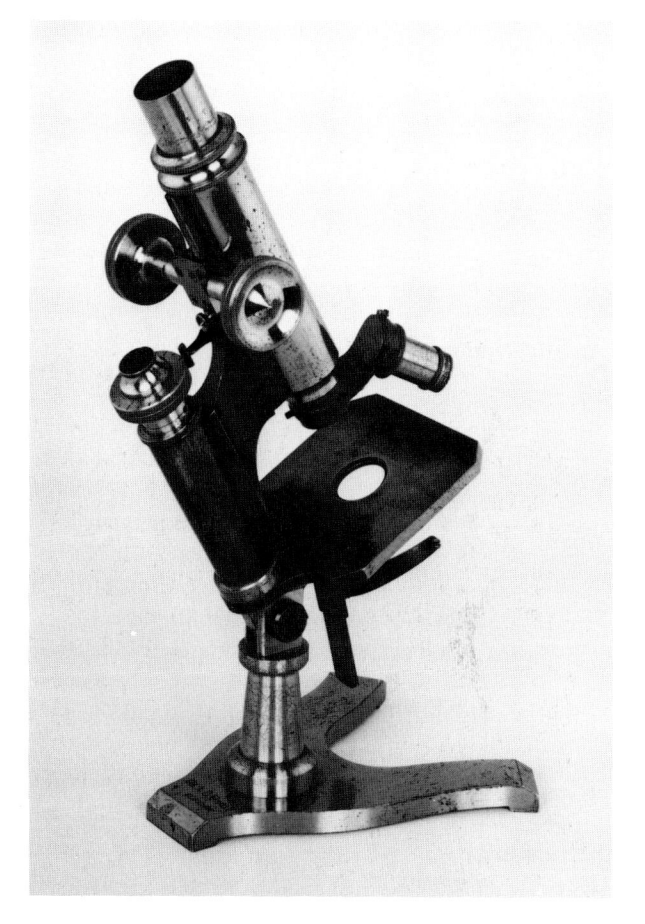

BAIRD & SON

Archibald Baird & Son was founded in 1870 as an engineers' ironmonger, at 17 and 19 Robertson Street, Glasgow. In 1872 it moved to 67 Robertson Street, expanding in 1880 to take in 46 Robertson Lane. Further expansion occurred in 1887 with the establishment of its workshop at the Clyde Steel and Engineering Works, Hamilton, and in 1888 showrooms at 105 Hope Street, Glasgow, were opened, and a warehouse at Peacock Cross, Hamilton, was operated until 1890. Between 1889 and 1893 the Glasgow shop was at 59 and 61 Waterloo Street, with another branch at 57 West Campbell Street from 1893 to 1898, and another at 41 Robertson Street from 1903 to 1905. After 1906 the firm operated solely from its Hamilton works until it ceased business in 1972.[1]

An article published in 1901 explained that

> 'Mr Archibald Baird, the founder of the firm, had a long experience both in coal mining and iron works. Entering the employment of Messrs. Colin Dunlop & Co., as a clerk at Clyde Iron Works, he rose step by step, and was finally appointed manager of Quarter Colliery. In 1870, Mr Baird relinquished his position as manager to start in business, along with his son, Mr Matthew B. Baird. The business was at first merely designed to supply the ironmongery required by collieries; but the practical knowledge of Mr Baird, sen., and the energy of the younger member of the firm quickly developed the business far beyond its original province.'[2]

By 1890, their range of stock manufactured in the works included 'all types of colliery, mining, railway, quarry and ironworks plant', as well as several types of fireproof cloths, winding and haulage ropes, and lighting equipment.[3]

A commercial account of 1888 speaks of 'the scope and magnitude of their gigantic business', which appears to have touched most aspects of colliery and factory work.[4] Although not involved in precision instrumentation and light engineering in quite the same direction as Dobbie McInnes Ltd. (q.v.), Baird & Son did produce instruments of comparable type. Among patents held by father and son individually, together or with others and manufactured by the firm, were Archibald Baird and G. McPherson's patent axle lubricator,[5] A. Baird's sight feed lubricator,[6] M.B. Baird and J.T. Pitcairn's patent greasing machine,[7] and most interesting in this context, Adams and Baird's self-recording anemometer for registering the variations of ventilating fans. This last item appears to have been of the type patented in 1879 by A. Baird and J. Adams,[8] which incorporated a recording drum, and is different from the later anemometer in this collection. M.B. Baird also took out a patent in 1889, for ventilating mines and tunnels.[9] He was elected a member of the Iron and Steel Institute, and also belonged to the Scottish Mining Institution.[10]

Baird & Son was evidently prospering from the early 1880s, as a result of the technical expertise and business flair of the two co-partners, Matthew B. Baird and Archibald Baird. In 1880 the firm's demand for castings became so large that they built their own foundry, and in 1887 they took over the Caledonian Crucible Steel Company, based in Glasgow, together with a number of other small businesses in the next few years.[11] In order to consolidate this diffuse empire, they moved their works out of Glasgow to neighbouring Hamilton, where they built premises at Peacock Cross, close to the railway station:

> 'The ground floor is a warehouse stored with all kinds of mining implements, from the miners' safety lamp to the colliery engine ... On the first floor are the counting-house, drawing office, chemical laboratory, equipped with scientific instruments, and partners' private room. Behind the warehouse is the timekeeper's office and workers' entrance, leading to an open courtyard, round which are ranged the works. To the right is the foundry, where ... locomotives ... by deft repairing are made fit for shunting operations, colliery sidings, works' railways, and even for heavy goods traffic ... To fulfil their function as colliery outfitters, the Messrs. Baird began the manufacture of fire-proof and tarred brattice-cloths, and have built up a good business for those requisites ... They are making the solid moulded gutta-percha pump buckets specialised by the firm ... It is a coterie of many industries, each complete in itself, and employs over 150 men.'[12]

Archibald Baird & Son exhibited at a number of industrial exhibitions. At the Mining Exhibition held in Glasgow 1885, they were awarded a silver medal, the following year in Edinburgh they received a gold medal, and in 1888 won a diploma from the Glasgow

Exhibition.[13] Doubtless the lucrative export trade in engineering products stimulated the improvement of particular types of equipment to meet competition, and helped the firm prosper. Clearly the Bairds were not in doubt about their business when their telegraphic address was 'Excello' Glasgow.[14]

REFERENCES

1. *Glasgow Post Office Directory* 1870-1972, *passim*.

2. Anon., 'Captains of Industry: Messrs. Archd. Baird & Sons, Clyde Steel and Engineering Works, Hamilton', *Glasgow Weekly Herald*, 26 January 1901. The author of this article has been identified by the Mitchell Library, Glasgow, as W.S. Murphy.

3. *Glasgow Post Office Directory* 1890.

4. *Glasgow of Today* (Glasgow, 1888), 103.

5. U.K. patent 2802, 10 July 1876.

6. U.K. patent 5053, 18 March 1884.

7. U.K. patent 5778, 28 April 1886.

8. U.K. patent 2378, 16 June 1879. Six months provisional protection was noted on 11 July 1879 in the *Engineer* 48 (1879), 37, where title was given as 'Indicating and Recording the Pressure of Air etc.', by J. Adams, Crossgates, and A. Baird, Glasgow.

9. U.K. Patent 16,273, 1889.

10. [Murphy], *op. cit.* (2).

11. *Ibid.*

12. *Ibid.*

13. *Ibid.*; *International Exhibition of Industry, Science & Art, Edinburgh, 1886. The Official Catalogue* (Edinburgh, 1886), 39; *International Exhibition in Glasgow 1888: Official Catalogue* (Glasgow, 1888), 6, 122.

14. *Glasgow of Today* (Glasgow, 1888), 103.

---

165. Airmeter: Baird & Son, Glasgow, c.1900 (T1980.217)

6-dial airmeter in brass and oxidised brass. Stamped on the silvered dial plate 'BIRAMS PATENT / BAIRD & SON GLASGOW 116D'. Vertical fan with 6 blades mounted on a stand beneath the dial work, and connected to it through a clutch which engages the hands. The 6 dials record air movement in feet passing the instrument per minute, the first dial graduated in tens of feet 1-[10] and identified by 'X', the others in rising multiples of 10, identified as 'C', 'M', 'XM', 'CM', 'M[M]', reading up to 10^7ft.

Fan diameter: 51mm.
Height: 91mm.
Diameter: 65mm.

Lacking the protective housing and cover.

Benjamin Biram's U.K. patent 9249 of 1842 covered the design of windmill sails, screw propellers, rotary engines, and 'apparatus for registering the velocities of bodies propelled through wind or water'. This instrument can be identified with one of several versions of Biram's anemometer offered for sale by the London wholesaler J.J. Hicks in his *Illustrated & Descriptive Wholesale Catalogue of Standard, Self-Recording, and other Meteorological Instruments ...* (London, c.1880), 111-2 and fig. 191, no. 534, 'Biram's Anemometer, 2-in. diameter, new, reading to 10,000,000 ft. £3 3 0'. Biram's instrument was described by Hicks as 'an ingenious and trustworthy "tell-tale" instrument for registering currents of air in mines, &c., and thus showing whether the furnace man, who is responsible for due ventilation, is mindful of his duty. ... In large gun or rifle practice the pocket size of this instrument yields trustworthy results.' Other variations of Biram's anemometer were still being discussed in 1924 (e.g. T. Bryson, *Mine Ventilation* (London, 1924), 209-211, and an advertisement by Short & Mason in M.H. Haddock, *Mine Ventilation and Ventilators* (London, 1924), endpaper).

ABRAHAM & CO.

Abraham & Co. was a substantial Liverpool firm of manufacturing optical, mathematical and scientific instrument makers established in about 1817 by Abraham Abraham, who in the early 1840s was also in partnership with the prominent microscope maker John Benjamin Dancer in Manchester.[1] In 1838 Abraham & Co. opened a retailing branch at 8 Exchange Square, Glasgow, moving to 82 Queen Street in 1841. This branch was taken over by one of the partners Simeon Phineas Cohen in 1843, and that partnership was dissolved the following year.[2] He traded as S.P. Cohen from 1844 at 82 Queen Street, at 105 Buchanan Street between 1845 and 1848, at 121 Buchanan Street in 1850, at 51 St. Vincent Street in 1851, and at 136 Buchanan Street in 1852.[3] In November 1853 his accumulated debts to Abraham (totalling £1,960)

65 S. P. Cohen's advertisement as successor to A. Abraham & Co., 1844, *Mitchell Library, Glasgow*

LATE
A. ABRAHAM & CO. 82 Queen Street,
GLASGOW.

S. P. COHEN,
OPTICAL, MATHEMATICAL, & PHILOSOPHICAL
INSTRUMENT MAKER,
82 QUEEN STREET, GLASGOW,

led him to petition for sequestration, describing himself as

> 'Simeon Phineas Cohen Optician and General Agent for Birmingham, Wolverhampton and Sheffield Manufactures in Glasgow; [petitioning] with concurrence of Abraham Abraham Optician in Liverpool·a creditor of the said Simeon Phineas Cohen.'[4]

Cohen was discharged from his bankruptcy on 23 January 1854.

Abraham & Co. sold and repaired all types of instruments at the Glasgow branch. An advertisement of 1840 reveals the range of stock, which was wider than, but mostly comparable to, that sold by others in this group such as Lizars, and by the Edinburgh firms Dunn (q.v.) and Lennie (q.v.). In addition to a standard range of microscopes, telescopes, barometers and thermometers, Abraham's stocked as sole agents 'improved Electrum Saccharometers and Sikes's Hydrometers' and a newly-invented pocket barometer.[5]

Abraham himself had been described as an optician in the Liverpool directories from 1819 and as an optician and mathematical instrument maker from 1841.[6] The business was called Abraham & Co. from 1851, and eventually in 1875 the name changed to that of the surviving partner George S. Wood.[7] Abraham had previously been in partnership with another optician, John Benjamin Dancer (1812-1887), whose Liverpool business was at a separate address between 1839 and 1845. Dancer moved to Manchester in 1841 where his partnership with Abraham lasted four years until 1845, with Dancer subsequently continuing alone.[8] Abraham & Co. produced at least one trade catalogue in 1855,[9] and appear to have been well-known for the quality of their microscopes.[10] The firm was one of the few non-London instrument makers to exhibit at the Crystal Palace in 1851.[11]

An indication of Abraham's business acumen is revealed in a letter of 1868 by his partner George S. Wood of Liverpool, in which he described how his 'late partner and friend' A. Abraham was among the first opticians to sell cheap achromatic microscopes outside London; as early as 1841 he sold large numbers of an instrument complete in a case with two lenses made by Nachet of Paris, for only £8 per set, capitalising on the popularity of the new instrument and reaching new markets by forcing down the price.[12]

REFERENCES

1. The most recent article which discusses Abraham, and his partnership with J.B. Dancer, is that by Stella Butler, 'Microscopes in Manchester', *Microscopy* 35 (1987), 570-572.

2. 'The Business carried on by the Subscribers, Abraham Abraham, and Simeon Phineas Cohen, in Glasgow, as Opticians, under the Firm of A. ABRAHAM & COMPANY, was DISSOLVED on the first day of May last, 1844, by mutual consent ...' (*Edinburgh Gazette*, 6 September 1844, 298).

3. *Glasgow Post Office Directory* 1838-1852.

4. Scottish Record Office, CS 280/40/40/4, Sequestration Papers, Simeon Phineas Cohen; there is no Sederunt Book and so details on this bankruptcy are scant. The Inventory of the Estate is an abstract, rather than a detailed breakdown of Cohen's affairs: his stock-in-trade was valued at £1034 9s 2d; his workshop materials and instruments at £25; there is no explanation for his discharge. Notices appeared in the *Edinburgh Gazette*: for the initial notice (18 November 1853, 917); for the announcement of Trustee and Commissioners (6 December 1853, 984); and the bankrupt's offer of composition at the creditors' meeting, to be considered at the next one to be held 6 January 1854 (27 January 1853, 1040). As Cohen was discharged shortly afterwards, it can be assumed that his offer was accepted.

5. *Glasgow Post Office Directory* 1840, appendix, 138-139.

6. E.G.R. Taylor, *Mathematical Practitioners of Hanoverian England* (Cambridge, 1966), 380 gives Abraham's dates as fl.1817-1849.

7. Information from J. Ravest, National Museums on Merseyside, May 1978. The firm's Liverpool address was 9 Lord Street 1829-1834, 76 Lord Street 1835-1836, 78 Lord Street 1837-1839, 20 Lord Street West 1840-1875, with an additional branch at 15 London Road East 1870-1875. J.B. Dancer was at 21 Pleasant Street, Liverpool 1839-1845; and the partnership's Manchester address was 43 Cross Street.

8. A recent résumé of the career of John Benjamin Dancer, inventor of the microphotograph and maker of instruments for John Dalton and James Prescott Joule, is that by Michael Hallett, 'John Benjamin Dancer 1812-1887: a perspective', *History of Photography* 10 (1986), 237-255.

9. *Descriptive and Illustrated Catalogue of Optical, Mathematical and Philosophical Instruments Manufactured by A. Abraham & Co., 20 Lord Street* (Liverpool, 1855). There is a copy at the Greater Manchester Museum of Science and Industry.

10. John Quekett wrote that 'report speaks well of the stand of the achromatic microscope constructed by Mr Abrahams, of Liverpool, which very much resembles that of Mr Ross ... the stage employed in this microscope has either a rack movement or is one after the plan of the author in which two levers, capable of being removed, are used to give motion in two opposite directions. Mr Abrahams also supplies a lenticular achromatic prism, as a substitute for the mirror and condensor.' (John Quekett, *A Practical Treatise on the Use of the Microscope* 2nd edition (London, 1852), 101). He devoted a paragraph to Abraham's achromatic prism and condenser, described as 'this very important instrument, answering the purpose both of mirror and achromatic condensor, was presented to the author by Mr Abrahams, optician, of Liverpool ...' (*ibid.*, 119). Jabez Hogg listed Messrs. Abraham of Liverpool among 'well-known makers

of microscopes ... all of whom have obtained a deservedly high reputation for their convenient forms of educational and other well-manufactured instruments.' (J. Hogg, *The Microscope Its History, Construction and Application* 6th edition (London, 1867), 111); and Hogg also described 'Abraham's Lenticular Prism' (*ibid.*, 172). Examples of Abraham's microscopes in the Royal Museum of Scotland are a pre-achromatic instrument NMS T1979.45 (described as item 14 in R.H. Nuttall, *Microscopes from the Frank Collection 1800-1860* (Jersey, 1979), 32); an achromatic microscope NMS T1980.43 (illustrated in R.H. Nuttall, 'An Early Microscope by John Benjamin Dancer of Manchester', *The Microscope* 28 (1980), 94) and an improved compound instrument on an iron stand, NMS T1983.182.

11. *Catalogue of the Great Exhibition* (London, 1851), I, 436. They exhibited improved multiple projection lanterns and a camera obscura. These are included in Abraham's prospectus: *Exhibition of the Works of Industry of All Nations 1851: Prospectuses of Exhibitors* vol. 8 (Class X, Part I: Philosophical Instruments), in the Science Museum Library, London.

12. Letter printed in *Quarterly Journal of the Microscopical Society* 8 (1868), 108.

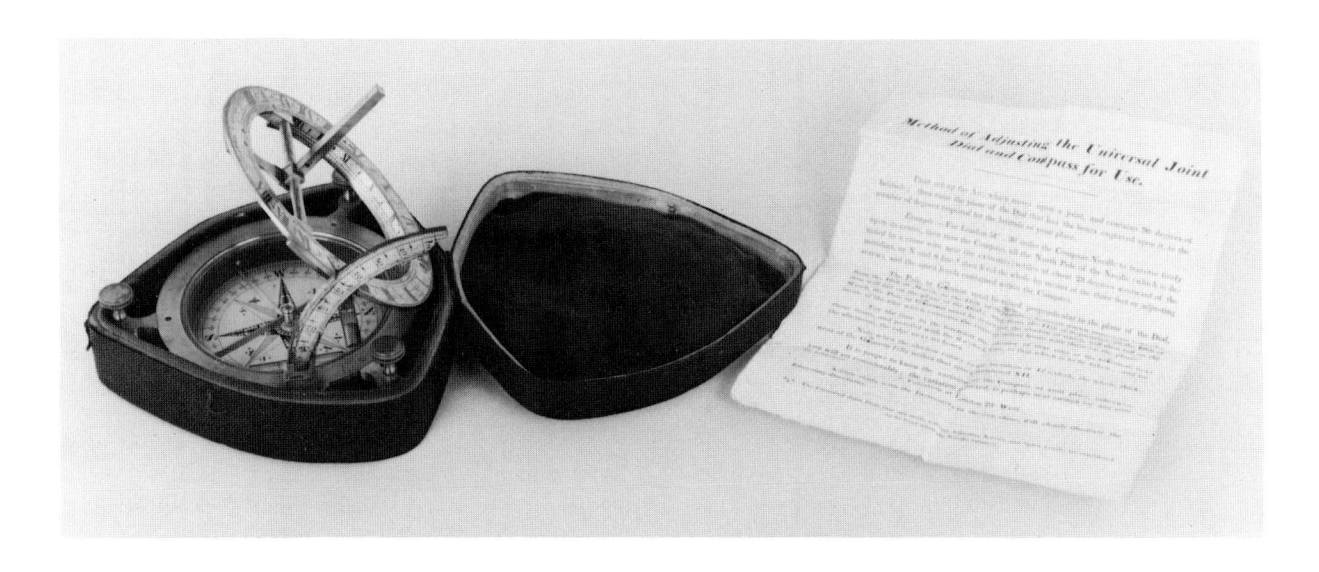

166. Portable sundial: Abraham & Co., Glasgow, c.1840 (T1980.167)

Folding universal equinoctial dial in brass, with silvered scales. Engraved on the chapter ring 'Abraham & Cº GLASGOW'. The spring-loaded stile, operating above or below the equator, is pivotted in a chapter ring, divided IIII-XII-VIII by 5 mins. on the upper and inner faces; hinged against an altitude arc, graduated in latitude 0-[85] by 1°, over a glazed compass dial incorporating crossed levels, and supported on 3 screw feet; the silvered dial with 8-point compass rose, divided [0]-[90]-[0]-[90]-[0] by 2°, with a needle on a relieved jewelled bearing. With a fitted velvet-lined shagreen-covered case, enclosing a printed sheet of instructions for use.

Diameter: 103mm.

The instruction sheet is headed 'Method of Adjusting the Universal Joint Dial and Compass for Use.': the spelling of gnomon is consistently given as 'guomon' and is corrected in MS.

JOHN LIZARS

John Lizars was born in Berwick-upon-Tweed in about 1810, the son of Isaac Lazarus or Lizars, and founded his optical business in 1830 at 12 Glassford Street, Glasgow, at about the age of twenty.[1] However, he was not entered in the street directory until 1858, when he appeared as an optical and mathematical instrument maker at 24 Glassford Street.[2] It is however possible that in the period before his Glasgow business became settled in the 1850s, that he moved around various population centres. He, or a relation, may be the I. Lazars, optician, of Edinburgh, who advertised his presence in 1841 in Cupar, Fife, and the 'Mr Lizar, optician' who did the same in Stirling in 1840 and again in 1853.[3]

Lizars remained at 24 Glassford Street, Glasgow, until 1875 when he moved briefly to 13 Wilson Street, before settling at 16 Glassford Street in 1877; he was based there until his removal in 1891 to 101-107 Buchanan Street. During the period 1888-1889 Lizars also traded from a second shop at 260 Sauchiehall Street.[4] Subsequently branches were opened in Edinburgh, Greenock, Paisley, Motherwell, Aberdeen, Liverpool, Belfast, Gourock and Strathaven.

Like other firms, too, the apparent broadening of the Lizars business, as recorded in entries in the street directories, continued steadily from the 1870s onwards. By 1880 for instance, the firm was listed as makers of chemical, mathematical, optical and philosophical instruments,[5] and by the 1880s it was making photographic apparatus and materials. By 1910 it is clear that business was concentrating on ophthalmic and photographic work, specialisations which have continued to this day.[6]

Unfortunately the Census entries recording John Lizars do not in his case record the number of his employees and, were it not for another piece of evidence, it would be tempting to dismiss Lizars as merely a retailer of scientific instruments and a maker of spectacles and the

66   John Lizars (1810-1879). *Messrs. Lizars Ltd.*

like, which he is known to have produced.[7] An account of Glasgow commerce in 1888 makes the interesting claim that 'Mr Lizars is the only manufacturer of microscopes in Scotland, and for this speciality has gained a widespread reputation'.[8] Allowing for the self-advertising nature of the entries in the book, and considering the output of other Scottish workshops at the time, this apparently grandiose claim should almost certainly be taken as accurate, and is a sharp indication of the domination of the Scottish microscope market by continental and English makers. The 1888 account also refers to the workshop at Glassford Street, where 'a number of hands are employed making and repairing optical instruments of every description'; to the warehouse; and to his trade which 'extends all over Scotland'. His 'beautifully illustrated catalogues' contained 'testimonials to his work from the most eminent scientific gentlemen'.[9]

John Lizars died in 1879, having survived both his first wife Ellen Solomon and his second wife Eliza Lyons.[10] His second daughter Rebecca is listed in the 1881 Census as 'Optician shopwoman',[11] and it is reasonable to suppose that she assisted the unidentified Mr Lizars who was running the firm in 1888, and whose 'thorough knowledge of the science of optics' and 'high principles of integrity' were adduced as the causes of his successful trade.[12] Matthew Ballantine, who married John Lizars' third and youngest daughter Juliet, entered the business in 1882. He was responsible for the expansion into photographic work, including the Liverpool branch, which was set up to sell off unshipped stock destined for an outlet in Cape Town, which fell through on the outbreak of the Boer War. The Edinburgh branch was formed by the acquisition of the optician Thomas Haddow's business. Since the 1880s Lizars has largely been in the hands of Ballantine and his descendants.[13]

REFERENCES

1. *A Century of Progress: J. Lizars, 1830-1930* (Glasgow, 1930) states that Lizars commenced business in 1830, but the commercial handbook *Glasgow of Today* (Glasgow, 1888), 170, gives 1835 as a start date. Nor are the conflicting ages given by Lizars in the 1861 Census, of 48, and in the 1871 Census, of 55, consistent with his age of 69 at his death in 1879, recorded in the Register of Deaths, although the latter age is probably the correct one (General Register Office (Scotland), Census of 1861 (Glasgow, Central), 644^1/28, p.3; Census of 1871 (Glasgow, Central), 644^1/10, p.14; Register of Deaths 1879 (Glasgow, Blythswood), 644^7/574).

2. *Glasgow Post Office Directory* 1858.

3. 'Until the Eleventh March I. LAZARS OPTICIAN, From 89, George St. Edinburgh, Respectfully begs to acquaint the Nobility, Gentry and Inhabitants of Cupar and its vicinity, that he has Opened that Shop in the Crossgate, next door but one to the George Inn, and, as his stay will be limited, he respectfully solicits an early call from those Ladies and Gentlemen intending to honour him with their Commands. The stock is of the most comprehensive and varied that has ever before been presented to the public, comprising OPTICAL, MATHEMATICAL, AND PHILOSOPHICAL INSTRUMENTS, VIZ:— Telescopes, Microscopes, Electrifying Machines and Apparatus, Air-Pumps, Theodolites, Spirit-Levels, Sun-Dials, Prisms; Globes of all diameters; a most extensive assortment of Cases of Mathematical Instruments, Engineers' Compasses, Ivory and Boxwood Rules and Scales, Opera Glasses; Barometers and Thermometers of the most approved and modern patterns; Storm-Glasses, Phantasmagoria and Magic Lanterns, with beautiful painted slides; also, a very extensive assortment of Spectacles in Gold, Silver and Shell, and Steel Mountings; Folding and Single Eye-Glasses, of very great varieties.' (*Fife Herald*, 4 March 1841). The Stirling announcements appeared in the *Stirling Journal and Advertiser*, 26 June 1840 and 18 March 1853.

4. *Glasgow Post Office Directory* 1858-1891, *passim*.

5. *Ibid.* 1880, 1890, 1910. [Lizars], *op. cit.* (1) emphasises the high quality of Lizars' ophthalmic products and services; to cope with demand the firm had a factory at Goldenacre Works, Craignestock Street, Glasgow and wholesale stores at 17 Melville Lane, Glasgow.

6. [Lizars], *op. cit.* (1), *passim*. In 1896 they announced that 'growth of business with Mr J. Lizars of Glasgow has necessitated the removal of his factory to enlarged premises in Craignestock Street, where considerable additions to the machinery have been made' (*Practical Photographer* 7 (1896), 253). We are grateful to Sara Stevenson for this reference. Lizars had a factory for producing their popular 'Challenge' camera. See also Jim Gilchrist, 'A Sharper Focus on the World', advertising feature in the *Scotsman*, 4 July 1980, 10.

7. *Glasgow of Today* (Glasgow, 1888), 170; similar to the business of James Brown (q.v.).

8. *Ibid.* The only other Scottish instrument maker who claimed to manufacture microscopes at about this time was Mathew Edwards (q.v.) in 1891. Apart from the microscopes signed by Lizars in the Frank collection (NMS T1980.223, T1980.224, T1980.225 and T1980.226), others have been recorded: a student microscope was offered for sale by Phillips Glasgow, 28 April 1983, Lot 90; a compound monocular microscope was offered by Christie's South Kensington, 17 April 1986, Lot 384, and another similar on 24 July 1986, Lot 254.

9. *Ibid.* None of these catalogues has been seen, nor are any commissions by 'scientific gentlemen' known to have survived.

10. G.R.O.(S.), Register of Deaths 1879 (Glasgow, Blythswood), 644^7/574.

11. G.R.O.(S.), Census of 1881 (Glasgow, St. Rollox), 644^6/7, p.12.

12. *Glasgow, op. cit.* (7), 170. Lizars appears to have only had three daughters, so his successor was probably a cousin or nephew.

13. Charles Gillies, *Lizars: 150 Years of Service and Progress* (Glasgow, 1980), 7-9.

167. Microscope: J. Lizars, Glasgow, c.1885 (T1980.224)

Compound microscope in brass, bronze and oxidised brass, on an inclining pillar and flat 2-toed tripod base. Engraved on the base 'J. LIZARS./ OPTICIAN./ GLASGOW./ 304'. Single-draw body tube, with rack and pinion coarse focus adjustment to the pillar; fine focus is by a spring-loaded sliding inner triangular-section bar, attached through the stage to the stand, and operated by a screw at the top of the pillar, graduated 0-[20]. The rectangular stage plate with aperture diaphragm inserted from below, and spring clips. Compass joint, and plane/concave mirror sliding on a swinging tail beneath the stage. Fine focus components stamped with the assembly mark '1'.

Body length (less eyepiece): 154mm.
Body diameter: 34mm.
Height (less eyepiece): 289mm.
Base size: 141 x 150mm.

Lacking objective and eyepiece. The quality of finish on the base does not match that of the rest of the instrument. The soldered joint preventing the triangular limb rotating about its attachment screw has parted. The bearing point of the fine focus adjustment has sheared from the screw arbor, and the thread of one of the coarse focus knobs has stripped. One screw from the diaphragm sleeve lacking.

The instrument bears a resemblance to item 164 (signed W.A.C. Smith) and the two may have a common source.

168. Microscope: J. Lizars, Glasgow, c.1875 (T1980.225)

Compound microscope in brass, bronze and oxidised brass, on an inclining pillar and flat 2-toed tripod base. Engraved on the base 'J. LIZARS / OPTICIAN / GLASGOW / 186'. Single-draw body tube (the draw tube with 2 settings marked), in a draw sleeve for coarse focus adjustment, and attached by a bar-limb to a pillar; fine focus is by a spring-loaded sliding inner triangular-section bar, attached through the stage to the stand, and operated by a screw at the top of the pillar. The rectangular stage plate with a rotating 3-aperture diaphragm, and spring clips. Compass joint and plane/concave mirror on a swinging tail beneath the stage. Components above the compass joint stamped with the assembly mark '8'.

Body length (less eyepiece): 147mm.
Body diameter: 25mm.
Height (less eyepiece): 238mm.
Base size: 94 x 98mm.

Lacking objective and eyepiece. The quality of finish on the base does not match that of the rest of the instrument. The microscope has been manufactured as a non-inclining instrument: it has been rather inexpertly converted to an inclining model by cutting a compass joint in the tapering bronze lower pillar, the height reduction being compensated for by adding a cylindrical unit to the base. Because the adapted instrument would be unstable if mounted in the attachment hole in the Lizars base, the hole has been filled and a new new one provided further forward.

167, 168

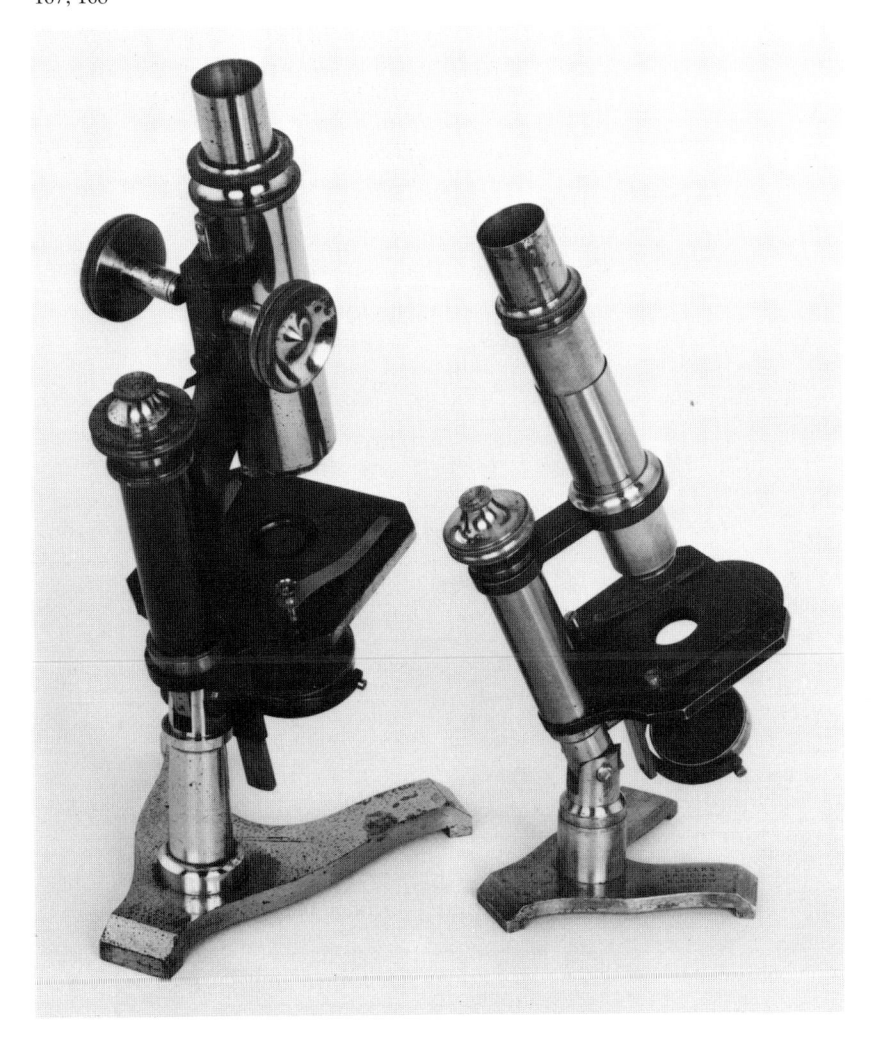

169. Microscope: J. Lizars, Glasgow, c.1880 (T1980.223)

Compound microscope in brass and oxidised brass, on a vertical pillar advanced by a screw over a flat base plate. Engraved on the base plate 'J. LIZARS / GLASGOW / 208'. The body tube mounted in a push-fit focusing sleeve screwed to a bar-limb attached to a slide between chamfered runners on a heavy rectangular base, with a screw to advance the slide so that the body tube traverses a fixed stage rising on pillars from the base. The stage with a large central aperture and spring clips, set over a bevelled edge circular aperture in the base: the length of adjustment of the body tube allows it to be lowered to focus at the level of the base plate. Four countersunk holes in the base allow attachment to a bench or other surface. Two untapped holes in the base may have supported illuminators or other accessories.

Body length (less eyepiece): 160mm.
Body diameter: 24mm.
Height (less eyepiece): 188mm.
Base size: 117 x 160mm.

Lacking objective, eyepiece and stage apertures. The sleeve for the body tube is distorted and the bar-limb has been bent and an attempt made to straighten it. Marks on the edge and underside of the base suggest it was unmounted or attached to a flat surface, but not raised on 4 legs.

The microscope appears to have been constructed as an inspection instrument for a special purpose. In some respects it resembles the 'Microscope a grand champ de vision pour l'examen des grandes surfaces', no. 198, offered by Nachet & Fils, *Instruments de Micrographie* (Paris, 1900): although Lizars' instrument is much simpler, the traversing motion is better engineered. It is possible that the instrument may have been fitted with standard-sized rectangular apertures used for an application such as thread counting in the fabric industry. (An example of such an instrument is the 'New Registered Silk-Mercers' Microscope' by J. Swift & Son of London, *A Catalogue of Microscopes* (London, 1880), 223).

170. Microscope: J. Lizars, Glasgow, c.1900 (T1980.226)

Compound bar-limb microscope, Society of Arts model, in brass and black-lacquered brass, on an inclining pillar and bent claw-footed tripod base. Engraved on the base 'J. LIZARS / GLASGOW. EDINBURGH. E^{TC}' The body tube screws into a bar-limb on a triangular-section slide within the pillar, a pinion engaging a rack cut into one edge for coarse focus adjustment; fine focus is by an external screw at the lower end of the body tube advancing a spring-loaded nosepiece for the objective. The rectangular stage hinged to the base, with a rotating 4-aperture diaphragm, and spring clips. A concave mirror sliding on an extension to the pillar beneath the stage. The base attached to a board, sliding into a fitted case, a brass name plate inside the door marked 'J. LIZARS,/ OPTICIAN & / PHOTOGRAPHIC DEALER,/ 101 & 107, BUCHANAN ST.,/ GLASGOW.'

Body length (less eyepiece): 170mm.
Body diameter: 28mm.
Height (less eyepiece): 295mm.
Base size (casting): 132 x 120mm.
Case size: 190 x 150 x 270mm.

Lacking objective and eyepiece.

A similar instrument, still called a 'Society of Arts' model, but with an iron base, was offered in 1894 by John J. Griffin & Sons, *Chemical Handicraft* (London, 1894), 302, no. 3202.

169, 170

171. Refracting telescope: J. Lizars, Glasgow, c.1870 (T1980.259)

1½" aperture single-draw hand-held refracting telescope with doublet objective (damaged), and 2-component eyepiece and erecting assemblies, the latter mounted at the division of the draw tube; in brass, with objective ray shade and attached 1⅛" aperture stop, and with leather-covered barrel. Engraved on the draw tube 'J, Lizars,/ 24 Glassford S.ᵗ. Glasgow,/ Day or Night,'. Scratched assembly mark 'X' on the draw tube and eye stop, but 'XI' on the eyelens mount. Sliding closures on external objective aperture and eye stop.

Length closed: 513mm.
Tube diameter: 59mm.
Aperture: 39mm.

The convex objective component has broken as a result of the two lenses being incorrectly inserted. Damage has been sustained to both ends of the instrument from falls or blows. The leather covering to the barrel has shrunk and the stitching has parted.

172. Projector: J. Lizars, Glasgow & Belfast, c.1895 (T1980.220)

Lantern slide projector in wood with red leather bellows and brass fittings, black-painted metal interior and cowling. Marked on 2 ivory name plates on the sides 'J. LIZARS,/ MANUFACTURING OPTICIAN,/ 101 & 107, BUCHANAN S.ᵀ / GLASGOW./ & AT 73 VICTORIA ST., BELFAST.' The illuminant chamber has metal lined side doors each with covered red-glass apertures. 2-component condenser lens assembly with ventilated cavity. Brass pressure plate attached to the front of the projector to locate the slide holder against the condenser mount, and projection lens mounting plate attached to 2 rods sliding in sleeves in the base of the projector and moved by a crank-operated screw; the 2 components connected by a removeable leather bellows. The lens sleeve focused by rack and pinion and with a swinging end cover plate and filter slot: stamped with the assembly mark '0'. Rear curtain to screen the light source.

Size of projector base: 365 x 202mm.
Height over cowling: 375mm.

Lacking the original projection lens, which would slide within the rack-focused sleeve. There is no light source: originally this would have been an oxy-hydrogen burner which would have been mounted from slides acting in the 2 rear sleeves in the base of the projector. Channels have been soldered to the base of the metal lining of the illuminant chamber to take a replacement source. The rear door is now missing and the curtain rail has been repositioned. The slide carrier for 3¼" square slides is modern.

NORTON & GREGORY

Norton & Gregory Ltd. was a London firm which first appeared in the London street directories as a tracing office at 3 Victoria Street in 1900; in 1902 they were described as photoprinters, at 24 Westminster Palace Gardens. The following year, the business had expanded, and diversified as 'drawing office stationers, photo paper makers, draughtsmen, map mounters and framers, lithographers, show-card manufacturers', with premises at 20, 21 and 24 Westminster Palace Gardens and works at 112 Grosvenor Road and 149 Lupus Street. However, these last two addresses were not listed in 1904, and in 1905 the business moved from Westminster Palace Gardens to Castle Lane, Buckingham Gate. In 1909 Norton & Gregory opened a Glasgow branch at 30 Robertson Street, and the following year this was given as 34 Robertson Street. In 1915, further premises were obtained at 6 Budge Row, London E.C. and by 1918 the firm was specialising in drawing-office stationery and equipment, but also advertised as makers of mathematical and surveying instruments.[1] An example of the type of apparatus which they produced was the unremarkable design for a parallel ruler, patented in 1915.[2]

In 1919 Norton & Gregory's Glasgow premises moved to 41 Queen Street and the following year to 71 Queen Street, by which time they were also advertising that they had factories at Harlesden and Willesden Green, London N.W.10. In 1926 the factory at Harlesden was replaced by one at Westminster. In 1939 the Glasgow branch moved to 199 Bath Street, and then to 17 Woodside Crescent in 1952; it eventually closed in 1962 after a final move to 461 Sauchiehall Street.[3] Further branches were opened, one in Broad Street, Birmingham in 1941, another at Manchester between 1949 and 1957 and a third in Bisley, Surrey between 1952 and 1954; the firm made its last appearance in the London directory in 1959.[4]

It would appear that much of the surveying equipment which Norton & Gregory sold would have been made by them. A trade catalogue, dating from 1906, lists the drawing and surveying instruments they offered, and states that 'if desired, instruments are sent to Kew for verification, the only extra expense being the cost of the certificate and carriage ... Theodolites constructed in aluminium with all wearing parts in bell or gun metal at an extra cost of 25 per cent. Larger sized theodolites, altazimuths, &c., made to any design.'[5] However, they were retailing surveyor's measuring chains and tapes manufactured by Chesterman of Sheffield.[6]

A second trade catalogue, dating from 1931, shows that the business had expanded greatly on the drawing and stationery sides, but that they were still selling their own surveying instruments: 'We can however supply every variety of instrument falling under this head and we welcome enquiries for other than listed patterns.'[7] Of the six sections into which the catalogue is divided, only two deal with instrumentation. An advertisement dating from 1938 proclaimed 'Everything for the Drawing Office From a Drawing Pin To A Photocopying Machine ... Norton & Gregory Ltd. founded 1899.'[8] A later advertisement implied that the firm had moved away from hardware towards photographic and other reproductive materials, yet still included the service 'suppliers of surveying equipment and all drawing office requirements.'[9] Another level with the firm's name engraved on it has been noted.[10]

REFERENCES

1. *Kelly's London Post Office Directory* 1899-1959, *passim*. We are grateful to Jane Insley and Tony Vincent, both of the Science Museum, London, for their assistance with this information. One of the firm's publications in this period was *Velography* (London, 1905), which deals with a form of high-speed photographic copying.

2. U.K. patent 18,210, 31 December 1915. Patented by Norton & Gregory and T.B. Pritchard.

3. *Glasgow Post Office Directory*, *passim*.

4. Kelly, *op. cit.* (1).

5. Norton & Gregory Limited, *Catalogue of Drawing and Surveying Instruments, Drawing materials, &c.* ([London], 1906), 2, 3. We are grateful to Dr Jon Darius for helping us with this reference.

6. *Ibid.*, 28-30.

7. Norton & Gregory Ltd., *Catalogue* (London, 1931), preface. We are grateful to Dr Jon Darius for this reference.

8. H. Fagg (ed.), *The Architects' Compendium and Annual Catalogue* (London, 1938), 579.

9. Anon., *The Architects' Standard Catalogues* 11th edition (London, 1950), I, 16.

10. Offered for sale at Sotheby's Belgravia, 28 May 1981, Lot 218.

---

173. Level: Norton & Gregory, London & Glasgow, 1930 (T1980.144)

12″ dumpy level in black-lacquered brass, with bubble tube (casing only) and crossed level over the telescope, rack and pinion focus to the objective, and objective ray shade divided for taking lateral inclines; the limb incorporating a ring compass divided [0]-360 by 30 mins. and with a prismatic eyeglass; mounted on a tribrach head with clamped micrometer azimuth motion. Engraved on the telescope tube with 'N&G' portcullis trade mark and 'NORTON & GREGORY L^{TD} / LONDON & GLASGOW / 1930'.

Telescope length: 292mm.
Aperture: 33mm.

The instrument can be identified with the 'Surveyor's Dumpy Level' advertised in various sizes, with accessories and extras, ranging from £18 to £22, in Norton & Gregory Ltd., *Catalogue* (London, 1931), 22.

174. Level: Norton & Gregory, London & Glasgow, c.1935 (T1980.145)

12″ dumpy level in black-lacquered brass, with bubble tube and crossed level over the telescope, rack and pinion focus to the objective, doublet objective, and objective ray shade divided for taking lateral inclines; the octagonal tube with cantilever limb attached by a ball-and-socket levelling adjustment to a tribrach head with clamped micrometer azimuth motion. Engraved on the telescope tube with 'N&G' portcullis trade mark and 'NORTON & GREGORY L^{TD} / LONDON & GLASGOW'.

Telescope length: 322mm.
Aperture: 41mm.

This instrument can be identified with the '"Engineer's" Permanent Collimation Level' advertised in various sizes, with accessories and extras, ranging from £19 to £23, in Norton & Gregory Ltd., *Catalogue* (London, 1931), 20: 'the telescope body, limb, and centre are in one gunmetal casting, eliminating screwed or soldered joints, thus ensuring rigidity and a fixed line of sight. As permanent collimation is thereby assured no alignment screws are necessary or provided, a great improvement on the ordinary Dumpy pattern.'

174, 173

CONCORDANCE

This concordance identifies items from the Frank collection which have appeared in previous publications. The first of these was Arthur Frank, *The Frank Collection of Early Scientific Instruments: an Interim List* (Glasgow, [1970]), and the second was the catalogue produced to accompany the special exhibition 'Tools of Science' in the Glasgow Art Gallery and Museum, Kelvingrove: R.H. Nuttall, *The Arthur Frank Loan Collection: Early Scientific Instruments* (Glasgow, 1973).

Here, we identify items by the catalogue number which they appear in this publication, their National Museums of Scotland inventory number, and the two numbers from the previous listings.

Catalogue Number	National Museums Inventory Number	Tools of Science Number	Interim List
1	T1981.27	700	TG1
2	T1981.28	701	
3	T1981.29	703	
4	T1981.30	704	TG5
5	T1981.31	705	
6	T1981.32	111	TG2
7	T1981.33	112	TG17
8	T1981.34		TG3
9	T1980.177		
10	T1980.245		
11	T1981.37		
12	T1980.124	706	SV7
13	T1980.125	707	SV6
14	T1980.221	708	
15	T1980.272	774	
16	T1980.163	711	
17	T1980.209	709	
18	T1981.35	710	
19	T1981.36	712	
20	T1980.126		
21	T1980.148	714	SV99
22	T1980.149		
23	T1980.150		
24	T1980.156		
25	T1980.157		
26	T1980.158		
27	T1980.169		
28	T1980.178		
29	T1980.210	715	
30	T1980.246		T85
31	T1980.273		
32	T1980.213		Misc.27
33	T1980.247	716	
34	T1980.248		
35	T1980.165	562	
36	T1980.206	567	N38
37	T1980.270		TL93
38	T1980.241	784	95
39	T1980.159		
40	T1980.166	799	
41	T1980.257	764	T34
42	T1980.182	789	OC22
43	T1980.181	786	OC11
44	T1980.265		
45	T1980.219	807	Misc.18
46	T1980.227	780	
47	T1980.228		
48	T1980.229		
49	T1980.242		
50	T1980.164	751	
51	T1980.236	750	
52	T1980.237		
53	T1980.261	752	T36
54	T1980.262		
55	T1980.263		
56	T1980.264		
57	T1980.277	772	TL142
58	T1980.278	778	
59	T1980.216	805	SV59
60	T1980.230		
61	T1980.243		Mac.36
62	T1980.183	787	OC17
63	T1980.184		
64	T1980.231	781	67
65	T1980.260		
66	T1981.42	777	
67	T1981.41	651	
68	T1980.189		
69	T1980.185	788	OC21
70	T1980.239	782	159
71	T1980.240	783	160
72	T1980.180		
73	T1980.146	717	SV107
74	T1980.147	721	SV83
75	T1980.127		
76	T1980.151	719	SV19
77	T1980.153		SV20
78	T1980.152	720	SV100
79	T1980.173	723	N17
80	T1980.222	718	
81	T1980.274	770	TL200
82	T1980.249		T158
83	T1980.250		
84	T1980.251		

309

Catalogue Number	National Museums Inventory Number	Tools of Science Number	Interim List	Catalogue Number	National Museums Inventory Number	Tools of Science Number	Interim List
85	T1980.275	771	TL145	130	T1980.134	735	
86	T1980.176	802	N18	131	T1980.132		
87	T1980.128	759		132	T1980.136		
88	T1980.129			133	T1980.131	733	SV1
89	T1980.170	760		134	T1980.133	734	
90	T1980.214	761	SV65	135	T1980.135	736	
91	T1980.215	762	SV58	136	T1980.154	738	SV18
92	T1980.232	758	152	137	T1980.155	739	SV16
93	T1980.276		TL115	138	T1980.160	737	SV93
94	T1981.38	776	TG18	139	T1980.161		
95	T1981.39	773		140	T1980.171		
96	T1981.40	775	TG15	141	T1980.174	746	N13
97	T1980.281	798		142	T1980.175	745	
98	T1980.252	765	T131	143	T1980.201	743	
99	T1980.253		T54	144	T1980.202	744	N15
100	T1980.212			145	T1980.233	740	
101	T1980.254	766		146	T1980.234	741	
102	T1980.168	801	D/C1	147	T1980.235	742	144
103	T1980.255	767		148	T1980.138		
104	T1980.200	791	N5	149	T1980.141		
105	T1980.196	790	N3	150	T1980.137	747	SV4
106	T1980.271		T72	151	T1980.139		
107	T1980.205	794	N41	152	T1980.140		
108	T1980.258	763		153	T1980.162		
109	T1980.194	796	S26	154	T1980.172	748	
110	T1980.208	797	CH12	155	T1980.195	749	
111	T1980.268			156	T1980.179	803	Misc.16
112	T1980.269		T4	157	T1980.142		SV2
113	T1980.188	728		158	T1980.143		
114	T1980.190	731		159	T1980.244	768	T81
115	T1980.193			160	T1980.198	792	
116	T1980.280			161	T1980.279	769	TL106
117	T1980.211			162	T1980.218	806	
118	T1980.186	730		163	T1980.256		T88
119	T1980.187	729	OC18	164	T1980.238	779	183
120	T1980.191			165	T1980.217	804	SV51
121	T1980.207	725	CH3	166	T1980.167	800	
122	T1980.266	726		167	T1980.224	755	161 or 184
123	T1980.267	727	T6	168	T1980.225	756	161 or 184
124	T1980.192		S49	169	T1980.223	754	184
125	T1980.204			170	T1980.226	757	157
126	T1980.197	724	N2	171	T1980.259	753	T19
127	T1980.199	793		172	T1980.220	693	
128	T1980.203	795		173	T1980.144		
129	T1980.130	732		174	T1980.145		

NAME INDEX

Main entries are entered in bold type. Occupations and titles are given, where known, particularly to distinguish between people bearing the same name. Corporate or business names are given in block capitals, and where firms and individuals share the same name (e.g. Alexander Adie, James White), the business entries are given a separate entry in block capitals.

Abraham, Abraham, instrument maker, 298, 299.
ABRAHAM & CO., instrument manufacturers, 99, 292, **298–300**.
Adair, William, sundial maker, 214.
Adam, John, architect, 54.
Adam, Margaret, 252.
Adam, Robert, architect, 55.
Adams, George, instrument maker, 25, 27, 29, 31, 48, 53.
Adams, J., patentee, 296, 297.
Adamson, Baillie –, 158.
Adamson, John, photographer, 149.
Adamson, Robert, photographer, 101, 149.
Adie, Agnes, 62.
ADIE, ALEXANDER, mathematical, optical and philosophical instrument makers, x, 5, 20, **25–74**, 81, 90, 245, 285.
Adie, Alexander James, 80.
Adie, Alexander James, [junior], civil engineer, 41, 47, 51, 59, 60, 62, 63, 64, 68, 75, 80.
Adie, Alexander (James), [senior], optician, 11, 13, 25, 26, 27, 30, 31, 32, **34–41**, 42, 43, 44, 47, 48, 49, 50, 51, 53, 54, 56, 57, 58, 60, 61, 62, 63, 64, 75, 80, 83, 89, 112, 113, 170, 191, 192, 285. See MILLER & ADIE.
Adie, Clement James Mellish, 82.
Adie, Elizabeth, 47, 59.
Adie, Elizabeth, 82.
Adie, Helen, 47.
Adie, Jane, 47, 62.
Adie, Janet, 47, 62.
Adie, John, 30.
Adie, John, instrument maker, **41–49**, 50, 51, 59, 60, 61, 62, 63, 75, 80, 89, 92, 94, 112, 113, 192, 226.
Adie, John, printer, 30, 53, 56.
Adie, Louise Jane, 82.
Adie, Marion, 62, 82.
Adie, Patrick, 82.
Adie, Patrick, instrument maker, 37, 41, 48, 50, 51, **75–84**.
Adie, Richard or Ritchie, optician, 41, 43, 47, 48, **49–51**, 53, 57, 62, 63, 64, 75.
Adie, Richard Haliburton, 82.
ADIE & SON, mathematical, optical and philosophical instrument makers, 22, 37, 39, **44–51**, 55, 57, 58, 59, 61, 62, 63, 64, 65, 69–73, 81, 85, 86, 89, 90, 93, 108, 113, 114, 115, 140, 177, 203, 245, 252.
Adie, Walter Sibbald, 82.
ADIE & WEDDERBURN, mathematical, optical and philosophical instrument makers, 48, **51–52**, 53, 64, 65, 73–74, 117.
Agnew of Lochnaw, Sir Andrew, landowner, 68–69.

Ainslie, John, surveyor and engraver, 30, 54.
Airth, Helen, 143.
Airy, George Biddell, Astronomer Royal, 59.
Aitken, John, meteorologist, 133, 135, 136.
Aiton, William, goldsmith and jeweller, 54.
Albert, J.W., instrument designer, 291.
Alemoor, Lord: see Pringle, Andrew.
Alexander, D.M., 277.
Alexander, George, watch and compass maker, 105.
Alexander, G.H., patentee, 266.
Alexander, Margaret, 159.
Allan, Alexander, instrument maker, 13, 206.
Allan, Isabella, 159.
Allan, R.S., director of D. McGregor & Co. Ltd., 240.
Allwood, George, compass maker, 232.
Amici, Giovanni Battista, scientist and microscope maker, 39, 58.
Anderson, Adam, schoolmaster and meteorologist, 40, 59.
Anderson, James, colliery manager, 268.
Anderson, James, schoolmaster, 18.
Anderson, John, professor of natural philosophy, 164.
Anderson, William, banker, 191.
Andrews, Elizabeth, 243.
Annan, Robert, photographer, 263, 265, 277.
Annan, Thomas, photographer, 215.
Archer, Helen, 109.
Argyll, 8th Duke of: see Campbell, George Douglas.
Armstrong, Andrew, cartographer, 29, 55.
Armstrong, Mostyn, cartographer, 29, 55.
ARROL, A. & SONS, engineers, 279.
Auld, William, clockmaker, 118.
Ayton, William, goldsmith, 26.

Baily, Francis, astronomer, 19, 25.
Baird, Alfred W., compass adjuster, 261, **262**, 265, 266. See KELVIN, BOTTOMLEY & BAIRD LTD.
Baird, Andrew H., instrument supplier, 135.
Baird, Archibald, ironmonger, **296–297**.
BAIRD, ARCHIBALD & SON, engineers' ironmongers, 188, 292, **296–297**.
Baird, Douglas Heriot, chairman of Baird & Tatlock (London) Ltd., 289.
Baird, George, principal of Edinburgh University, 59.
Baird, Hugh Harper, instrument manufacturer, **289–291**.
Baird, John, wholesale and manufacturing optician, 244.
Baird, Matthew B., ironmonger, **296–297**. See BAIRD, ARCHIBALD & SON.
BAIRD & TATLOCK, instrument manufacturers, 135, 278, **289–291**.
BAIRD & TATLOCK LTD., instrument manufacturers, 290.
BAIRD & TATLOCK, LONDON AND GLASGOW, instrument manufacturers, 289.
BAIRD & TATLOCK (LONDON) LTD., instrument manufacturers, 289, 290.
Baker, Edward(?), chronometer maker, 159.
Balerno, Domenico, barometer maker, 139, 147.
Ballantine, Matthew, instrument maker, 302.
Barazoni, Anthony, optician, 153.
Barclay, Andrew, engineer and amateur telescope maker, 188–189, **197–202**, 292.

311

Barclay, John, millwright, 197.
Barr, John Haddin, optician, 252, 253, 261, 262.
BARR & STROUD, optical instrument manufacturers, 82, 188.
BARRAUD & LUND, chronometer makers, 228.
Barron, Elizabeth, 63.
Barron, George, lawyer, 47.
Barry, James, surveyor, 164.
BASSNETT, THOMAS & CO., nautical instrument makers, 239, 242, 243.
BATTISTESSA & CO., carvers and gilders, 102.
Batty, William, compass designer, 257, 264.
Beattie, James F., surveyor, 158.
BECK, R. & J., microscope manufacturers, 87, 135, 277, 278.
BECKER, F.E., instrument suppliers, 290, 291.
Bedford, E.J., naval officer, 236.
Begbie, Christina, 63.
Bell, Henry, nautical and optical instrument maker, 217, **224**.
Bell, Isaac, brassfounder and compass adjuster, 224.
Bell, James, philosophical instrument maker, 112, 118, 125.
Berry, Andrew, pharmacist, 159.
Berry, George Allan, watchmaker, 156, 157, 158, 159, 160.
Berry, George Francis, watchmaker and compass adjuster, 157.
Berry, James, chronometer, nautical and optical instrument maker, 106, 153, **156–157, 158–159**, 160.
Berry, James, shipmaster, 156.
Berry, James, produce merchant, 159.
BERRY, JAMES & SON, chronometer, nautical and optical instrument makers, 156, 160.
Berry, John Philip, minister, 159.
BERRY & MACKAY, chronometer, nautical and optical instrument makers, 109, 152, **156–161**, 162.
Berry, William, doctor, 159.
Beverley, Arthur, apprentice, 159.
Billcliff, J., camera manufacturer, 130.
Biram, Benjamin, patentee, 297.
Bird, John, instrument maker, 3, 6.
Bisset, Christina, 156.
Black, D., sailor, 247.
Black, Joseph, professor of chemistry, 27, 54, 105.
Black, William Galt, surgeon and meteorologist, 133, 136.
Blackie, William, optical lapidary, 13, 14, 38, 58.
Blackwood, James, carpet manufacturer, 193.
BLAIKIE BROTHERS, engineers, 154.
Blair, Bryce, turner and wheelwright, 190.
Blair, Jean, 219.
Blakeney, John W., nautical instrument maker, 225, **226**. See CAMERON & BLAKENEY.
Blanc, Hippolyte, architect, 120.
Blood, W.B., professor of civil engineering, 77.
Blythswood, Lord: see Campbell, Sir Archibald Campbell.
Boath, Robert, 72.
Bon, John, chronometer, watch and nautical instrument maker, 139, 245.
Bon, Mrs John, 139.
Bonar, John, sundial maker, 211, 214.
Bottomley, James Thomson, university demonstrator, 254, 255, **260–262**, 263, 265. See KELVIN, BOTTOMLEY & BAIRD LTD.
Bough, Samuel, artist, 148.
Boulton, Matthew, industrialist, 164.
BOULTON & WATT, engineers, 168.
Boyack, Alexander, photographic goods merchant, 170.
Boyle, George Frederick, 6th Earl of Glasgow, 213.
Bradford, W., coin scale designer, 164.
Braham, Frederick J., optician, 97.
BRAHAM, L. & CO., opticians, 97.
BRAHAM, WALTER & CO., instrument engineers, 278.
Brand, –, railway contractor, 158.
Brebner, James, advocate, 158.
Brewster, Sir David, polymath, 13, 14, 18, 19, 20, 21, 22, 23, 24, 31, 32, 37, 38, 39, 40, 50, 55, 56, 58, 59, 83, 89, 94, 97, 98, 101, 107, 112, 119, 147, 149.

Brisbane, Sir Thomas Makdougall, soldier and scientist, 19, 22, 26, 41, 75, 80, 114, 119.
Bristow, Fanny, 157.
BROCKBANK & ATKINS, chronometer makers, 159.
Broun, John Allan, magnetician and meteorologist, 75, 80, 82.
Brown, Alexander Crum, professor of chemistry, 120.
Brown, Ann, 162.
BROWN & CHALMERS, clock, watch and nautical instrument makers, 105.
Brown, George B., clock, watch and nautical instrument maker, 105.
Brown, Hugh, ironmonger, maker of tools, guns and mathematical instruments, 184.
Brown, Hugh, missionary, 183.
Brown, James, optician, 125, 151, 168, 170, 179, **183–187**, 271, 278, 289.
BROWN, JAMES & SON, nautical stationers and publishers, 184.
Brown, Mary Wilkie, 228.
Brown, William Henry, rule maker, 253.
Browning, John, instrument maker, 120.
Browning, William Spencer, instrument wholesaler, 57.
Bryce, Alexander, minister, 26, 27, 29.
Bryson, Alexander, geologist, 54, 61, 92, 112, 113, **114–115**, 118, 119.
Bryson, Archibald Gillespie, 119, 120.
Bryson, David Dunn, 120.
Bryson, James Mackay, instrument maker, 48, 51, 54, 64, 92, **112–115**, 118, 119, 252, 260, 263, 292.
Bryson, Jessie Gillespie, 120.
Bryson, Margaret Bannatyne, 120.
Bryson, Mary, 120.
Bryson, Robert, 120.
Bryson, Robert, [junior], clockmaker, 112, **115–116**, 117, 118, 120.
Bryson, Robert, [senior], clockmaker, 94, 112, 113, 114, **115**, 118.
Bryson, William Gillespie, factor, 118.
Buccleuch, 3rd Duke of: see Scott, Henry.
Buchan, Hugh, brassfounder and gasfitter, 226.
Buchanan, George, civil engineer, 49, 62, 93, 94, 113, 118.
Buird, Edward, ship chandler and supplier of navigational instruments, 104.
Burges, William, architect, 213.
Burns, George, businessman(?), 220–221.
Bute, 3rd Earl of: see Stuart, John.
Butterfield, William, architect, 213.
BUTTI & CO., carvers and gilders, 102.
Butti, James A., carver, gilder and looking glass manufacturer, 102.
Butti, Louis Joseph, looking glass manufacturer, **102–103**. See ZENONE & BUTTI.

Cairne, John, artist, 148.
Caldecott, John, astronomer, 40, 49.
Calwell, Robert, merchant, 212, 214.
CAMBRIDGE SCIENTIFIC INSTRUMENT COMPANY, instrument manufacturers, 188.
Cameron, Alexander, jeweller, clock, watch and nautical instrument maker, 58, 106, 139, **140–141**, 154, 162.
CAMERON & BLAKENEY, mathematical, nautical and optical instrument makers, 217, **225–227**.
Cameron, J.R., instrument maker, 231.
Cameron, Paul, mathematical, optical and philosophical instrument maker, **225–227**. See HOUSTON & CAMERON.
CAMERON, PAUL & CO., mathematical and nautical instrument makers, **226**.
Campbell, Sir Archibald Campbell, 1st Baron Blythswood, politician and scientist, 200, 201.
Campbell, George Douglas, 8th Duke of Argyll, politician and scientist, 120.
CAMPBELL, J. & W. & CO., textile merchants, 183.

Campbell, Janet, 200.
Campbell, John, 4th Earl of Loudoun, soldier, 27, 29, 53, 55.
Carlaw, Bernard, warper, 180.
Carlaw, David, 168, 170, **180–182**, 188, 253, 260, 261, 262.
CARLAW, DAVID & SONS, engineers, 181.
Caroline, Queen, consort of George II, 1.
Carrick, John Wilson, book keeper to D. McGregor & Co., 239, 242, 243.
Carter, John, chronometer maker, 158, 159.
Cary, William, instrument maker, 32, 34, 49.
Casartelli, Joseph, instrument maker, 81.
CASELLA, LOUIS PASCAL, instrument wholesaler, 77, 83, 130, 170, 252.
Cetti, Edward, barometer maker and wholesaler, 238, 252.
CHADBURN BROTHERS, instrument wholesalers and manufacturers, 278.
Chalmers, Charles, experimenter, 90.
Chalmers, Patrick, antiquary and bookseller, 89.
CHAMBERLAIN, C.J. & CO., photographers, 170.
Chambers, W.H., government safety officer, 115, 119.
CHANCE BROTHERS, optical glass manufacturers, 200.
Charlton, –, 19.
CHESTERMAN, JAMES & CO., manufacturers of steel tape measures, 238, 278, 306.
Chetwynd, Hon. Louis Wentworth P., naval officer and compass designer, 260, 265.
Chickie, Francis, barometer maker, 153.
Chiene, John, professor of surgery, 87.
Chisholm, –, 80.
Christie, Andrew, instrument maker and compass adjuster, 239, 240, **250–251**, 283.
Christie, Andrew, law clerk, 250.
CHRISTIE, ANDREW, nautical instrument makers, 239.
CHRISTIE & LEE, nautical instrument makers, 239.
Christie, Mary, 250.
CHRISTIE & WILSON, nautical opticians and compass adjusters, 239, 240, **250–251**, 283, 284.
Christison, Sir Robert, toxicologist, 49, 62, 77, 81.
CICERI & CO., carvers and gilders, 103.
Ciceri, Joshua, 103.
CICERI, MANTICHA & TORRE, carvers and gilders, 103.
CICERI & PINI, carvers and gilders, 102, 103.
CICERI, PINI & CO., carvers and gilders, 102.
Clark, –, schoolmaster, 17.
Clark, Athole, compass adjuster, 262.
Clark, Frank Wood, patentee, 262, 263, 283.
Clark, James, instrument maker(?), 139.
Clark, William, patent licensee, 79, 80.
Clarke, Edward, instrument wholesalers, 243.
Clarke, George Thomas, engineer and archaeologist, 97.
Clarke, James, instrument maker, 97.
Clerk of Penicuik, Sir John, antiquary and advocate, 1.
Clutton, Henry, architect, 213.
CLYDESIDE CONSTRUCTION CO. LTD., public works contractors, 274.
Cohen, Simeon Phineas, instrument maker, 298, 299.
Cockburn, Henry Thomas, Lord Cockburn, judge, 94.
Coddington, Henry, mathematician, 61.
Combe, George, phrenologist, 94.
Congreve, Sir William, soldier and inventor, 118.
Connell, Arthur, professor of chemistry, 133.
Connor, Benjamin, railway engineer, 270.
Conrady, A.E., patentee, 266.
COOKE, THOMAS & SONS, engineers and instrument manufacturers, 77, 81, 97.
Copland, Patrick, professor of natural philosophy, 152, 153.
Corti, Antoni, glassworker, 167, 172.
Coulter, William, stocking maker, 30.
Cox, Frederick J., instrument maker, 97.
Craig, John, architect, 164.
Craig, William, optician, **111**, 130.
Crichton, James, instrument wholesaler, 228, 231.

CRICHTON, J. & SON, opticians and instrument manufacturers, 97.
Crisp, Sir Frank, microscopist and collector, 149.
Crouch, Henry, microscope manufacturer, 276, 277.
Cumberland, William Augustus, Duke of, soldier, 1.
CUNNINGHAM SHEARER GROUP, 230, 231.
Cuthbert, John, microscope maker, 39, 58.
CUTTS, J.P., instrument wholesalers, 167.
CUTTS, J.P., SUTTON & SON, instrument wholesalers and manufacturers, 97, 253.

Dallmeyer, John Henry, optical instrument manufacturer, 81.
Dalrymple, William, instrument maker, 168, 184.
Dalton, John, chemist and natural philosopher, 299.
Dalyell, –, coach driver, 158.
Dancer, John Benjamin, instrument maker, 298, 299, 300.
Darkes, –, 242.
DARLING, BROWN & SHARPE, tool manufacturers, 81.
DARTON, F. & CO., instrument makers and manufacturing opticians, 278.
Davidson, Thomas, optician, 45, 99, 101, 119.
Davis, Jacob, wholesale furrier and skin merchant, 100.
Davis, John, instrument supplier, 97, **99–101**.
DENNY & CO., engineers, 181.
Desaguliers, J.T., lecturer and author in natural philosophy, 30, 55.
Dicas, John, hydrometer designer, 206.
Dick, John, apprentice instrument maker(?), 228.
Dick, Robert, professor of natural philosophy, 164.
Dick, Thomas, scientific writer, 139.
Dickenson, –, patentee, 260.
Dickman, John, chronometer, clock watch and nautical instrument maker, 105.
Dobbie, Alexander, chronometer and nautical instrument maker, 217, **228–234**, 235, 256.
DOBBIE, ALEXANDER & SON, chronometer and nautical instrument makers, 228, **229**, 233, 238, 239, 240, 243, 264, 292.
DOBBIE, ALEXANDER & SON LTD., chronometer and nautical instrument makers, 230.
Dobbie, Alexander Brown, engineer, 230, 232.
Dobbie, Andrew, watchmaker, 231.
DOBBIE, HUTTON & GEBBIE LTD., chronometer and nautical instrument makers, 229, 230.
Dobbie, Sir James Johnston, chemist, 229, 230, 231, 232.
Dobbie, John, watchmaker, 228, 230, 231.
Dobbie, John Clark, nautical instrument maker, **228–232**.
DOBBIE, McINNES LTD., manufacturers of engineering and navigation instruments, **229**, 230, 231, 234, 243, 281, 296.
DOBBIE, McINNES & CLYDE LTD., chronometer and nautical instrument makers, 230, 231.
Dobbie, Mary W., 232.
DOBBIE, SON & HUTTON, chronometer and nautical instrument makers, 229, 230.
Dobbie, Thomas, watch and clockmaker(?), 231.
Dobbie, William, instrument maker(?), 228, 230, 231, 232.
Dobbie, William Wilson, 232.
Dobbin, William, 212.
Dods, –, teacher, 80.
Döbereiner, Johann Wolfgang, chemist, 40.
DOLLOND & AITCHISON, ophthalmic opticians, 168, 184.
Dollond, George, optical and mathematical instrument maker, 47, 97, 99.
Dollond, John, optical and mathematical instrument maker, 3, 5, 27, 55.
Donald, Isabella, 162.
Donaldson, James, bookseller and printer, 56.
Douglas of Cavers, James, landowner, 19.
Douglas, James, 14th Earl of Morton, politician and scientist, 1, 3, 4, 5, 6, 27.
Douglas, Sholto Charles, Lord Aberdour, 3.
Driffield, V.C., photographic experimenter, 134, 136.
Duff, Adam, sheriff depute, 60.

313

Duff, James, clockmaker, 53.
Duncan, William, mathematical, optical and philosophical instrument maker, 153.
DUNLOP, COLIN & CO., ironmasters, 296.
Dunn, David, 117.
Dunn, Hamilton, builder, 89.
Dunn, John, mathematical, optical and philosophical instrument maker, 49, 62, **89–95**, 100, 112, 113, 118.
Dunn, Mary, 117.
Dunn, Thomas, optician, mathematical and philosophical instrument maker, 62, 89, 90, 92, **93**, 219, 299.
DUTHIE, ALEX. & CO., shipowners, 154, 158.

EASTMAN PHOTO. CO., photographic suppliers, 278.
Easton, Alexander R., watch, nautical and optical instrument maker, 153.
Easton, Robert, surveyor, 18.
Edington, Alexander, bacteriologist, 87.
EDWARD, GEORGE & SON, jewellers, 170.
Edwards, Frederick, tanner, 276.
Edwards, Mathew, instrument maker, 252, 254, 257, 260, 261, 263, 265, **276–277**, 280, 292, 302.
Edwards, Matthew, tobacconist and hardware merchant, 276.
Elliot, Archibald, architect, 34.
Elliot, Admiral Sir George, naval officer, 20.
Elliot, Sir Gilbert, 1st Earl of Minto, diplomat, 19, 20, 22.
Elliot, James, experimenter, 118.
Elliot, Admiral John, naval officer, 17, 19, 20.
Elliott, Major –, 19.
ELLIOTT BROTHERS, instrument manufacturers and wholesalers, 77, 83, 97, 116, 242, 252.
ELLIOTT & SONS, instrument makers, 83.
Ellis of Otterburn, James, antiquary and solicitor, 21.

Fair of Langton, James, 18.
Farquhar, William, chronometer maker, 159.
Feathers, James, instrument maker, 144.
Feathers, James, shipowner and coal merchant, 143.
Feathers, John Murray Mitchell, flax merchant, 144, 145.
FEATHERS, P. & SON, **142–145**, 256.
Feathers, Peter, photographic supplier, 133, 144.
Feathers, Peter Airth, chronometer and nautical instrument maker, 109, 139, **142–145**.
Fergus, Andrew, engineer, 185.
Ferguson, James, astronomer and author, 6, 152, 210, 214.
Ferguson, James, electrician, 260, 286.
Ferguson, Robert, compass adjuster, 262.
Fergusson, George, Lord Hermand, judge, 19, 22.
Field, –, 241.
Field, M.B., patentee, 266.
FIELD & SON, opticians, 97.
Finlay, Robert, mathematical, optical and philosophical instrument maker, 225.
Finlayson, James, doctor, 185.
Fisher, William, pocket book maker, 165.
Fitzwygram, Sir Frederick Wellington John, army veterinary surgeon, 121.
Fleming, James, optician and chronometer maker, 219.
Fleming, John, minister and naturalist, 119.
Fleming, William Malcolm, agriculturalist, 215.
Flintoff, Thomas, army veterinary surgeon, 121.
Forbes, –, coppersmith, 5.
Forbes, George, professor of natural philosophy, 149.
Forbes, James David, professor of natural philosophy, 20, 23, 35, 40, 41, 43, 45, 47, 49, 50, 57, 60, 61, 62, 63, 114, 118.
Forgan, William, optician, 133.
Forrest, Samuel, apprentice instrument maker(?), 228.
FORRESTER & NICHOL, lithographers, 94.
Foucart, Virginia, 206.
FRANK, CHARLES LTD., opticians, xi, 48, 85, 233.
Fraser, William, 94.
Frazer, Alexander, optician, 48, 51, 52, 64, 133.

FRITH, PETER & CO., instrument wholesalers and retailers, 167, 170, 253.
Froccatt, Thomas, optician, 97.
Frodsham, Charles, chronometer maker, 141, 237, 242.
Fulton, John, shoemaker and maker of orreries, 191, 194.
Fyfe, Dr Andrew, chemist, 59, 89.
FYFE & BLACK, mathematical instrument makers, 222.
Fyfe, John, ship chandler, 220, 222.
Fyfe, John, surveyor, 222.
Fyfe, Margery, 220.
Fyfe, Robert, ship chandler, 220, 222.
FYFE, R. & W., mathematical instrument makers, 220, 222.
Fyfe, Samuel Holborn, ship chandler and instrument retailer, 217, 220, 221, **222–223**. See MACALISTER & FYFE.
Fyfe, William Holborn, ship chandler, 219, **220–221**, 222.
FYFE, WILLIAM H. & CO., ship chandlers and paint manufacturers, **220**.
FYFE & YEO, ironmongers and ship chandlers, **220–221**.

Gairdner, William, physician, 114, 115, 119.
Galbraith, William, teacher of mathematics, 41, 43, 49, 60, 62.
Gall, John, minister, 159.
Galletti, Antoni, carver and gilder, 167, **205–207**.
Galletti, Charles, print seller, 205, 206.
Galletti, John, instrument maker, **205–207**.
Garden, Alexander, instrument maker, 40.
GARDNER & CO., instrument makers, xi, 20, 50, 73, 129, 151, 168, **169**, 170, 174–179, 183, 184, 187, 193, 203, 245, 248, 252, 261, 289.
GARDNER, J. & J., instrument makers, **165**, 168, 171.
GARDNER, JAMIESON & CO., **165**, 168, 171.
Gardner, John, gardener and maltman, 164.
Gardner, John, [junior], instrument maker, 125, **165**, 168.
Gardner, John, [senior], instrument maker, **164–165**, 168, 171.
Gardner, John, [younger], instrument maker, 165, 168, 170, 171.
GARDNER & LAURIE, instrument makers, **164**, 168, 173.
GARDNER & LYLE, instrument makers, 168, **170**, 173.
Gardner, Margaret, 125, 165, 167, 168, 170, 171.
GARDNER, M. & CO., **165–169**, 171, 172.
GARDNER, M. & SONS, **165–169**, 171, 172.
Gardner, Robert, chronometer springer and balance maker, 228, 232.
Gardner, Thomas Rankine, [junior], instrument maker, 168, **170**, 173.
Gardner, Thomas Rankine, [senior], instrument maker, 165, 167, 168, **169–170**, 172, 173, 180, 226, 252.
Gardner, William, instrument maker, 165, 167, 168, **170**, 172, 253.
Gartly, John, watch and clockmaker, 154.
Gaskin, George Augustus Samuel, instrument maker, 97.
Gebbie, Hugh, compass adjuster, 230, 232.
Gebbie, W.W., compass adjuster, 232, 237, 238, 239, 240, 243.
Geikie, Archibald, geologist, 115, 119.
George III, King of Great Britain and Ireland, 25, 27, 54.
George IV, King of Great Britain and Ireland, 37.
GEORGE & BECKER LTD., W. & J., instrument suppliers, 290.
Gerletti, Charles, looking glass manufacturer, **208**.
Gerletti, Dominick, optician and firework artist, **208**.
Gerletti, John, barometer and thermometer maker, 208.
Gerrard, Alexander, instrument maker, 226.
Gibb, John, civil engineer, 152.
Gibson, John, clockmaker and instrument maker, 20, 23, 55.
Gibson, Thomas, instrument maker(?), 228.
Gibson–Craig, Sir James, architect, 93.
Giesecké, Sir Charles, mineralogist, 23.
Gilchrist, James, photographer, 130.
Gilchrist, James, specialist constructor of compasses and binnacles, 283, 284.
Glaisher, James, meteorologist, 81.
Glanford, Robert, salesman to D. McGregor & Co., 242.
Glasgow, 6th Earl of: see Boyle, George Frederick.

Glendinning, Alexander, manager to D. McGregor & Co., 242.
Gordon, George, 9th Marquess of Huntly, soldier and politician, 158.
Gordon, John, landowner, 212.
Gordon, Peter, clockmaker, 160.
Goring, C.R., scientific author, 37, 38, 39, 58.
Graham of Blagowan, Lord Cathcart, agriculturalist, 17.
Graham, George, clock, watch and instrument maker, 3, 6.
Graham, John, compass adjuster, 283, 284, 286.
Graham, John, printer, 207.
Graham, Thomas, chemist, 89, 93.
Grant, J., instrument retailer, 105.
Grant, John, clockmaker, 161.
Grant, John, nautical instrument maker and optician, 153.
Granton, Lord: see Hope, Charles.
Grassick, James, clerk with Adie & Son, 40, 48, 50, 63.
Gray, James, ironmaster, 26, 53.
Gray, Robert, teacher of mathematics, 156, 160.
Gray, William, optician, 260.
Green, George, personal assistant to Lord Kelvin, 262.
Green, William, instrument maker, 225.
GREGORY, THOMSON & CO., engineers, 197.
Gregg, Elizabeth, 26, 53.
GRIFFIN & CO., R., manufacturers of chemical apparatus, 252.
GRIFFIN & GEORGE LTD., instrument suppliers, 290, 291.
Griffin, John Joseph, writer on technical subjects, 290.
GRIFFIN, JOHN JOSEPH & SONS, instrument wholesalers, 135, 271, 278, 290, 304.
GRIFFIN & SONS LTD., JOHN JOSEPH, instrument wholesalers, 290.
GRIFFIN & TATLOCK, instrument suppliers, 290.
GRIFFIN & TATLOCK LTD., suppliers of chemicals and chemical apparatus, 290.
GRIMOLDI & CO., barometer manufacturers, 157.
Grubb, Sir Howard, astronomical instrument maker, 80, 188, 189.
Grubb, Thomas, instrument maker, 80, 188, 189.
Gurney, Jason, 82.
Gwilt, Hannah Jackson, 63.

Haddow, Thomas, photographic and instrument supplier, 48, 52, 65, 302.
Hadley, James, mathematician and scientific mechanist, 20, 23.
HALL, Messrs., shipbuilders, 152, 154.
Hall of Dunglass, Sir James, geologist, 32, 57.
Hall, Robert, minister, 18.
Hamilton, Hill, merchant, 213.
HAMILTON & INCHES, watch and clockmakers, 115.
Hamilton, James, director of D. McGregor & Co. Ltd., 241.
Hammersby, Joseph, telescope and opera glass manufacturer, 168, 180, 181, 253, 261.
Hannan, Robert, brewer, 209.
HARRIS, PHILIP & CO., instrument manufacturers and wholesalers, 163.
Harris, William, instrument maker, 32, 58, 107.
Harrison, John, chronometer maker, 5, 27.
Hart, William, instrument maker, 253.
Hartnack, Edmund, microscope manufacturer, 51, 122, 135, 277.
HARTNACK & PRAZMOWSKI, microscope manufacturers, 121, 122.
Hassard, W.J., instrument maker, 261, 276, 277.
Hauksbee, Francis, [the elder], instrument maker, 152.
Haworth, Sir Norman, professor of chemistry, 230, 231.
Hay, John, carver and gilder, 148.
Hay, John, landscape gardener, 69.
Hay of Drumelzier, Robert, landowner, 55.
Hay Craft, W.T., experimenter, 40, 59.
Hearne, George, instrument maker, 152.
HEATH & CO., nautical instrument makers, 238, 243.
Hellaby, Clementina, 80.
Henderson, Angus, instrument maker, 48, 51, 52, 115.

Henderson, Thomas, Astronomer Royal for Scotland, 47, 60, 61, 62.
Herd, David, antiquary, 30, 56.
Hermand, Lord: see Fergusson, George.
HERON & CO., ship chandlers and nautical warehouse, **281**.
Heron, David, instrument maker, 221, 235, 239, 241, **281**, 284, 285, 286.
HERON, DAVID, ship chandler, 239, **281**, 284, 287.
HERON, DAVID & CO., nautical instrument makers, 239, **281**, 284.
Heron, James, instrument maker, 235, 285.
Heron, Jessie, 281, 284.
Heron, John, chronometer maker, watchmaker and jeweller, 235, 239, 281, 284, 285.
HERON & JOHNSTON, ironmongers, ship chandlers and flag makers, **281**, 284, 286.
HERON & SON, watchmakers, 235.
Heron, William, instrument retailer, 37, 57, 285.
Heron, William, lecturer in natural philosophy, 281, 285.
Herschel, Sir William, astronomer, 23.
Hewitt, Thomas, chronometer maker, 156, 159.
HICKS, JAMES JOSEPH, instrument wholesaler, 80, 82, 83, 109, 110, 130, 131, 186, 208, 246, 297.
Higginbotham, Samuel, printer, 214.
Hilger, Adam, instrument maker, 201.
Hilger, Otto, instrument maker, 200, 201.
Hill, David Octavius, artist and photographer, 101.
Hill, George, smith and beam maker, 12, 13, 14.
Hill, James, 12, 14.
Hill, Peter, mathematical instrument maker, **11–15**, 38, 89.
Hill, Peter, [senior], smith, 11, 12, 14.
Hogg, –, observer, 80.
Hogg, J., instrument retailer, 105.
Hogg, Jabez, ophthalmic surgeon and scientific author, 299, 300.
Holborn, Elizabeth, 220, 222.
HOMAN, sundial makers, 123.
HOOD, Messrs., shipbuilders, 153.
Hooke, Robert, natural philosopher, 35.
Hope, Charles, Lord Granton, judge, 19, 22.
Hope, Sir John, 4th Earl of Hopetoun, soldier, 19.
Hope, Thomas Charles, professor of chemistry, 89.
Hopetoun, 4th Earl of: see Hope, Sir John.
Horner, Leonard, geologist and educationalist, 112, 113.
HOUSTON & CAMERON, mathematical and nautical instrument makers, **226**.
How, James, optician, 97.
Howden, James, watchmaker, 34, 57, 83.
Howie, James, optician and philosophical instrument maker, 139.
Howie, Peter, optician, 139.
Hoy, James, clerk, 27, 55.
HUGHES, HENRY & SON, nautical instrument makers, 97, 238, 252, 257, 261, 264.
Hughes, Joseph, optician, 97.
Hume, David, philosopher, 26.
Hume, David, smith, 13.
Hume, Elliott, 135.
Hume, Elizabeth, 133.
Hume, Hall, 135.
Hume, John, 135.
Hume, Scott, 135.
Hume, William, commercial flour traveller, 133.
Hume, William, philosophical instrument maker, **132–137**, 147.
Hume, William, [junior], 135.
Hunter of Thurston, James, landowner and amateur scientist, 34, 57, 83.
Huntly, 9th Marquess of: see Gordon, George.
Hutcheson, George, benefactor of Hutcheson's Hospital, 267.
Hutcheson, Thomas, benefactor of Hutcheson's Hospital, 267.
Hutchison, George, [junior], ophthalmic optician, 48, **85–87**.
Hutchison, George, [senior], instrument maker, 48, 53, 58, 62, **85–87**.

HUTCHISON, G. & SONS, opticians and surveying instrument makers, 48, 71, **85–87**.
Hutchison, John R., optician and instrument collector, 39, 48, 53, 58, 69, **85–87**.
Hutton, James, geologist, 32.

Inglis, John R., compass adjuster, 283, 284, 286.
IMRAY, JAMES & SON, instrument wholesalers, 238.
Ivory, Sir James, mathematician, 139.

Jardine, James, civil engineer, 41, 43, 50, 60, 94, 191, 194.
James IV, King of Scotland, 104.
Jamieson, Robert, instrument maker(?), 165, 168, 171, 285. See GARDNER, JAMIESON & CO.
Jenkin, Fleeming, engineer and electrician, 256.
Jennings, John, optician, 97.
Johnston, Alexander Keith, cartographer, 119.
Johnston, James Meldrum, ironmonger and ship chandler, 284, 286. See HERON & JOHNSTON.
Johnston, Thomas, surveyor, 56.
Jones, –, instrument maker, 34.
Jones, Thomas, instrument maker, 35, 57, 165.
JONES, W. & S., instrument makers, 99.
Joule, James Prescott, physicist, 264, 299.

Kean, J., patentee, 266.
Keith of Ravelston, Alexander, antiquary and scientific patron, 29, 55.
KELVIN, BOTTOMLEY & BAIRD LTD., navigational and electrical instrument makers, 252, 259, 261, **262**, 266.
Kelvin, Lord: see Thomson, William.
KELVIN & HUGHES DIVISION of SMITH'S INDUSTRIES LTD., 252, 261.
KELVIN & HUGHES LTD., instrument makers, 252, 261.
KELVIN & JAMES WHITE LTD., opticians and instrument makers, x, 86, 158, 188, 252, 254, 256, **260–262**, 264, 265, 272–277, 280, 283, 285, 286.
KELVIN, WHITE & HUTTON, instrument makers, 261, 262.
KELVIN & WILFRED O. WHITE, instrument makers, 261, 262.
Kemp, Andrew Meikle, senior staff member at James White, 260, 261.
KEMP & CO., philosophical instrument makers, 133, 135.
Kenton, Frances, 148.
KEYZOR & BENDON, wholesale opticians, 157.
Kidd, John, nautical instrument maker, 139.
Kinloch, James, compass adjuster, 230, 231.
King, Alexander, nautical instrument maker, 139.
King, F.A., patentee, 266.
King, John, clockmaker, 152, 153.
Kinnaird, George William Fox, 9th Baron Kinnaird, 147.
Kirkwood, Robert, cartographer, 56.
Kitchin, Thomas, engraver and publisher, 55.
Kullberg, Victor, chronometer maker, 238, 242.
Kyle, William, land surveyor, 12, 13, 14, 31, 34, 56.
Kynynmound, Hugh Murray, advocate, 1.

LADD & OERTLING, philosophical instrument makers, 252.
Laing, –, chronometer maker, 228.
Laird, David Laird, watchmaker and nautical instrument maker, 105, **106–107**.
Laird, Mary, 107.
Laird, Sarah, 107.
Lamont, Colin, mathematics teacher, 217.
Lauchlan, Margaret, 276.
LAURIE & HAMILTON, merchants, 165, 171.
Laurie, James, surveyor, **164**, 171. See GARDNER & LAURIE.
LANCASTER & SON, J., optical instrument and camera manufacturers, 136.

Lazars, I.: see Lizars, John.
Lee, James Gilchrist, shipbroker, 239, 250.
Lee, John, astronomer and antiquarian, 197, 200.
Lee, Thomas, mathematics and astronomy teacher, 191, 194.
Lees, George, teacher of natural philosophy, 59, 90.
Légé, A., instrument maker, 255, 256.
Leitz, Ernst, microscope manufacturer, 277.
LEJEUNE & PERKEN, importers of optical instruments, 170.
LENNIE, instrument suppliers, 97, **123–129**, 133.
LENNIE, E., instrument suppliers, 123, 125, 152, 179, 187, 287, 292, 299.
LENNIE, E.& J., instrument suppliers, 123, 125, 130.
Lennie, Eliza, businesswoman, **123**.
Lennie, George, toy warehouseman, 125.
LENNIE, J., instrument suppliers, 123.
LENNIE, J.& J., instrument suppliers, 123.
Lennie, Jack, optician, 123, 125.
Lennie, James, [junior], optician, 123.
Lennie, James, [senior], jeweller and optician, **123**, 125.
Lennie, John, optician, 123.
Lennie, Joseph C., optician, 123, 125.
LENNIE & McCALL, ophthalmic opticians, 125.
Lennie, William, apprentice optician, 123.
Leslie, General, 20.
Leslie, Sir John, professor of natural philosophy, 32, 33, 34, 35, 40, 43, 47, 49, 57, 59, 60, 62.
LEVI, JOSEPH & CO., instrument wholesalers and retailers, 157, 238, 243, 253.
Lind, James, physician and natural philosopher, 27, 29, 30, 53, 54, 55.
Lindsay, James Bowman, electrician and philologist, 148, 149.
LILLEY, J. & SON, chronometer and instrument makers, 256.
Lipton, Sir Thomas, grocer, 231.
Lizars, Isaac, 301.
Lizars, John, optical and mathematical instrument maker, 123, 133, 147, 253, 292, 295, 299, **301–305**.
Lizars, Juliet, 302.
Lizars, Rebecca, optician's shopwoman, 302.
Lloyd, –, foreman with Patrick Adie, 48, 80.
Lloyd, Richard, agent, 81.
LONDON & PARIS OPTICAL CO., instrument suppliers, 238.
Loudoun, 4th Earl of: see Campbell, John.
Lovi, Angelo, barometer and specific gravity bead maker, 205.
LOWDON BROTHERS, electrical engineers, 147.
Lowdon, Edward Joseph Bonar, electrical engineer, 147, 148.
Lowdon, George, [junior], optician, 117, 129, 138, 139, 140, 141, **146–151**, 170, 179, 187, 287.
Lowdon, George, [senior], grocer, 146, 148.
Lowdon, John, electrician, 147.
Lunan, Charles, watch and clockmaker, 152, 153.
Lunan, William, watch clock and philosophical instrument maker, 60, 153.
Lyle, James, 168, **170**, 173. See GARDNER & LYLE.
Lyon, Peter, mathematical, nautical and optical instrument maker, 105.
Lyons, Eliza, 302.

Mabon, Alexander Fernie, mathematical instrument maker, 48, 52, 65.
MABON, ALEX. & SONS, 48, 65, 274.
Macalister, Daniel, painter and decorator, 220, 221, 222.
MACALISTER & FYFE, ship chandlers and instrument retailers, 217, **220–221**, 222, 241.
McCALL & LENNIE, ophthalmic opticians, 125.
Macclesfield, 2nd Earl of: see Parker, George.
McCracken, Robert, dentist, 254, 261, 263.
McCulloch, Thomas, smith, 197.
McDougall, Sir Henry, landowner, 19.
McFarlane, J.W., sergeant in the Lanarkshire Rifle Volunteers, 248.
McGregor, Annie, 243.
McGregor, Bessie, 243.

McGREGOR, D. & CO., chronometer and nautical instrument makers, 108, 110, 226, 228, 229, 230, 232, **235–249**, 250, 256, 257, 281, 283, 286, 292.
McGREGOR, D. & CO. LTD., 239, 240, 241.
McGregor, Duncan, 243.
McGREGOR, DUNCAN, chronometer and nautical instrument makers, 58, 73, 177, 217, 220, 232, **235–249**, 284.
McGregor, Duncan, cooper, 241.
McGregor, Duncan, [junior], optician and nautical instrument maker, **235–249**.
McGregor, Duncan, [senior], chronometer and nautical instrument maker, **235**, 239, 241, 242, 281, 284, 285.
McGregor, John, 243.
McGregor, Malcom, 235, 242.
McGregor, Marion, 243.
McGregor, Robert, instrument retailer, 239, **240–241**, 243, 244.
McINNES & CAIRNS, engineers, 230.
McINNES, T.S. & CO. LTD., manufacturers of mathematical and engineering instruments, 229, 230, 231.
McINNES, THOMAS S., engineers, 230. See DOBBIE McINNES LTD.
Mackay, Alexander Spence, watchmaker and optician, **157**, 159, 160. See BERRY & MACKAY.
Mackay, James T., jeweller, 157, 158.
MACKAY, JAMES T. & CO., working silversmiths, 157.
MACKAY, JAMES T. & SON, opticians, 157.
Mackay, Margaret Alexander, 159.
Mackenzie, Miss –, 103.
Mackenzie of Delvine, Sir Alexander Muir, landowner, 19, 22.
Mackenzie, Donald Falconer, factor, 52, 64, 65.
Mackenzie, Frederick, 277.
Mackenzie of Coul, Sir George Steuart, mineralogist, 119.
Mackenzie, Henry, novelist and writer, 27, 29, 54.
Mackenzie, James Stuart, politician, 29, 55.
Mackenzie, Murdoch, senior, surveyor, 6.
Mackie, A.D., cutler, 48, 52.
Mackie, Alexander, toolmaker, 48, 52, 53.
Mackie, James, compass adjuster, 230, 232.
Mackie, William James, optician, 48, 52, 53.
McKilliam, John, watch, clock and nautical instrument maker, 153.
Mackintosh, L., shopman to D. McGregor & Co., 242.
McLaggan, –, chronometer maker, 159.
Maclaurin, Colin, professor of mathematics, 1, 4, 5, 6.
McLennan, Duncan, compass adjuster, 262.
McLeod, J.M., director of D. McGregor & Co. Ltd., 240, 241.
McMichael, James, [junior], 277.
McMillan, Peter, 153, **154–155**.
McMillan, William, watch, clock and mathematical instrument maker, 154.
McMILLAN, WILLIAM & CO., watch, clock and nautical instrument makers, 153, 154.
McNair, Angus, compass adjuster, 250.
McNally, J., sundial maker(?), 214.
McNeilage, John, optician and nautical instrument maker, 219.
McPherson, G., patentee, 296.
McQueen, Hugh, compass adjuster, 230, 232, 239, 243.
McVean, Effie, 200.
Mann, John, mathematician and mathematical instrument maker, 104.
Mann, William, 212.
Marlborough, 3rd Duke of: see Spencer, Charles.
Marr, –, poor law inspector, 158.
Marshall, James, 47.
Martin, Benjamin, instrument maker and populariser of science, 66, 127, 210, 214.
Maskelyne, Nevil, Astronomer Royal, 29, 54.
Mason, George, chemical and optical instrument maker, 293.
MASON, GEORGE & CO., optical, mathematical, chemical, philosophical and photographic instrument makers, 115, 293.
Maxwell, James Clerk, professor of experimental physics, 114, 118, 152.

Maxwell of Pollok, Sir John, landowner, 212.
Maxwell, John Hall, advocate and agriculturalist, 212, 215.
Mayall, John, photographer, 75.
Mayo, Thomas, ship chandler, 104.
Meikle, Henry, scientist, 90, 94.
Mein, Thomas, optician, 48, 52, 65.
MEISER & MERTIG, instrument manufacturers, 135.
Melville, Richard, sundial maker, **210–216**.
Melvin, Agnes, 212.
Melvin, Richard, sundial maker: see Melville, Richard.
Menzies, Alexander James, optician, 48, 52, 64, 65.
Menzies, Charles D., 65.
Menzies, William John, solicitor, 47.
Mercer, Thomas, chronometer maker, 228, 232, 237, 238, 242.
Merz, Georg, instrument maker, 114.
MERZ & MAHLER, instrument manufacturers, 114.
MIDDLETON & TENNENT, merchants, 165.
Millar, George, merchant, 54.
Millar, John, goldsmith's apprentice, 26, 54.
Millar, Richard, clock, watch and nautical instrument maker, 105, 107.
Millar, Robert, bank teller, 147.
MILLER & ADIE, mathematical, optical and philosophical instrument makers, **30–34**, 35, 43, 48, 54, 56, 57, 66, 67, 72, 75, 89.
Miller, Alexander, clockmaker, 30, 53, 54.
Miller, Betty, 30, 53, 56.
Miller, Charles, turner, 26.
Miller, Hugh, geologist, 47, 62.
Miller, Isabel, 17.
Miller, John, [junior], instrument maker, 11, 25, 26, **27–30**, 31, 32, 34, 48, 49, 50, 53, 54, 55, 56, 65, 75.
Miller, John, [senior], turner, **25–27**, 30, 53, 48, 54.
Milne, James, brassfounder, 34, 50, 113.
MILNE, JAMES & SON, brassfounders and gas engineers, 62, 75, 80.
Milne, John, brassfounder, 62, 94.
Minto, 1st Earl of: see Elliot, Sir Gilbert.
Mitchell, Elizabeth, 102.
Mitchell, Kate, 80.
Moffat, John, photographer, 252.
Molinari, Charles, glassblower, 205.
MOLTENI, ZERBONI & CO., looking glass and picture frame manufacturers, 102.
Monro, Alexander, [primus], professor of medicine, 1, 26.
Moodie, Margaret, 108.
Moore, F.M., nautical instrument maker, 257, 264.
More, Alexander, engineer, 279.
More, David, engineer and millwright, 278.
More, James, optician, 260, 261, **278–279**.
MORE, JAMES & CO., opticians, 269, **278–279**.
More, Margaret, 279.
Moreau, –, microscope manufacturer, 126.
Morgan, John, instrument maker, 27, 164.
Morrison, Skelton, ship's mate, 158.
Morton, 14th Earl of: see Douglas, James.
Morton, Alexander, carpet machine manufacturer and amateur telescope maker, **193**, 194, 198, 253.
Morton, Alexander, textile manufacturer, 193.
Morton, Charles, carpet machine manufacturer, 193.
Morton, Gavin, carpet manufacturer, 193.
Morton, James, textile manufacturer, 193.
Morton, John, compass adjuster, 239, 240.
MORTON, JOHN & CO., nautical instrument makers and publishers, 239, 240.
Morton, Robert James, compass adjuster, 230, 232.
Morton, Thomas, carpet machine manufacturer and amateur telescope maker, 20, 43, 61, 188–189, **190–196**, 197, 198, 292.
Morton, Thomas, [senior], brickmaker, 190.
MORTON, WILLIAM & SONS, cutlers, 48, 53, 65.
Mossman, –, surveyor, 45.
MOTION SMITH & SON LTD., instrument repairers, 233.

317

Muir, John, mariner, supplier of navigation instruments, 104.
Muir, William, engineer, 194.
Muirhead, D., patentee, 286.
Munro, R.W., engineer, 188, 189.
Murdoch, Laurie, instrument maker with Adie & Son, 48, 50.
Murray, D., author, 60.
Murray, John, compass maker, 139.
Murray, John, watch, clock and nautical instrument maker, 153.
Murray, Walter, business proprietor, 157, 159.
Murray, Walter, compass adjuster, 157.
Murray of Ochtertyre, Sir William Keith, amateur astronomer, 44.

Nachet, Camille Sebastien, microscope maker, 51, 64, 299.
NACHET ET FILS, microscope manufacturers, 304.
Napier, James R., engineer, 226.
Napier, Robert, marine engineer, 270.
Negretti, Enrico, glassblower, 75.
NEGRETTI & ZAMBRA, instrument wholesalers, 77, 81, 130, 252.
Neish, Sheriff –, 244.
Nelson, E.M., designer of microscope apparatus, 137.
Nevatt, Eliza, 243.
Newcombe, Frederick, philosophical instrument maker, 253.
Newton, Sir Isaac, natural philosopher, 2, 23.
Nicol, Catherine, 162.
Nicol, William, teacher of natural philosophy, 58, 61, 114, 115, 119.
Noble, George, poet, 18.
Noble, William, scientific journalist, 199, 201.
Norremberg, –, optical mineralogist, 291.
NORIE & WILSON, nautical instrument makers, 238.
NORTON & GREGORY LTD., mathematical and surveying instrument makers, 292, **306–307**.

Oberhaeuser, Georges, microscope manufacturer, 51.
Oerstedt, Hans Christian, scientist, 43.
Ogg, William, watch, clock and nautical instrument maker, 154.
OGG & McMILLAN, watch, clock and nautical instrument makers, 153.
Ord, George, 167.
Owenson, Sydney, Lady Morgan, 103.

Park, Robert, nautical instrument maker, 217, **219**.
Park, Robert, [senior], wheelwright, 219.
Parker, George, 2nd Earl of Macclesfield, astronomer, 1.
PARKES, JAMES & SON, instrument wholesalers, 73, 74, 83, 116, 163, 179, 187, 248, 253, 271, 293.
Parkes, S.H., patentee, 187.
PASTORELLI, J., glassblowers, 253.
Paterson, James, almanack maker, 104.
Peden, John, designer of carpet machinery, 194.
Pennant, Thomas, traveller and naturalist, 29.
PERKEN, SON & RAYMENT, photographic suppliers, 173.
Perreaux, –, engineer, 77.
Petitdidier, O.L., astronomer, 199.
Petrie, A., public servant, 13.
PETRIE & SON, clockmakers, 161.
Philips, James, sundial maker(?), 214.
Philips, Robert, 277.
Phillips, John, geologist and instrument designer, 119.
PILLISCHER, MORRICE, optical instrument manufacturer, 135.
Pitcairn, J.T., patentee, 296.
Playfair, –, gunmaker, 158.
Playfair, John, professor of natural philosophy, 19, 22, 35, 57.
Pole, William, professor of civil engineering, 114.
Poole, John, chronometer manufacturer, 236, 237.
Ponton, Mungo, advocate and amateur scientist, 90, 94.
Porthouse, Thomas, chronometer maker, 159.
Potter, J.D., nautical instrument maker, 238, 252.

POWELL & LEALAND, optical instrument manufacturers, 121, 253, 277.
Prazmowski, Adam, microscope maker, 121, 277.
Prendergast, Helen, 206.
PRESTON, JOSEPH & SONS, chronometer manufacturers, 242, 247.
Pringle, Andrew, Lord Alemoor, judge, 27, 29, 54, 55.
Pringle, Walter, 56.
Pritchard, Andrew, instrument maker, 14, 39, 58.
Pritchard, T.B., patentee, 306.
Purves, Elizabeth, 89.
Purves, William Laidlaw, 82.

Quekett, John Thomas, histologist and author, 299.

RAE BROTHERS, photographers, 261, 278.
Ramage, Charles, optical and nautical instrument maker, 152.
Ramage, John, optical instrument maker, 23, 153.
Ramsay, Sir William, professor of chemistry, 290.
Ramsden, Jesse, instrument maker, 29, 49, 55.
Ranken, George, instrument maker with Adie & Son, 48, 50, 63.
Rattray, J.W.A., ophthalmic optician, 130, 131.
Reichert, Carl, microscope manufacturer, 135, 277.
Reid, –, schoolmaster, 158.
Reid, Alexander, optician, 97, 98.
Reid, Amos, 18.
Reid, David, manager of James White, 254, 255, 257, 260, 261, 262, 263, 264, 276, 277, 280.
Reid, David Hay, muslin manufacturer, 263, 264, 277.
Reid, Jane, 260.
Reid, John W., optician, 261, 262, **280**.
Reid, Thomas, watchmaker, 118.
REID & YOUNG, opticians and mathematical instrument makers, 261, 262, **280**.
Renfrew, D., patentee, 266.
Renton, Lord –, 19, 22.
REPSOLD, ADOLF & GEORG, instrument makers, 114.
RICHARDS FRÈRES, instrument wholesalers, 242.
RICHARDSON, ADIE & CO., instrument suppliers, 48, **53**, 65.
Riddel of Riddel, Sir John, agriculturalist, 17.
Ritchie, Edward Samuel, instrument maker, 242.
Ritchie, John, slater, 31, 54, 56.
Ritchie, Marion, 31, 41, 56.
RIVA, A. & CO., carvers and gilders, 209.
RIVA, J. & M., carvers and gilders, **209**.
Riva, Michael, 209.
Robertson, Lieutenant –, naval officer, 37.
Robertson, David, smith, 53.
Robertson, James, watchmaker, 106, 107.
Robertson, Jean, 53.
Robertson, Lewis, surveyor, 158.
Robison, John, professor of natural philosophy, 26, 27, 30, 38, 54, 55.
Robison, Sir John, natural philosopher, 49, 62.
Robinson, Thomas Charles, philosophical instrument maker, 39, 58.
Ronalds, Sir Francis, electrician and meteorologist, 77, 81.
Ross, Andrew, optical instrument maker, 39, 147, 149, 151, 277, 299.
Ross, Captain John, naval officer, 35.
Ross, Sir John, polar explorer, 43, 192, 194, 197, 200.
Ross, Thomas, instrument manufacturer, 81.
Roy, William, military engineer and surveyor, 29, 55.
Russell, –, observer, 80.
Rust, Richard, instrument maker, 97.
Rutherford, John, 92.
Rutherfurd of Egerston, –, landowner, 19.
Rutherfurd of Hunthill, Sir John, landowner, 17.
Ruthven, John, inventor, 31, 97, 98.

Salleron, J., instrument maker, 256.

Salmon, Benjamin, optician, 99.
Sandwich, Edward, tenant, 214.
Sandwich, John, 212, 214.
Sang, Edward, polymath, 45, 49, 60, 61, 62, 63, 92, 94.
Sang, John, land surveyor, 45, 62.
Sangster, Alexander, clockmaker, 160.
Saxton, Joseph, inventor, 83.
Schumacher, Heinrich, astronomer, 19, 22, 23.
Scoresby, William, scientist, 225, 226.
Scott, Alexander, 185.
Scott of Fala, Alexander, draper and amateur mechanic, 18.
Scott, Henry, 3rd Duke of Buccleuch, president of the Royal Society of Edinburgh, 55.
Scott, James, minister, 18.
Scott, John, shopman, 219.
Scott, R., diesinker, 245.
Scott, Captain Robert Falcon, polar explorer, 138.
Scott, Thomas, 18.
Scott, Sir Walter, novelist and poet, 19, 21, 22.
SEAGROVE & WOODS, instrument wholesalers, 243.
Selby of Twizel Castle, –, 19.
Seller, William, physician, 62.
Sewill, F., chronometer and nautical instrument maker, 256.
Shairp, Robert, watchmaker, 20.
Shaw, R. Norman, inventor, 133.
Shepherd, Charles, chronometer maker, 159.
SHEW, J.F. & CO., camera manufacturers, 149, 278.
Short, James, telescope maker, x, **1–10**, 20, 23, 26, 27, 29, 105, 164.
SHORT & MASON, instrument suppliers, 297.
Short, Thomas, instrument maker, 3, 4, 27, 105.
Shortrede, John, 19.
Shortrede, Robert, soldier and surveyor, 19, 22.
Sibbald, Janet, 31, 56.
Sillars, John D. clerk to M. Walker & Son, 242.
Simpson, W., patentee, 81.
Sinclair, James, merchant, 59.
Sinclair, Sir John, agriculturalist, 17.
Sinclair, Louisa, 60, 80.
Sisson, J., instrument maker, 60.
Sivright of Meggetland, Thomas, amateur optician, 13.
Slater, John, solicitor, 172.
Sloper, Alfred, mathematical instrument maker, 253.
Small, James, ploughwright, 17, 21.
Smeaton, John, civil engineer, 54, 152.
SMITH, A. & J., jewellers, watchmakers and opticians, 157, **162–163**.
Smith, Alexander, shoemaker, 162.
Smith, Alexander, working jeweller, 162, 163.
Smith, Alexander George Nicol, watchmaker and jeweller, 162.
Smith, Alfred, watchmaker and jeweller, 162.
Smith, Andrew, manufacturer, 90.
SMITH & BECK, microscope manufacturers, 253.
SMITH, BECK & BECK, manufacturing opticians, 97.
Smith, Charles, 152, 153.
Smith, Charles A., 263, 265, 277.
Smith, David, land surveyor, 167.
Smith, Harold, watchmaker and jeweller, 162.
Smith, James, instrument maker, 50, 63.
Smith, John, working jeweller, 162, 163.
SMITH & RAMAGE, nautical and optical instrument makers, 118, 153.
Smith, Robert, author, 5.
Smith, Thomas, civil engineer, 111.
Smith, William, architect, 158.
Smith, William A.C., mechanical engineer, 292, **294–295**, 303.
Smollett, Tobias George, novelist, 2.
Smyth, Charles Piazzi, Astronomer Royal for Scotland, 43, 49, 51, 59, 61, 62, 64, 115, 120.
Solomon, Ellen, 302.
SOMALVICO, JOSEPH & CO., instrument wholesalers, 170.
Somerville, Mary, scientific writer, 19, 21, 63.

Somerville, Thomas, minister and historian, 19.
Sorby, Henry Clifton, metallurgist and mineralogist, 115.
SORLEY, R.& W., chronometer makers, 228, 232.
Sorrell, S.C., ophthalmic optician, 130.
South, Sir James, astronomer, 63.
Spark, William, watch and clockmaker, 156, 158.
Spear, Richard, mathematical instrument maker, 109.
SPENCER & BROWNING, instrument wholesalers, 245.
SPENCER, BROWNING & CO., instrument wholesalers, 241, 245.
SPENCER, BROWNING & RUST, instrument wholesalers, 29, 37, 55, 167, 245, 286.
Spencer, Charles, 3rd Duke of Marlborough, soldier and statesman, 6.
Spencer, J.L., jeweller, 293.
Spencer, John, optician, 97, 170, 253, **293**.
Spencer, William, hardware merchant, 293.
Stalker, David, optician, 105, **108–110**, 111, 130, 144.
Stalker, William, sailmaker, 108.
Stampa, Dominick, carver and gilder, 205.
STANDLEY, BELCHER & MASON LTD., instrument suppliers, 290.
Stanley, W. F., instrument maker, 268.
STEPHEN & FORBES, shipbuilders, 158.
Stevenson, Alan, civil engineer, 45, 50, 61.
Stevenson, David, civil engineer, 62, 128.
Stevenson, J., watch and optical instrument maker, 153.
Stevenson, Peter, instrument maker, 226, 253.
Stevenson, Robert, civil engineer, 34, 35, 41, 45, 59, 63, 74, 94, 111, 128.
Stevenson, Thomas, civil engineer, 45, 62.
Stewart, Alexander N.M., shop assistant, 168, 184.
Stewart, Dr. Charles, benefactor to Edinburgh University, 30.
Stewart, John, professor of natural philosophy, 26, 27, 53.
Stewart, John, clerk to D. McGregor & Co., 242.
Stirling, William, agriculturalist, 212, 215.
Stokes, George Gabriel, professor of mathematics, 254, 263.
Stopani, John, barometer maker, 153.
Strachan, Alexander, nautical and optical instrument maker, 153.
Stuart, John, 3rd Earl of Bute, Prime Minister, 6, 29.
Sturrock, –, Provost of Kilmarnock, 200.
Swan, William, professor of natural philosophy, 44, 45, 61, 133.
Swart, Jacob, author, 58.
Swift, James, microscope manufacturer, 122, 176.
SWIFT, JAMES & SON, microscope manufacturers, 135, 137, 272, 277, 304.
Sykes, Bartholomew, hydrometer designer, 206.
Sym, James [junior], instrument maker, 164.
Sym, James [senior], instrument maker, 164.

Tarone, Antony, barometer maker, 139, 147.
Tatlock, J. Douglas, instrument manufacturer, 290.
Tatlock, John, laboratory assistant and instrument maker, **289–291**. See BAIRD & TATLOCK.
Tatlock, Robert, analytical chemist, 289, 290.
TAYLOR & CO., nautical instrument makers, 139.
Taylor, James, optician, 123.
Taylor, John, amateur astronomer, 63.
TAYLOR, TAYLOR & HOBSON, photographic lens manufacturers, 130.
Telford, Thomas, engineer, 152.
Tennent, Hugh, 171.
Tennent, Marion, 171.
Thompson, H. A., instrument retailer, 172.
Thomson, Elizabeth, 276.
Thomson, James, compass adjuster, 250.
Thomson, James, [junior], compass adjuster, **283**, 284. See WHYTE, THOMSON & CO.
Thomson, James, [senior], compass adjuster, 237, 239, **281–283**, 284.
Thomson, Robert, 191, 194.

319

Thomson, Thomas, chemist, 13, 206.
Thomson, William, 1st Baron Kelvin of Largs, professor of natural philosophy, x, 94, 180, 181, 199, 235, **252–275**, 285, 289, 290. See JAMES WHITE and KELVIN & JAMES WHITE LTD.
Thurot, François, privateer, 17.
TODD & HIGGINBOTHAM, printers, 214.
TORRE, P.D. & CO., carvers and gilders, 103.
Torrie, Thomas James, correspondent, 60.
Torry, Alexander, clockmaker, 160.
TOWNSON & MERCER, instrument wholesalers, 271.
Traill, Thomas Stewart, professor of medical jurisprudence, 63.
Trotter, John, 41.
Trotter, John, instrument maker, 115.
Trotter, Robert, watch and compass maker, 105.
Troughton, Edward, instrument maker, 35, 72.
TROUGHTON & SIMMS, instrument manufacturers, 50, 81, 158.
Tulley, Charles, instrument maker, 23, 38.
TURNBULL & CO., opticians, 97, 109, 111, 123, 125, **130–131**.
Turnbull, Frederick Brodie, compass adjuster and optician, 130.
Turnbull, Hilda B., ophthalmic optician, 130, 131.
Turnbull, John Miller, optician and instrument maker, **130–131**.
Turnbull, Robert, fishing rod and tackle manufacturer, 131.
Turnbull, William, optician, 130.
Turner, Edward, professor of chemistry, 90.
Twaddell, William, glassblower and hydrometer maker, 167.

UNITED COLLIERIES LTD., 274.
Ure, Andrew, chemist and scientific writer, 31.

Van Heurck, Henri, microscopist and author, 87.
Veitch, James, farmer, 17.
Veitch, James, ploughwright and optician, 13, **16–24**, 38, 56, 58, 60, 197.
Veitch, John, watchmaker, 20.
Veitch, William, farmer, 23.
Veitch, William, wright, 17.
Veitch, William, [junior], ploughwright, 18, 19, 21.
Victoria, Queen of Great Britain and Ireland, 170.
Vogel, Hermann Carl, astronomer and spectroscopist, 80.
Vulliamy, Joseph, clockmaker, 55.

Wallace, Robert, 180.
Wallace, William, professor of mathematics, 12, 26, 27, 49, 50, 54, 62, 72, 90, 94.
Walker, James, clockmaker, 160.
Walker, Jessie, 239, 243.
Walker, John Gray, nautical instrument maker, 235, 237, 239, 240, 242, 243.
WALKER, M. & SON, chronometer and nautical instrument makers, 229, 230, 232, **237–240**, 242, 243, 256. See DUNCAN McGREGOR.
WALKER & SONS, M., nautical instrument makers, 238.
Walker, Malcolm McNeil, nautical instrument maker, 226, 235, **237–240**, 242, 243.
Wasserlein, R., instrument maker, 121.
Wastell, Rev. –, minister and amateur meteorologist, 41.
WATERSON, A. & CO., carvers and gilders, 102.
WATKINS & HILL, instrument makers, 83, 252.
Watson, Sheriff –, 158.
Watson, Marion, 135.
WATSON, W. & SONS LTD., microscope manufacturers, 85, 87, 266, 279, 295.
Watt, James, engineer, 5, 27, 164, 168, 170, 217.
Wedderburn, Thomas, optician, 48, **51–52**, 64. See ADIE & WEDDERBURN.

Weir, James, chronometer maker, 228, 232.
Weir, John, 262.
Weir, Robert, compass adjuster, 237, 238, 239, 240.
Weir, Thomas, instrument maker, 115.
Welsh, John, meteorologist, 43, 75, 76, 77, 80, 81, 82.
Wesley, John, minister, 199.
West, Hon. Frederick Richard, Member of Parliament, 213.
Wheeler, Thomas, watch casemaker, 186.
Whitbread, George, instrument maker, 109.
White, Alexander Adam, 158.
White, Barbara, 11, 14.
WHITE & BARR, instrument makers, 168, 252, 261, 262.
WHITE, JAMES, instrument makers, x, 20, 58, 97, 116, 168, 170, 180, 185, 217, 235, 237, 238, **252–275**, 276, 278, 279, 285, 292.
White, James, optician, 168, **252–258**, 260, 261, 262, 263, 264. See KELVIN & JAMES WHITE LTD.
WHITE, JOSEPH & SONS, wholesale suppliers, 243.
White, William, yarn merchant, 252.
Whitehead, John, Lanarkshire Artillery Volunteers, 248.
Whitworth, Sir Joseph, mechanical engineer, 194.
WHYTE & CO., nautical instrument makers, 256, **281**, 284, 286.
WHYTE, JAMES, nautical instrument maker, 284.
Whyte, James, [junior], instrument maker, **283**, 284.
Whyte, James, [senior], instrument maker, **281**, 284, 285.
WHYTE, THOMSON & CO., 57, 129, 151, 179, 187, 217, 228, 231, 232, 235, 237, 238, 239, 250, 257, 264, 265, **281–287**.
Whyte, W.D., compass adjuster, 231, 283, 284, **285**, 286.
Williams, William, professor of veterinary studies, 121.
Wilson, Agnes, 193.
Wilson, George Washington, photographer, 148, 149, 278.
Wilson, Hugh, engraver, 170.
Wilson, James, compass adjuster, 239, 240, **250**, 283, 284, 286. See CHRISTIE & WILSON.
Winning, Andrew, colliery employee, 268.
Wise, Dr –, 147.
Wood, Alexander S., compass maker, 232.
Wood, E.G., optician, 97.
Wood, George S., optician, 299.
Wood, William, merchant, 165.
Wollaston, Francis John Hyde, professor of chemistry and natural philosophy, 83.
Wollaston, William Hyde, polymath, 61.
WRAY, W., instrument wholesalers, 130.
Wrench, John, instrument supplier, 170.
Wright, William, spectacle manufacturer, 170.

Yeaman, John, gunsmith, 53.
Yeaman, John, instrument maker, **25–26**, 27, 29, 48, 53, 54.
YEATES & SON, instrument manufacturers, 116.
Yeo, Daniel D., painter and decorator, 220, 221. See FYFE & YEO.
YORK & SON, photographic publishers, 170.
YOUNG & CUNNINGHAM LTD., remote control and gauging instruments, 230, 231.
Young, David, compass adjuster, 283, 284, 286.
Young, Robert, nautical instrument maker, 239, 242.
Young, Thomas, [junior], optician, 261, 262, **280**. See REID & YOUNG.
Young, Thomas, [senior], compass adjuster, 261, 262, 280.
Yuill, John C., 238.

ZEISS, CARL, optical instrument manufacturer, 135.
ZENONE & BUTTI, carvers and gilders, 97, **102–103**.
Zenone, John, carver and gilder, **102–103**.
Zenone, Joseph, importer of French flowers, 102, 103.